Intelligent Systems
for Information Processing
From Representation to Applications

Cover figure taken from: "Normal forms for fuzzy relations and their contribution to universal approximation", by I. Perfilieva,, Figure 3 (pp 391) in this book.

Intelligent Systems for Information Processing
From Representation to Applications

Edited by

Bernadette Bouchon-Meunier
CNRS-UPMC, LIP6
Paris, France

Laurent Foulloy
University of Savoie
LISTIC/ESIA
Annecy, France

Ronald R. Yager
Machine Intelligence Institute
Iona College
New Rochelle, New York, USA

2003

ELSEVIER

Amsterdam • Boston •Heidelberg • London • New York • Oxford • Paris
San Diego • San Francisco • Singapore • Sydney • Tokyo

ELSEVIER SCIENCE B.V.
Sara Burgerhartstraat 25
P.O. Box 211, 1000 AE Amsterdam, The Netherlands

© 2003 Elsevier Science B.V. All rights reserved.

This work is protected under copyright by Elsevier Science, and the following terms and conditions apply to its use:

Photocopying
Single photocopies of single chapters may be made for personal use as allowed by national copyright laws. Permission of the Publisher and payment of a fee is required for all other photocopying, including multiple or systematic copying, copying for advertising or promotional purposes, resale, and all forms of document delivery. Special rates are available for educational institutions that wish to make photocopies for non-profit educational classroom use.

Permissions may be sought directly from Elsevier's Science & Technology Rights Department in Oxford, UK: phone: (+44) 1865 843830, fax: (+44) 1865 853333, e-mail: permissions@elsevier.com. You may also complete your request on-line via the Elsevier Science homepage (http://www.elsevier.com), by selecting 'Customer Support' and then 'Obtaining Permissions'.

In the USA, users may clear permissions and make payments through the Copyright Clearance Center, Inc., 222 Rosewood Drive, Danvers, MA 01923, USA; phone: (+1) (978) 7508400, fax: (+1) (978) 7504744, and in the UK through the Copyright Licensing Agency Rapid Clearance Service (CLARCS), 90 Tottenham Court Road, London W1P 0LP, UK; phone: (+44) 207 631 5555; fax: (+44) 207 631 5500. Other countries may have a local reprographic rights agency for payments.

Derivative Works
Tables of contents may be reproduced for internal circulation, but permission of Elsevier Science is required for external resale or distribution of such material.
Permission of the Publisher is required for all other derivative works, including compilations and translations.

Electronic Storage or Usage
Permission of the Publisher is required to store or use electronically any material contained in this work, including any chapter or part of a chapter.

Except as outlined above, no part of this work may be reproduced, stored in a retrieval system or transmitted in any form or by any means, electronic, mechanical, photocopying, recording or otherwise, without prior written permission of the Publisher.
Address permissions requests to: Elsevier's Science & Technology Rights Department, at the phone, fax and e-mail addresses noted above.

Notice
No responsibility is assumed by the Publisher for any injury and/or damage to persons or property as a matter of products liability, negligence or otherwise, or from any use or operation of any methods, products, instructions or ideas contained in the material herein. Because of rapid advances in the medical sciences, in particular, independent verification of diagnoses and drug dosages should be made.

The article "Toward a perception-based theory of probabilistic reasoning with imprecise probabilities" by Lotfi A. Zadeh has been previously published in the Journal of Statistical Planning and Inference 105 (2002), pp. 233-264.

First edition 2003

Library of Congress Cataloging in Publication Data
A catalog record from the Library of Congress has been applied for.

British Library Cataloguing in Publication Data

```
International Conference on Information Processing and
Management of Uncertainty in Knowledge Based Systems (9th :
Annecy, France)
  Intelligent systems for information processing : from
representation to applications
  1.Artificial intelligence - Congresses 2.Intelligent agents
(Computer software) - Congresses
  I.Title II.Bouchon-Meunier, B. (Bernadette) III.Foulloy,
Laurent IV.Yager, R. R. (Ronald R), 1941-
  006.3

  ISBN 0444513795
```

ISBN: 0 444 51379 5

∞ The paper used in this publication meets the requirements of ANSI/NISO Z39.48-1992 (Permanence of Paper).
Printed in The Netherlands.

Preface

One of the earliest applications of the computer was to the processing of information. These applications led to a revolution and created a whole discipline. During the recent decade we saw a second revolution in information processing motivated by the introduction of the Internet, which is essentially a confluence of the computing and communications technologies.

We are now at the early stage of a new revolution in information processing technology. This revolution is focused on the introduction of intelligence to the processing of information. In large part this revolution is the next step in the development of Internet technology. We need intelligence to search the Internet in an efficient way, we need intelligence to build agents to prowl the Internet and do our bidding and of course we need intelligence to help us in understanding the multi-media environment of the Internet. However, the Internet is not the only place where this revolution is developing. The control systems, used in industrial processes, are getting more intelligent.

The focus of the International Conference on Information Processing and Management of Uncertainty in Knowledge Based System (IPMU) is on the development of technology needed for the construction of intelligent systems. The ninth IPMU conference held in Annecy France, organized by the University of Savoie, brought together some of the world's leading experts in these technologies. In this volume, we have collected and organized a selection of papers from this conference.

The book starts with an introduction to perception-based information processing by Prof. L.A. Zadeh. This paper is reprinted from the Journal of Statistical Planning and Inference (2002) by Elsevier. Here Prof. Zadeh continues his development of a framework for computing with words by investigating perception-based probabilities, which are needed for the development of intelligent decision systems.

An important task in the development of intelligent systems is the representation of knowledge in a manner that is rich enough to capture the subtlety of human intelligence, but still formal enough to allow computer manipulation. In the second section of this volume we present a number of papers on representing knowledge.

The task of retrieving information is central to many activities involved in the processing of information. The third section addresses this important issue. A number of the selected papers focus on the development of aggregation operators.

Reasoning is an important part in human intelligent activities. Modeling human activities by means of computers is a real challenge for the implementation of intelligent systems. Section 4 of this volume is concerned with these topics.

Intelligent systems have to take into account the quality of the information they process. The representation and handling of uncertain information is one of the interesting aspects in the construction of intelligent systems and is tackled in the fifth section.

The mining of information from the large amount of data available is one of the most active areas in information technology. Many new technologies and paradigms for the task are being developed. In the sixth section we provide some papers on data mining and the related issue of learning.

The last two sections are respectively devoted to formal foundations of the technologies used for constructing intelligent systems and applications illustrating the wide spectrum of domains where the tools are applied.

B. Bouchon-Meunier
L. Foulloy
R. R. Yager

Contents

Preface v

Perception-based Information Processing 1

Toward a Perception-Based Theory of Probabilistic Reasoning with
Imprecise Probabilities 3
L.A. Zadeh

Representing Knowledge 35

Rough Set Uncertainty in an Object Oriented Data Model 37
T. Beaubouef, F.E. Petry

On the Representation of Fuzzy Spatial Relations in Robot Maps 47
I. Bloch, A. Saffiotti

Fuzzy "Between" Operators in the Framework of Fuzzy Orderings 59
U. Bodenhofer

A Step towards Conceptually Improving Takagi-Sugeno's Approximation 71
S. Guadarrama, E. Trillas, J. Gutiérrez, F. Fernández

Combining Heterogeneous Information in Group Decision Making 81
F. Herrera, E. Herrera-Viedma, L. Martínez, P.J. Sánchez

Function Approximation by Fuzzy Rough Sets 93
M. Inuiguchi, T. Tanino

Retrieving Information 105

Query Aggregation Reflecting Domain-knowledge 107
T. Andreasen

Modelling of Fuzzy and Uncertain Spatio-Temporal Information in
Databases: A Constraint-based Approach 117
G. de Tré, R. de Caluwe, A. Hallez, J. Verstraete

A General Framework for Meta-search based on Query-weighting and
Numerical Aggregation Operators 129
M. Gómez, C. Abásolo

On the Comparison of Aggregates over Fuzzy Sets. 141
P. Bosc, O. Pivert, L. Liétard

Towards an Intelligent Text Categorization for Web Resources:
An Implementation 153
S. Zadrozny, K. Lawcewicz, J. Kacprzyk

Reasoning **165**

Prototype Based Reasoning and Fuzzy Modeling. 167
R.R. Yager

Gradual Handling of Contradiction in Argumentation Frameworks 179
C. Cayrol, M.C. Lagasquie

Coherent Conditional Probability as a Tool for Default Reasoning. 191
G. Coletti, R. Scozzafava, B. Vantaggi

Detecting Conflict-Free Assumption-Based Knowledge Bases. 203
R. Haenni

Uncertainty **211**

Theory of Belief Functions: History and Prospects 213
A.P. Dempster

Towards Another Logical Interpretation of Theory of Evidence and
a New Combination Rule 223
L. Cholvy

Uncertainty, Type-2 Fuzzy Sets, and Footprints of Uncertainty 233
J. M. Mendel

Rough Sets, Bayes' Theorem and Flow Graphs. 243
Z. Pawlak

Belief Revision as Combinatorial Optimization. 253
A. Ramer

Showing why Measures of Quantified Beliefs are Belief Functions 265
Ph. Smets

Extension of Coherent Lower Previsions to Unbounded Random Variables . . . 277
M. C.M. Troffaes, G. de Cooman

Learning and Mining ... **289**

Clustering of Proximity Data using Belief Functions ... 291
Th. Denoeux, M. Masson

A Hierarchical Linguistic Clustering Algorithm for Prototype Induction ... 303
L. González Rodríguez, J. Lawry, J.F. Baldwin

A Multiobjective Genetic Algorithm for Feature Selection and Data Base
Learning in Fuzzy-Rule Based Classification Systems ... 315
O. Cordón, F. Herrera, M.J. del Jesus, L. Magdalena, A.M. Sánchez, P. Villar

Mining Implication-Based Fuzzy Association Rules in Databases ... 327
E. Hüllermeier, J. Beringer

Learning Graphical Models by Extending Optimal Spanning Trees ... 339
C. Borgelt, R. Kruse

Clustering Belief Functions Based on Attracting and Conflicting
Metalevel Evidence ... 349
J. Schubert

Foundations ... **361**

Models and Submodels of Fuzzy Theories ... 363
V. Novák

Numerical Representations of Fuzzy Relational Systems ... 375
S. Ovchinnikov

Normal Forms for Fuzzy Relations and Their Contribution to Universal
Approximation ... 381
I. Perfilieva

Associative Operators Based on t-Norms and t-Conorms ... 393
M. Mas, R. Mesiar, M. Monserrat, J. Torrens

Applications ... **405**

Non-Analytical Approaches to Model-Based Fault Detection and Isolation ... 407
P.M. Frank

A Hybrid Fuzzy-Fractal Approach for Time Series Analysis and
Prediction and its Applications to Plant Monitoring ... 419
O. Castillo, P. Melin

Linguistic Modeling of Physical Task Characteristics ... 431
S. Visa, A. Ralescu, S. Yeung, A. Genaidy

Validation of Diagnostic Models Using Graphical Belief Networks 443
O. Kipersztok, H. Wang

Adjustment of Parallel Queuing Processes by Multi-Agent Control 453
R. Palm, T.A. Runkler

Is she gonna like it? Automated Inspection System Using Fuzzy Aggregation. . . 465
A. Soria-Frisch, M. Köppen, Th. Sy

Author Index **477**

Perception-based Information Processing

B. Bouchon-Meunier, L. Foulloy and R.R. Yager (Editors)
Intelligent Systems for Information Processing:
From Representation to Applications
© 2003 Published by Elsevier Science B.V.

Toward a perception-based theory of probabilistic reasoning with imprecise probabilities[☆]

Lotfi A. Zadeh[*]

Computer Science Division and the Electronics Research Laboratory, Department of EECS, University of California, Berkeley, CA 94720-1776, USA

To Herbert Robbins, Hung T. Nguyen and the memory of Professor Kampe de Feriet

Abstract

The perception-based theory of probabilistic reasoning which is outlined in this paper is not in the traditional spirit. Its principal aim is to lay the groundwork for a radical enlargement of the role of natural languages in probability theory and its applications, especially in the realm of decision analysis. To this end, probability theory is generalized by adding to the theory the capability to operate on perception-based information, e.g., "Usually Robert returns from work at about 6 p.m." or "It is very unlikely that there will be a significant increase in the price of oil in the near future". A key idea on which perception-based theory is based is that the meaning of a proposition, p, which describes a perception, may be expressed as a generalized constraint of the form X isr R, where X is the constrained variable, R is the constraining relation and is$r$ is a copula in which r is a discrete variable whose value defines the way in which R constrains X. In the theory, generalized constraints serve to define imprecise probabilities, utilities and other constructs, and generalized constraint propagation is employed as a mechanism for reasoning with imprecise probabilities as well as for computation with perception-based information.
© 2002 Elsevier Science B.V. All rights reserved.

MSC: primary 62A01; secondary 03B52

Keywords: Perception-based information; Fuzzy set theory; Fuzzy logic; Generalized constraints; Constraint languages

1. Introduction

Interest in probability theory has grown markedly during the past decade. Underlying this growth is the ballistic ascent in the importance of information technology. A related cause is the concerted drive toward automation of decision-making in a wide variety of fields ranging from assessment of creditworthiness, biometric authentication,

---

[☆] Program of UC Berkeley.
[*] Tel.: +1-510-642-4959; fax: +1-510-642-1712.
E-mail address: zadeh@cs.berkeley.edu (L.A. Zadeh).

and fraud detection to stock market forecasting, and management of uncertainty in knowledge-based systems. Probabilistic reasoning plays a key role in these and related applications.

A side effect of the growth of interest in probability theory is the widening realization that most real-world probabilities are far from being precisely known or measurable numbers. Actually, reasoning with imprecise probabilities has a long history (Walley, 1991) but the issue is of much greater importance today than it was in the past, largely because the vast increase in the computational power of information processing systems makes it practicable to compute with imprecise probabilities—to perform computations which are far more complex and less amenable to precise analysis than computations involving precise probabilities.

Transition from precise probabilities to imprecise probabilities in probability theory is a form of generalization and as such it enhances the ability of probability theory to deal with real-world problems. The question is: Is this mode of generalization sufficient? Is there a need for additional modes of generalization? In what follows, I argue that the answers to these questions are, respectively, No and Yes. In essence, my thesis is that what is needed is a move from imprecise probabilities to perception-based probability theory—a theory in which perceptions and their descriptions in a natural language play a pivotal role.

The perception-based theory of probabilistic reasoning which is outlined in the following is not in the traditional spirit. Its principal aim is to lay the groundwork for a radical enlargement in the role of natural languages in probability theory and its applications, especially in the realm of decision analysis.

For convenience, let PT denote standard probability theory of the kind found in textbooks and taught in courses. What is not in dispute is that standard probability theory provides a vast array of concepts and techniques which are highly effective in dealing with a wide variety of problems in which the available information is lacking in certainty. But alongside such problems we see many very simple problems for which PT offers no solutions. Here are a few typical examples:

1. What is the probability that my tax return will be audited?
2. What is the probability that my car may be stolen?
3. How long does it take to get from the hotel to the airport by taxi?
4. Usually Robert returns from work at about 6 p.m. What is the probability that he is home at 6:30 p.m.?
5. A box contains about 20 balls of various sizes. A few are small and several are large. What is the probability that a ball drawn at random is neither large nor small?

Another class of simple problems which PT cannot handle relates to commonsense reasoning (Kuipers, 1994; Fikes and Nilsson, 1971; Smithson, 1989; Shen and Leitch, 1992; Novak et al., 1992; Krause and Clark, 1993) exemplified by

6. Most young men are healthy; Robert is young. What can be said about Robert's health?

7. Most young men are healthy; it is likely that Robert is young. What can be said about Robert's health?
8. Slimness is attractive; Cindy is slim. What can be said about Cindy's attractiveness?

Questions of this kind are routinely faced and answered by humans. The answers, however, are not numbers; they are linguistic descriptions of fuzzy perceptions of probabilities, e.g., not very high, quite unlikely, about 0.8, etc. Such answers cannot be arrived at through the use of standard probability theory. This assertion may appear to be in contradiction with the existence of a voluminous literature on imprecise probabilities (Walley, 1991). In may view, this is not the case.

What are the sources of difficulty in using PT? In Problems 1 and 2, the difficulty is rooted in the basic property of conditional probabilities, namely, given $P(X)$, all that can be said about $P(X|Y)$ is that its value is between 0 and 1, assuming that Y is not contained in X or its complement. Thus, if I start with the knowledge that 1% of tax returns are audited, it tells me nothing about the probability that my tax return will be audited. The same holds true when I add more detailed information about myself, e.g., my profession, income, age, place of residence, etc. The Internal Revenue Service may be able to tell me what fraction of returns in a particular category are audited, but all that can be said about the probability that my return will be audited is that it is between 0 and 1. The tax-return-audit example raises some non-trivial issues which are analyzed in depth in a paper by Nguyen et al. (1999).

A closely related problem which does not involve probabilities is the following.

Consider a function, $y = f(x)$, defined on an interval, say [0, 10], which takes values in the interval [0, 1]. Suppose that I am given the average value, a, of f over [0, 10], and am asked: What is the value of f at $x = 3$? Clearly, all I can say is that the value is between 0 and 1.

Next, assume that I am given the average value of f over the interval [2, 4], and am asked the same question. Again, all I can say is that the value is between 0 and 1. As the length of the interval decreases, the answer remains the same so long as the interval contains the point $x = 3$ and its length is not zero. As in the previous example, additional information does not improve my ability to estimate $f(3)$.

The reason why this conclusion appears to be somewhat counterintuitive is that usually there is a tacit assumption that f is a smooth function. In this case, in the limit the average value will converge to $f(3)$. Note that the answer depends on the way in which smoothness is defined.

In Problem 3, the difficulty is that we are dealing with a time series drawn from a nonstationary process. When I pose the question to a hotel clerk, he/she may tell me that it would take approximately 20–25 min. In giving this answer, the clerk may take into consideration that it is raining lightly and that as a result it would take a little longer than usual to get to the airport. PT does not have the capability to operate on the perception-based information that "it is raining lightly" and factor-in its effect on the time of travel to the airport.

In problems 4–8, the difficulty is more fundamental. Specifically, the problem is that PT—as stated above—has no capability to operate on perceptions described in a natural language, e.g., "usually Robert returns from work at about 6 p.m.", or "the box contains several large balls" or "most young men are healthy". This is a basic shortcoming that will be discussed in greater detail at a later point.

What we see is that standard probability theory has many strengths and many limitations. The limitations of standard probability theory fall into several categories. To see them in a broad perspective, what has to be considered is that a basic concept which is immanent in human cognition is that of partiality. Thus, we accept the reality of partial certainty, partial truth, partial precision, partial possibility, partial knowledge, partial understanding, partial belief, partial solution and partial capability, whatever it may be. Viewed through the prism of partiality, probability theory is, in essence, a theory of partial certainty and random behavior. What it does not address—at least not explicitly—is partial truth, partial precision and partial possibility—facets which are distinct from partial certainty and fall within the province of fuzzy logic (FL) (Zadeh, 1978; Dubois and Prade, 1988; Novak, 1991; Klir and Folger, 1988; Reghis and Roventa, 1998; Klir and Yuan, 1995; Grabisch et al., 1995). This observation explains why PT and FL are, for the most part, complementary rather than competitive (Zadeh, 1995; Krause and Clark, 1993; Thomas, 1995).

A simple example will illustrate the point. Suppose that Robert is three-quarters German and one-quarter French. If he were characterized as German, the characterization would be imprecise but not uncertain. Equivalently, if Robert stated that he is German, his statement would be partially true; more specifically, its truth value would be 0.75. Again, 0.75 has no relation to probability.

Within probability theory, the basic concepts on which PT rests do not reflect the reality of partiality because probability theory is based on two-valued Aristotelian logic. Thus, in PT, a process is random or not random; a time series is stationary or not stationary; an event happens or does not happen; events A and B are either independent or not independent; and so on. The denial of partiality of truth and possibility has the effect of seriously restricting the ability of probability theory to deal with those problems in which truth and possibility are matters of degree.

A case in point is the concept of an event. A recent Associated Press article carried the headline, "Balding on Top Tied to Heart Problems; Risk of disease is 36 percent higher, a study finds". Now it is evident that both "balding on top", and "heart problems", are matters of degree or, more concretely, are fuzzy events, as defined in Zadeh (1968), Kruse and Meyer (1987) and Wang and Klir (1992). Such events are the norm rather than exception in real-world settings. And yet, in PT the basic concept of conditional probability of an event B given an event A is not defined when A and B are fuzzy events.

Another basic, and perhaps more serious, limitation is rooted in the fact that, in general, our assessment of probabilities is based on information which is a mixture of

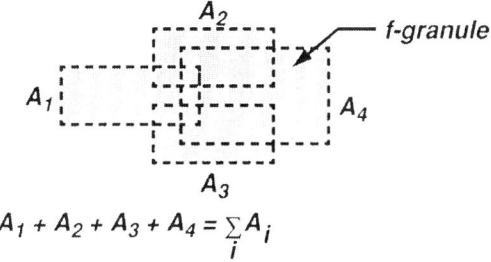

f-granularity is a reflection of the bounded ability of sensory organs and, ultimately, the brain, to resolve detail and store information

Fig. 1. f-Granularity (fuzzy granularity).

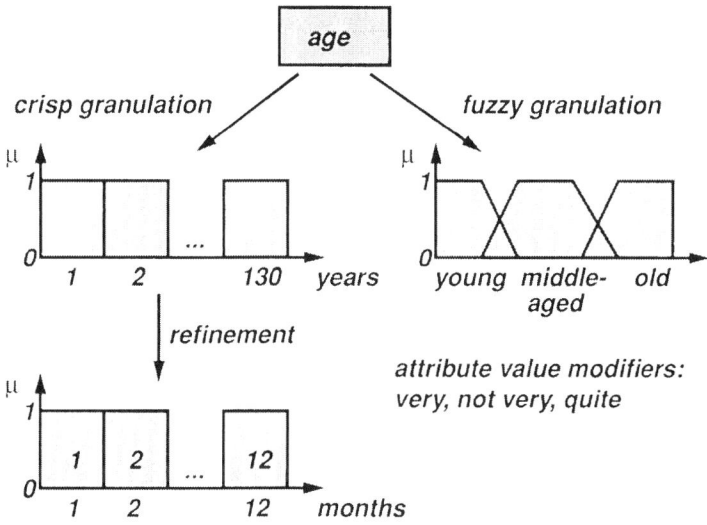

Fig. 2. Crisp and fuzzy granulation of *Age*.

measurements and perceptions (Vallee, 1995; Barsalou, 1999). Reflecting the bounded human ability to resolve detail and store information, perceptions are intrinsically imprecise. More specifically, perceptions are f-granular (Zadeh, 1979, 1997), that is: (a) perceptions are fuzzy in the sense that perceived values of variables are not sharply defined and (b) perceptions are granular in the sense that perceived values of variables are grouped into granules, with a granule being a clump of points drawn together by indistinguishability, similarity, proximity or functionality (Fig. 1). For example, the fuzzy granules of the variable Age might be young, middle-aged and old (Fig. 2). Similarly, the fuzzy granules of the variable Probability might be likely, not likely, very unlikely, very likely, etc.

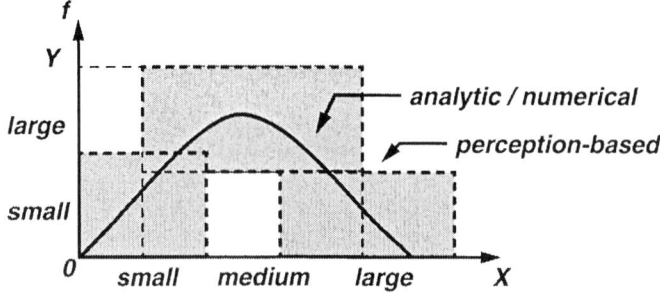

Fig. 3. Coarse description of a function by a collection of linguistic rules. Linguistic representation is perception-based.

Perceptions are described by propositions expressed in a natural language. For example

- Dana is young,
- it is a warm day,
- it is likely to rain in the evening,
- the economy is improving,
- a box contains several large balls, most of which are black.

An important class of perceptions relates to mathematical constructs such as functions, relations and counts. For example, a function such as shown in Fig. 3 may be described in words by a collection of linguistic rules (Zadeh, 1973, 1975, 1996). In particular, a probability distribution, e.g., discrete-valued probability distribution of Carol's age, P^*, may be described in words as

$\quad$ Prob{Carol is *young*} is *low*,
$\quad$ Prob{Carol is *middle-aged*} is *high*,
$\quad$ Prob{Carol is *old*} is *low*

or as a linguistic rule-set

$\quad$ if *Age* is *young* then P^* is *low*,
$\quad$ if *Age* is *middle-aged* then P^* is *high*,
$\quad$ if *Age* is *old* then P^* is *low*.

For the latter representation, using the concept of a fuzzy graph (Zadeh, 1996, 1997), which will be discussed later, the probability distribution of Carol's age may be represented as a fuzzy graph and written as

$\quad P^* = young \times low + middle\text{-}aged \times high + old \times low$

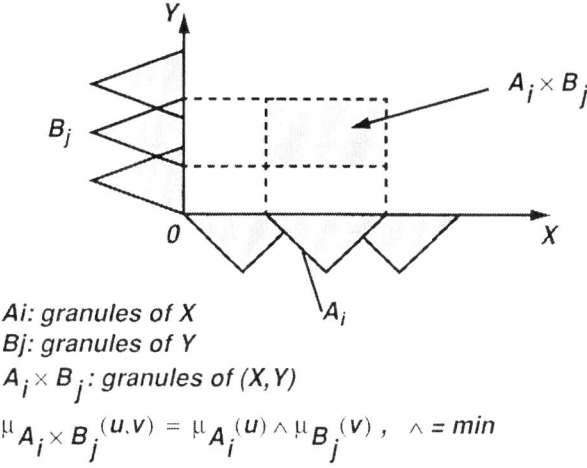

Fig. 4. Cartesian granulation. Granulation of X and Y induces granulation of (X, Y).

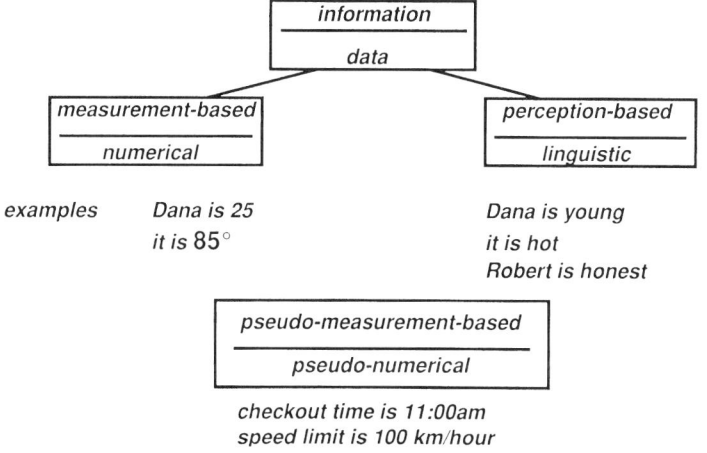

Fig. 5. Structure of information: measurement-based, perception-based and pseudo-measurement-based information.

which, as shown in Fig. 4, should be interpreted as a disjunction of cartesian products of linguistic values of *Age* and *Probability* (Zadeh, 1997; Pedrycz and Gomide, 1998).

An important observation is in order. If I were asked to estimate Carol's age, it would be unrealistic to expect that I would come up with a numerical probability distribution. But I would be able to describe my perception of the probability distribution of Carol's age in a natural language in which *Age* and *Probability* are represented—as described above—as linguistic, that is, granular variables (Zadeh, 1973, 1975, 1996, 1997).

Information which is conveyed by propositions drawn from a natural language will be said to be perception-based (Fig. 5). In my view, the most important

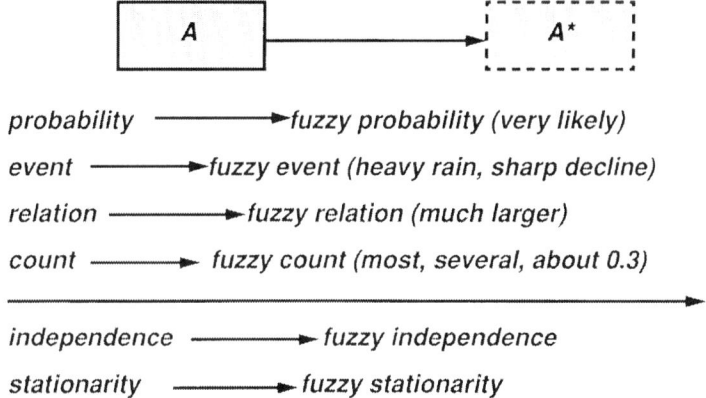

Fig. 6. f-Generalization (fuzzification). Fuzzification is a mode of generalization from crisp concepts to fuzzy concepts.

shortcoming of standard probability theory is that it does not have the capability to process perception-based information. It does not have this capability principally because there is no mechanism in PT for (a) representing the meaning of perceptions and (b) computing and reasoning with representations of meaning.

To add this capability to standard probability theory, three stages of generalization are required.

The first stage is referred to as f-generalization (Zadeh, 1997). In this mode of generalization, a point or a set is replaced by a fuzzy set. f-generalization of standard probability theory, PT, leads to a generalized probability theory which will be denoted as PT+. In relation to PT, PT+ has the capability to deal with

1. fuzzy numbers, quantifiers and probabilities, e.g., about 0.7, most, not very likely,
2. fuzzy events, e.g., warm day,
3. fuzzy relations, e.g., much larger than,
4. fuzzy truths and fuzzy possibilities, e.g., very true, quite possible.

In addition, PT+ has the potential—as yet largely unrealized—to fuzzify such basic concepts as independence, stationarity and causality. A move in this direction would be a significant paradigm shift in probability theory.

The second stage is referred to as f.g-generalization (fuzzy granulation) (Zadeh, 1997). In this mode of generalization, a point or a set is replaced by a granulated fuzzy set (Fig. 6). For example, a function, f, is replaced by its fuzzy graph, f^* (Fig. 7). f.g-generalization of PT leads to a generalized probability theory denoted as PT++.

PT++ adds to PT+ further capabilities which derive from the use of granulation. They are, mainly

1. linguistic (granular) variables,
2. linguistic (granular) functions and relations,

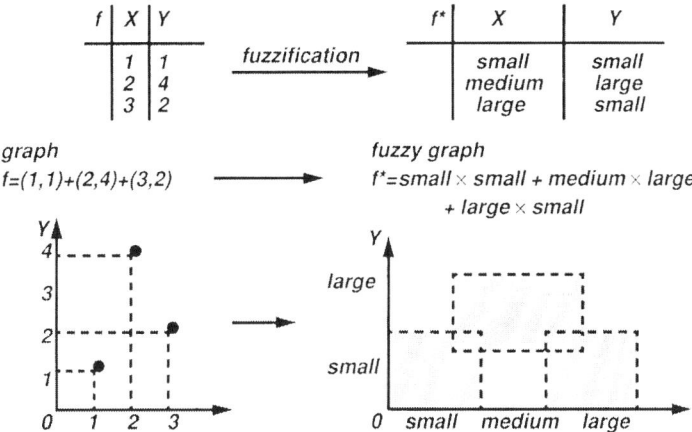

Fig. 7. Fuzzy graph of a function. A fuzzy graph is a generalization of the concept of a graph of a function.

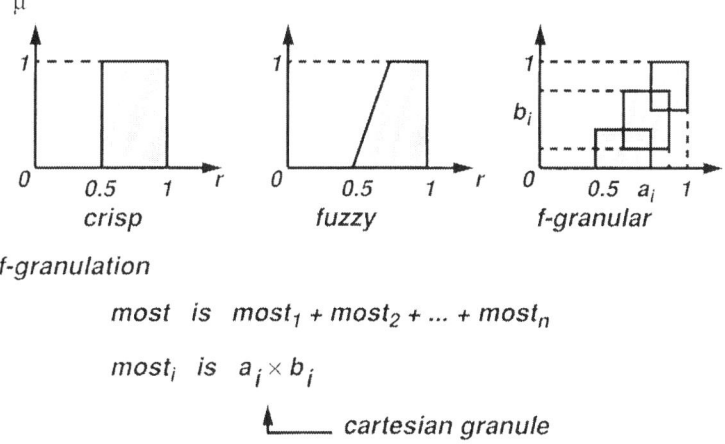

Fig. 8. Representation of *most*. Crisp, fuzzy and f-granular.

3. fuzzy rule-sets and fuzzy graphs,
4. granular goals and constraints,
5. granular probability distributions.

As a simple example, representation of the membership function of the fuzzy quantifier *most* (Zadeh, 1983) in PT, PT+ and PT++ is shown in Fig. 8.

The third stage is referred to a p-generalization (perceptualization). In this mode of generalization, what is added to PT++ is the capability to process perception-based information through the use of the computational theory of perceptions (CTP) (Zadeh, 1999, 2000). p-generalization of PT leads to what will be referred to as perception-based probability theory (PT_P).

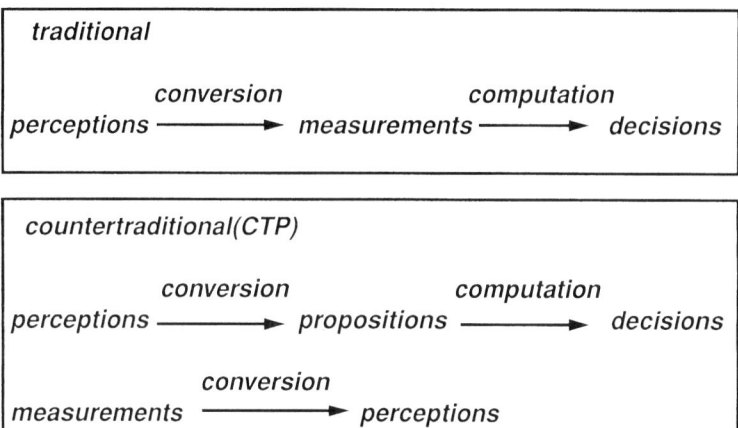

Fig. 9. Countertraditional conversion of measurements into perceptions. Traditionally, perceptions are converted into measurements.

The capability of PT_p to process perception-based information has an important implication. Specifically, it opens the door to a major enlargement of the role of natural languages in probability theory. As a simple illustration, instead of describing a probability distribution, P, analytically or numerically, as we normally do, P could be interpreted as a perception and described as a collection of propositions expressed in a natural language. A special case of such description is the widely used technique of describing a function via a collection of linguistic if–then rules (Zadeh, 1996). For example, the function shown in Fig. 7 may be described coarsely by the rule-set

f: if X is *small* then Y is *small*,
if X is *medium* then Y is *large*,
if X is *large* then Y is *small*,

with the understanding that the coarseness of granulation is a matter of choice.

In probability theory, as in other fields of science, it is a long-standing tradition to deal with perceptions by converting them into measurements. PT_p does not put this tradition aside. Rather, it adds to PT a countertraditional capability to convert measurements into perceptions, or to deal with perceptions directly, when conversion of perceptions into measurements is infeasible, unrealistic or counterproductive (Fig. 9).

There are three important points that are in need of clarification. First, when we allude to an enlarged role for natural languages in probability theory, what we have in mind is not a commonly used natural language but a subset which will be referred to as a precisiated natural language (PNL). In essence, PNL is a descriptive language which is intended to serve as a basis for representing the meaning of perceptions in a way that lends itself to computation. As will be seen later, PNL is a subset of a natural language which is equipped with constraint-centered semantics and is translatable into what is referred to as the generalized constraint language (GCL). At this point, it will suffice to observe that the descriptive power of PNL is much higher than that of the subset of a natural language which is translatable into predicate logic.

The second point is that in moving from measurements to perceptions, we move in the direction of lesser precision. The underlying rationale for this move is that precision carries a cost and that, in general, in any given situation there is a tolerance for imprecision that can be exploited to achieve tractability, robustness, lower cost and better rapport with reality.

The third point is that perceptions are more general than measurements and PT_p is more general that PT. Reflecting its greater generality, PT_p has a more complex mathematical structure than PT and is computationally more intensive. Thus, to exploit the capabilities of PT, it is necessary to have the capability to perform large volumes of computation at a low level of precision.

Perception-based probability theory goes far beyond standard probability theory both in spirit and in content. Full development of PT_p will be a long and tortuous process. In this perspective, my paper should be viewed as a sign pointing in a direction that departs from the deep-seated tradition of according more respect to numbers than to words.

Basically, perception-based probability theory may be regarded as the sum of standard probability theory and the computational theory of perceptions. The principal components of the computational theory of perceptions are (a) meaning representation and (b) reasoning. These components of CTP are discussed in the following sections.

2. The basics of perception-based probability theory; the concept of a generalized constraint

As was stated already, perception-based probability theory may be viewed as a p-generalization of standard probability theory. In the main, this generalization adds to PT the capability to operate on perception-based information through the use of the computational theory of perceptions. What follows is an informal precis of some of the basic concepts which underlie this theory.

To be able to compute and reason with perceptions, it is necessary to have a means of representing their meaning in a form that lends itself to computation. In CTP, this is done through the use of what is called constraint-centered semantics of natural languages (CSNL) (Zadeh, 1999).

A concept which plays a key role in CSNL is that of a generalized constraint (Zadeh, 1986). Introduction of this concept is motivated by the fact that conventional crisp constraints of the form $X \in C$, where X is a variable and C is a set, are insufficient to represent the meaning of perceptions.

A generalized constraint is, in effect, a family of constraints. An unconditional constraint on a variable X is represented as

$$X \text{ isr } R, \qquad (2.1)$$

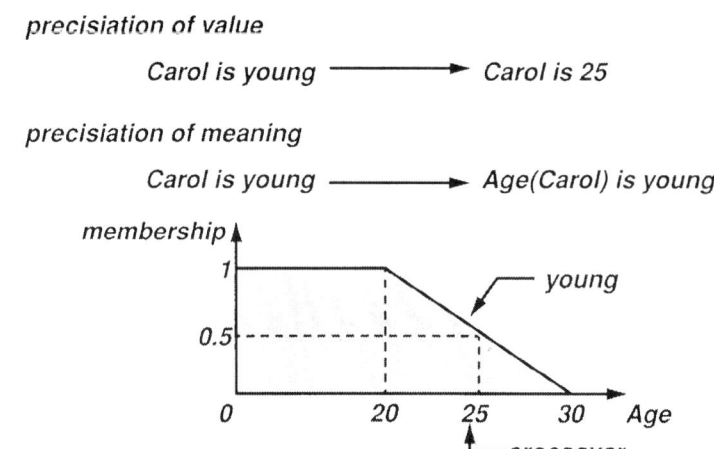

Fig. 10. Membership function of *young* (context-dependent). Two modes of precisiation.

where R is the constraining relation and is$r$, pronounced as ezar, is a variable copula in which the discrete-valued variable r defines the way in which R constrains X.

The principal constraints are the following:

$r :=$	equality constraint; $X = R$
r : blank	possibilistic constraint; X is R; R is the possibility distribution of X (Zadeh, 1978; Dubois and Prade, 1988)
$r : v$	veristic constraint; X is$v$ R; R is the verity distribution of X (Zadeh, 1999)
$r : p$	probabilistic constraint; X is$p$ R; R is the probability distribution of X
$r : pv$	probability-value constraint; X ispv R; X is the probability of a fuzzy event (Zadeh, 1968) and R is its value
$r : rs$	random set constraint; X isrs R; R is the fuzzy-set-valued probability distribution of X
$r : fg$	fuzzy graph constraint; X isfg R; X is a function and R is its fuzzy graph
$r : u$	usuality constraint; X is$u$ R; means: usually (X is R).

As an illustration, the constraint

Carol is *young*

in which *young* is a fuzzy set with a membership function such as shown in Fig. 10, is a possibilistic constraint on the variable X: $Age(Carol)$. This constraint defines the possibility distribution of X through the relation

$$\text{Poss}\{X = u\} = \mu_{young}(u),$$

where u is a numerical value of Age; μ_{young} is the membership function of *young*; and $\text{Poss}\{X = u\}$ is the possibility that Carol's age is u.

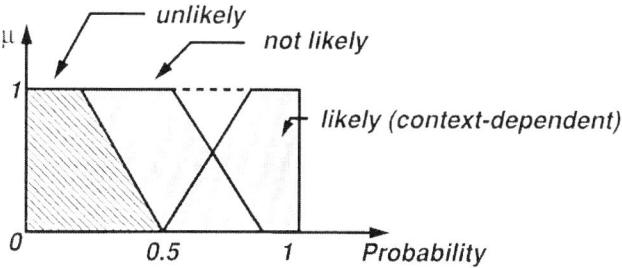

Fig. 11. Membership function of *likely* (context-dependent).

The veristic constraint

$$X \text{ isv } R \tag{2.2}$$

means that the verity (truth value) of the proposition $\{X = u\}$ is equal to the value of the verity distribution R at u. For example, in the proposition "Alan is half German, quarter French and quarter Italian", the verity of the proposition "Alan is German" is 0.5. It should be noted that the numbers 0.5 and 0.25 are not probabilities.

The probabilistic constraint

$$X \text{ is} p\ N(m, \sigma^2) \tag{2.3}$$

means that X is a normally distributed random variable with mean m and variance σ^2.

The proposition

$$p: \quad \text{it is likely that Carol is young} \tag{2.4}$$

may be expressed as the probability-value constraint

$$\text{Prob}\{Age(Carol) \text{ is } young\} \text{ is } likely. \tag{2.5}$$

In this expression, the constrained variable is X: $\text{Prob}\{Age(Carol) \text{ is } young\}$ and the constraint

$$X \text{ is } likely \tag{2.6}$$

is a possibilistic constraint in which *likely* is a fuzzy probability whose membership function is shown in Fig. 11.

In the random-set constraint, X is a fuzzy-set-valued random variable. Assuming that the values of X are fuzzy sets $\{A_i, i = 1, \ldots, n\}$ with respective probabilities $p_1, \ldots, p_n$, the random-set constraint on X is expressed symbolically as

$$X \text{ isrs } (p_1 \backslash A_1 + \cdots + p_n \backslash A_n). \tag{2.7}$$

It should be noted that a random-set constraint may be viewed as a combination of (a) a probabilistic constraint, expressed as

$$X \text{ is} p\ (p_1 \backslash u_1 + \cdots + p_n \backslash u_n), \quad u_i \in U \tag{2.8}$$

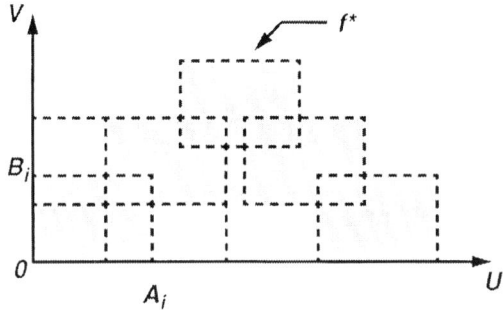

Fig. 12. Fuzzy-graph constraint. f^* is a fuzzy graph which is an approximate representation of f.

and a possibilistic constraint expressed as

$$(X, Y) \text{ is } R, \tag{2.9}$$

where R is a fuzzy relation defined on $U \times V$, with membership function $\mu_R : U \times V \to [0, 1]$.

If A_i is a section of R, defined as in Zadeh (1997) by

$$\mu_{A_i}(v) = \mu_R(u_i, v), \tag{2.10}$$

then the constraint on Y is a random-set constraint expressed as

$$Y \text{ isrs } (p_1 \backslash A_1 + \cdots + p_n \backslash A_n). \tag{2.11}$$

Another point that should be noted is that the concept of a random-set constraint is closely related to the Dempster–Shafer theory of evidence (Dempster, 1967; Shafer, 1976) in which the focal sets are allowed to be fuzzy sets (Zadeh, 1979).

In the fuzzy-graph constraint

$$X \text{ is} fg \ R, \tag{2.12}$$

the constrained variable, X, is a function, f, and R is a fuzzy graph (Zadeh, 1997) which plays the role of a possibility distribution of X. More specifically, if $f : U \times V \to [0, 1]$ and A_i, $i = 1, \ldots, m$ and B_j, $j = 1, \ldots, n$, are, respectively, fuzzy granules in U and V (Fig. 12), then the fuzzy graph of f is the disjunction of cartesian products (granules) $U_i \times V_j$, expressed as

$$f^* = \sum_{i=1, j=1}^{m,n} U_i \times V_j, \tag{2.13}$$

with the understanding that the symbol $\sum$ should be interpreted as the union rather than as an arithmetic sum, and U_i and V_j take values in the sets $\{A_1, \ldots, A_m\}$ and $\{B_1, \ldots, B_n\}$, respectively.

A fuzzy graph of f may be viewed as an approximate representation of f. Usually, the granules A_i and B_j play the role of values of linguistic variables. Thus, in the case of the function shown in Fig. 7, its fuzzy graph may be expressed as

$$f^* = small \times small + medium \times large + large \times small. \tag{2.14}$$

Equivalently, if f is written as $Y = f(X)$, then f^* may be expressed as the rule-set

f^*: if X is *small* then Y is *small*,
if X is *medium* then Y is *large*,
if X is *large* then Y is *small*. (2.15)

This rule-set may be interpreted as a description—in a natural language—of a perception of f.

The usuality constraint is a special case of the probability-value constraint. Thus,

X isu A (2.16)

should be interpreted as an abbreviation of

usually $(X$ is $A)$, (2.17)

which in turn may be interpreted as

Prob$\{X$ is $A\}$ is *usually*, (2.18)

with *usually* playing the role of a fuzzy probability which is close to 1. In this sense, A is a *usual* value of X. More generally, A is a *usual* value of X if the fuzzy probability of the fuzzy event $\{X$ is $A\}$ is close to one and A has high specificity, that is, has a tight possibility distribution, with tightness being a context-dependent characteristic of a fuzzy set. It is important to note that, unlike the concept of the expected value, the usual value of a random variable is not uniquely determined by its probability distribution. What this means is that the usual value depends on the calibration of the context-dependent natural language predicates "close to one" and "high specificity".

The difference between the concepts of the expected and usual values goes to the heart of the difference between precise and imprecise probability theories. The expected value is precisely defined and unique. The usual value is context-dependent and hence is not unique. However, its definition is precise if the natural language predicates which occur in its definition are defined precisely by their membership functions. In this sense, the concept of the usual value has a flexibility that the expected value does not have. Furthermore, it may be argued that the concept of the usual value is closer to our intuitive perception of "expected value" than the concept of the expected value as it is defined in PT.

In the foregoing discussion, we have focused our attention on unconditional generalized constraints. More generally, a generalized constraint may be conditional, in which case it is expressed in a generic form as an if–then rule

if X isr R then Y iss S (2.19)

or, equivalently, as

Y iss S if X isr R. (2.20)

Furthermore, a generalized constraint may be exception-qualified, in which case it is expressed as

X isr R unless Y iss S. (2.21)

A generalized rule-set is a collection of generalized if–then rules which collectively serve as an approximate representation of a function or a relation. Equivalently, a generalized rule-set may be viewed as a description of a perception of a function or a relation.

As an illustration, consider a function, $f: (U \times V) \to [0,1]$, expressed as $Y = f(X)$, where U and V are the domains of X and Y, respectively. Assume that U and V are granulated, with the granules of U and V denoted, respectively, as A_i, $i = 1, \ldots, m$, and B_j, $j = 1, \ldots, n$. Then, a generic form of a generalized rule set may be expressed as

$$f^*: \quad \{\text{if } X \text{ is} r \ U_i \text{ then } Y \text{ is} s \ V_j\} \quad i = 1, \ldots, m, \ j = 1, \ldots, n, \qquad (2.22)$$

where U_i and V_j take values in the sets $\{A_1, \ldots, A_m\}$ and $\{B_1, \ldots, B_n\}$, respectively. In this expression, f^* represents a fuzzy graph of f.

A concept which plays a key role in the computational theory of perceptions is that of the Generalized Constraint Language, GCL (Zadeh, 1999). Informally, GCL is a meaning-representation language in which the principal semantic elements are generalized constraints. The use of generalized constraints as its semantic elements makes a GCL a far more expressive language than conventional meaning-representation languages based on predicate logic.

3. Meaning-representation: constraint-centered semantics of natural languages

In perception-based probability theory, perceptions—and, in particular, perceptions of likelihood, dependency, count and variations in time and space—are described by propositions drawn from a natural language. To mechanize reasoning with perceptions, it is necessary to have a method of representing the meaning of propositions in a way that lends itself to computation. In the computational theory of perceptions, a system that is used for this purpose is called the constraint-centered semantics of natural language (CSNL) (Zadeh, 1999).

Meaning-representation is a central part of every logical system. Why, then, is it necessary to introduce a system that is significantly different from the many meaning-representation methods that are in use? The reason has to do with the intrinsic imprecision of perceptions and, more particularly, with their f-granularity. It is this characteristic of perceptions that puts them well beyond the expressive power of conventional meaning-representation methods, most of which are based on predicate logic.

To illustrate, consider the following simple perceptions:

- Ann is much younger than Mary.
- A box contains black and white balls of various sizes. Most are large. Most of the large balls are black.
- Usually it is rather cold in San Francisco during the summer.
- It is very unlikely that there will be a significant increase in the price of oil in the near future.

Conventional meaning-representation methods do not have the capability to represent the meaning of such perceptions in a form that lends itself to computation.

A key idea which differentiates CSNL from conventional methods is that the meaning of a proposition, p, drawn from a natural language, is represented as a generalized constraint, with the understanding that the constrained variable and the constraining relation are, in general, implicit rather than explicit in p. For example, in the proposition

p: it is likely that Kate is young,

the constraint is possibilistic; the constrained variable is the probability that Kate is young; and the constraining relation is *likely*.

The principal ideas and assumptions which underlie CSNL may be summarized as follows:

1. Perceptions are described by propositions drawn from a natural language.
2. A proposition, p, may be viewed as an answer to a question.
   In general, the question is implicit and not unique. For example, the proposition "Carol is young" may be viewed as an answer to the question: "How old is Carol", or as the answer to "Who is young?"
3. A proposition is a carrier of information.
4. The meaning of a proposition, p, is represented as a generalized constraint which defines the information conveyed by p.
5. Meaning-representation is viewed as translation from a language into the GCL.

In CSNL, translation of a proposition, p, into GCL is equated to explicitation of the generalized constraint which represents the meaning of p. In symbols

$$p \; \frac{\text{translation}}{\text{explicitation}} \; X \text{ isr } R. \tag{3.1}$$

The right-hand member of this relation is referred to as a canonical form of p, written as $CF(p)$. Thus, the canonical form of p places in evidence (a) the constrained variable which, in general, is implicit in p; (b) the constraining relation, R; and (c) the copula variable r which defines the way in which R constrains X.

The canonical form of a question, q, may be expressed as

$$CF(q): \quad X \text{ isr } ?R \tag{3.2}$$

and read as "What is the generalized value of X?"

Similarly, the canonical form of p, viewed as an answer to q, is expressed as

$$CF(p): \quad X \text{ isr } R \tag{3.3}$$

and reads "The generalized value of X isr R".

As a simple illustration, if the question is "How old is Carol?", its canonical form is

$$CF(q): \quad Age(Carol) \text{ is } ?R. \tag{3.4}$$

Correspondingly, the canonical form of

$$p: \quad \text{Carol is young} \tag{3.5}$$

is

$$CF(p): \quad Age(Carol) \text{ is } young. \tag{3.6}$$

If the answer to the question is

$$p: \quad \text{it is likely that Carol is young} \tag{3.7}$$

then

$$CF(p): \quad \text{Prob}\{Age(Carol) \text{ is } young\} \text{ is } likely. \tag{3.8}$$

More explicitly, if *Age(Carol)* is a random variable with probability density g, then the probability measure (Zadeh, 1968) of the fuzzy event "Carol is young" may be expressed as

$$\int_0^{120} \mu_{young}(u) g(u) \, du, \tag{3.9}$$

where μ_{young} is the membership function of *young*. Thus, in this interpretation the constrained variable is the probability density g, and, as will be seen later, the membership function of the constraining relation is given by

$$\mu_R(g) = \mu_{likely} \left(\int_0^{120} \mu_{young}(u) g(u) \, du \right). \tag{3.10}$$

A concept which plays an important role in CSNL is that of cardinality, that is, the count of elements in a fuzzy set (Zadeh, 1983; Ralescu, 1995; Hajek, 1998). Basically, there are two ways in which cardinality can be defined: (a) crisp cardinality and (b) fuzzy cardinality (Zadeh, 1983; Ralescu et al., 1995; Ralescu, 1995). In the case of (a), the count of elements in a fuzzy set is a crisp number; in the case of (b) it is a fuzzy number. For our purposes, it will suffice to restrict our attention to the case where a fuzzy set is defined on a finite set and is associated with a crisp count of its elements.

More specifically, consider a fuzzy set A defined on a finite set $U = \{u_1, \ldots, u_n\}$ through its membership function $\mu_A: U \to [0, 1]$. The sigma-count of A is defined as

$$\sum Count(A) = \sum_{i=1}^{n} \mu_A(u_i). \tag{3.11}$$

If A and B are fuzzy sets defined on U, then the relative sigma-count, $\sum Count(A/B)$, is defined as

$$\sum Count(A/B) = \frac{\sum_{i=1}^{n} \mu_A(u_i) \wedge \mu_B(u_i)}{\sum_{i=1}^{n} \mu_B(u_i)}, \tag{3.12}$$

where $\wedge = \min$, and summations are arithmetic.

As a simple illustration, consider the perception

   p: most Swedes are tall.

In this case, the canonical form of p may be expressed as

$$CF(p): \quad \sum Count(tall \cdot Swedes/Swedes) \text{ is } \frac{1}{n}\sum_{i=1}^{n} \mu_{tall \cdot Swede}(u_i), \tag{3.13}$$

where u_i is the height of the ith Swede and $\mu_{tall \cdot Swede}(u_i)$ is the grade of membership of the ith Swede in the fuzzy set of tall Swedes.

In a general setting, how can a given proposition, p, be expressed in its canonical form? A framework for translation of propositions drawn from a natural language into GCL is partially provided by the conceptual structure of test-score semantics (Zadeh, 1981). In this semantics, X and R are defined by procedures which act on an explanatory database, ED, with ED playing the role of a collection of possible worlds in possible world semantics (Cresswell, 1973). As a very simple illustration, consider the proposition (Zadeh, 1999)

   p: Carol lives in a small city near San Francisco

and assume that the explanatory database consists of three relations:

$$\begin{aligned} ED = {} & POPULATION[Name; Residence] \\ & + SMALL[City; \mu] \\ & + NEAR[City1; City2; \mu]. \end{aligned} \tag{3.14}$$

In this case,

$$X = Residence(Carol) = {}_{Residence} POPULATION[Name = Carol], \tag{3.15}$$

$$R = SMALL[City; \mu] \cap_{City1} NEAR[City2 = San\_Francisco]. \tag{3.16}$$

In R, the first constituent is the fuzzy set of small cities; the second constituent is the fuzzy set of cities which are near San Francisco; and $\cap$ denotes the intersection of these sets. Left subscripts denote projections, as defined in Zadeh (1981).

There are many issues relating to meaning-representation of perception-based information which go beyond the scope of the present paper. The brief outline presented in this section is sufficient for our purposes. In the following section, our attention will be focused on the basic problem of reasoning based on generalized constraint propagation. The method which will be outlined contains as a special case a basic idea suggested in an early paper of Good (1962). A related idea was employed in Zadeh (1955).

4. Reasoning based on propagation of generalized constraints

One of the basic problems in probability theory is that of computation of the probability of a given event from a body of knowledge which consists of information about the relevant functions, relations, counts, dependencies and probabilities of related events.

As was alluded to earlier, in many cases the available information is a mixture of measurements and perceptions. Standard probability theory provides a vast array of tools for dealing with measurement-based information. But what is not provided is a machinery for dealing with information which is perception-based. This limitation of PT is exemplified by the following elementary problems—problems in which information is perception-based.

1. X is a normally distributed random variable with small mean and small variance.
   Y is much larger than X.
   What is the probability that Y is neither small nor large?
2. Most Swedes are tall.
   Most Swedes are blond.
   What is the probability that a Swede picked at random is tall and blond?
3. Consider a perception-valued times series

   $$T = \{t_1, t_2, t_3, \ldots\},$$

   in which the t_i's are perceptions of, say temperature, e.g., warm, very warm, cold,.... For simplicity, assume that the t_i's are independent and identically distributed. Furthermore, assume that the t_i's range over a finite set of linguistic values, $A_1, A_2, \ldots, A_n$, with respective probabilities $P_1, \ldots, P_n$. What is the average value of T?

To be able to compute with perceptions, it is necessary, as was stressed already, to have a mechanism for representing their meaning in a form that lends itself to computation. In the computational theory of perceptions, this purpose is served by the constraint-centered semantics of natural languages. Through the use of CSNL, propositions drawn from a natural language are translated into the GCL.

The second stage of computation involves generalized constraint propagation from premises to conclusions. Restricted versions of constraint propagation are considered in Zadeh (1979), Bowen et al. (1992), Dubois et al. (1993), Katai et al. (1992) and Yager (1989). The main steps in generalized constraint propagation are summarized in the following. As a preliminary, a simple example is analyzed.

Assume that the premises consist of two perceptions:

p_1:   most Swedes are tall,
p_2:   most Swedes are blond.

and the question, q, is: What fraction of Swedes are tall and blond? This fraction, then, will be the linguistic value of the probability that a Swede picked at random is tall and blond.

To answer the question, we first convert p_1, p_2 and q into their canonical forms:

$$CF(p_1): \quad \sum Count(tall \, . \, Swedes/Swedes) \text{ is } most, \tag{4.1}$$

$$CF(p_2): \quad \sum Count(blond \, . \, Swedes/Swedes) \text{ is } most, \tag{4.2}$$

$$CF(q): \quad \sum Count(tall \cap blond \, . \, Swedes/Swedes) \text{ is } ?Q, \tag{4.3}$$

where Q is the desired fraction.

Next, we employ the identity (Zadeh, 1983)

$$\sum Count(A \cap B) + \sum Count(A \cup B) = \sum Count(A) + \sum Count(B), \quad (4.4)$$

in which A and B are arbitrary fuzzy sets. From this identity, we can readily deduce that

$$\sum Count(A) + \sum Count(B) - 1 \leqslant \sum Count(A \cap B)$$
$$\leqslant \min(\sum Count(A), \sum Count(B)), \quad (4.5)$$

with the understanding that the lower bound is constrained to lie in the interval $[0, 1]$. It should be noted that the identity in question is a generalization of the basic identity for probability measures

$$P(A \cap B) + P(A \cup B) = P(A) + P(B). \quad (4.6)$$

Using the information conveyed by canonical forms, we obtain the bounds

$$2most - 1 \leqslant \sum Count(tall \cap blond \, . \, Swedes/Swedes) \leqslant most, \quad (4.7)$$

which may be expressed equivalently as

$$\sum Count(tall \cap blond \, . \, Swedes/Swedes) \text{ is } \leqslant most \cap \geqslant (2most - 1). \quad (4.8)$$

Now

$$\leqslant most = [0, 1] \quad (4.9)$$

and

$$\geqslant (2most - 1) = 2most - 1, \quad (4.10)$$

in virtue of monotonicity of *most* (Zadeh, 1999).

Consequently,

$$\sum Count(tall \cap blond \, . \, Swedes/Swedes) \text{ is } 2most - 1 \quad (4.11)$$

and hence the answer to the question is

$$a: \quad (2most - 1) \text{ Swedes are } tall \text{ and } blond. \quad (4.12)$$

In a more general setting, the principal elements of the reasoning process are the following.

1. Question (query), q. The canonical form of q is assumed to be

$$X \text{ isr } ?Q. \quad (4.13)$$

2. Premises. The collection of premises expressed in a natural language constitutes the initial data set (IDS).
3. Additional premises which are needed to arrive at an answer to q. These premises constitute the external data set (EDS). Addition of EDS to IDS results in what is referred to as the augmented data set (IDS+).

Example. Assume that the initial data set consists of the propositions

p_1: Carol lives near Berkeley,

p_2: Pat lives near Palo Alto.

Suppose that the question is: How far is Carol from Pat? The external data set in this case consists of the proposition

> distance between Berkeley and Palo Alto is approximately 45 miles. (4.14)

4. Through the use of CSNL, propositions in IDS+ are translated into the GCL. The resulting collection of generalized constraints is referred to as the augmented initial constraint set ICS+.
5. With the generalized constraints in ICS+ serving as antecedent constraints, the rules which govern generalized constraint propagation in CTP are applied to ICS+, with the goal of deducing a set of generalized constraints, referred to as the terminal constraint set, which collectively provide the information which is needed to compute q.

The rules governing generalized constraint propagation in the computational theory of perceptions coincide with the rules of inference in fuzzy logic (Zadeh, 1999, 2000). In general, the chains of inference in CTP are short because of the intrinsic imprecision of perceptions. The shortness of chains of inference greatly simplifies what would otherwise be a complex problem, namely, the problem of selection of rules which should be applied in succession to arrive at the terminal constraint set. This basic problem plays a central role in theorem proving in the context of standard logical systems (Fikes and Nilsson, 1971).

6. The generalized constraints in the terminal constraint set are re-translated into a natural language, leading to the terminal data set. This set serves as the answer to the posed question. The process of re-translation is referred to as linguistic approximation (Pedrycz and Gomide, 1998). Re-translation will not be addressed in this paper.

The basic rules which govern generalized constraint propagation are of the general form

$$\frac{\begin{array}{c} p_1 \\ p_2 \\ \cdot \\ \cdot \\ \cdot \\ p_k \end{array}}{p_{k+1},} \quad (4.15)$$

where $p_1, \ldots, p_k$ are the premises and p_{k+1} is the conclusion. Generally, $k = 1$ or 2.

In a generic form, the basic constraint-propagation rules in CTP are expressed as follows (Zadeh, 1999):

1. *Conjunctive rule* 1:

$$\frac{\begin{array}{ll} X & \text{isr} \quad R \\ X & \text{iss} \quad S \end{array}}{X \quad \text{ist} \quad T.} \tag{4.16}$$

The different symbols r, s, t in constraint copulas signify that the constraints need not be of the same type.

2. *Conjunctive rule* 2:

$$\frac{\begin{array}{ll} X & \text{isr} \quad R \\ Y & \text{iss} \quad S \end{array}}{(X,Y) \quad \text{ist} \quad T.} \tag{4.17}$$

3. *Disjunctive rule* 1:

$$\frac{\begin{array}{ll} & X \quad \text{isr} \quad R \\ \text{or} & X \quad \text{iss} \quad S \end{array}}{X \quad \text{ist} \quad T.} \tag{4.18}$$

4. *Disjunctive rule* 2:

$$\frac{\begin{array}{ll} & X \quad \text{isr} \quad R \\ \text{or} & Y \quad \text{iss} \quad S \end{array}}{(X,Y) \quad \text{ist} \quad T.} \tag{4.19}$$

5. *Projective rule*:

$$\frac{(X,Y) \quad \text{isr} \quad R}{Y \quad \text{iss} \quad S.} \tag{4.20}$$

6. *Surjective rule*:

$$\frac{X \quad \text{isr} \quad R}{(X,Y) \quad \text{iss} \quad S.} \tag{4.21}$$

7. *Inversive rule*:

$$\frac{f(X) \quad \text{isr} \quad R}{X \quad \text{iss} \quad S,} \tag{4.22}$$

where $f(X)$ is a function of X.

From these basic rules the following frequently used rules may be derived:

8. *Compositional rule*:

$$\frac{\begin{array}{ll} X & \text{isr} \quad R \\ (X,Y) & \text{iss} \quad S \end{array}}{Y \quad \text{ist} \quad T.} \tag{4.23}$$

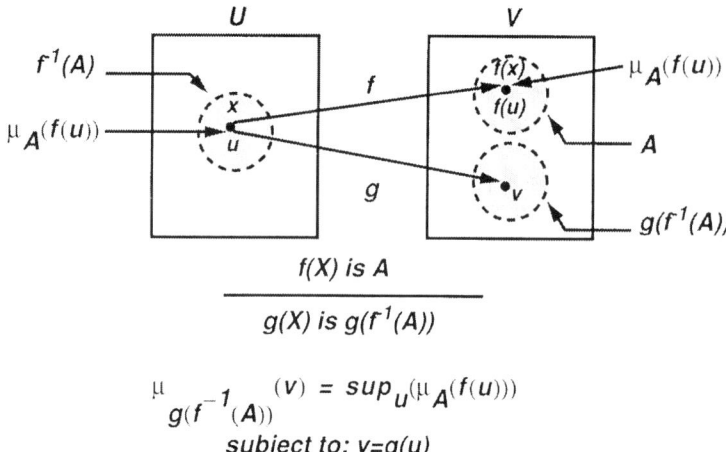

Fig. 13. Generalized extension principle. Constraint on $f(X)$ induces a constraint on $g(X)$.

9. *Generalized extension principle*:

$$\frac{f(X) \quad \text{isr} \quad R}{g(X) \quad \text{iss} \quad S,} \tag{4.24}$$

where f and g are given functions. The generalized extension principle is the principal rule of inference in fuzzy logic.

The generic rules lead to specialized rules for various types of constraints. In particular, for possibilistic constraints we have, for example (Pedrycz and Gomide, 1998)

Conjunctive rule 1:

$$\frac{\begin{array}{ll} X & \text{is} \quad R \\ X & \text{is} \quad S \end{array}}{X \quad \text{is} \quad R \cap S,} \tag{4.25}$$

where R and S are fuzzy sets and $R \cap S$ is their intersection.

Compositional rule:

$$\frac{\begin{array}{ll} X & \text{is} \quad R \\ (X,Y) & \text{is} \quad S \end{array}}{Y \quad \text{is} \quad R \bullet S,} \tag{4.26}$$

where $R \bullet S$ is the composition of R and S. If conjunction and disjunction are identified with min and max, respectively, then

$$\mu_{R \bullet S}(v) = \max_u(\min(\mu_R(u), \mu_S(u,v))), \tag{4.27}$$

where μ_R and μ_S are the membership functions of R and S.

Generalized extension principle (Fig. 13):

$$\frac{f(X) \text{ is } R}{g(X) \text{ is } g(f^{-1}(R)),} \tag{4.28}$$

where

$$\mu_{g(f^{-1}(R))}(v) = \max_{u|v=g(u)} \mu_R(f(u)). \qquad (4.29)$$

Compositional rule for probabilistic constraints (Bayes' rule):

$$\frac{\begin{array}{ll} X & \text{is}p \ \ R \\ Y|X & \text{is}p \ \ S \end{array}}{Y \ \ \text{is}p \ \ R \bullet S,} \qquad (4.30)$$

where $Y|X$ denotes Y conditioned on X, and $R \bullet S$ is the composition of the probability distributions R and S.

Compositional rule for probabilistic and possibilistic constraints (random-set constraint):

$$\frac{\begin{array}{ll} X & \text{is}p \ \ R \\ (X,Y) & \text{is} \ \ S \end{array}}{Y \ \ \text{is}rs \ \ T,} \qquad (4.31)$$

where T is a random set. As was stated at an earlier point, if X takes values in a finite set $\{u_1,\ldots,u_n\}$ with respective probabilities $p_1,\ldots,p_n$, then the constraint X is$p$ R may be expressed compactly as

$$X \text{ is}p \ \left(\sum_{i=1}^{n} p_i \backslash u_i \right). \qquad (4.32)$$

When X takes a value u_i, the possibilistic constraint (X,Y) is S induces a constraint on Y which is given by

$$Y \text{ is } S_i, \qquad (4.33)$$

where S_i is a fuzzy set defined by

$$S_i = S(u_i, Y). \qquad (4.34)$$

From this it follows that when X takes the values $u_1,\ldots,u_n$ with respective probabilities $p_1,\ldots,p_n$, the fuzzy-set-valued probability distribution of Y may be expressed as

$$Y \text{ is}p \ \left(\sum_{i=1}^{n} p_i \backslash S_i \right). \qquad (4.35)$$

This fuzzy-set-valued probability distribution defines the random set T in the random-set constraint

$$Y \text{ is}rs \ T. \qquad (4.36)$$

Conjunctive rule for random set constraints: For the special case in which R and S in the generic conjunctive rule are random fuzzy sets as defined above, the rule assumes a more specific form:

$$X \text{ isrs } \sum_{i=1}^{m} p_i \backslash R_i$$

$$X \text{ isrs } \sum_{j=1}^{n} q_j \backslash S_j \qquad (4.37)$$

$$\overline{X \text{ isrs } \sum_{i=1,j=1}^{m,n} p_i q_j \backslash (R_i \cap S_j).}$$

In this rule, R_i and S_i are assumed to be fuzzy sets. When R_i and S_i are crisp sets, the rule reduces to the Dempster rule of combination of evidence (Dempster, 1967; Shafer, 1976). An extension of Dempster's rule to fuzzy sets was described in a paper dealing with fuzzy information granularity (Zadeh, 1979). It should be noted that in (4.37) the right-hand member is not normalized, as it is in the Dempster–Shafer theory (Strat, 1992).

The few simple examples discussed above demonstrate that there are many ways in which generic rules can be specialized, with each specialization leading to a distinct theory in its own right. For example, possibilistic constraints lead to possibility theory (Zadeh, 1978; Dubois and Prade, 1988); probabilistic constraints lead to probability theory; and random-set constraints lead to the Dempster–Shafer theory of evidence. In combination, these and other specialized rules of generalized constraint propagation provide the machinery that is needed for a mechanization of reasoning processes in the logic of perceptions and, more particularly, in a perception-based theory of probabilistic reasoning with imprecise probabilities.

As an illustration, let us consider a simple problem that was stated earlier—a typical problem which arises in situations in which the decision-relevant information is perception-based. Given the perception: Usually Robert returns from work at about 6 p.m.; the question is: What is the probability that he is home at 6:30 p.m.?

An applicable constraint-propagation rule in this case is the generalized extension principle. More specifically, let g denote the probability density of the time at which Robert returns from work. The initial data set is the proposition

p: usually Robert returns from work at about 6 p.m.

This proposition may be expressed as the usuality constraint

$$X \text{ isu } 6^*, \qquad (4.38)$$

where 6^* is an abbreviation for "about 6 p.m.", and X is the time at which Robert returns from work. Equivalently, the constraint in question may be expressed as

p: Prob$\{X \text{ is } 6^*\}$ is *usually*.   (4.39)

Using the definition of the probability measure of a fuzzy event (Zadeh, 1968), the constraint on g may be expressed as

$$\int_0^{12} g(u) \mu_{6^*}(u) \, du \text{ is } usually, \qquad (4.40)$$

where $\mu_{6^*}(u)$ is the membership function of 6^* (Fig. 14).

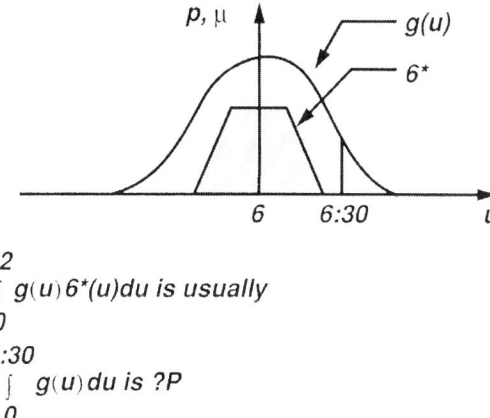

$$\int_0^{12} g(u)6^*(u)du \text{ is usually}$$

$$\int_0^{6:30} g(u)du \text{ is } ?P$$

Fig. 14. Application of the generalized extension principle. P is the probability that Robert is at home at 6:30 p.m.

Let $P(g)$ denote the probability that Robert is at home at 6:30 p.m. This probability would be a number if g were known. In our case, information about g is conveyed by the given usuality constraint. This constraint defines the possibility distribution of g as a functional:

$$\mu(g) = \mu_{usually}\left(\int_0^{12} g(u)\mu_{6^*}(u)\,du\right). \quad (4.41)$$

In terms of g, the probability that Robert is home at 6:30 p.m. may be written as a functional:

$$P(g) = \int_0^{6:30} g(u)\,du. \quad (4.42)$$

The generalized extension principle reduces computation of the possibility distribution of P to the solution of the variational problem

$$\mu_P(v) = \max_g \left(\mu_{usually}\left(\int_0^{12} g(u)\mu_{6^*}(u)\,du\right)\right) \quad (4.43)$$

subject to

$$v = \int_0^{6:30} g(u)\,du.$$

The reduction of inference to solution of constrained variational problems is a basic feature of fuzzy logic (Zadeh, 1979).

Solution of variational problems of form (4.43) may be simplified by a discretization of g. Thus, if u is assumed to take values in a finite set $U = \{u_1, \ldots, u_n\}$, and the respective probabilities are $p_1, \ldots, p_n$, then the variational problem (4.43) reduces to the nonlinear program

$$\mu_P(v) = \max_P \left(\mu_{usually}\left(\sum_{i=1}^n p_i\mu_{6^*}(u_i)\right)\right) \quad (4.44)$$

subject to

$$v = \sum_{j=1}^{m} P_j,$$

$$0 \leqslant P_j \leqslant 1,$$

$$\sum_{i=1}^{n} P_i = 1,$$

where $p = (p_1, \ldots, p_n)$, and m is such that $u_m = 6{:}30$.

In general, probabilities serve as a basis for making a rational decision. As an illustration, assume that I want to call Robert at home at 6:30 p.m. and have to decide on whether I should call him person-to-person or station-to-station. Assume that we have solved the variational problem (4.43) and have in hand the value of P defined by its membership function $\mu_P(v)$. Furthermore, assume that the costs of person-to-person and station-to-station calls are a and b, respectively.

Then the expected cost of a person-to-person call is

$$A = aP,$$

while that of a station-to-station call is

$$B = b,$$

where A is a fuzzy number defined by (Kaufmann and Gupta, 1985)

$$\mu_A(v) = a\mu_P(v).$$

More generally, if X is a random variable taking values in the set of numbers $U = \{a_1, \ldots, a_n\}$ with respective imprecise (fuzzy) probabilities $P_1, \ldots, P_n$, then the expected value of X is the fuzzy number (Zadeh, 1975; Kruse and Meyer, 1987)

$$E(X) = \sum_{i=1}^{n} a_i P_i. \tag{4.45}$$

The membership function of $E(X)$ may be computed through the use of fuzzy arithmetic (Kaufmann and Gupta, 1985; Mares, 1994). More specifically, if the membership functions of P_i are μ_i, then the membership function of $E(X)$ is given by the solution of the variational problem

$$\mu_{E(X)}(v) = \max_{u_1, \ldots, u_n} (\mu_{P_1}(u_1) \wedge \cdots \wedge \mu_{P_n}(u_n)) \tag{4.46}$$

subject to the constraints

$$0 \leqslant u_i \leqslant 1,$$

$$\sum_{i=1}^{n} u_i = 1,$$

$$v = \sum_{i=1}^{n} a_i u_i.$$

Returning to our discussion of the Robert example, if we employ a generalized version of the principle of maximization of expected utility to decide on how to place the call, then the problem reduces to that of ranking the fuzzy numbers A and B. The problem of ranking of fuzzy numbers has received considerable attention in the literature (see Pedrycz and Gomide, 1998), and a number of ranking algorithms have been described.

Our discussion of the Robert example is aimed at highlighting some of the principal facets of the perception-based approach to reasoning with imprecise probabilities. The key point is that reasoning with perception-based information may be reduced to solution of variational problems. In general, the problems are computationally intensive, even for simple examples, but well within the capabilities of desktop computers. Eventually, novel methods of computation involving neural computing, evolutionary computing, molecular computing or quantum computing may turn out to be effective in computing with imprecise probabilities in the context of perception-based information.

As a further illustration of reasoning with perception-based information, it is instructive to consider a perception-based version of a basic problem in probability theory.

Let X and Y be random variables in U and V, respectively. Let f be a mapping from U to V. The basic problem is: Given the probability distribution of $X, P(X)$, what is the probability distribution of Y?

In the perception-based version of this problem it is assumed that what we know are perceptions of f and $P(X)$, denoted as f^* and $P^*(X)$, respectively. More specifically, we assume that X and f are granular (linguistic) variables and f^* is described by a collection of granular (linguistic) if–then rules:

$$f^*: \quad \{\text{if } X \text{ is } A_i \text{ then } Y \text{ is } B_i\}, \quad i = 1, \ldots, m, \tag{4.47}$$

where A_i and B_i are granules of X and Y, respectively (Fig. 12). Equivalently, f^* may be expressed as a fuzzy graph

$$f^* = \sum_{i=1}^{m} A_i \times B_i, \tag{4.48}$$

where $A_i \times B_i$ is a cartesian granule in $U \times V$. Furthermore, we assume that the perception of $P(X)$ is described as

$$P^*(X) \text{ is } \sum_{j=1}^{n} p_j \backslash C_j, \tag{4.49}$$

where the C_j are granules of U, and

$$p_j = \text{Prob}\{X \text{ is } C_j\}. \tag{4.50}$$

Now, let $f^*(C_j)$ denote the image of C_j. Then, application of the extension principle yields

$$f^*(C_j) = \sum_{i=1}^{m} m_{ij} \wedge B_i, \tag{4.51}$$

where the matching coefficient, m_{ij}, is given by

$$m_{ij} = \sup(A_i \cap C_j), \tag{4.52}$$

with the understanding that

$$\sup(A_i \cap C_j) = \sup_u (\mu_{A_i}(u) \wedge \mu_{C_j}(u)), \qquad (4.53)$$

where $u \in U$ and μ_{A_i} and μ_{C_j} are the membership functions of A_i and C_j, respectively. In terms of $f^*(C_j)$, the probability distribution of Y may be expressed as

$$P^*(Y) \text{ is } \sum_{j=1}^{n} p_j \backslash f^*(C_j) \qquad (4.54)$$

or, more explicitly, as

$$P^*(Y) \text{ is } \sum_{j=1}^{n} p_j \backslash \left(\sum_i m_{ij} \wedge B_i \right). \qquad (4.55)$$

What these examples show is that computation with perception-based functions and probability distribution is both more general and more complex than computation with their measurement-based counterparts.

5. Concluding remarks

The perception-based theory of probabilistic reasoning which is outlined in this paper may be viewed as an attempt to add to probability theory a significant capability—a capability to operate on information which is perception-based. It is this capability that makes it possible for humans to perform a wide variety of physical and mental tasks without any measurements and any computations.

Perceptions are intrinsically imprecise, reflecting a fundamental limitation on the cognitive ability of humans to resolve detail and store information. Imprecision of perceptions places them well beyond the scope of existing meaning-representation and deductive systems. In this paper, a recently developed computational theory of perceptions is used for this purpose. Applicability of this theory depends in an essential way on the ability of modern computers to perform complex computations at a low cost and high reliability.

Natural languages may be viewed as systems for describing perceptions. Thus, to be able to operate on perceptions, it is necessary to have a means of representing the meaning of propositions drawn from a natural language in a form that lends itself to computation. In this paper, the so-called constraint-centered semantics of natural languages serves this purpose.

A conclusion which emerges from these observations is that to enable probability theory to deal with perceptions, it is necessary to add to it concepts and techniques drawn from semantics of natural languages. Without these concepts and techniques, there are many situations in which probability theory cannot answer questions that arise when everyday decisions have to be made on the basis of perception-based information. Examples of such questions are given in this paper.

A related point is that, in perception-based theory of probabilistic reasoning, imprecision can occur on may different levels—and not just on the level of imprecise probabilities. In particular, imprecision can occur on the level of events, counts and relations. More basically, it can occur on the level of definition of such basic concepts as random variable, causality, independence and stationarity. The concept of precisiated natural language may suggest a way of generalizing these and related concepts in a way that would enhance their expressiveness and operationality.

The confluence of probability theory and the computational theory of perceptions opens the door to a radical enlargement of the role of natural languages in probability theory. The theory outlined in this paper is merely a first step in this direction. Many further steps will have to be taken to develop the theory more fully. This will happen because it is becoming increasingly clear that real-world applications of probability theory require the capability to process perception-based information as a basis for rational decisions in an environment of imprecision, uncertainty and partial truth.

Acknowledgements

Research supported in part by ONR Contract N00014-99-C-0298, NASA Contract NCC2-1006, NASA Grant NAC2-1177, ONR Grant N00014-96-1-0556, ONR Grant FDN0014991035, ARO Grant DAAH 04-961-0341 and the BISC Program.

References

Barsalou, L.W., 1999. Perceptual symbol systems. Behav. Brain Sci. 22, 577–660.
Bowen, J., Lai, R., Bahler, D., 1992. Fuzzy semantics and fuzzy constraint networks. Proceedings of the First IEEE Conference on Fuzzy Systems, San Francisco, pp. 1009–1016.
Cresswell, M.J., 1973. Logic and Languages. Methuen, London.
Dempster, A.P., 1967. Upper and lower probabilities induced by a multivalued mapping. Ann. Math. Statist. 38, 325–339.
Dubois, D., Prade, H., 1988. Possibility Theory. Plenum Press, New York.
Dubois, D., Fargier, H., Prade, H., 1993. The calculus of fuzzy restrictions as a basis for flexible constraint satisfaction. Proceedings of the Second IEEE International Conference on Fuzzy Systems, San Francisco, pp. 1131–1136.
Fikes, R.E., Nilsson, N.J., 1971. STRIPS: a new approach to the application of theorem proving to problem solving. Artif. Intell. 2, 189–208.
Good, I.J., 1962. Subjective probability as the measure of a non-measurable set. In: Nagel, E., Suppes, P., Tarski, A. (Eds.), Logic, Methodology and Philosophy of Science. Stanford University Press, Stanford, pp. 319–329.
Grabisch, M., Nguyen, H.T., Walker, E.A., 1995. Fundamentals of Uncertainty Calculi with Applications to Fuzzy Inference. Kluwer, Dordrecht.
Hajek, P., 1998. Metamathematics of Fuzzy Logic. Kluwer, Dordrecht.
Katai, O., Matsubara, S., Masuichi, H., Ida, M., et al., 1992. Synergetic computation for constraint satisfaction problems involving continuous and fuzzy variables by using occam. In: Noguchi, S., Umeo, H. (Eds.), Transputer/Occam, Proceedings of the Fourth Transputer/Occam International Conference. IOS Press, Amsterdam, pp. 146–160.
Kaufmann, A., Gupta, M.M., 1985. Introduction to Fuzzy Arithmetic: Theory and Applications. Von Nostrand, New York.
Klir, G., Folger, T.A., 1988. Fuzzy Sets, Uncertainty, and Information. Prentice-Hall, Englewood Cliffs, NJ.
Klir, G., Yuan, B., 1995. Fuzzy Sets and Fuzzy Logic. Prentice-Hall, Englewood Cliffs, NJ.
Krause, P., Clark, D., 1993. Representing Uncertain Knowledge. Kluwer, Dordrecht.

Kruse, R., Meyer, D., 1987. Statistics with Vague Data. Kluwer, Dordrecht.
Kuipers, B.J., 1994. Qualitative Reasoning. MIT Press, Cambridge.
Mares, M., 1994. Computation over Fuzzy Quantities. CRC Press, Boca Raton.
Nguyen, H.T., Kreinovich, V., Wu, B., 1999. Fuzzy/probability-fractal/smooth. Internat. J. Uncertainty Fuzziness Knowledge-based Systems 7, 363–370.
Novak, V., 1991. Fuzzy logic, fuzzy sets, and natural languages. Int. J. Gen. Systems 20, 83–97.
Novak, V., Ramik, M., Cerny, M., Nekola, J. (Eds.), 1992. Fuzzy Approach to Reasoning and Decision-Making. Kluwer, Boston.
Pedrycz, W., Gomide, F., 1998. Introduction to Fuzzy Sets. MIT Press, Cambridge.
Ralescu, A.L., Bouchon-Meunier, B., Ralescu, D.A., 1995. Combining fuzzy quantifiers. Proceedings of the International Conference of CFSA/IFIS/SOFT'95 on Fuzzy Theory and Applications, Taipei, Taiwan.
Ralescu, D.A., 1995. Cardinality, quantifiers and the aggregation of fuzzy criteria. Fuzzy Sets and Systems 69, 355–365.
Reghis, M., Roventa, E., 1998. Classical and Fuzzy Concepts in Mathematical Logic and Applications. CRC Press, Boca Raton.
Shafer, G., 1976. A Mathematical Theory of Evidence. Princeton University Press, Princeton.
Shen, Q., Leitch, R., 1992. Combining qualitative simulation and fuzzy sets. In: Faltings, B., Struss, P. (Eds.), Recent Advances in Qualitative Physics. MIT Press, Cambridge.
Smithson, M., 1989. Ignorance and Uncertainty. Springer, Berlin.
Strat, T.M., 1992. Representative applications of fuzzy measure theory: decision analysis using belief functions. In: Wang, Z., Klir, G.J. (Eds.), Fuzzy Measure Theory. Plenum Press, New York, pp. 285–310.
Thomas, S.F., 1995. Fuzziness and Probability. ACG Press, Wichita, KS.
Vallee, R., 1995. Cognition et Systeme. l'Interdisciplinaire Systeme(s), Paris.
Walley, P., 1991. Statistical Reasoning with Imprecise Probabilities. Chapman & Hall, London.
Wang, Z., Klir, G., 1992. Fuzzy Measure Theory. Plenum Press, New York.
Yager, R.R., 1989. Some extensions of constraint propagation of label sets. Internat. J. Approx. Reasoning 3, 417–435.
Zadeh, L.A., 1955. General filters for separation of signals and noise. Proceedings of the Symposium on Information Networks, Polytechnic Institute of Brooklyn, New York, pp. 31–49.
Zadeh, L.A., 1968. Probability measures of fuzzy events. J. Math. Anal. Appl. 23, 421–427.
Zadeh, L.A., 1973. Outline of a new approach to the analysis of complex systems and decision processes. IEEE Trans. Systems Man Cybernet. SMC-3, 28–44.
Zadeh, L.A., 1975. The concept of a linguistic variable and its application to approximate reasoning. Part I: Inf. Sci. 8, 199–249; Part II: Inf. Sci. 8, 301–357; Part III: Inf. Sci. 9, 43–80.
Zadeh, L.A., 1978. Fuzzy sets as a basis for a theory of possibility. Fuzzy Sets and Systems 1, 3–28.
Zadeh, L.A., 1979. Fuzzy sets and information granularity. In: Gupta, M., Ragade, R., Yager, R. (Eds.), Advances in Fuzzy Set Theory and Applications. North-Holland, Amsterdam, pp. 3–18.
Zadeh, L.A., 1981. Test-score semantics for natural languages and meaning representation via PRUF. In: Rieger, B. (Ed.), Empirical Semantics. Brockmeyer, Bochum, W. Germany, pp. 281–349.
Zadeh, L.A., 1983. A computational approach to fuzzy quantifiers in natural languages. Comput. Math. 9, 149–184.
Zadeh, L.A., 1986. Outline of a computational approach to meaning and knowledge representation based on a concept of a generalized assignment statement. In: Thoma, M., Wyner, A. (Eds.), Proceedings of the International Seminar on Artificial Intelligence and Man–Machine Systems. Springer, Heidelberg, pp. 198–211.
Zadeh, L.A., 1995. Probability theory and fuzzy logic are complementary rather than competitive. Technometrics 37, 271–276.
Zadeh, L.A., 1996. Fuzzy logic and the calculi of fuzzy rules and fuzzy graphs: a precis. Multivalued Logic 1, 1–38.
Zadeh, L.A., 1997. Toward a theory of fuzzy information granulation and its centrality in human reasoning and fuzzy logic. Fuzzy Sets and Systems 90, 111–127.
Zadeh, L.A., 1999. From computing with numbers to computing with words—from manipulation of measurements to manipulation of perceptions. IEEE Trans. Circuits Systems 45, 105–119.
Zadeh, L.A., 2000. Outline of a computational theory of perceptions based on computing with words. In: Sinha, N.K., Gupta, M.M. (Eds.), Soft Computing and Intelligent Systems: Theory and Applications. Academic Press, London, pp. 3–22.

Representing Knowledge

B. Bouchon-Meunier, L. Foulloy and R.R. Yager (Editors)
Intelligent Systems for Information Processing:
From Representation to Applications
© 2003 Elsevier Science B.V. All rights reserved.

Rough Set Uncertainty in an Object Oriented Data Model

Theresa Beaubouef
Southeastern Louisiana University
Computer Science Department
Hammond, LA 70402
Email: tbeaubouef@selu.edu

Frederick E. Petry
Center for Intelligent and Knowledge-Based Systems
Tulane University
New Orleans, LA 70118
Email: fep@eecs.tulane.edu

Abstract
We introduce a rough set model of uncertainty for object oriented databases (OODB). This model is formally defined, consistent with the notation of both rough set theory and object oriented database formalisms. Uncertainty and vagueness, which are inherent in all real world applications, are incorporated into the database model through the indiscernibility relation and approximation regions of rough sets. Because spatial database and geographic information systems (GIS) have particular needs for uncertainty management in spatial data, examples from this type of application are used to illustrate the benefits of this rough OODB approach.

Keywords: object-oriented database, rough sets, uncertainty, spatial data

1 Introduction

A database semantic model aims to capture the meaning of some enterprise in the real world, and is a high level, conceptual model that must then be implemented. At a lower, more practical level, the database is simply a collection of data and constraints stored in some schema, which attempts to model this enterprise in the real world. The real world abounds in uncertainty, and any attempt to model aspects of the world must therefore include some mechanism for incorporating uncertainty. There may be uncertainty in the understanding of the enterprise or in the quality or meaning of the data. There may be uncertainty in the modeling of the enterprise, which leads to uncertainty in entities, the attributes describing them, or the relationships that exist between various entities. There may also be uncertainty about the uncertainty itself, the degree or types of uncertainty present in the data. Because there is a particular need for uncertainty management in spatial data [10] and in the relationships among various spatial entities and vague regions, we illustrate the formalisms of the rough object-oriented database with examples taken from a geographic information systems (GIS) perspective.

It is well established that rough set techniques as part of an underlying relational database model effectively manages uncertainty in relational databases [1, 2]. Additionally, the rough querying of crisp data [3] allows for rough set uncertainty to be applied to existing ordinary relational databases. In this paper we define a rough set model for an object-oriented database.

2 Rough Sets

Rough set theory, introduced by Pawlak [11] and discussed in greater detail in [8,9,12], is a technique for dealing with uncertainty and for identifying cause-effect relationships in databases as a form of database learning [13]. It has also been used for improved information retrieval [14] and for uncertainty management in relational databases [2, 3].

Rough sets involve the following:

> U is the *universe*, which cannot be empty,
> R is the *indiscernibility (equivalence) relation*,
> $A = (U,R)$ is the *approximation space*,
> $[x]_R$ is the equivalence class of R containing x,
> *elementary sets* in A are the equivalence classes,
> *definable set* in A is any finite union of elementary sets in A.

Therefore, for any given approximation space defined on some universe U and having an equivalence relation R imposed upon it, U is partitioned into equivalence classes called elementary sets which may be used to define other sets in A. Given that $X \subseteq U$, X can be defined in terms of the definable sets in A by the following:

lower approximation of X in A: $\underline{R}X = \{x \in U \mid [x]_R \subseteq X\}$

upper approximation of X in A: $\overline{R}X = \{x \in U \mid [x]_R \cap X \neq \emptyset\}$.

Another way to describe the set approximations is as follows. Given the upper and lower approximations $\overline{R}X$ and $\underline{R}X$, of X a subset of U, the R-positive region of X is $POS_R(X) = \underline{R}X$, the R-negative region of X is $NEG_R(X) = U - \overline{R}X$, and the boundary or R-borderline region of X is $BN_R(X) = \overline{R}X - \underline{R}X$. X is called R-definable if and only if $\underline{R}X = \overline{R}X$. Otherwise, $\underline{R}X \neq \overline{R}X$ and X is rough with respect to R. In Figure 1 the universe U is partitioned into equivalence classes denoted by the rectangles. Those elements in the lower approximation of X, $POS_R(X)$, are denoted with the letter P and elements in the R-negative region by the letter N. All other classes belong to the boundary region of the upper approximation.

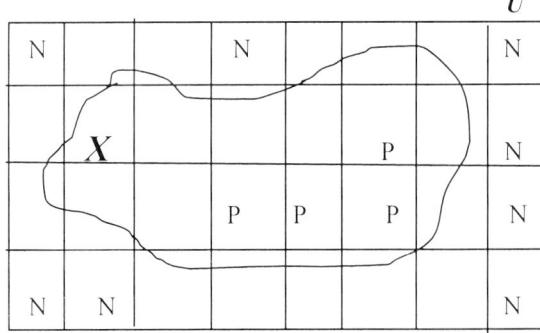

Figure 1. Example of a rough set X.

Consider the following example:

Let U = {tower, stream, creek, river, forest, woodland, pasture, meadow}.

Let the equivalence relation R^1 be defined as follows:

R^1 = {[tower], [stream, creek, river], [forest, woodland], [pasture, meadow]}.

Given that some set X = {tower, stream, creek, river, forest, pasture}, we can define it in terms of its lower and upper approximations:

$\underline{R}X$ = {tower, stream, creek, river}, and
$\overline{R}X$ = {tower, stream, creek, river, forest, woodland, pasture, meadow}.

A *rough set in A* is the group of subsets of U with the same upper and lower approximations. For this example, the rough set is

{{tower, stream, creek, river, forest, pasture}
{tower, stream, creek, river, forest, meadow}
{tower, stream, creek, river, woodland, pasture}
{tower, stream, creek, river, woodland, meadow}}.

The major rough set concepts of interest are the use of an indiscernibility relation to partition domains into equivalence classes and the concept of lower and upper approximation regions to allow the distinction between certain and possible, or partial, inclusion in a rough set.

The indiscernibility relation allows us to group items based on some definition of 'equivalence' as it relates to the application domain. We may use this partitioning to increase or decrease the granularity of a domain, to group items together that are considered indiscernible for a given purpose, or to "bin" ordered domains into range groups.

In order to allow *possible* results, in addition to the obvious, certain results encountered in querying an ordinary spatial database system, we may employ the use of the boundary region information in addition to that of the lower approximation region. The results in the lower approximation region are certain. These correspond to exact matches. The boundary region of the upper approximation contains those results that are possible, but not certain.

3 Object Oriented Databases

The object-oriented programming paradigm has become quite popular in recent years, both as a modeling tool and for code development for databases and other applications. Often objects can more realistically model an enterprise, enabling developers to easily transition from a conceptual design to the implementation. The concepts of *classes* and *inheritance* allow for code reuse through specialization and generalization. A *class hierarchy* is designed such that classes at the top of the hierarchy are the most general and those nearer the bottom more specialized. A class *inherits* data and behavior from classes at higher levels in the class hierarchy. This promotes reuse of existing functionality, which can save valuable programming time. If code is already available for a task and that code has been tested, it is often better to use that code, perhaps with some slight modification, than to develop and test code from scratch. The concept of *polymorphism* allows the same name to be used for methods differing in functionality for different object types.

Essentially, an *object* is an instance of a *class* in a class hierarchy. Each class defines a particular type of object including its public and private variables and operations associated with the functionality of the object, which are called *methods*. An object method is invoked by the passing of a *message* to the object in which the method is defined. The data variables and methods are *encapsulated* in the object that defines them, which means that they are packaged in, and can only be accessed through, the object. Encapsulation enables a component of the system to be extended or modified with minimal impact on other parts of the system.

There are many advantages of using an object-oriented database approach as compared to a relational database approach. A major advantage is that objects can be defined to represent very complex data structures and relationships in the data, as is often the case in spatial data. According to Fayad and Tsai [7], object-oriented technology provides several other benefits. These include reusability, extensibility, robustness, reliability, and scalability. Object modeling helps in requirements understanding and collaboration of group members and the use of object-oriented techniques leads to high quality systems that are easy to modify and to maintain.

Spatial data is particularly suited to object oriented modeling and implementation. Often this data is more complex than that of typical database applications requiring simple values and strings. The complex data types, data structures, and relationships between data objects in a spatial database can be modeled quite effectively with object oriented techniques, and the advantages discussed previously can also be realized. It is necessary, therefore, that we develop a mechanism for integrating uncertainty management into the OODB model.

4 Rough Object-Oriented Database (ROODB) Model

In this section we develop the rough object-oriented database model. We follow the formal framework and type definitions for generalized object-oriented databases proposed by [6], which conforms to the standards set forth by the Object Database Management Group [5]. We extend this framework, however, to allow for rough set indiscernibility and approximation regions for the representation of uncertainty as we have previously done for relational databases [1,2]. The rough object database scheme is formally defined by the following type system and constraints.

The type system, *ts*, contains literal types $T_{literal}$, which can be a base type, a collection literal type, or a structured literal type. It also contains T_{object}, which specifies object types, and $T_{reference}$, the set of specifications for reference types. In the type system, each domain $dom_{ts} \in D_{ts}$, the set of domains. This domain set, along with a set of operators

O_{ts} and a set of axioms A_{ts}, capture the semantics of the type specification. The type system is then defined based on these type specifications, the set of all programs P, and the implementation function mapping each type specification for a domain onto a subset of the powerset of P that contains all the implementations for the type system.

We are particularly interested in object types. Following [6], we may specify a class t of object types as

$$\text{Class } id(id_1:s_1; ...; id_n:s_n) \quad \text{or} \quad \text{Class } id: \overline{id}_1, ..., \overline{id}_n(id_1:s_1; ...; id_n:s_n)$$

where id, an identifier, names an object type, $\{\overline{id}_i \mid 1 \leq i \leq m\}$ is a finite set of identifiers denoting parent types of t, and $\{id_i:s_i \mid 1 \leq i \leq n\}$ is the finite set of characteristics specified for object type t within its syntax. This set includes all the attributes, relationships and method signatures for the object type. The identifier for a characteristic is id_i and the specification is s_i for each of the $id_i:s_i$.

Consider a GIS which stores spatial data concerning water and land forms, structures, and other geographic information. If we have simple types defined for string, set, geo, integer, etc., we can specify an object type

Class *ManMadeFeature* (
    Location: geo;
    Name: string;
    Height: integer;
    Material: Set(string));

Some example instances of the object type *ManMadeFeature* might include

[oid1, Ø, *ManMadeFeature*, Struct(0289445, "KXYZ radio tower", 60, Set(steel, plastic, aluminum))]

or

[oid2, Ø, *ManMadeFeature*, Struct(01122345, "Ourtown water tower", 30, Set(steel, water, iron))],

following the definition of instance of an object type [6], the quadruple o = [*oid, N, t, v*] consisting of a unique object identifier, a possibly empty set of object names, the name of the object type, and for all attributes, the values ($v_i \in dom_{s_i}$) for that attribute, which represent the state of the object. The object type t is an instance of the type system *ts* and is formally defined in terms of the type system and its implementation function $t = [ts, f_{impl}^{type}(ts)]$.

In the rough set object-oriented database, indiscernibility is managed through classes. Every domain is implemented as a class hierarchy, with the lowest elements of the hierarchy representing the equivalence classes based on the finest possible partitioning for the domain as it pertains to the application. Consider, for example, a GIS, where objects have an attribute called *landClass*. Of the many different classifications for land area features, some are those categorized by water. This part of the hierarchy is depicted in Figure 2.

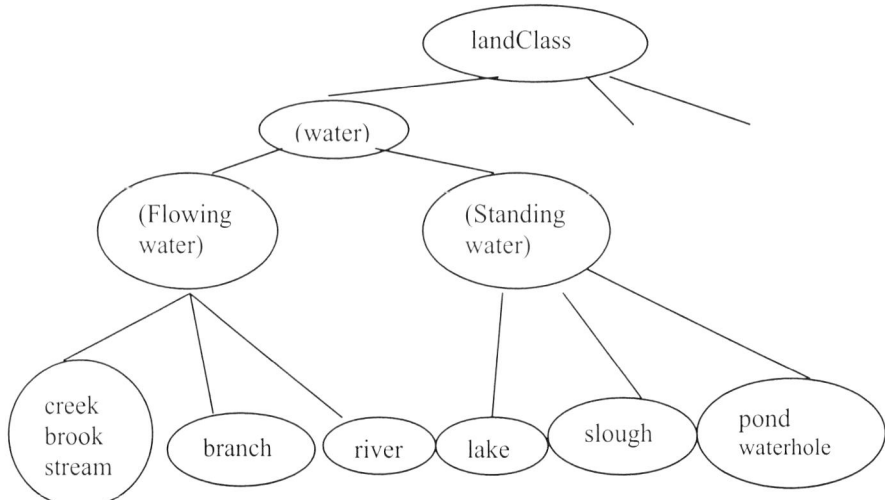

Figure 2. Part of the class hierarchy of *landClass* features involving water.

Ignoring the non-water parts of the *landClass* domain, and focusing on the water-related parts, we see that the domain set

$$dom_{landClass} = \{creek, brook, stream, branch, river, lake, pond, waterhole, slough\}$$

can be partitioned in several different ways. One partitioning, which represents the finest partitioning (more, but smaller, equivalence classes) is given by

$$R1 = \{[creek, brook, stream], [branch], [river], [lake], [pond, waterhole], slough]\}.$$

This can be discerned from the lowest level of the hierarchy.

An object type (domain class) for landClass may be defined as

    Class landClass (
        numEquivClass: integer;
        name: string;
        indiscernibility: Set(Ref(equivClass)))

At this lowest level, each landClass object has only one reference in its attribute for indiscernibility, the object identifier for the particular equivalence class. These reference individual equivalence class objects defined by

Class equivClass(
    element: Set(string);
    N: integer;
    Name: string).

In this case, we have six separate equivalence classes, three of which are shown below:

[oid56, Ø, *equivClass*, Struct(Set("creek," "brook," "stream"), 3, "creek")]

[oid57, Ø, *equivClass*, Struct(Set("branch"), 1, "branch")]

[oid61, Ø, *equivClass*, Struct(Set("pond," "waterhole"), 2, "pond")]

Note that the name of the class can be set equal to any of the values within the class.

Let the other water classes be similarly defined with oid58 denoting the "river" class, oid59 denoting "lake", and oid60 denoting "slough." If we want to change the partitioning, such that our application only distinguishes between flowing and standing water, for example, our equivalence classes become

R^2 = {[creek, brook, river, stream, branch], [lake, pond, waterhole, slough]}.

We would then have the landClass objects

[oid101, Ø, *landClass*,
Struct(3, "Flowing water,"
Set(oid56, oid57, oid58))]   and

[oid102, Ø, *landClass*,
Struct(3, "Standing water,"
Set(oid59, oid60, oid61))].

Lastly, if the application only requires that a feature be categorized as water or non-water for land classification, an even coarser partitioning may be used that includes all the possible water values in one equivalence class:

R^3 = {[creek, brook, stream, branch, river, lake, pond, waterhole, slough]}.

An instance of this class would be defined in a manner similar to those above.

Each domain class *i* in the database, $dom_i \in D_i$, has methods for maintaining the current level of granulation, changing the partitioning, adding new domain values to the hierarchy, and for determining equivalence based on the current indiscernibility relation imposed on the domain class.

Every domain class, then, must be able to not only store the legal values for that domain, but to maintain the grouping of these values into equivalence classes. This can be achieved through the type implementation function and class methods, and can be specified through the use of generalized constraints as in [6] for a generalized OODB.

An ordinary (non-indiscernibility) object class in this database, having one of its attributes *landClass*, may be defined as follows:

>Class *RuralProperty* (
>   Location: geo;
>   Name: string;
>   Owner: string;
>   landClass: Set(string));

having particular instances of the class, for example:

>[oid24, Ø, *RuralProperty,* Struct(01987345, "Elm Plantation", "Bob Owner", Set("waterhole," "pasture"))],

>[oid27, Ø, *RuralProperty,* Struct(01987355, Ø , "Betty Owner", Set("forest," "lake"))],

>[oid31, Ø, *RuralProperty,* Struct(01987390, "Hodge Mill Runoff Lagoon", "Hodge Mill", Set("waterhole"))],

>[oid32, Ø, *RuralProperty,* Struct(01987394, "Heart Lake", "Blackham County", Set("lake"))].

>[oid26, Ø, *RuralProperty,* Struct(01987358, Ø , "Brown County", Set("pond"))].

Now let us assume that we are trying to sell fish fingerlings and want to retrieve the names of all landowners that own land that contains ponds. Our query may look something like this:

>SELECT Owner
>FROM RuralProperty
>WHERE landClass = "pond".

If our goal is selling fish for stocking small ponds, we may want our indiscernibility relation to be defined with a very fine partitioning as discussed previously:

>R^1 = {[creek, brook, stream],[branch], [river], [lake], [pond, waterhole], [slough]}.

Here "pond" and "waterhole" are considered indiscernible and the query will match either as a certain result. Possible results will in addition contain those objects that have "pond" or "waterhole" as one of the values in the set for landClass. (See [1,2] for complete semantics of rough database operations.)

For the partitioning R^1 and the five sample objects above, our rough set result would include the following:

$\underline{R}X$ = {Brown County, Hodge Mill}
$\overline{R}X$ = {Brown County, Hodge Mill, Bob Owner}

Here, oid26 (Owner is "Brown County") provides an exact match with "pond" so it is included in the lower approximation region, which represents certain data. Because "waterhole" is in the same equivalence class as "pond", oid31 (Owner is "Hodge Mill") is also included in the lower approximation region. The upper approximation region contains those objects where at least one of the values in the set of values for that attribute match the request. Hence, oid21 (Owner is "Bob Owner") is returned since "pond" and "waterhole" both belong to the class [pond] and this value is included in the set of values {waterhole, pasture}.

If we had decided that all standing water locations are likely candidates for fish stocking, then we might have coarsened the partitioning, using the equivalence relation:

R^2 = {[creek, brook, river, stream, branch], [lake, pond, waterhole, slough]}.

In this case, oid32 (Owner is Blackham County) also belongs to the lower approximation region since "lake", "pond", "waterhole", and "slough" are now indiscernible. Likewise, oid27 (Owner is Betty Owner) becomes part of the upper approximation since "lake" is a subset of {lake, forest}. Now the rough set results are given as:

$\underline{R}X$ = {Brown County, Hodge Mill, Blackham County}
$\overline{R}X$ = {Brown County, Hodge Mill, Blackham County, Bob Owner, Betty Owner}.

The semantics of rough set operations discussed for relational databases in [1,2] apply similarly for the object database paradigm. However, the implementation of these operations is done via methods associated with the individual object classes.

5 Conclusion and Related Work

In this paper we used rough set formalisms to define a model for object-oriented databases. We showed that the rough set concepts of indiscernibility and approximation regions can be integrated into a rough object-oriented framework, resulting in a model that allows for the management of uncertainty. A geographic information system example illustrated the usefulness of this rough object oriented database model.

We have further extended the rough object oriented database model to incorporate fuzzy set uncertainty as well as rough set uncertainty [4]. In [6], the generalized object database model presented allowed the values of attributes to contain fuzzy sets. Our fuzzy rough model allows for the incorporation of these fuzzy membership values, as done for the fuzzy rough relational database in [1]. Having both rough and fuzzy set uncertainty makes possible the management of uncertainty through the use of indiscernibility relations and approximation regions, and also the ability to quantify the degree of uncertainty through the use of fuzzy sets.

References

[1] T. Beaubouef and F. Petry (2000). Fuzzy Rough Set Techniques for Uncertainty Processing in a Relational Database. *International Journal of Intelligent Systems*, vol. 15, Issue 5, pp. 389-424, April, 2000.

[2] T. Beaubouef, F. Petry, and B. Buckles (1995). Extension of the Relational Database and its Algebra with Rough Set Techniques. *Computational Intelligence*, Vol. 11, No. 2, pp. 233-245, May 1995.

[3] T. Beaubouef and F. Petry (1994). Rough Querying of Crisp Data in Relational Databases. *Proc. Third International Workshop on Rough Sets and Soft Computing (RSSC'94)*, San Jose, California, pp. 368-375, November 1994.

[4] T. Beaubouef and F. Petry (2002). Fuzzy Set Uncertainty in a Rough Object Oriented Database. *Proc. North American Fuzzy Information Society (NAFIPS)*, New Orleans, LA, pp. 365-370, June, 2002.

[5] R. Cattell, D. Barry, D. Bartels, M. Berler, J. Eastman, S. Gamerman, D. Jordan, A. Springer, H. Strickland, and D. Wade (1997). *The Object Database Standard: ODMG2.0*, Morgan Kaufmann Publishers, Inc., San Francisco, CA.

[6] G. De Tré and R. De Caluwe (1999). A Generalized Object-Oriented Database Model with Generalized Constraints. *In Proc. of NAFIPS'99*, pp. 381-386, New York, 1999.

[7] M. Fayad, and Tsai, W. (1995). Object-Oriented Experiences. *Communications of the ACM*, **38**, pp. 51-53.

[8] J. Grzymala-Busse (1991). *Managing Uncertainty in Expert Systems*, Kluwer Academic Publishers, Boston.

[9] J. Komorowski, Z. Pawlak, L. Polkowski, and A. Skowron (1999). Rough Sets: A Tutorial. *in Rough Fuzzy Hybridization: A New Trend in Decision-Making* (ed. S. K. Pal and A. Skowron), Springer-Verlag, Singapore, pp. 3-98, 1999.

[10] R. Laurini and D. Thompson (1992). *Fundamentals of Spatial Information Systems*, Academic Press, London.

[11] Z. Pawlak (1984). Rough Sets. *International Journal of Man-Machine Studies*, vol. 21, 1984, pp. 127-134.

[12] Z. Pawlak (1991). *Rough Sets: Theoretical Aspects of Reasoning About Data*, Kluwer Academic Publishers, Norwell, MA, 1991.

[13] R. Slowinski (1992). A Generalization of the Indiscernibility Relation for Rough Sets Analysis of Quantitative Information. *First International Workshop on Rough Sets: State of the Art and Perspectives*, Poland, September 1992.

[14] P. Srinivasan (1991). The importance of rough approximations for information retrieval. *International Journal of Man-Machine Studies*, 34, pp. 657-671, 1991.

On the Representation of Fuzzy Spatial Relations in Robot Maps

Isabelle Bloch

Ecole Nat. Sup. des Télécommunications
Dept. TSI - CNRS URA 820
46 rue Barrault, 75013 Paris, France
Isabelle.Bloch@enst.fr

Alessandro Saffiotti

Applied Autonomous Sensor Systems
Dept. of Technology, Örebro University
S-70182 Örebro, Sweden
Alessandro.Saffiotti@aass.oru.se

Abstract

Spatial directional relations, like "north of," play an important role in the modeling of the environment by an autonomous robot. We propose an approach to represent spatial relations grounded in fuzzy set theory and fuzzy mathematical morphology. We show how this approach can be applied to robot maps, and suggest that these relations can be used for self-localization and for reasoning about the environment. We illustrate our approach on real data collected by a mobile robot in an office environment.

Keywords: autonomous robots, occupancy grids, topological maps, fuzzy spatial relations, fuzzy mathematical morphology.

1 Introduction

Autonomous robots need the ability to perceive their environment, build a model of it, and use this model to effectively navigate and operate in that environment. One important aspect of these models is the ability to incorporate spatial directional relations, like "north of." These relations are inherently vague, since they depend on how much of an object is in the specified direction with respect to the reference object.

Relative directional relations have not been extensively studied in the mobile robotics literature. The field of image processing contains a comparatively larger body of work on spatial relations, although directional positions have received much less attention in that field than topological relations like set relationships, part-whole relationships, and adjacency. Most non-fuzzy approaches use a set of basic relations based on Allen's interval relations [1] (e.g., [22]) or on simplifications of objects (e.g. [10]). Some approaches use intervals to represent qualitative expressions about angular positions [15]. Stochastic approaches have also been proposed for representing spatial uncertainty in robotics, e.g., [24]. Most of the above approaches, however, suffer from a somehow simplified treatment of the uncertainty and vagueness which is intrinsic in spatial relations. Concepts related to directional relative position are rather ambiguous, and defy precise definitions. However, humans have a rather intuitive and common way of understanding and interpreting them. From our everyday experience, it is clear that any "all-or-nothing" definition of these concepts leads to unsatisfactory results in several situations of even moderate complexity. Fuzzy set theory appears then as an appropriate tool for such modeling since it allows to integrate both quantitative and qualitative knowledge, using semiquantitative interpretation of fuzzy sets. As noted by Freeman in [9], this allows us to provide a computational representation and interpretation of imprecise spatial relations, expressed in a linguistic way, possibly including quantitative knowledge.

In this paper, we show how fuzzy mathematical morphology can be used to compute approximate spatial relations between objects in a robot map. The key step is to represent the space in the robot's environment by an occupancy grid [7, 20], and to treat this grid as a grey-scale image. This allows us to apply techniques from the field of image processing to extract spatial information from this grid. In particular, we are interested in the spatial relations between rooms and corridors in the environment.

In the rest of this paper, we briefly introduce fuzzy mathematical morphology and we show how it can be used to define fuzzy spatial relations. We then discuss the use of this approach in the context of one particular type of robot maps, called *topology-based maps*, which are built from occupancy grids [8]. We illustrate our approach on real data collected by a mobile robot in an office environment. Finally, we discuss a few possible applications of fuzzy spatial relations to robot navigation.

2 Fuzzy mathematical morphology

Mathematical morphology is originally based on set theory. Introduced in 1964 by Matheron [16] to study porous media, mathematical morphology has rapidly evolved into a general theory of shape and its transformations, and it has found wide applications in image processing and pattern recognition [21].

The four basic operations of mathematical morphology are dilation, erosion, opening and closing. The *dilation* of a set X of an Euclidean space $\mathcal{S}$ (typically $\mathbb{R}^n$ or $\mathbb{Z}^n$) by a set B is defined by [21]:

$$D_B(X) = \{x \in \mathcal{S} \mid B_x \cap X \neq \emptyset\}, \qquad (1)$$

where B_x denotes the translation of B at x. Similarly the *erosion* of X by B is defined by:

$$E_B(X) = \{x \in \mathcal{S} \mid B_x \subseteq X\}. \qquad (2)$$

The set B, called *structuring element*, defines the neighborhood that is considered at each point. It controls the spatial extension of the operations: the result at a point only depends on the neighborhood of this point defined by B.

From these two operators, *opening* and *closing* are defined respectively by: $O(X) = D_{\check{B}}[E_B(X)]$, and $C(X) = E_{\check{B}}[D_B(X)]$, where $\check{B}$ denotes the symmetrical of B with respect to the origin of the space.

The above operators satisfy a number of algebraic properties [21]. Among the most important ones are commutativity of dilation (respectively erosion) with union or sup (respectively intersection or inf), increasingness[1] of all operators, iteration properties of dilation and erosion, idempotency of opening and closing, extensivity[2] of dilation (if the origin belongs to the structuring element) and of closing, anti-extensivity of erosion (if the origin belongs to the structuring element) and of opening.

Mathematical morphology has been extended in many ways. In the following, we make use of *fuzzy morphology*, where operations are defined on fuzzy sets (representing spatial entities along with their imprecision) with respect to fuzzy structuring elements. Several definitions of fuzzy mathematical morphology have been proposed (e.g. [3, 5, 23]). Here, we define dilation and erosion of a fuzzy set μ by a structuring element ν for all $x \in \mathcal{S}$ by, respectively:

$$D_\nu(\mu)(x) = \sup_y\{t[\nu(y-x),\mu(y)]\},$$
$$E_\nu(\mu)(x) = \inf_y\{T[c(\nu(y-x)),\mu(y)]\}$$

where y ranges over the Euclidean space $\mathcal{S}$ where the objects are defined, t is a t-norm, and T its associated t-conorm with respect to the complementation c [27]. In these equations, fuzzy sets are assimilated to their membership functions. These definitions extend classical morphology in a natural way, providing similar properties as in the crisp case [3, 19].

Through the notion of structuring element, mathematical morphology can deal with local or regional spatial context. It also has some features that allow us to include more global information, which is particularly important when the spatial arrangement of objects in a scene has to be assessed. This fact is exploited in the following.

3 Spatial relations from fuzzy mathematical morphology

Spatial relationships between the objects in the environment carry structural information about the environment, and provide important information for object recognition and for self localization [11]. Fuzzy mathematical morphology can be used here to represent and compute in a uniform setting several types of relative position information, like distance, adjacency and directional relative position. In this section, we explain how we can use it to deal with directional relations.

A few works propose fuzzy approaches for assessing the directional relative position between objects, which is an intrinsically vague relation [2, 12, 13, 17, 18]. The approach used here and described in more details in [2] relies on a fuzzy dilation that provides a map (or fuzzy landscape) where the membership value of each point represents the degree of the satisfaction of the relation to the reference object. This approach has interesting features: it works directly in the image space, without reducing the objects to points or histograms, and it takes the object shape into account.

We consider a (possibly fuzzy) object R in the space $\mathcal{S}$, and denote by $\mu_\alpha(R)$ the fuzzy subset of $\mathcal{S}$ such that points of areas which satisfy to a high degree the relation "to be in the direction $\vec{u}_\alpha$ with respect to object R" have high membership values, where $\vec{u}_\alpha$ is a vector making an angle α with respect to a reference axis.

---
[1] An operation ψ is increasing if $\forall X, Y \ X \subseteq Y \Rightarrow \psi(X) \subseteq \psi(Y)$.
[2] An operation ψ is extensive if $\forall X, \ X \subseteq \psi(X)$ and anti-extensive if $\forall X, \psi(X) \subseteq X$.

The form of $\mu_\alpha(R)$ may depend on the application domain. Here, we use the definition proposed in [2], which considers those parts of the space that are visible from a reference object point in the direction $\vec{u}_\alpha$. This can be expressed formally as the fuzzy dilation of μ_R by ν, where ν is a fuzzy structuring element depending on α: $\mu_\alpha(R) = D_\nu(\mu_R)$ where μ_R is the membership function of the reference object R. This definition applies both to crisp and fuzzy objects and behaves well even in case of objects with highly concave shape [2]. In polar coordinates, ν is defined by: $\nu(\rho,\theta) = f(\theta - \alpha)$ and $\nu(0,\theta) = 1$, where $\theta - \alpha$ is defined modulo π and f is a decreasing function. In the experiments reported here, we have used $f(x) = \max[0, \cos x]^2$ for $x \in [0, \pi]$ — see Figure 1. Techniques for reducing the computation cost have been proposed in [2].

Figure 1: Structuring element ν for $\alpha = 0$ (high grey values correspond to high membership values).

Once we have defined $\mu_\alpha(R)$, we can use it to define the degree to which a given object A is in direction $\vec{u}_\alpha$ with respect to R. Let us denote by μ_A the membership function of the object A. The evaluation of relative position of A with respect to R is given by a function of $\mu_\alpha(R)(x)$ and $\mu_A(x)$ for all x in S. The histogram of $\mu_\alpha(R)$ conditionally to μ_A is such a function. If A is a binary object, then the histogram of $\mu_\alpha(R)$ in A is given by:

$$h(z) = \text{Card}\left(\{x \in A \mid \mu_\alpha(R)(x) = z\}\right),$$

where $z \in [0, 1]$. This extends to the fuzzy case by:

$$h(z) = \sum_{x \,:\, \mu_\alpha(R)(x) = z} \mu_A(x).$$

While this histogram gives the most complete information about the relative spatial position of two objects, it is difficult to reason in an efficient way with it. A summary of the contained information could be more useful in practice. An appropriate tool for defining this summary is the fuzzy pattern matching approach [6]. Following this approach, the matching between two possibility distributions is summarized by two numbers, a necessity degree N (a pessimistic evaluation) and a possibility degree Π (an optimistic evaluation), as often used in the fuzzy set community. In our application, they take the following forms:

$$\Pi_\alpha^R(A) = \sup_{x \in S} t[\mu_\alpha(R)(x), \mu_A(x)], \tag{3}$$

$$N_\alpha^R(A) = \inf_{x \in S} T[\mu_\alpha(R)(x), 1 - \mu_A(x)], \tag{4}$$

where t is a t-norm and T a t-conorm. The possibility corresponds to a degree of intersection between the fuzzy sets A and $\mu_\alpha(R)$, while the necessity corresponds to a degree of

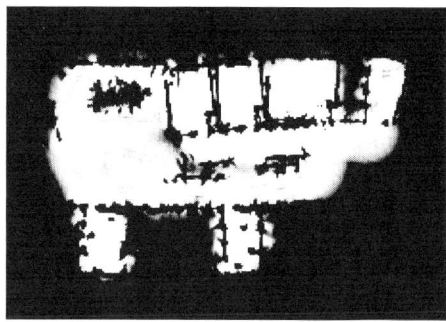

Figure 2: The occupancy grid built by the robot from sensor data in a test environment.

inclusion of A in $\mu_\alpha(R)$. These operations can also be interpreted in terms of fuzzy mathematical morphology, since $\Pi_\alpha^R(A)$ is equal to the dilation of μ_A by $\mu_\alpha(R)$ at the origin of $\mathcal{S}$, while $N_\alpha^R(A)$ is equal to the erosion at the origin [3]. The set-theoretic and the morphological interpretations indicate how the shape of the objects is taken into account.

It should be emphasized that, since the aim of these definitions is not to find only the dominant relationship, an object may satisfy several different relationships, for different angles, with high degrees. Therefore, "to be to the right of R" does not mean that the object should be completely to the right of the reference object, but only that it is at least to the right of some part of it.

The defined directional relations are symmetrical (only for Π), invariant with respect to translation, rotation and scaling, both for crisp and for fuzzy objects, and when the distance between the objects increases, the shape of the objects plays a smaller and smaller role in the assessment of their relative position [2].

4 Robot maps

We now study how fuzzy spatial relations can be used to enrich the spatial representations used by a mobile robot, or *robot maps*. A number of different representations of space have been proposed in the literature on mobile robotics. Most of these fall into two categories: *metric maps*, which represent the environment according to the absolute geometric position of objects (or places); and *topological maps*, which represent the environment according to the relationships among objects (or places) without an absolute reference system (e.g., [14, 25]).

In this work, we consider robot maps in the form of digital grids ($\mathcal{S}$ is therefore a 2D discrete space) on which certain objects, corresponding to the sub-spaces of interest (rooms and corridors), have been isolated. The reason for this is that we can directly apply the above methods to these representations.

More precisely, we consider the particular type of maps, called *topology-based maps*, proposed by Fabrizi and Saffiotti [8]. These maps represent the environment as a graph of rooms and corridors connected by doors and passages. The authors use image processing techniques to automatically extract regions that correspond to large open spaces (rooms and corridors) from a fuzzy occupancy grid that represents the free space in the

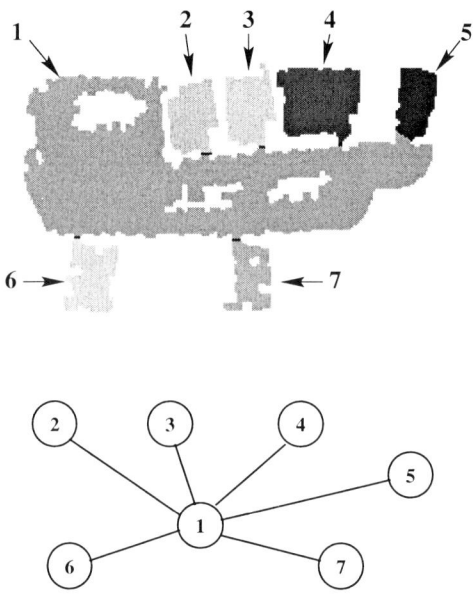

Figure 3: (top) regions extracted from the above occupancy grid; (bottom) the corresponding topology-based map.

environment. This grid is built by the robot itself using the technique described in [20].

Figure 2 shows a fuzzy occupancy grid built by a Nomad 200 robot in an office environment of 21×14 meters using sonar sensors. The environment consists of six rooms connected to a large corridor, which expands to a hall on the left hand side of the map. The dark areas in the corridor correspond to pieces of furniture. Each cell in the grid represents a square of side 10 cm, and its value, in the $[0, 1]$ interval, represents the degree of necessity of that space being empty. White cells have received sensor evidence of being empty; darker cells have not—they are either occupied or unexplored. (A dual grid, not used here, represents the occupied space.)

In order to extract the desired rooms and corridors, the authors in [8] regard this occupancy grid as a grey-scale image and process it using a technique based on fuzzy mathematical morphology. The open spaces can be extracted from the grid by performing a morphological opening by a fuzzy structuring element of a conic shape that represents the fuzzy concept of a large space. The result of the opening is then segmented by a watershed algorithm [26] in order to separate these spaces. Figure 3 (top) shows the result obtained by applying this procedure to our occupancy grid. The extracted regions correspond to the open spaces in the environment. These regions, together with the adjacency relation, constitute a topology-based map for our environment, summarized in graph form in Figure 3 (bottom). This graph provides an abstract representation that captures the structure of the space with a reduced number of parameters.

5 Adding fuzzy spatial relations to a robot map

Once we have segmented the environment into regions (rooms and corridors) we can use the technique described in the previous section to compute directional spatial relationships between these regions. These relations provide important information for object recognition and for self localization [11].

Figure 4: Fuzzy landscapes for being West, North, East and South of fuzzy region 4.

Figure 4 shows the *fuzzy landscapes* for the fuzzy notions of being, respectively, West, North, East, and South of the fuzzy region number 4 in Figure 3. These landscapes represent the $\mu_\alpha(R)$ fuzzy sets (see Section 3 above) with R being the fuzzy occupancy grid restricted to region number 4, and α taking the values 0, $\frac{1}{2}\pi$, π and $\frac{3}{2}\pi$, respectively.

We can use these landscapes to compute the relative directional position of any other region in our map with respect to region 4. For instance, Figure 5 shows the histograms of these fuzzy landscapes computed conditionally to region 1. These histograms represent the satisfaction of the relationships "region 1 is to the West (respectively, North) of region 4".

It should be noted that the direct computation of $\mu_\alpha(R)$ can be very expensive. Interestingly, the interpretation of that definition as a fuzzy dilation may suggest a few ways to reduce the computation time by reducing the precision of $\mu_\alpha(R)$: e.g., we can perform the dilation with a limited support for the structuring element, which corresponds to using a rough quantification of angles.

The above histograms can give the robot important information about the environment. In practice, however, storing and manipulating the whole histograms for each pair of regions may be prohibitive, and in real applications it is convenient to summarize the information contained in the histograms by a few parameters. A common choice is to use a pair of necessity and possibility degrees, computed according to equations (3) and (4) above.

The following table shows, for each region in our example, the degrees of necessity and possibility of being West, East, South and North of region 4. Degrees are written as a $[N, \Pi]$ interval.

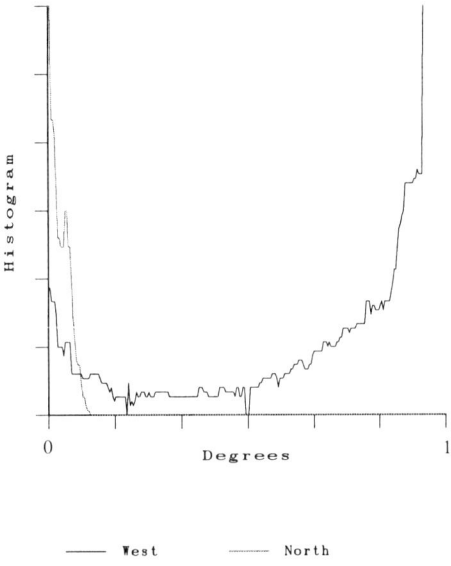

Figure 5: Histograms of the fuzzy landscapes of region 4 (west and north) conditionally to region 1.

	West	East	South	North
1	[0.0, 1.0]	[0.00, 0.99]	[0.0, 1.0]	[0.0, 0.11]
2	[0.99, 1.0]	[0.0, 0.0]	[0.0, 0.36]	[0.0, 0.24]
3	[0.92, 1.0]	[0.0, 0.0]	[0.00, 0.85]	[0.0, 0.83]
4	[0.55, 1.0]	[0.51, 1.0]	[0.50, 1.0]	[0.55, 1.0]
5	[0.0, 0.0]	[0.98, 1.0]	[0.02, 0.40]	[0.00, 0.59]
6	[0.65, 0.87]	[0.0, 0.0]	[0.30, 0.56]	[0.0, 0.0]
7	[0.17, 0.54]	[0.0, 0.0]	[0.86, 0.99]	[0.0, 0.0]

These results correspond well to intuition. For instance, regions 2 and 3 are found to be fully West of region 4, and totally not East of it; while region 5 is fully East of it and totally not West. Region 1 offers an interesting example. This region surrounds region 4 on the West and South side, and extends further East from it. Correspondingly, it has full possibility of being considered West, South and East of region 4, although no one of these relations is necessary. Its possibility of being considered North of region 4 is, however, neglectable, which is consistent with intuition. This can also be seen in the histogram, where no high degrees are obtained for the North direction, while many points satisfy the West relation to a degree close to 1. Finally, regions 6 and 7 are, at different degrees, both South and West of region 4, again conforming with intuition.

6 Discussion and conclusions

The proposed approach to represent directional relations has several interesting features. The interval representation allows us to capture the ambiguity of some relations, like in the case of the relation between region 1 and region 4 in the above example. The formal properties listed at the end of Section 3 are also of direct interest for applications in autonomous robotics. For instance, the invariance with respect to geometrical transformations is needed to guarantee that localization and recognition are independent of the frame of reference used to define directions. The fact that the shape of an object plays a smaller and smaller role as the distance of that object increases is useful when considering relationships to the robot itself: far away objects are seen by the robot as points, which is consistent with the idea that the spatial extent of these objects becomes irrelevant. The behavior of our definitions in case of concave objects agrees with intuition: an object can satisfy several relations with respect to a concave one at a high degree. In the above example, regions 2, 3, 4, 5 are all both East and North of region 1 to a high degree, which expresses that they are in the upper-right concave area of region 1. This is a way to express more complex relationships.

The computed fuzzy directional information can be used in several ways during autonomous navigation. Perhaps the most direct application is to improve the self-localization ability of the robot. The robot can perform coarse self-localization on the topological map by estimating, at every moment, the node (room) in which it is. Markov techniques can be used to update this estimate when the robot detects a transition from one node to the next: directional information can then be used to produce an expectation about the next node, by comparing the direction of travel with the distribution of possible directions associated to the outgoing links from the current node.

The ability to produce a fuzzy landscape for a given direction with respect to a node opens the possibility of additional applications. For instance, the robot can use linguistic directional information to identify important areas in the environment. As an example, we can tell the robot that the door to a given room is North with respect to the room where it currently is: the corresponding fuzzy landscape limits the area where the door should be looked for. Alternatively, we can tell the robot that the area North of a given corridor is dangerous (e.g., there is a staircase) and it should be avoided. A similar use of fuzzy logic to incorporate linguistic information in a robot map has been proposed in [11].

The proposed method to define directional information and fuzzy directional landscapes is not limited to a fixed set of directions (e.g., North, South, West, East), but can be applied to any desired angle. Also, we can tune the f function used in the definition of the structuring element ν in order to define directions which are more or less vague, depending on the application needs. The definition of fuzzy landscapes makes it easy to define complex directional relations by combining elementary relations using fuzzy operations. For instance, we can define a landscape for "North but not East" by fuzzy intersection of the landscape for North and the complement of the one for East.

Finally, it should be noted that fuzzy mathematical morphology can be used to solve several other problems in mobile robot navigation, including self-localization and spatial object processing (see [4]).

While the initial results reported in this paper show the viability of our technique, more experiments on real robotic applications are needed in order to establish the actual utility of this technique, for instance for robot self-localization or for human-robot interaction by linguistic expressions. These experiments are part of our current work.

Acknowledgments

This work was partially supported by the Swedish KK Foundation.

References

[1] J. Allen. Maintaining Knowledge about Temporal Intervals. *Comunications of the ACM*, 26(11):832–843, 1983.

[2] I. Bloch. Fuzzy Relative Position between Objects in Image Processing: a Morphological Approach. *IEEE Transactions on Pattern Analysis and Machine Intelligence*, 21(7):657–664, 1999.

[3] I. Bloch and H. Maître. Fuzzy Mathematical Morphologies: A Comparative Study. *Pattern Recognition*, 28(9):1341–1387, 1995.

[4] I. Bloch and A. Saffiotti. Why Robots should use Fuzzy Mathematical Morphology. In *Neuro Fuzzy Technologies NF2002*, La Havana, Cuba, 2002. Online at http://www.aass.oru.se/~asaffio/.

[5] B. de Baets. Fuzzy Morphology: a Logical Approach. In B. Ayyub and M. Gupta, editors, *Uncertainty in Engineering and Sciences: Fuzzy Logic, Statistics and Neural Network Approach*, pages 53–67. Kluwer Academic, 1997.

[6] D. Dubois and H. Prade. Weighted Fuzzy Pattern Matching. *Fuzzy Sets and Systems*, 28:313–331, 1988.

[7] A. Elfes. Using Occupancy Grids for Mobile Robot Perception and Navigation. *IEEE Computer*, 26(6):46–57, 1989.

[8] E. Fabrizi and A. Saffiotti. Extracting Topology-Based Maps from Gridmaps. In *IEEE International Conference on Robotics and Automation (ICRA-2000)*, San Francisco, CA, 2000. Online at http://www.aass.oru.se/~asaffio/.

[9] J. Freeman. The Modelling of Spatial Relations. *Computer Graphics and Image Processing*, 4(2):156–171, 1975.

[10] K. P. Gapp. Basic Meanings of Spatial Relations: Computation and Evaluation in 3D Space. In *12th National Conference on Artificial Intelligence, AAAI-94*, pages 1393–1398, Seattle, Washington, 1994.

[11] J. Gasós and A. Saffiotti. Using Fuzzy Sets to Represent Uncertain Spatial Knowledge in Autonomous Robots. *Spatial Cognition and Computation*, 1(3):205–226, 1999. Online at http://www.aass.oru.se/~asaffio/.

[12] J. M. Keller and X. Wang. Comparison of Spatial Relation Definitions in Computer Vision. In *ISUMA-NAFIPS'95*, pages 679–684, College Park, MD, September 1995.

[13] L. T. Koczy. On the Description of Relative Position of Fuzzy Patterns. *Pattern Recognition Letters*, 8:21–28, 1988.

[14] B. J. Kuipers. Modeling Spatial Knowledge. *Cognitive Science*, 2:129–153, 1978.

[15] J. Liu. A Method of Spatial Reasoning based on Qualitative Trigonometry. *Artificial Intelligence*, 98:137–168, 1998.

[16] G. Matheron. *Eléments pour une théorie des milieux poreux*. Masson, Paris, 1967.

[17] P. Matsakis and L. Wendling. A New Way to Represent the Relative Position between Areal Objects. *IEEE Trans. on Pattern Analysis and Machine Intelligence*, 21(7):634–642, 1999.

[18] K. Miyajima and A. Ralescu. Spatial Organization in 2D Segmented Images: Representation and Recognition of Primitive Spatial Relations. *Fuzzy Sets and Systems*, 65:225–236, 1994.

[19] M. Nachtegael and E. E. Kerre. Classical and Fuzzy Approaches towards Mathematical Morphology. In E. E. Kerre and M. Nachtegael, editors, *Fuzzy Techniques in Image Processing*, Studies in Fuzziness and Soft Computing, chapter 1, pages 3–57. Physica-Verlag, Springer, 2000.

[20] G. Oriolo, G. Ulivi, and M. Vendittelli. Fuzzy Maps: A New Tool for Mobile Robot Perception and Planning. *Journal of Robotic Systems*, 14(3):179–197, 1997.

[21] J. Serra. *Image Analysis and Mathematical Morphology*. Academic Press, London, 1982.

[22] J. Sharma and D. M. Flewelling. Inferences from Combined Knowledge about Topology and Directions. In *Advances in Spatial Databases*, volume 951 of *LNCS*, pages 279–291. Springer, 1995.

[23] D. Sinha, P. Sinha, E. R. Dougherty, and S. Batman. Design and Analysis of Fuzzy Morphological Algorithms for Image Processing. *IEEE Trans. on Fuzzy Systems*, 5(4):570–584, 1997.

[24] R. Smith, M. Self, and P. Cheeseman. Estimating Uncertain Spatial Relationships in Robotics. In I. J. Cox and G. T. Wilfong, editors, *Autonomous Robot Vehicles*, pages 167–193. Springer, New York, 1990.

[25] S. Thrun. Learning Metric-Topological Maps for Indoor Mobile Robot Navigation. *Artificial Intelligence*, 99(1):21–71, 1999.

[26] L. Vincent and P. Soille. Watersheds in Digital Spaces: An Efficient Algorithm Based on Immersion Simulations. *IEEE Ttransactions on Pattern Analysis and Machine Intelligence*, 13(6):583–598, 1991.

[27] S. Weber. A General Concept of Fuzzy Connectives, Negations and Implications Based on t-Norms and t-Conorms. *Fuzzy sets and systems*, 11:115–134, 1983.

Fuzzy "Between" Operators in the Framework of Fuzzy Orderings

Ulrich Bodenhofer
Software Competence Center Hagenberg
A-4232 Hagenberg, Austria
ulrich.bodenhofer@scch.at

Abstract

Ordering-based modifiers have fruitful applications in fuzzy rule-based systems. In continuation of ongoing research work on this topic, this paper is concerned with the construction of two binary ordering-based modifiers that model a concept of fuzzy *'between'*, both in an inclusive and a non-inclusive setting.

Keywords: between operator, fuzzy ordering, fuzzy relation, interpretability, ordering-based modifier.

1 Introduction

Fuzzy systems have always been regarded as appropriate methodologies for controlling complex systems and for carrying out complicated decision processes [21]. The compactness of rule bases, however, is still a crucial issue—the surveyability and interpretability of a rule base decreases with its number of rules. In particular, if rule bases are represented as complete tables, the number of rules grows exponentially with the number of variables. Therefore, techniques for reducing the number of rules in a rule base while still maintaining the system's behavior and improving surveyability and interpretability should receive special interest. In this paper, we deal with operators which are supposed to serve as a key to rule base reduction—*ordering-based modifiers*.

Almost all fuzzy systems involving numerical variables implicitly use orderings. It is standard to decompose the universe of a linearly ordered system variable into a certain number of fuzzy sets by means of the ordering of the universe—typically resulting in labels like *'small'*, *'medium'*, or *'large'*.

Let us consider a simple example. Suppose that we have a system with two real-valued input variables x_1, x_2 and a real-valued output variable y, where all domains are divided into five fuzzy sets with the linguistic labels *'Z'*, *'S'*, *'M'*, *'L'*, and *'V'* (standing for *'approx. zero'*, *'small'*, *'medium'*, *'large'*, and *'very large'*, respectively).

$x_1 \backslash x_2$	Z	S	M	L	V
Z	Z	S	M	L	V
S	S	S	M	L	L
M	S	M	L	M	M
L	S	S	M	M	S
V	Z	S	M	S	Z

It is easy to see that, in the above table, there are several adjacent rules having the same consequent value. Assuming that we had a unique and unambiguous computational methodology to compute *'at least A'*, *'at most A'*, or *'between A and B'*, it would be possible to group and replace such neighboring rules. For instance, the three rules

IF x_1 is *'S'* AND x_2 is *'Z'* THEN y is *'S'*
IF x_1 is *'M'* AND x_2 is *'Z'* THEN y is *'S'*
IF x_1 is *'L'* AND x_2 is *'Z'* THEN y is *'S'*

could be replaced by the following rule[1] (adopting an inclusive view on the adverb *'between'*):

IF x_1 is *'between S and L'* AND x_2 is *'Z'* THEN y is *'S'*

Of course, there is actually no need to do so in such a simple case. Anyway, grouping neighboring rules in such a way could help to reduce the size of larger high-dimensional rule bases considerably.

It is considered as another opportunity for reducing the size of a rule base to store only some representative rules and to "interpolate" between them [15], where, in this context, we understand interpolation as a computational method that is able to obtain a meaningful conclusion even if an observation does not match any antecedent in the rule base [14]. In any case, it is indispensable to have criteria for determining between which rules the interpolation should take place. Beside distance, orderings play a fundamental role in this selection. As an alternative to distance-based methods [15], it is possible to fill the gap between the antecedents of two rules using a non-inclusive concept of fuzzy *'between'*.

In [1,4], a basic framework for defining the unary modifiers ATL and ATM (short for *'at least'* and *'at most'*, respectively) by means of image operators of fuzzy orderings has been introduced. This general approach has the following advantages: it is applicable to any kind of fuzzy set, it can be used for any kind of fuzzy ordering without any restriction to linearly ordered or real-valued domains, and it even allows to take a domain-specific context of indistinguishability into account.

This paper is concerned with an extension of this framework by two binary ordering-based modifiers named BTW and SBT which both represent fuzzy *'between'* operators, where BTW stands for the inclusive and SBT ("strictly between") stands for the non-inclusive interpretation.

2 Preliminaries

Throughout the whole paper, we will not explicitly distinguish between fuzzy sets and their corresponding membership functions. Consequently, uppercase letters will be used for both synonymously. The set of all fuzzy sets on a domain X will be denoted with $\mathcal{F}(X)$. As usual, we call a fuzzy set A *normalized* if there exists an $x \in X$ such that $A(x) = 1$ holds.

In general, *triangular norms* [13], i.e. associative, commutative, and non-decreasing binary operations on the unit interval (i.e. a $[0,1]^2 \to [0,1]$ mappings) which have 1 as

---

[1]It depends on the underlying inference scheme whether the result is actually the same; we leave this aspect aside for the present paper, since this is not its major concern.

neutral element, will be considered as our standard models of logical conjunction. In this paper, assume that T denotes a left-continuous triangular norm, i.e. a t-norm whose partial mappings $T(x,.)$ and $T(.,x)$ are left-continuous.

Definition 1. Let T be a t-norm. The *T-intersection* of two fuzzy sets $A, B \in \mathcal{F}(X)$ is defined by means of the following membership function:
$$(A \cap_T B)(x) = T(A(x), B(x))$$

For $T = \min$, we will simply use the notation $A \cap B$. Correspondingly, the max-union of two fuzzy sets $A, B \in \mathcal{F}(X)$ is defined as
$$(A \cup B)(x) = \max(A(x), B(x)).$$

So-called residual implications will be used as the concepts of logical implication [7–9, 13].

Definition 2. For any left-continuous t-norm T, the corresponding *residual implication* $\vec{T}$ is defined as
$$\vec{T}(x, y) = \sup\{u \in [0,1] \mid T(u, x) \leq y\}.$$

The residual implication can be used to define a logical negation which logically fits to the t-norm and its implication.

Definition 3. The *negation* corresponding to a left-continuous t-norm T is defined as
$$N_T(x) = \vec{T}(x, 0).$$

Lemma 4. *N_T is a left-continuous non-increasing $[0,1] \to [0,1]$ mapping. Moreover, the so-called law of contraposition holds*
$$\vec{T}(x, y) \leq \vec{T}(N_T(y), N_T(x))$$
which also implies $x \leq N_T(N_T(x))$.

Note that the reverse inequality does not hold in general (unlike the Boolean case, where $p \Rightarrow q$ is equivalent to $\neg q \Rightarrow \neg p$).

Definition 5. The *T-complement* of a fuzzy set $A \in \mathcal{F}(X)$ is defined as
$$(\complement_T A)(x) = N_T(A(x)).$$

Lemma 6. *The following holds for all fuzzy sets $A, B \in \mathcal{F}(X)$:*

1. $A \cap_T \complement_T A = \emptyset$
2. $A \subseteq \complement_T \complement_T A$
3. $A \subseteq B$ *implies* $\complement_T A \supseteq \complement_T B$

Lemma 7. *As long as only min-intersections and max-unions are considered, the so-called De Morgan laws hold:*
$$\complement_T(A \cup B) = (\complement_T A) \cap (\complement_T B)$$
$$\complement_T(A \cap B) = (\complement_T A) \cup (\complement_T B)$$

As usual, we call a fuzzy set on a product space $X \times X$ *binary fuzzy relation*. The following two kinds of binary fuzzy relations will be essential.

Definition 8. A binary fuzzy relation E on a domain X is called *fuzzy equivalence relation* with respect to T, for brevity T-*equivalence*, if and only if the following three axioms are fulfilled for all $x, y, z \in X$:

Reflexivity: $\quad E(x,x) = 1$

Symmetry: $\quad E(x,y) = E(y,x)$

T-transitivity: $\quad T(E(x,y), E(y,z)) \leq E(x,z)$

In contrast to previous definitions of fuzzy orderings [7, 18, 20], we consider a general concept of fuzzy orderings taking a given context of indistinguishability into account which is modeled by a fuzzy equivalence relation [2, 10].

Definition 9. Let $L: X^2 \to [0,1]$ be a binary fuzzy relation. L is called *fuzzy ordering* with respect to T and a T-equivalence E, for brevity T-E-*ordering*, if and only if it is T-transitive and fulfills the following two axioms for all $x, y \in X$:

E-reflexivity: $\quad E(x,y) \leq L(x,y)$

T-E-antisymmetry: $\quad T(L(x,y), L(y,x)) \leq E(x,y)$

A subclass which will be of special importance in the following are so-called direct fuzzifications.

Definition 10. A T-E-ordering L is called a *direct fuzzification* of a crisp ordering $\preceq$ if and only if it admits the following resolution:

$$L(x,y) = \begin{cases} 1 & \text{if } x \preceq y \\ E(x,y) & \text{otherwise} \end{cases}$$

It is worth to mention that there is a one-to-one correspondence between direct fuzzifications of crisp linear orderings and so-called fuzzy weak orderings, i.e. reflexive and T-transitive binary fuzzy relations which fulfill strong completeness (i.e., for all $x, y \in X$, $\max(L(x,y), L(y,x)) = 1$) [1, 2].

3 Unary Ordering-Based Modifiers

Throughout the remaining paper, assume that we are given a T-E-ordering L (for some left-continuous t-norm T and a given T-equivalence E). Then the unary ordering-based modifiers ATL and ATM are defined as follows [1, 4]:

$$\text{ATL}(A)(x) = \sup\{T(A(y), L(y,x)) \mid y \in X\}$$
$$\text{ATM}(A)(x) = \sup\{T(A(y), L(x,y)) \mid y \in X\}$$

In the case that L coincides with a crisp ordering $\preceq$, we will explicitly indicate that by using the notations LTR and RTL (short for "left-to-right" and "right-to-left continuations") instead of ATL and ATM, respectively. It is easy to verify that the following simplified representation holds in such a case:

$$\mathrm{LTR}(A)(x) = \sup\{A(y) \mid y \preceq x\}$$
$$\mathrm{RTL}(A)(x) = \sup\{A(y) \mid y \succeq x\}$$

Moreover, for a given fuzzy set A, LTR is the smallest superset of A with a non-decreasing membership function and RTL is the smallest superset of A with a non-increasing membership function. For convenience, let us use the notation EXT for the so-called *extensional hull* operator of the T-equivalence E:

$$\mathrm{EXT}(A)(x) = \sup\{T(A(y), E(y,x)) \mid y \in X\}$$

Note that, for an arbitrary fuzzy set A, $\mathrm{EXT}(A)$ is the smallest superset fulfilling the property

$$T(A(x), E(x,y)) \leq A(y)$$

for all $x, y \in X$. This property is usually called *extensionality* [6, 5, 11, 12, 16].

Lemma 11. [4, 11, 16] *Let E be a T-equivalence and let $A, B \in \mathcal{F}(X)$ be two extensional fuzzy sets. Then $A \cap B$, $A \cup B$, and $\complement_T A$ are also extensional. Moreover, $\mathrm{EXT}(A \cup B) = \mathrm{EXT}(A) \cup \mathrm{EXT}(B)$ holds.*

Lemma 12. [1, 4] *The operators ATL and ATM are non-decreasing with respect to the inclusion of fuzzy sets and the following holds for all fuzzy sets $A, B \in \mathcal{F}(X)$:*

1. $\mathrm{ATL}(A \cup B) = \mathrm{ATL}(A) \cup \mathrm{ATL}(B)$
2. $\mathrm{ATM}(A \cup B) = \mathrm{ATM}(A) \cup \mathrm{ATM}(B)$
3. $\mathrm{ATL}(A \cap B) \subseteq \mathrm{ATL}(A) \cap \mathrm{ATL}(B)$
4. $\mathrm{ATM}(A \cap B) \subseteq \mathrm{ATM}(A) \cap \mathrm{ATM}(B)$
5. $\mathrm{ATL}(\mathrm{ATL}(A)) = \mathrm{ATL}(A)$
6. $\mathrm{ATM}(\mathrm{ATM}(A)) = \mathrm{ATM}(A)$

Theorem 13. [1, 4] *If L is a direct fuzzification of some crisp ordering $\preceq$, the following equalities hold:*

$$\mathrm{ATL}(A) = \mathrm{EXT}(\mathrm{LTR}(A)) = \mathrm{LTR}(\mathrm{EXT}(A)) = \mathrm{EXT}(A) \cup \mathrm{LTR}(A)$$
$$\mathrm{ATM}(A) = \mathrm{EXT}(\mathrm{RTL}(A)) = \mathrm{RTL}(\mathrm{EXT}(A)) = \mathrm{EXT}(A) \cup \mathrm{RTL}(A)$$

Moreover, $\mathrm{ATL}(A)$ is the smallest fuzzy superset of A which is extensional and has a non-decreasing membership function. Analogously, $\mathrm{ATM}(A)$ is the smallest fuzzy superset of A which is extensional and has a non-increasing membership function.

The notion of convexity and convex hulls will also be essential in the following.

Definition 14. Provided that the domain X is equipped with some crisp ordering $\preceq$ (not necessarily linear), a fuzzy set $A \in \mathcal{F}(X)$ is called *convex* (compare with [17, 19]) if and only if, for all $x, y, z \in X$,

$$x \preceq y \preceq z \text{ implies } A(y) \geq \min(A(x), A(z)).$$

Lemma 15. *Fuzzy sets with non-increasing/non-decreasing membership function are convex. For any two convex fuzzy sets $A, B \in \mathcal{F}(X)$, $A \cap B$ is convex.*

Lemma 16. *Assume that $\preceq$ is an arbitrary, not necessarily linear ordering on a domain X. Then the fuzzy set*
$$\mathrm{CVX}(A) = \mathrm{LTR}(A) \cap \mathrm{RTL}(A)$$
is the smallest convex fuzzy superset of A.

Theorem 17. [1,4] *With the assumptions of Theorem 13 and the definition*
$$\mathrm{ECX}(A) = \mathrm{ATL}(A) \cap \mathrm{ATM}(A),$$
the following representation holds:
$$\mathrm{ECX}(A) = \mathrm{EXT}(\mathrm{CVX}(A)) = \mathrm{CVX}(\mathrm{EXT}(A)) = \mathrm{EXT}(A) \cup \mathrm{CVX}(A)$$
Furthermore, $\mathrm{ECX}(A)$ is the smallest fuzzy superset of A which is extensional and convex.

4 The Inclusive Operator

Finally, we can now define an operator representing an inclusive version of *'between'* with respect to a fuzzy ordering.

Definition 18. Given two fuzzy sets $A, B \in \mathcal{F}(X)$, the binary operator BTW is defined as
$$\mathrm{BTW}(A, B) = \mathrm{ECX}(A \cup B).$$

Note that it can easily be inferred from basic properties of ATL and ATM (cf. Lemma 12) that the following alternative representation holds:
$$\mathrm{BTW}(A, B) = (\mathrm{ATL}(A) \cup \mathrm{ATL}(B)) \cap (\mathrm{ATM}(A) \cup \mathrm{ATM}(B)) \tag{1}$$

This representation is particularly helpful to prove the following basic properties of the BTW operator.

Theorem 19. *The following holds for all fuzzy sets $A, B \in \mathcal{F}(X)$:*

1. $\mathrm{BTW}(A, B) = \mathrm{BTW}(B, A)$
2. $A \subseteq \mathrm{BTW}(A, B)$
3. $\mathrm{BTW}(A, \emptyset) = \mathrm{BTW}(A, A) = \mathrm{ECX}(A)$
4. $\mathrm{BTW}(A, B)$ *is extensional*

If L is a direct fuzzification of a crisp ordering $\preceq$, then $\mathrm{BTW}(A, B)$ is convex as well and $\mathrm{BTW}(A, B)$ is the smallest convex and extensional fuzzy set containing both A and B.

Proof. Trivial by elementary properties of maximum union and the operator ECX (cf. Theorem 17). □

It is, therefore, justified (in particular due to Point 2. above) to speak of an inclusive interpretation. Moreover, it is even possible to show that BTW is an associative operation; hence $(\mathcal{F}(X), \mathrm{BTW})$ is a commutative semigroup.

5 The Non-Inclusive Operator

Now let us study how a *'strictly between'* operator can be defined. It seems intuitively clear that *'strictly between A and B'* should be a subset of $\text{BTW}(A,B)$ which should not include any relevant parts of A and B.

Definition 20. The *'strictly between'* operator is a binary connective on $\mathcal{F}(X)$ which is defined as

$$\text{SBT}(A,B) = \text{BTW}(A,B) \cap \complement_T((\text{ATL}(A) \cap \text{ATL}(B)) \cup (\text{ATM}(A) \cap \text{ATM}(B))).$$

Note that Lemma 7 yields the following alternative representation:

$$\text{SBT}(A,B) = \text{BTW}(A,B) \cap \complement_T((\text{ATL}(A) \cap \text{ATL}(B)) \cap \complement_T(\text{ATM}(A) \cap \text{ATM}(B))) \quad (2)$$

Theorem 21. *The following holds for all fuzzy sets $A, B \in \mathcal{F}(X)$:*

1. $\text{SBT}(A,B) = \text{SBT}(B,A)$
2. $\text{SBT}(A,B) \subseteq \text{BTW}(A,B)$
3. $\text{SBT}(A,\emptyset) = \text{ECX}(A)$
4. $\text{SBT}(A,B)$ *is extensional*

If L is a direct fuzzification of a crisp ordering $\preceq$, $\text{SBT}(A,B)$ is convex as well. If we assume that L is a direct fuzzification of a crisp linear ordering $\preceq$ and that A and B are normalized, the following holds:

5. $\text{ECX}(A) \cap_T \text{SBT}(A,B) = \emptyset$

Proof. The first two assertions follow trivially from the definition of SBT and elementary properties. For proving the third assertion, consider the following:

$$\text{SBT}(A,\emptyset) = \text{BTW}(A,\emptyset) \cap \complement_T((\text{ATL}(A) \cap \text{ATL}(\emptyset)) \cup (\text{ATM}(A) \cap \text{ATM}(\emptyset)))$$
$$= \text{ECX}(A) \cap \complement_T(\emptyset) = \text{ECX}(A)$$

The extensionality of $\text{SBT}(A,B)$ follows from Lemma 11 and the fact that $\text{BTW}(A,B)$, $\text{ATL}(A)$, $\text{ATL}(B)$, $\text{ATM}(A)$, and $\text{ATM}(B)$ are all extensional.

Assume that L is a direct fuzzification. As $\text{ATL}(A)$ and $\text{ATL}(B)$ have non-decreasing membership functions, also $\text{ATL}(A) \cap \text{ATL}(B)$ has a non-decreasing membership function. Consequently, $\complement_T(\text{ATL}(A) \cap \text{ATL}(B))$ has a non-increasing membership function and, by Lemma 15, this fuzzy set is convex. Following analogous arguments, it can be proved that $\complement_T(\text{ATM}(A) \cap \text{ATM}(B))$ is convex. Then we see from the alternative representation (2) that $\text{SBT}(A,B)$ is an intersection of convex fuzzy sets. Therefore, by Lemma 15, $\text{SBT}(A,B)$ is convex.

The following follows easily from the distributivity of minimum and maximum:

$$(\text{ATL}(A) \cap \text{ATL}(B)) \cup (\text{ATM}(A) \cap \text{ATM}(B))$$
$$= (\text{ATL}(A) \cup \text{ATM}(A)) \cap (\text{ATL}(A) \cup \text{ATM}(B))$$
$$\cap (\text{ATL}(B) \cup \text{ATM}(A)) \cap (\text{ATL}(B) \cup \text{ATM}(B)) = (*)$$

Let A and B now be normalized and let L be a direct fuzzification of a crisp linear ordering $\preceq$. It is easy to see that, in such a setting, $\mathrm{ATL}(A) \cup \mathrm{ATM}(A) = X$ holds. Therefore,

$$(*) = \bigl(\mathrm{ATL}(A) \cup \mathrm{ATM}(B)\bigr) \cap \bigl(\mathrm{ATL}(B) \cup \mathrm{ATM}(A)\bigr).$$

As the fuzzy sets $\bigl(\mathrm{ATL}(A) \cup \mathrm{ATM}(B)\bigr)$ and $\bigl(\mathrm{ATL}(B) \cup \mathrm{ATM}(A)\bigr)$ are both supersets of $\mathrm{ECX}(A)$, we obtain that $\bigl(\mathrm{ATL}(A) \cap \mathrm{ATL}(B)\bigr) \cup \bigl(\mathrm{ATL}(A) \cap \mathrm{ATM}(B)\bigr)$ is a superset of $\mathrm{ECX}(A)$. Then we can infer the following:

$\mathrm{ECX}(A) \cap_T \mathrm{SBT}(A,B)$
$\quad = \mathrm{ECX}(A) \cap_T \bigl(\mathrm{BTW}(A,B) \cap \complement_T((\mathrm{ATL}(A) \cap \mathrm{ATL}(B)) \cup (\mathrm{ATM}(A) \cap \mathrm{ATM}(B)))\bigr)$
$\quad = \bigl(\mathrm{ECX}(A) \cap_T \mathrm{BTW}(A,B)\bigr)$
$\qquad \cap \bigl(\mathrm{ECX}(A) \cap_T \complement_T((\mathrm{ATL}(A) \cap \mathrm{ATL}(B)) \cup (\mathrm{ATM}(A) \cap \mathrm{ATM}(B)))\bigr)$
$\quad \subseteq \bigl(\mathrm{ECX}(A) \cap_T \mathrm{BTW}(A,B)\bigr) \cap \bigl(\mathrm{ECX}(A) \cap_T \complement_T(\mathrm{ECX}(A))\bigr)$
$\quad = \bigl(\mathrm{ECX}(A) \cap_T \mathrm{BTW}(A,B)\bigr) \cap \emptyset = \emptyset$ $\square$

Note that the last equality particularly implies

$$A \cap_T \mathrm{SBT}(A,B) = \emptyset$$

which justifies to speak of an non-inclusive concept.

6 Ordering Properties

Despite basic properties that have already been presented in the previous two sections, it remains to be clarified whether the results $\mathrm{BTW}(A,B)$ and $\mathrm{SBT}(A,B)$ obtained by the two operators are really *lying between* A and B. We will approach this question from an ordinal perspective. It is straightforward to define the following binary relation on $\mathcal{F}(X)$:

$$A \preceq_L B \text{ if and only if } \mathrm{ATL}(A) \supseteq \mathrm{ATL}(B) \text{ and } \mathrm{ATM}(A) \subseteq \mathrm{ATM}(B)$$

This relation is reflexive, transitive, and antisymmetric up to the following equivalence relation:
$$A \sim_L B \text{ if and only if } \mathrm{ECX}(A) = \mathrm{ECX}(B)$$

Moreover, if we restrict ourselves to fuzzy numbers and to the natural ordering of real numbers, it is relatively easy to see that $\preceq_L$ coincides with the interval ordering of fuzzy numbers induced by the extension principle. It is, therefore, justified to consider $\preceq_L$ as a meaningful general concept of ordering of fuzzy sets with respect to a given fuzzy ordering L [1, 3].

The following theorem gives a clear justification that we may consider the definitions of the operators BTW and SBT as appropriate.

Theorem 22. *Suppose that we are given two normalized fuzzy sets $A, B \in \mathcal{F}(X)$ such that $A \preceq_L B$ holds. Then the following inequality holds:*

$$A \preceq_L \mathrm{BTW}(A,B) \preceq_L B$$

Now let us assume that L is strongly complete (therefore, a direct fuzzification of a crisp linear ordering $\preceq$) and that there exists a value $x \in X$ which separates A and B in the way that (for all $y, z \in X$) $A(y) > 0$ implies $y \prec x$ and $B(z) > 0$ implies $x \prec z$. Then the following inequality holds too:

$$A \preceq_L \mathrm{SBT}(A,B) \preceq_L B$$

Proof. Assume that $A \preceq_L B$ holds, i.e. $\mathrm{ATL}(A) \supseteq \mathrm{ATL}(B)$ and $\mathrm{ATM}(A) \subseteq \mathrm{ATM}(B)$. Then we obtain using the alternative representation (1) and Lemma 12:

$$\mathrm{ATL}(\mathrm{BTW}(A,B)) = \mathrm{ATL}((\mathrm{ATL}(A) \cup \mathrm{ATL}(B)) \cap (\mathrm{ATM}(A) \cup \mathrm{ATM}(B)))$$
$$= \mathrm{ATL}(\mathrm{ATL}(A) \cap \mathrm{ATM}(B)) \subseteq \mathrm{ATL}(\mathrm{ATL}(A)) = \mathrm{ATL}(A)$$

Moreover, $\mathrm{ATM}(A) \subseteq \mathrm{ATM}(\mathrm{BTW}(A,B))$ must hold, since A is a subset of $\mathrm{BTW}(A,B)$ (non-decreasingness of ATM with respect to inclusion; cf. Lemma 12), and we have proved that $A \preceq_L \mathrm{BTW}(A,B)$ holds. The inequality $\mathrm{BTW}(A,B) \preceq_L B$ can be proved analogously.

We have proved above that $\mathrm{ATL}(\mathrm{BTW}(A,B)) \subseteq \mathrm{ATL}(A)$. As, trivially, $\mathrm{SBT}(A,B) \subseteq \mathrm{BTW}(A,B)$ holds, we obtain $\mathrm{ATL}(\mathrm{SBT}(A,B)) \subseteq \mathrm{ATL}(A)$. Now fix the value $x \in X$ as described above. All values y with $A(y) > 0$ are below x, i.e. $y \prec x$, and all values z with $B(z) > 0$ are above x, i.e. $x \prec z$. That entails the four equalities $\mathrm{ATL}(A)(x) = 1$, $\mathrm{ATM}(A)(x) = 0$, $\mathrm{ATL}(B)(x) = 0$, and $\mathrm{ATM}(B)(x) = 1$, and we obtain the following:

$$\mathrm{BTW}(A,B)(x) = (\mathrm{ATL}(A) \cup \mathrm{ATL}(B)) \cap (\mathrm{ATM}(A) \cup \mathrm{ATM}(B))(x)$$
$$= \min(\max(\mathrm{ATL}(A)(x), \mathrm{ATL}(B)(x)), \max(\mathrm{ATM}(A)(x), \mathrm{ATM}(B)(x)))$$
$$= \min(\max(1,0), \max(0,1)) = 1$$

Moreover,

$$(\mathrm{ATL}(A) \cap \mathrm{ATL}(B)) \cup (\mathrm{ATM}(A) \cap \mathrm{ATM}(B))(x) = \max(\min(1,0), \min(1,0)) = 0.$$

Therefore, $\mathrm{SBT}(A,B)(x) = 1$. As all values of A having non-zero membership degrees are below x, it follows that

$$\mathrm{ATM}(A) \subseteq \mathrm{ATM}(\mathrm{SBT}(A,B)).$$

Hence, $A \preceq_L \mathrm{SBT}(A,B)$ holds. The second inequality $\mathrm{SBT}(A,B) \preceq_L B$ can again be proved analogously. □

7 Examples

In order to underline these rather abstract results with an example, let us consider two fuzzy subsets of the real numbers:

$$A(x) = \max(1 - 3 \cdot |1 - x|, 0.7 - 2 \cdot |1.5 - x|, 0)$$
$$B(x) = \max(1 - |4 - x|, 0)$$

It is easy to see that both fuzzy sets are normalized; B is convex, while A is not convex.

The natural ordering of real numbers $\leq$ is a fuzzy ordering with respect to any t-norm and the crisp equality. No matter which t-norm we choose, we obtain the fuzzy sets BTW(A,B) and SBT(A,B) as shown in Figure 1.

Now let us consider the following two fuzzy relations:

$$E(x,y) = \max(1-|x-y|,0)$$

$$L(x,y) = \begin{cases} 1 & \text{if } x \leq y \\ E(x,y) & \text{otherwise} \end{cases}$$

One easily verifies that E is indeed a T_L-equivalence on the real numbers and that L is a T_L-E-ordering which directly fuzzifies the linear ordering of real numbers [1,2]. Figure 2 shows the results of computing BTW(A,B) and SBT(A,B) for A and B from above. It is a routine matter to show that B is extensional and that A is not extensional. This means that A contains parts that are defined in an unnaturally precise way. Since the operators BTW and SBT have been designed to take the given context of indistinguishability into account, they try to remove all uncertainties arising from the non-extensionality of A. This is reflected in the fact that BTW(A,B) also contains some parts to the left of A that are potentially indistinguishable from A. In the same way, SBT(A,B) does not include those parts to the right of A that are potentially indistinguishable from elements in A.

8 Conclusion

This paper has been concerned with the definition of two binary ordering-based modifiers BTW and SBT. The operator BTW has been designed for computing the fuzzy set of all objects lying between two fuzzy sets including both boundaries. The purpose of SBT is to extract those objects which are lying strictly between two fuzzy sets—not including the two boundaries. We have shown several basic properties of the two operators and, from the viewpoint of orderings of fuzzy sets, that the two operators indeed yield meaningful results. Therefore, we conclude that the two operators are appropriate as modifiers for fuzzy systems applications and rule interpolation.

Acknowledgements

The author gratefully acknowledges support of the K*plus* Competence Center Program which is funded by the Austrian Government, the Province of Upper Austria and the Chamber of Commerce of Upper Austria.

References

[1] U. Bodenhofer. *A Similarity-Based Generalization of Fuzzy Orderings*, volume C 26 of *Schriftenreihe der Johannes-Kepler-Universität Linz*. Universitätsverlag Rudolf Trauner, 1999.

[2] U. Bodenhofer. A similarity-based generalization of fuzzy orderings preserving the classical axioms. *Internat. J. Uncertain. Fuzziness Knowledge-Based Systems*, 8(5):593–610, 2000.

[3] U. Bodenhofer. A general framework for ordering fuzzy sets. In B. Bouchon-Meunier, J. Guitiérrez-Ríoz, L. Magdalena, and R. R. Yager, editors, *Technologies for Constructing Intelligent Systems 1: Tasks*, volume 89 of *Studies in Fuzziness and Soft Computing*, pages 213–224. Physica-Verlag, Heidelberg, 2002.

[4] U. Bodenhofer, M. De Cock, and E. E. Kerre. Openings and closures of fuzzy preorderings: Theoretical basics and applications to fuzzy rule-based systems. *Int. J. General Systems*. (to appear).

[5] D. Boixader, J. Jacas, and J. Recasens. Fuzzy equivalence relations: Advanced material. In D. Dubois and H. Prade, editors, *Fundamentals of Fuzzy Sets*, volume 7 of *The Handbooks of Fuzzy Sets*, pages 261–290. Kluwer Academic Publishers, Boston, 2000.

[6] R. Bělohlávek. *Fuzzy Relational Systems. Foundations and Principles*. IFSR Int. Series on Systems Science and Engineering. Kluwer Academic/Plenum Publishers, New York, 2002.

[7] J. Fodor and M. Roubens. *Fuzzy Preference Modelling and Multicriteria Decision Support*. Kluwer Academic Publishers, Dordrecht, 1994.

[8] S. Gottwald. *Fuzzy Sets and Fuzzy Logic*. Vieweg, Braunschweig, 1993.

[9] P. Hájek. *Metamathematics of Fuzzy Logic*, volume 4 of *Trends in Logic*. Kluwer Academic Publishers, Dordrecht, 1998.

[10] U. Höhle and N. Blanchard. Partial ordering in L-underdeterminate sets. *Inform. Sci.*, 35:133–144, 1985.

[11] F. Klawonn and J. L. Castro. Similarity in fuzzy reasoning. *Mathware Soft Comput.*, 3(2):197–228, 1995.

[12] F. Klawonn and R. Kruse. Equality relations as a basis for fuzzy control. *Fuzzy Sets and Systems*, 54(2):147–156, 1993.

[13] E. P. Klement, R. Mesiar, and E. Pap. *Triangular Norms*, volume 8 of *Trends in Logic*. Kluwer Academic Publishers, Dordrecht, 2000.

[14] L. T. Kóczy and K. Hirota. Ordering, distance and closeness of fuzzy sets. *Fuzzy Sets and Systems*, 59(3):281–293, 1993.

[15] L. T. Kóczy and K. Hirota. Size reduction by interpolation in fuzzy rule bases. *IEEE Trans. Syst. Man Cybern.*, 27(1):14–25, 1997.

[16] R. Kruse, J. Gebhardt, and F. Klawonn. *Foundations of Fuzzy Systems*. John Wiley & Sons, New York, 1994.

[17] R. Lowen. Convex fuzzy sets. *Fuzzy Sets and Systems*, 3:291–310, 1980.

[18] S. V. Ovchinnikov. Similarity relations, fuzzy partitions, and fuzzy orderings. *Fuzzy Sets and Systems*, 40(1):107–126, 1991.

[19] L. A. Zadeh. Fuzzy sets. *Inf. Control*, 8:338–353, 1965.

[20] L. A. Zadeh. Similarity relations and fuzzy orderings. *Inform. Sci.*, 3:177–200, 1971.

[21] L. A. Zadeh. Outline of a new approach to the analysis of complex systems and decision processes. *IEEE Trans. Syst. Man Cybern.*, 3(1):28–44, 1973.

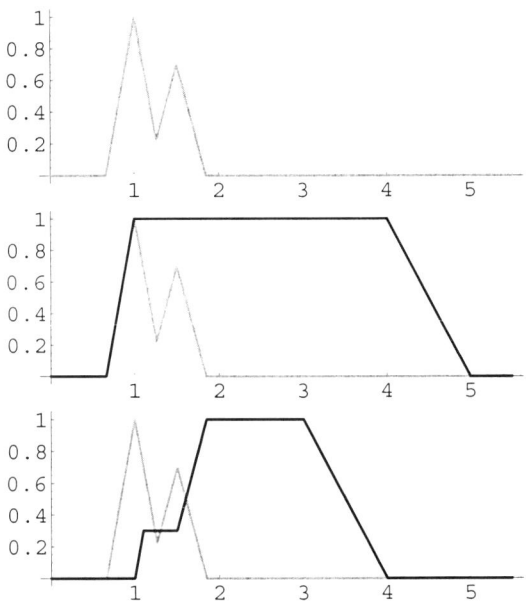

Figure 1: Two fuzzy sets A, B (top) and the results of BTW(A,B) (middle) and SBT(A,B) (bottom), using the crisp ordering of real numbers

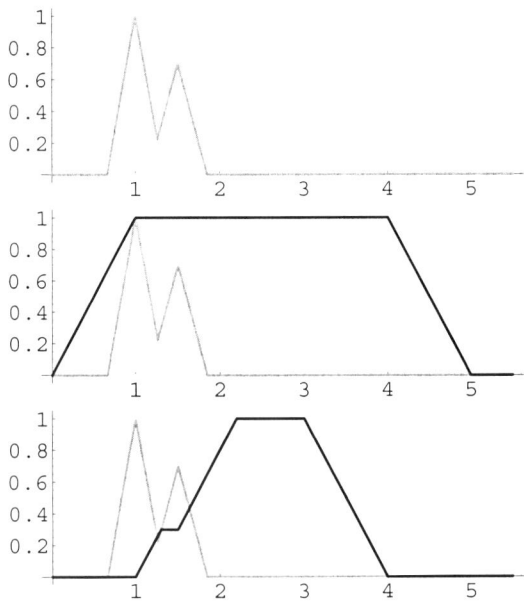

Figure 2: Two fuzzy sets A, B (top) and the results of BTW(A,B) (middle) and SBT(A,B) (bottom) with respect to a fuzzy ordering on $\mathbb{R}$

A step towards conceptually improving Takagi-Sugeno's Approximation[*]

S. Guadarrama[1]   E. Trillas[2]
Dept. Inteligencia Artificial
Universidad Politécnica de Madrid
sguada@isys.dia.fi.upm.es[1] etrillas@fi.upm.es[2]

J. Gutiérrez[3]   F. Fernández[4]
Dept. Tecnología Fotónica
Universidad Politécnica de Madrid
jgr@dtf.fi.upm.es[3] felipe.fernandez@es.bosch.com[4]

Abstract

This paper tries to explore some lines to improve Takagi-Sugeno's Approximation from a point of view, joining both the logical rationale of Fuzzy Control as Intelligent Control and the numerical results' accuracy.

Keywords: Fuzzy Control, Takagi-Sugeno, Condictional Functions, Convolution.

1 Introduction

In the last years we have seen increasing interest in fuzzy systems research and applications. This is mainly due to the success of fuzzy technologies in many fields of engineering including consumer products, transportation, manufacturing, medical, control and signal processing systems.

Classical models try to avoid vague, imprecise or uncertain information because it is considered as having a negative influence in the corresponding technique. However, fuzzy systems take advantage of this kind of information because it leads to simpler and more suitable models, which are both easier to handle and more familiar to human thinking.

During decades Fuzzy Control using Takagi-Sugeno' Approximation has been successfully applied to a wide range of control problems and have demonstrated significant advances in non-linear control.

A zero-order Takagi-Sugeno (T-S) model [4], specifies a fuzzy system by a set of rules of the form:

" If x_1 is P_1^1 and $\cdots$ and x_n is P_1^n then y is q_1"
$$\vdots$$
" If x_1 is P_i^1 and $\cdots$ and x_n is P_i^n then y is q_i"

---

[*] This paper is partially supported by CICYT (Spain) under project TIC2000-1420

where x_j are input variables, P_i^n are linguistic labels on $\mathbb{X}$ represented by a fuzzy set $\mu_{P_i^n}$, y is the output variable and q_i are constant values.
The global output of a T-S model, given an input vector of values $[x_1, ..., x_n]$ is:

$$y = \frac{\sum_i (\omega_i \cdot q_i)}{\sum_i (\omega_i)}$$

where $\omega_i = T(\mu_{P_1^i}(x_1), \mu_{P_2^i}(x_2), \cdots, \mu_{P_n^i}(x_n))$ is the matching degree between the inputs and the antecedent of the rule, usually $T = Prod$.
The other model widely used is Mamdani-Larsen (M-L) model [6], composed by a set of rules of the form:

" If x_1 is P_1^1 and $\cdots$ and x_n is P_1^n then y is Q_1 "
$$\vdots \qquad \vdots$$
" If x_1 is P_i^1 and $\cdots$ and x_n is P_i^n then y is Q_i "

where x_1 are input variables, P_i^n are linguistic labels on $\mathbb{X}$ represented by a fuzzy set $\mu_{P_i^n}$, Q_i are linguistic labels on $\mathbb{Y}$ represented by a fuzzy set μ_{Q_i}. And each rule is represented using a conditional function $J : [0,1] \times [0,1] \to [0,1]$.
The fuzzy output of each single rule, given an input vector of values $[x_1, ..., x_n]$ follows from Zadeh's Compositional Rule of Inference, and is:

$$\mu_{Q_i^*}(y) = J(T(\mu_{P_i^1}(x_1), \cdots, \mu_{P_i^n}(x_n)), \mu_{Q_i}(y))).$$

To obtain a numerical output it is needed to defuzzify this fuzzy output using either centre of gravity, or centre of area, or left or right most value, *et cetera*.
Numerical functions $J : [0,1] \times [0,1] \to [0,1]$ such that for some continuous t-norm T verify the **Modus Ponens** inequality $T(a, J(a,b)) \leq b$, for all a,b in $[0,1]$, are called T-Conditional Functions. They are used in Fuzzy Logic to represent conditional statements or rules *"If x is P, then y is Q"* as $J(\mu_P(x), \mu_Q(y))$, provided x varies in a referential $\mathbb{X}$, y in a referential $\mathbb{Y}$, and P, Q are linguistic labels on $\mathbb{X}$ and $\mathbb{Y}$, respectively.
Fuzzy control begins with the linguistic rules describing the expert's knowledge of the system's behaviour, and in that kind of control it is usual to explicitly or implicitly represent the rules by means of either the Min-Conditional function $J(a,b) = Min(a,b)$ - Mamdani -, or Prod-Conditional function $J(a,b) = Prod(a,b)$ - Larsen -. Both functions are particular cases of the general expression $J(a,b) = T(\varphi(a), \psi(b))$ [1] [2], where T is a continuous t-norm, $\varphi : [0,1] \to [0,1]$ any continuous function verifying $\varphi(0) = 0$ and $\varphi(1) = 1$ and $\psi : [0,1] \to [0,1]$ a contractive function, that is, such that $\psi(b) \leq b$ for all b in $[0,1]$.
From:
$$T(a, T(\varphi(a), \psi(b))) \leq Min(a, Min(\varphi(a), \psi(b)))$$
$$= Min((Min(a, \varphi(a)), \psi(b)) \leq \psi(b) \leq b$$

because of $T \leq Min$, it follows that $J(a,b) = T(\varphi(a), \psi(b))$ is always a T-Conditional Function.
It should be pointed out that given a rule *"If $\mu_P(x)$, then $\mu_Q(y)$"*, to pass from the representation as $J(\mu_P(x), \mu_Q(y)) = T(\mu_P(x), \mu_Q(y))$ - with, for example, $T = Min$ or

$T = Prod$ - to a more general one $J(\mu_P(x), \mu_Q(y)) = T(\varphi(\mu_P(x)), \psi(\mu_Q(y)))$, **is equivalent to** change the given rule to the new one "*If $\varphi(\mu_P(x))$, then $\psi(\mu_Q(y))$*" and to maintain for this rule the old representation by means of $J(\varphi(\mu_P(x)), \psi(\mu_Q(y))) = T(\varphi(\mu_P(x)), \psi(\mu_Q(y)))$.

A simple and useful type of this functions are obtained by taking $\varphi(a) = a^r$ with $r \in \mathbb{R}$ and $\psi = id$: $J(a,b) = T(a^r, b)$. With $r = 1$ it is $J(a,b) = T(a,b)$.

It should be also noticed that given a system of several rules it can happen that each one of them can be more adequately represented by a different T-Conditional Function $T(\varphi(a), \psi(b))$. For example, given two rules "*If x is P_1, then y is Q_1*" and "*If x is P_2, then y is Q_2*", some respective representation $T_1(\mu_{P_1}(x)^{r_1}, \mu_{Q_1})$ and $T_2(\mu_{P_2}(x)^{r_2}, \mu_{Q_2})$, with $T_1 \neq T_2$ and $r_1 \neq r_2$, can fit better their respective meanings or semantics in a concrete problem than the single representation obtained with $T_1 = T_2$ and $r_1 = r_2$.

2 Improving T-S model in two directions

This paper takes the position of considering that if we can get a better representation of the rules and a better control surface then we can get get a better T-S model and therefore a better approximation.

2.1 By using a better representation of the rules

By using the new representation of the rules means of operators $J(a,b) = T(a^r, b)$ then we can try to adjust each representation properly to obtain a better representation of the given rule. Taking into account that we can modify each exponent independently, in what follows we were modifying one exponent until the output was not closer to the expected output, and then did the same with the next one.

2.2 Looking for a better behaviour for control

As T-S model is applied to control machines it is desirable for the correct work of a machine that the changes in the output be smooth enough. Following this idea the convolution technique transforms the input labels from triangular functions to smoother functions, without sudden changes. And thanks to that, the output of the T-S model is also smoother. The transform F, was defined in [3], consists on the convolution of a function ϕ with a fuzzy set μ, which yields a new fuzzy set μ^*:

$$\mu^*(x) = (\phi * \mu)(x) = \int_U \mu(u) \cdot \phi(u-x) du$$

Some important properties of this transform are (see [3]):

- The transform of a partition of the unity is another partition of the unity.

- The transform preserves the area if ϕ has unity area.

- The transform is a smoothing one: if μ and ϕ have smoothness of order m and n respectively ($\mu \in C^{m-2}$ and $\phi \in C^{n-2}$), then $\mu^* = \mu * \phi$ has smoothness of order $m+n$. Smoothness of order k means that the derivative of order k of the function becomes impulsive, and the differentiability class C^l is the space of functions that are l times continuously differentiable.

- The transforms change a convex fuzzy set in another convex fuzzy set, but non in a necessarily normal one.

The function ϕ must be chosen for each case taking into account the smoothness wished for the surface output.

3 Case Examples

To illustrate how this changes in the representation of the rules and in the representation of the input labels can improve the result of T-S model in two ways, first, reducing the mean square error (MSE) between the known function and the approximating function, and second, obtaining a smoother function, we present the following two examples.

3.1 A simple example

We chose a non symmetrical function to show how we can reduce the MSE of the approximation and how we can increase the smoothness of the approximating function:

$$y = \frac{sin(x)}{x}$$

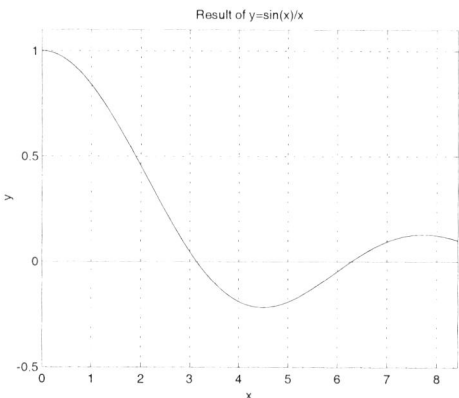

Figure 1: Simple example

The T-S model composed by the following seven rules can approximate well enough the function:

" If x is Cto 0 then $y = 1$"
" If x is Cto 4/3 then $y = \frac{3}{4}sin(4/3)$"
" If x is Cto 8/3 then $y = \frac{3}{8}sin(8/3)$"
" If x is Cto 4 then $y = \frac{1}{4}sin(4)$"
" If x is Cto 16/3 then $y = \frac{3}{16}sin(16/3)$"
" If x is Cto 20/3 then $y = \frac{3}{20}sin(20/3)$"
" If x is Cto 8 then $y = \frac{1}{8}sin(8)$"

Suppose that predicate Close-to (Cto) is represented by the fuzzy sets in the figure 2.

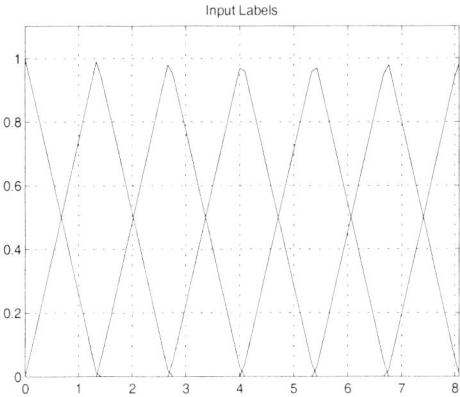

Figure 2: Input Labels

This model allows us to obtain the next approximating function (see left side of figure 3) with mean square error $MSE = 0.00070459$.

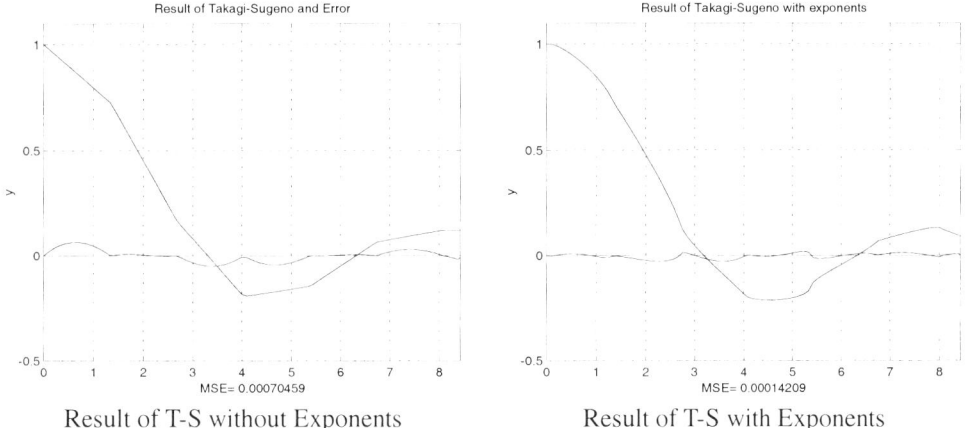

Result of T-S without Exponents   Result of T-S with Exponents

Figure 3: Result of Takagi-Sugeno

If we use exponents:

" If x is $(Cto\ 0)^{(0.9)}$ then $y = 1$ "
" If x is $(Cto\ 4/3)^{(0.915)}$ then $y = \frac{3}{4}sin(4/3)$ "
" If x is $(Cto\ 8/3)^{(0.915)}$ then $y = \frac{3}{8}sin(8/3)$ "
" If x is $(Cto\ 4)^{(0.6)}$ then $y = \frac{1}{4}sin(4)$ "
" If x is $(Cto\ 16/3)^{(0.8)}$ then $y = \frac{3}{16}sin(16/3)$ "
" If x is $(Cto\ 20/3)^{(0.45)}$ then $y = \frac{3}{20}sin(20/3)$ "
" If x is $(Cto\ 8)^{(0.95)}$ then $y = \frac{1}{8}sin(8)$ "

we obtain the next approximating function (see right side of figure 3) with a considerable reduction of the error $MSE = 0.00014209$.

Taking ϕ as a triangular function defined by:

$$\phi = \begin{cases} 1 - \frac{x}{5} & -5 \leq x \leq 0 \\ \frac{x}{5} - 1 & 0 \leq x \leq 5 \end{cases}$$

We use this function to transform, applying convolution, the input labels to obtain the new fuzzy sets (see figure 4), which are smother

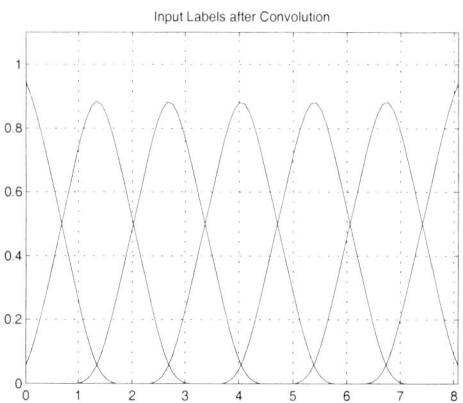

Figure 4: Input labels after convolution

and the following approximating function (see figure 5) with a slight increase of the error $MSE = 0.00075886$, but it is much smoother function.

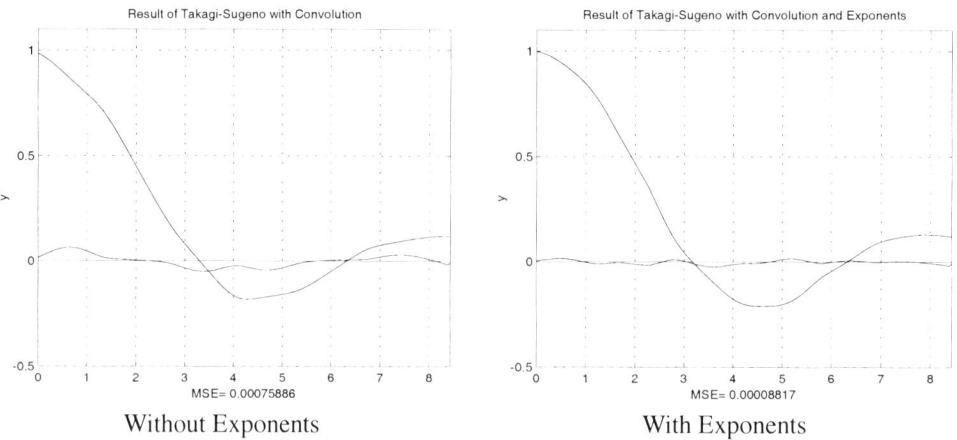

Figure 5: Result of T-S with convolution

If we apply convolution plus exponents they work better together than separate as it can be seen in figure 5 $MSE = 0.00008817$.

3.2 A complex example

Let's consider a more complex problem, that is, an approximation of the surface:

$$z = \frac{sin(x^2)e^{-x} + sin(y^2)e^{-y} + 0.2338}{0.8567}, \text{ for } x \in [0,3] \text{ and } y \in [0,3].$$

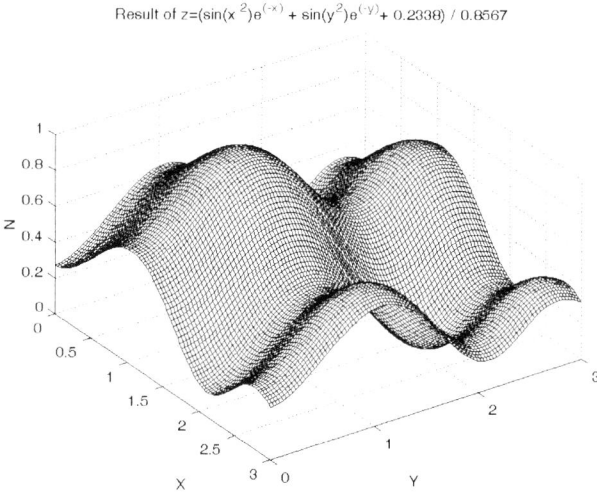

Figure 6: Complex example

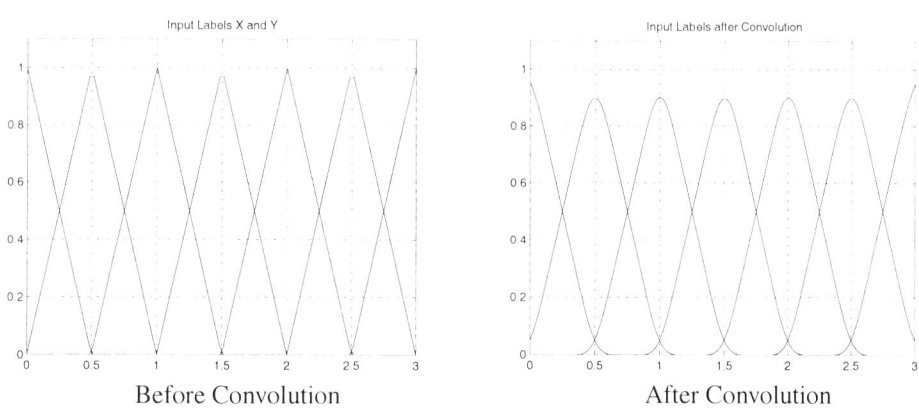

Figure 7: Input labels

Using a T-S model (see [5]) composed by 49 rules, because seven input labels for each variable are used, we can model the above target function.

$$\text{" If } x \text{ is } Cto\ x_1 \text{ and } y \text{ is } Cto\ y_1 \text{ then } z = q_1 \text{"}$$
$$\vdots \qquad \vdots$$
$$\text{" If } x \text{ is } Cto\ x_7 \text{ and } y \text{ is } Cto\ y_7 \text{ then } z = q_{49} \text{"}$$

In the approximation obtained using Takagi-Sugeno $MSE = 0.20451039$ (see figure 8).

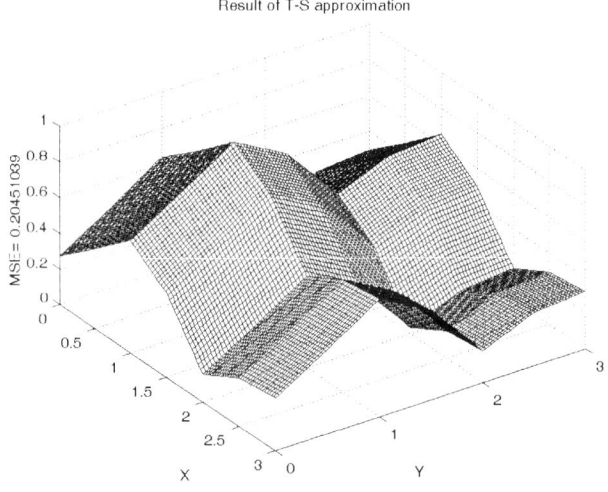

Figure 8: Result of Takagi-Sugeno

In the approximation obtained using Takagi-Sugeno with exponents, in this case we use the same exponent for all rules, and the error is reduced up to a 10% until $MSE = 0.18357553$ (see figure 9).

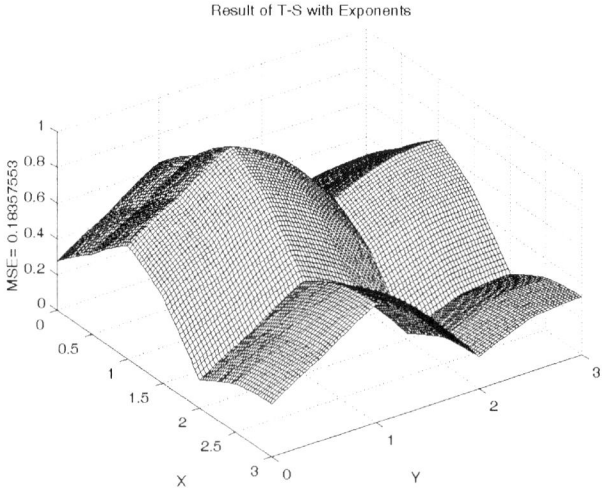

Figure 9: Result of Takagi-Sugeno with Exponents

Input labels after convolution, applying same ϕ from previous example, can been see in figure 7.

In the approximation obtained using Takagi-Sugeno with convolution, the error is slight increased up to a 5% until $MSE = 0.21149813$, but the smoothness is clearly improved, as can seen in figure 10.

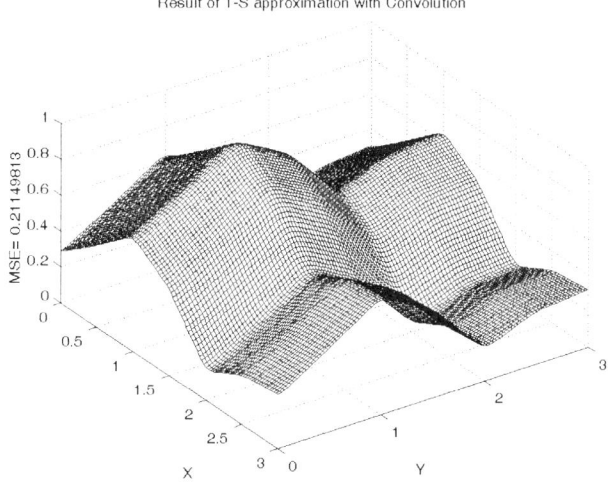

Figure 10: Result of Takagi-Sugeno after convolution

In the approximation using Takagi-Sugeno with convolution plus exponents, the error is reduced up to a 25% $MSE = 0.15090679$. And also smoothness is clearly improved (see figure 11).

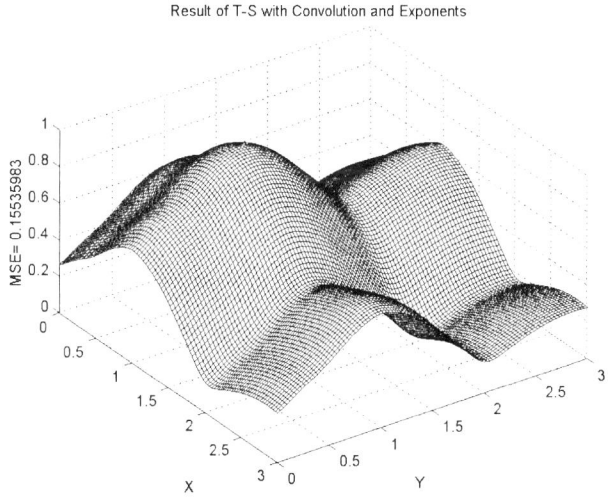

Figure 11: Result of T-S with convolution and exponents

4 Conclusion

It has been shown that the benefits of the improvement in the representation of the rules and the improvement in the smoothness of the inputs can contribute to obtain a better approximation of a given function.

These ideas need to be deeply explored to get a better understanding of the way they help to the T-S model to give a better approximation and the theoretical explanation of this improvement.

References

[1] C. del Campo and E. Trillas. On Mamdani-Larsen's Type Fuzzy Implications. In *Proceedings IPMU 2000*, volume II, pages 712–716, Madrid, 2000.

[2] E. Trillas and S. Guadarrama. On Some Logical Aspects of Numerical Fuzzy Inference. In *Proceedings IFSA/NAFIPS*, pages 998–1002, Vancouver, 2001.

[3] F. Fernandez and J. Gutierrez. Specification and optimization of fuzzy systems using convolution techniques. In *Signal Procesing*, 80 pages 935–949, 2000.

[4] M. Sugeno and T. Takagi. Multi-Dimensional Fuzzy Reasoning. In *Fuzzy Sets and Systems*, 9, pages 313–325, 1983.

[5] P. Carmona, J.L. Castro and J.M. Zurita. Learning Maximal Structure Fuzzy Rules with Exceptions. In *Proceedings EUSFLAT 2001*, pages 113–117, 2001.

[6] Driankov, Hellendoorn and Reinfrank. An Introducction to Fuzzy Control. *Springer-Verlag*, 1996.

B. Bouchon-Meunier, L. Foulloy and R.R. Yager (Editors)
Intelligent Systems for Information Processing:
From Representation to Applications
© 2003 Elsevier Science B.V. All rights reserved.

Combining Heterogeneous Information in Group Decision Making

F. Herrera[a], E. Herrera-Viedma[a], L. Martínez[b], P.J. Sánchez[b] *

[a]Dept. of Computer Science and A.I. University of Granada,
18071 - Granada, Spain. e-mail: herrera,viedma@decsai.ugr.es

[b]Dept. of Computer Science University of Jaén,
23071 - Jaén, Spain. e-mail: martin,pedroj@ujaen.es

Abstract

Decision processes for solving group decision making problems are composed of two phases: (i) *aggregation* and, (ii) *exploitation*. When experts that participate in the group decision making process are not able to express their opinions using a same expression domain, then the use of information assessed in different domains, i.e., *heterogeneous information*, is necessary. In these cases, the information can be assessed in domains with different nature as linguistic, numerical and interval-valued. The aim of this contribution is to present an aggregation process to manage *heterogeneous information contexts* in the case of linguistic, numerical and interval-valued information. To do this, we take as representation base the 2-tuple fuzzy linguistic representation model [5].

Keywords: decision making, aggregation, linguistic 2-tuples, heterogeneous information.

1 Introduction

Group Decision Making (GDM) problems have a finite set of alternatives $X = \{x_1, ..., x_n\}$ $n \geq 2$, as well as a finite set of experts $E = \{e_1, ..., e_m\}$ $m \geq 2$. Usually, each expert e_k provides his/her preferences on X by means of a preference relation P_{e_k}, being $P_{e_k}(x_i, x_j) = p_{ij}^k$ the degree of preference of alternative x_i over x_j.

It seems difficult that the nature of the preference values, p_{ij}^k, provided by the experts be the same. Because it depends on the knowledge of them over the alternatives (usually it is not precise). Therefore, the preference values have been expressed in different domains. Early in DM problems, the uncertainty were expressed in the preference values by means of real values assessed in a predefined range [11, 16], soon other approaches based on interval valued [12, 15] and linguistic one [4, 17] were proposed. The most of the proposals for solving GDM problems are focused on cases where all the experts provide their preferences in a unique domain, however, the experts could work in different knowledge fields and could express their preferences with different types of information depending on their knowledge. We shall call this type of information as *Heterogeneous Information*. Hence, the GDM problem is defined in a heterogeneous information context.

A solution for a GDM problem is derived either from the individual preference relations, without constructing a social preference relation, or by computing first a social fuzzy preference relation and then using it to find a solution [10]. In any of the above

---

*This work is supported by Research Project TIC2002-03348 and FEDER fonds

approaches called direct and indirect approaches respectively the process for reaching a solution of the GDM problems is composed by two steps [14]:

- *Aggregation phase*: that combines the expert preferences, and
- *Exploitation one*: that obtains a solution set of alternatives from a preference relation.

The main difficulty for managing GDM problems defined in heterogeneous information contexts is the *aggregation phase*, i.e., *how to aggregate this type of information?*. Because of, there not exist standard operators or processes for combining this type of information.

The 2-tuple fuzzy linguistic representation model presented in [5] has shown itself as a good choice to manage non-homogeneous information in aggregation processes [6, 8, 9]. In this paper, we propose an aggregation process based on the 2-tuple model that is able to deal with heterogeneous information contexts.

Our proposal for aggregating heterogeneous information follows a scheme comprised of three phases:

1. **Unification:** The heterogeneous information is unified in an unique expression domain by means of fuzzy sets. Different transformation functions will be defined to transform the input information into fuzzy sets.

2. **Aggregation:** The fuzzy sets will be aggregated by means of an aggregation operator to obtain collective preference values expressed by fuzzy sets.

3. **Transformation:** The collective preference values expressed by means of fuzzy sets will be transformed into linguistic 2-tuples.

The exploitation phase of the decision process is carried out over the collective linguistic 2-tuples, to obtain the solution for the GDM problem.

In order to do so, this paper is structured as follows: in Section 2 we shall review different basic concepts; in Section 3 we shall propose the aggregation process for combining heterogeneous information; in Section 4 we shall solve an example of a GDM problem defined in a heterogeneous information context and finally, some concluding remarks are pointed out.

2 Preliminaries

We have just seen that in GDM problems the experts express their preferences depending on their knowledge over the alternatives by means of preference relations. Here, we review different approaches to express those preferences. And afterwards, we shall review the 2-tuple fuzzy linguistic representation model.

2.1 Approaches for Modelling Preferences

2.1.1 Fuzzy Binary Relations

A valued (fuzzy) binary relation R on X is defined as a fuzzy subset of the direct product $X \times X$ with values in $[0, 1]$, i.e, $R : X \times X \rightarrow [0, 1]$. The value, $R(x_i, x_j) = p_{ij}$, of a valued relation R denotes the degree to which $x_i R x_j$. In preference analysis, p_{ij} denotes the *degree to which an alternative x_i is preferred to x_j*. These were the first type of relations used in decision making [10, 11].

2.1.2 Interval-valued Relations

About the fuzzy binary approach has been argued that the most experts are unable to make a fair estimation of the inaccuracy of their judgements, making far larger estimation errors that the boundaries accepted by them as feasible [2].

A first approach to overcome this problem is to add some flexibility to the uncertainty representation problem by means of interval-valued relations:

$$R: X \times X \to \wp([0,1]).$$

Where $R(x_i, x_j) = p_{ij}$ denotes the interval-valued preference degree of the alternative x_i over x_j. In these approaches [12, 15], the preferences provided by the experts consist of interval values assessed in $\wp([0,1])$, where the preference is expressed as $[\underline{a}, \overline{a}]_{ij}$, with $\underline{a} \leq \overline{a}$.

2.1.3 Linguistic Approach

Usually, we work in a quantitative setting, where the information is expressed by means of numerical values. However, many aspects of different activities in the real world cannot be assessed in a quantitative form, but rather in a qualitative one, i.e., with vague or imprecise knowledge. In that case, a better approach may be to use linguistic assessments instead of numerical values. The fuzzy linguistic approach represents qualitative aspects as linguistic values by means of linguistic variables [18].

To use the linguistic approach we have to choose the appropriate linguistic descriptors for the term set and their semantics. In the literature, several possibilities can be found (see [7] for a wide description). An important aspect to analyze is the *"granularity of uncertainty"*, i.e., the level of discrimination among different counts of uncertainty. The *"granularity of uncertainty"* for the linguistic term set $S = \{s_0, ..., s_g\}$ is $g+1$, while its *"interval of granularity"* is $[0, g]$.

One possibility of generating the linguistic term set consists of directly supplying the term set by considering all terms distributed on a scale on which a total order is defined [17]. For example, a set of seven terms S, could be given as follows:

$$S = \{s_0 : N, s_1 : VL, s_2 : L, s_3 : M, s_4 : H, s_5 : VH, s_6 : P\}$$

Usually, in these cases, it is required that in the linguistic term set satisfy the following additional characteristics:

1. There is a negation operator: $\text{Neg}(s_i) = s_j$, with, $j = g - i$ (g+1 is the cardinality).

2. $s_i \leq s_j \iff i \leq j$. Therefore, there exists a *min* and a *max* operator.

The semantics of the linguistic terms are given by fuzzy numbers defined in the [0,1] interval. A way to characterize a fuzzy number is to use a representation based on parameters of its membership function [1]. The linguistic assessments given by the users are just approximate ones, some authors consider that linear trapezoidal membership functions are good enough to capture the vagueness of those linguistic assessments. The parametric representation is achieved by the 4-tuple (a, b, d, c), where b and d indicate the interval in which the membership value is 1, with a and c indicating the left and right limits of the definition domain of the trapezoidal membership function [1]. A particular case of this type of representation are the linguistic assessments whose membership functions are triangular, i.e., $b = d$, then we represent this type of membership functions by a 3-tuple (a, b, c). A possible semantics for the above term set, S, may be the following (Figure 1):

$P = (.83, 1, 1)$ $VH = (.67, .83, 1)$ $H = (.5, .67, .83)$ $M = (.33, .5, .67)$
$L = (.17, .33, .5)$ $VL = (0, .17, .33)$ $N = (0, 0, .17)$

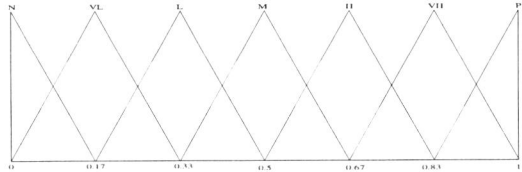

Figure 1: A set of seven linguistic terms with its semantics

2.2 The 2-Tuple Fuzzy Linguistic Representation Model

This model was presented in [5], for overcoming the drawback of the loss of information presented by the classical linguistic computational models [7]: (i) The model based on the Extension Principle [1], (ii) and the symbolic one [3]. The 2-tuple fuzzy linguistic representation model is based on symbolic methods and takes as the base of its representation the concept of Symbolic Translation.

Definition 1. *The Symbolic Translation of a linguistic term $s_i \in S = \{s_0, ..., s_g\}$ is a numerical value assessed in $[-.5, .5)$ that support the "difference of information" between a counting of information $\beta \in [0, g]$ and the closest value in $\{0, ..., g\}$ that indicates the index of the closest linguistic term in $S(s_i)$, being $[0,g]$ the interval of granularity of S.*

¿From this concept a new linguistic representation model is developed, which represents the linguistic information by means of 2-tuples (r_i, α_i), $r_i \in S$ and $\alpha_i \in [-.5, .5)$. r_i represents the linguistic label center of the information and α_i is the Symbolic Translation.

This model defines a set of functions between linguistic 2-tuples and numerical values.

Definition 2. *Let $S = \{s_0, ..., s_g\}$ be a linguistic term set and $\beta \in [0, g]$ a value supporting the result of a symbolic aggregation operation, then the 2-tuple that expresses the equivalent information to β is obtained with the following function:*

$$\Delta : [0, g] \longrightarrow S \times [-0.5, 0.5)$$
$$\Delta(\beta) = (s_i, \alpha), \text{ with } \begin{cases} s_i & i = round(\beta) \\ \alpha = \beta - i & \alpha \in [-.5, .5) \end{cases}$$

where $round(\cdot)$ is the usual round operation, s_i has the closest index label to "β" and "α" is the value of the symbolic translation.

Proposition 1. *Let $S = \{s_0, ..., s_g\}$ be a linguistic term set and (s_i, α) be a linguistic 2-tuple. There is always a Δ^{-1} function, such that, from a 2-tuple it returns its equivalent numerical value $\beta \in [0, g]$ in the interval of granularity of S.*

Proof. It is trivial, we consider the following function:

$$\Delta^{-1} : S \times [-.5, .5) \longrightarrow [0, g]$$
$$\Delta^{-1}(s_i, \alpha) = i + \alpha = \beta$$

Remark 1. From Definitions 1 and 2 and Proposition 1, it is obvious that the conversion of a linguistic term into a linguistic 2-tuple consist of adding a value 0 as symbolic translation: $s_i \in S \Longrightarrow (s_i, 0)$

3 Aggregation Process for Heterogeneous Information in a GDM Problem

In this section we propose a method to carry out the *aggregation step* of a GDM process defined in a heterogeneous information context. We focus on GDM problems in which the preference relations provided, can be:

- Fuzzy preference relations [11].
- Interval-valued preference relation [15].
- Linguistic preference relation assessed in a pre-established label set [4].

Our proposal for combining the heterogeneous information is composed of the following phases:

1. *Making the information uniform.* The heterogeneous information will be unified into a specific linguistic domain, called *Basic Linguistic Term Set* (BLTS) and symbolized as S_T. Each numerical, interval-valued and linguistic performance value is transformed into a fuzzy set in S_T, $F(S_T)$. The process is carried out in the following order:

    (a) Transforming numerical values in $[0, 1]$ into $F(S_T)$.
    (b) Transforming linguistic terms into $F(S_T)$.
    (c) Transforming interval-valued into $F(S_T)$.

2. *Aggregating individual performance values.* For each alternative, a collective performance value is obtained by means of the aggregation of the above fuzzy sets on the BLTS that represents the individual performance values assigned by the experts according to his/her preference.

3. *Transforming into 2-tuple.* The collective performance values (fuzzy sets) are transformed into linguistic 2-tuples in the BLTS and obtained a collective 2-tuple linguistic preference relation.

Following, we shall show in depth each phase of the aggregation process.

3.1 Making the Information Uniform

In this phase, we have to choose the domain, S_T, to unify the heterogeneous information and afterwards, the input information will be transformed into fuzzy sets in S_T.

3.1.1 Choosing the Basic Linguistic Term Set

The heterogeneous information is unified in a unique expression domain. In this case, we shall use fuzzy sets over a BLTS, denoted as $F(S_T)$. We study the linguistic term set S used in the GDM problem. If:

1. *S is a fuzzy partition*,
2. *and the membership functions of its terms are triangular*, i.e., $s_i = (a_i, b_i, c_i)$

Then, we select S as BLTS due to the fact that, these conditions are necessary and sufficient for the transformation between values in $[0, 1]$ and 2-tuples, being them carried out without loss of information [6].

If the linguistic term set S, used in the definition context of the problem, does not satisfy the above conditions then we shall choose as BLTS a term set with a larger number of terms than the number of terms that a person is able to discriminate (normally 11 or 13, see [1]) and satisfies the above conditions. We choose the BLTS with 15 terms symmetrically distributed, with the following semantics (graphically, Figure 2).

s_0	(0,0,0.07)	s_1	(0,0.07,0.14)	s_2	(0.07,0.14,0.21)
s_3	(0.14,0.21,0.28)	s_4	(0.21,0.28,0.35)	s_5	(0.28,0.35,0.42)
s_6	(0.35,0.42,0.5)	s_7	(0.42,0.5,0.58)	s_8	(0.5,0.58,0.65)
s_9	(0.58,0.65,0.72)	s_{10}	(0.65,0.72,0.79)	s_{11}	(0.72,0.79,0.86)
s_{12}	(0.79,0.86,0.93)	s_{13}	(0.86,0.93,1)	s_{14}	(0.93,1,1)

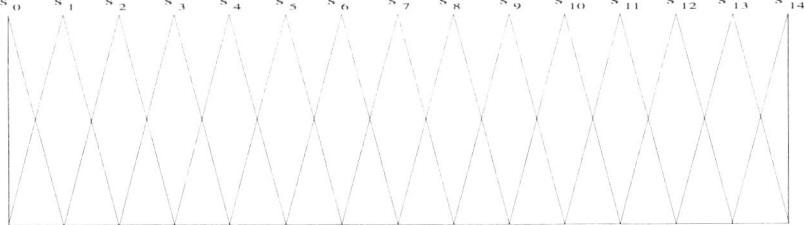

Figure 2: A BLTS with 15 terms symmetrically distributed

3.1.2 Transforming the Input Information Into $F(S_T)$

Once chosen the BLTS, we shall define the *transformation functions* that will be necessary to unify the heterogeneous information. The process of unifying the heterogeneous information involves in any occasions the comparison between fuzzy sets. These comparisons are usually carried out by means of a measure of comparison. We focus on measures of comparison which evaluate the resemblance or likeness of two objects (fuzzy sets in our case). These type of measures are called *measures of similitude* [13]. For simplicity, in this paper we shall choose a measure of similitude based on a possibility function $S(A, B) = \max_x \min(\mu_a(x), \mu_B(x))$, where μ_A and μ_B are the membership function of the fuzzy sets A and B respectively.

3.1.2.1. Transforming numerical values in $[0, 1]$ into $F(S_T)$.

Let $F(S_T)$ be the set of fuzzy sets in $S_T = \{s_0, \ldots, s_g\}$, we shall transform a numerical value $\vartheta \in [0, 1]$ into a fuzzy set in $F(S_T)$ computing the membership value of ϑ in the membership functions associated with the linguistic terms of S_T.

Definition 3. [6] *The function τ transforms a numerical value into a fuzzy set in S_T:*

$$\tau : [0, 1] \rightarrow F(S_T)$$
$$\tau(\vartheta) = \{(s_0, \gamma_0), \ldots, (s_g, \gamma_g)\}, s_i \in S_T \text{ and } \gamma_i \in [0, 1]$$

$$\gamma_i = \mu_{s_i}(\vartheta) = \begin{cases} 0, & if \vartheta \notin Support(\mu_{s_i}(x)) \\ \frac{\vartheta - a_i}{b_i - a_i}, & if a_i \leq \vartheta \leq b_i \\ 1, & if b_i \leq \vartheta \leq d_i \\ \frac{c_i - \vartheta}{c_i - d_i}, & if d_i \leq \vartheta \leq c_i \end{cases}$$

Remark 2. We consider membership functions, $\mu_{s_i}(\cdot)$, for linguistic labels, $s_i \in S_T$, that achieved by a parametric function (a_i, b_i, d_i, c_i). A particular case are the linguistic assessments whose membership functions a triangular, i.e., $b_i = d_i$.

Example 1.
Let $\vartheta = 0.78$ be a numerical value to be transformed into a fuzzy set in $S = \{s_0, ..., s_4\}$. The semantics of this term set is:

$s_0 = (0, 0, 0.25)$ $s_1 = (0, , 0.25, 0.5)$ $s_2 = (0.25, 0.5, 0.75)$ $s_3 = (0.5, 0.75, 1)$ $s_4 = (0.75, 1, 1)$

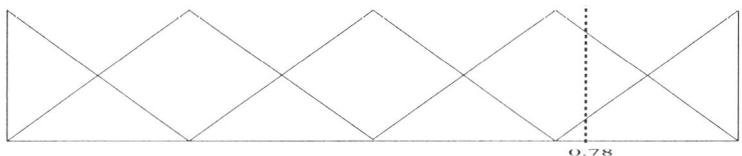

Figure 3: Transforming a numerical value into a fuzzy set in S

Then, the fuzzy set obtained is (See Fig. 3):

$$\tau(0.78) = \{(s_0, 0), (s_1, 0), (s_2, 0), (s_3, 0.88), (s_4, 0.12)\}.$$

3.1.2.2. Transforming Linguistic Terms in S into $F(S_T)$.

Definition 4.[9] Let $S = \{l_0, \ldots, l_p\}$ and $S_T = \{s_0, \ldots, s_g\}$ be two linguistic term sets, such that, $g \geq p$. Then, a multi-granularity transformation function, τ_{SS_T}, is defined as:

$$\tau_{SS_T} : A \to F(S_T)$$
$$\tau_{SS_T}(l_i) = \{(c_k, \gamma_k^i) / k \in \{0, ..., g\}\}, \forall l_i \in S$$
$$\gamma_k^i = \max_y \min\{\mu_{l_i}(y), \mu_{c_k}(y)\}$$

where $F(S_T)$ is the set of fuzzy sets defined in S_T, and $\mu_{l_i}(\cdot)$ and $\mu_{c_k}(\cdot)$ are the membership functions of the fuzzy sets associated with the terms l_i and c_k, respectively.

Therefore, the result of τ_{SS_T} for any linguistic value of S is a fuzzy set defined in the BLTS, S_T.

Example 2.
Let $S = \{l_0, l_1, \ldots, l_4\}$ and $S_T = \{s_0, s_1, \ldots, s_6\}$ be two term sets, with 5 and 7 labels, respectively, and with the following semantics associated:

$l_0 = (0, 0, 0.25)$ $l_1 = (0, , 0.25, 0.5)$ $s_0 = (0, 0, 0.16)$ $s_1 = (0, 0.16, 0.34)$
$l_2 = (0.25, 0.5, 0.75)$ $l_3 = (0.5, 0.75, 1)$ $s_2 = (0.16, 0.34, 0.5)$ $s_3 = (0.34, 0.5, 0.66)$
$l_4 = (0.75, 1, 1)$ $s_4 = (0.5, 0.66, 0.84)$ $s_5 = (0.66, 0.84, 1)$
$s_6 = (0.84, 1, 1)$

The fuzzy set obtained after applying τ_{SS_T} for l_1 is (see Fig. 4):

$$\tau_{SS_T}(l_1) = \{(s_0, 0.39), (s_1, 0.85), (s_2, 0.85), (s_3, 0.39), (s_4, 0), (s_5, 0), (s_6, 0)\}.$$

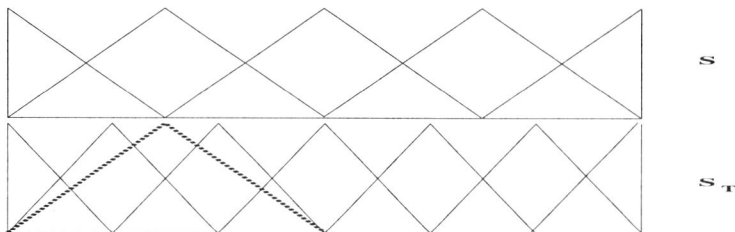

Figure 4: Transforming $l_1 \in S$ into a fuzzy set in S_T

3.1.2.3 Transforming Interval-Valued into $F(S_T)$.

Let $I = [\underline{i}, \overline{i}]$ be an interval-valued in $[0,1]$, to carry out this transformation we assume that the interval-valued has a representation, inspired in the membership function of a fuzzy set [12] as follows:

$$\mu_I(\vartheta) = \begin{cases} 0, & if\ \vartheta < \underline{i} \\ 1, & if\ \underline{i} \leq \vartheta \leq \overline{i} \\ 0, & if\ \overline{i} < \vartheta \end{cases}$$

where ϑ is a value in $[0,1]$. In Figure 5 can be observed the graphical representation of an interval.

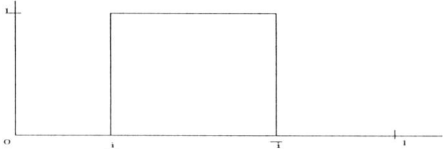

Figure 5: Membership function of $I = [\underline{i}, \overline{i}]$

Definition 5. Let $S_T = \{s_0, \ldots, s_g\}$ be a BLTS. Then, the function τ_{IS_T} transforms a interval-valued I in $[0,1]$ into a fuzzy set in S_T as follows

$$\tau_{IS_T} : I \to F(S_T)$$
$$\tau_{IS_T}(I) = \{(c_k, \gamma_k^i) / k \in \{0, \ldots, g\}\},$$
$$\gamma_k^i = \max_y \min\{\mu_I(y), \mu_{c_k}(y)\}$$

where $F(S_T)$ is the set of fuzzy sets defined in S_T, and $\mu_I(\cdot)$ and $\mu_{c_k}(\cdot)$ are the membership functions associated with the interval-valued I and terms c_k, respectively.

Example 3.

Let $I = [0.6, 0.78]$ be an interval-valued to be transformed into $F(S_T)$. The semantic of this term set is the same of Example 1. The fuzzy set obtained applying τ_{IS_T} is (see Fig. 6):

$$\tau_{IS_T} = \{(s_0, 0), (s_1, 0), (s_2, 0.6), (s_3, 1), (s_4, 0.2)\}$$

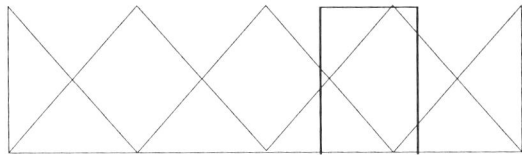

Figure 6: Transforming $[0.6, 0.78]$ into a fuzzy set in S_T

3.2 Aggregating Individual Performance Values

Using the above transformation functions we express the input information by means of fuzzy sets on the BLTS, $S_T = \{s_0, \ldots, s_g\}$. Now we use an aggregation function for combining the fuzzy sets on the BLTS to obtain a collective performance for each alternative that will be a fuzzy set on the BLTS.

For the heterogeneous GDM the preference relations are expressed by means of fuzzy sets on the BLTS, as in the Table 1. Where p_{ij}^k is the preference degree of the alternative x_i over x_j provides by the expert e_k.

Table 1: The preference relation

$$P_{e_k} = \begin{pmatrix} p_{11}^k = \{(s_0, \gamma_{k_0}^{11}), \ldots, (s_g, \gamma_{k_g}^{11})\} & \cdots & p_{1n}^k = \{(s_0, \gamma_{k_0}^{1n}), \ldots, (s_g, \gamma_{k_g}^{1n})\} \\ \vdots & \cdots & \vdots \\ p_{n1}^k = \{(s_0, \gamma_{k_0}^{n1}), \ldots, (s_g, \gamma_{k_g}^{n1})\} & \cdots & p_{nn}^k = \{(s_0, \gamma_{k_0}^{nn}), \ldots, (s_g, \gamma_{k_g}^{nn})\} \end{pmatrix}$$

We shall represent each fuzzy set, p_{ij}^k, as $r_{ij}^k = (\gamma_{k_0}^{ij}, \ldots, \gamma_{k_g}^{ij})$ being the values of r_{ij}^k their respective membership degrees. Then, the collective performance value of the preference relation according to all preference relations provided by experts $\{r_{ij}^k, \forall e_k\}$ is obtained aggregating these fuzzy sets. These collective performance values are denoted as r_{ij}, form a new preference relation of fuzzy sets defined in S_T, i.e.,

$$r_{ij} = (\gamma_0^{ij}, \ldots, \gamma_g^{ij})$$

characterized by the following membership function:

$$\gamma_v^{ij} = f(\gamma_{1_v}^{ij}, \ldots, \gamma_{k_v}^{ij}),$$

where f is an "aggregation operator" and k is the number of experts.

3.3 Transforming into Linguistic 2-Tuples

In this phase we transform the fuzzy sets on the BLTS into linguistic 2-tuples over the BLTS. In [9] was presented a function χ that transforms a fuzzy set in a linguistic term set into a numerical value in the interval of granularity of S_T, $[0, g]$:

$$\chi : F(S_T) \to [0, g]$$

$$\chi(\tau(\vartheta)) = \chi(\{(s_j, \gamma_j), j = 0, \ldots, g\}) = \frac{\sum_{j=0}^{g} j\gamma_j}{\sum_{j=0}^{g} \gamma_j} = \beta.$$

Therefore, applying the Δ function to β we shall obtain a collective preference relation whose values are linguistic 2-tuples.

4 A GDM Problem Defined in a Heterogeneous Information Context

Let's suppose that a company want to renew its computers. There exist four models of computers available, {HP, IBM, COMPAQ and DELL} and three experts provide his/her preference relations over the four cars. The first expert expresses his/her preference relation using numerical values in $[0, 1]$, P_1^n. The second one expresses the preferences by means of linguistic values in a linguistic term set S (see Figure 1), P_2^S. And the third expert can express them using interval-valued in $[0, 1]$, P_3^I. The three experts attempt to reach a collective decision.

Table 2: Preference relations

$$P_1^n \begin{pmatrix} - & .5 & .8 & .4 \\ .5 & - & .9 & .5 \\ .8 & .9 & - & .4 \\ .4 & .5 & .4 & - \end{pmatrix} \quad P_2^S \begin{pmatrix} - & H & VH & M \\ H & - & H & VH \\ VH & H & - & VH \\ M & VH & VH & - \end{pmatrix} \quad P_3^I \begin{pmatrix} - & [.7,.8] & [.65,.7] & [.8,.9] \\ [.7,.8] & - & [.6,.7] & [.8,.85] \\ [.8,.9] & [.6,.7] & - & [.7,.9] \\ [.8,.9] & [.8,.85] & [.7,.9] & - \end{pmatrix}$$

4.1 Decision Process

We shall use the following decision process to solve this problem:

A) Aggregation Phase

We use the aggregation process presented in Section 3.

1. **Making the information uniform**

   (a) *Choose the BLTS.* It will be S, due to the fact, it satisfies the conditions showed in Section 3.1.1.

   (b) *Transforming the input information into $F(S_T)$.* (e.g., see Table 3).

Table 3: Fuzzy sets in a BLTS

$$P_1^n = \begin{pmatrix} - & (0,0,0,1,0,0,0) & (0,0,0,0,.19,.81,0) & (0,0,.59,.41,0,0,0) \\ (0,0,0,1,0,0,0) & - & (0,0,0,0,0,.59,.41) & (0,0,0,1,0,0,0) \\ (0,0,0,0,.19,.81,0) & (0,0,0,0,0,.59,.41) & - & (0,0,.59,.41,0,0,0) \\ (0,0,.59,.41,0,0,0) & (0,0,0,1,0,0,0) & (0,0,.59,.41,0,0,0) & - \end{pmatrix}$$

   (c) *Aggregating individual performance values.* In this example we use as aggregation operator, f, the arithmetic mean obtaining the collective preference relation:

Table 4: The collective Preference relation.

$$\begin{pmatrix} - & (0,0,0,0,.6,.27,0) & (0,0,0,.04,.4,.67,0) & (0,0,.2,.47,.27,.33,.14) \\ (0,0,0,0,.6,.27,0) & - & (0,0,0,.14,.67,.26,.14) & (0,0,0,.33,.06,.67,.04) \\ (0,0,0,.04,.4,.67,0) & (0,0,0,.14,.67,.26,.14) & - & (0,0,.2,.14,.27,.67,.14) \\ (0,0,.2,.47,.27,.33,.14) & (0,0,0,.33,.06,.67,.04) & (0,0,.2,.14,.27,.67,.14) & - \end{pmatrix}$$

2. **Transforming into linguistic 2-tuples**. The result of this transformation is the following:

$$\begin{pmatrix} - & (H,.31) & (VH,-.43) & (H,-.18) \\ (H,.31) & - & (H,.33) & (H,.38) \\ (VH,-.43) & (H,.33) & - & (H,.29) \\ (H,-.18) & (H,.38) & (H,.29) & - \end{pmatrix}$$

B) Exploitation Phase

To solve the GDM problem, finally we calculate the dominance degree for the alternative x_i over the rest of alternatives. To do so, we shall use the following function:

$$\Lambda(x_i) = \frac{1}{n-1} \sum_{j=0 \mid j \neq i}^{n} \beta_{ij}$$

where n is the number of alternatives and $\beta_{ij} = \Delta^{-1}(p_{ij})$ being p_{ij} a linguistic 2-tuple. In this phase we shall calculate the dominance degree for this preference relation:

Table 5: Dominance degree of the alternatives

HP	IBM	**COMPAQ**	$DELL$
$(H,.23)$	$(H,.34)$	**(H,.4)**	$(H,.16)$

Then, dominance degrees rank the alternatives and we choose the best alternative(s) as solution set of the GDM problem, in this example the solution set is {**COMPAQ**}.

5 Concluding Remarks

We have presented an aggregation process for aggregating heterogeneous information in the case of numerical, interval-valued and linguistic values. This aggregation process is based on the transformation of the heterogeneous information into fuzzy sets assessed in a unique basic linguistic term set. And afterwards, these fuzzy sets are converted into linguistic 2-tuples. The aggregation process has been applied to a GDM problem defined in a heterogeneous information context.

In the future, we shall apply this aggregation process to other types of information used in the literature to express preference values as Interval-Valued Fuzzy Sets and Intuitionistic Fuzzy Sets.

References

[1] P.P. Bonissone and K.S. Decker. *Selecting Uncertainty Calculi and Granularity: An Experiment in Trading-Off Precision and Complexity*. Uncertainty in A. I. North-Holland in: L.H. Kanal and J.F. Lemmer, Eds., 1986.

[2] R. López de Mántaras and L. Godó. From intervals to fuzzy truth-values: Adding flexibility to reasoning under uncertainty. *International Journal of Uncertainty, Fuzziness and Knowledge-Based Systems*, 5(3):251–260, 1997.

[3] M. Delgado, J.L. Verdegay, and M.A. Vila. Linguistic decision making models. *International Journal of Intelligent Systems*, 7:479–492, 1993.

[4] F. Herrera, E. Herrera-Viedma, and J.L. Verdegay. A linguistic decision process in group decision making. *Group Decision and Negotiation*, 5:165–176, 1996.

[5] F. Herrera and L. Martínez. A 2-tuple fuzzy linguistic representation model for computing with words. *IEEE Transactions on Fuzzy Systems*, 8(6):746–752, 2000.

[6] F. Herrera and L. Martínez. An approach for combining linguistic and numerical information based on 2-tuple fuzzy representation model in decision-making. *International Journal of Uncertainty, Fuzziness and Knowledge-Based Systems*, 8(5):539–562, 2000.

[7] F. Herrera and L. Martínez. The 2-tuple linguistic computational model. Advantages of its linguistic description, accuracy and consistency. *International Journal of Uncertainty, Fuzziness and Knowledge-Based Systems*, 9:33–49, 2001.

[8] F. Herrera and L. Martínez. A model based on linguistic 2-tuples for dealing with multigranularity hierarchical linguistic contexts in multiexpert decision-making. *IEEE Trans. on Systems, Man and Cybernetics. Part B: Cybernetics*, 31(2):227–234, 2001.

[9] F. Herrera, L. Martínez, E. Herrera-Viedma, and F. Chiclana. Fusion of multigranular linguistic information based on the 2-tuple fuzzy linguistic representation method. In *Proceedings of Ninth International Conference IPMU 2002*, pages 1155–1162, Annecy (France), 2002.

[10] J. Kacprzyk and M. Fedrizzi. Group decision making with a fuzzy linguistic majority. *Fuzzy Sets and Systems*, 18:105–118, 1986.

[11] J. Kacprzyk and M. Fedrizzi. *Multi-Person Decision Making Using Fuzzy Sets and Possibility Theory*. Kluwer Academic Publishers, Dordrecht, 1990.

[12] D. Kuchta. Fuzzy capital budgeting. *Fuzzy Sets and Systems*, 111:367–385, 2000.

[13] M. Rifqi, V. Berger, and B. Bouchon-Meunier. Discrimination power of measures of comparison. *Fuzzy Sets and Systems*, 110:189–196, 2000.

[14] M. Roubens. Fuzzy sets and decision analysis. *Fuzzy Sets and Systems*, 90:199–206, 1997.

[15] J.F. Le Téno and B. Mareschal. An interval version of PROMETHEE for the comparison of building products' design with ill-defined data on environmental quality. *European Journal of Operational Research*, 109:522–529, 1998.

[16] R.R. Yager. On ordered weighted averaging aggregation operators in multicriteria decision making. *IEEE Trans. on Systems, Man, and Cybernetics*, 18:183–190, 1988.

[17] R.R. Yager. An approach to ordinal decision making. *International Journal of Approximate Reasoning*, 12:237–261, 1995.

[18] L.A. Zadeh. The concept of a linguistic variable and its applications to approximate reasoning. *Information Sciences, Part I, II, III*, 8,8,9:199–249,301–357,43–80, 1975.

Function Approximation by Fuzzy Rough Sets

Masahiro Inuiguchi
Graduate School of Engineering
Osaka University
2-1 Yamadaoka, Suita
Osaka 565-0871, Japan
inuiguti@eie.eng.osaka-u.ac.jp

Tetsuzo Tanino
Graduate School of Engineering
Osaka University
2-1 Yamadaoka, Suita
Osaka 565-0871, Japan
tanino@eie.eng.osaka-u.ac.jp

Abstract

In this paper, we show a possible application of fuzzy rough sets to function approximation based on decision tables. Through this study, we demonstrate that continuous attribute values can be treated by fuzzy rough sets. A general approach for reasoning of decision attribute values from condition attribute values based on a given decision table is presented. A specific method in one dimensional case is discussed and some properties of the method are shown. Some modifications are applied for getting a better approximation and a smaller body of rules.

Keywords: Fuzzy rough sets, approximation, continuous attribute, inference

1 Introduction

Rough sets [1] have been well applied to analyze information tables. By the methodologies based on rough sets, we can find the reduced information tables without losing the accuracy of the object classification, and the minimum descriptive decision rules [2]. These methodologies are effective mainly when attribute values are discrete. To treat continuous attribute values, some discretization is necessary. Thus these methodologies are not good at extraction of continuous functions implicit in given information tables.

Recently rough sets are generalized in various ways. Yao and Lin [3] has shown the correspondence between rough sets and Kripke model of modal logic, and generalized rough sets by extending equivalence relations to various relations from viewpoint of the correspondence. Słowinski and Vanderpooten [4] have discussed rough sets under a similarity relation which satisfies the reflexivity only. Greco et al. [5][6] have proposed rough sets under a dominance relation. Yao [7][8] and Inuiguchi and Tanino [9] have also investigated the extensions of rough sets.

On the other hand, rough sets under fuzzy relations were discussed from relatively long ago [10][11]. However fuzzy rough sets have never been applied to analysis of information tables so far. Recently a new type of fuzzy rough sets has been proposed by Inuiguchi and Tanino [12]. It is shown that this new fuzzy rough sets give better approximations than the previous fuzzy rough sets [12].

Those generalizations reveal that several interpretations can be applied to rough sets. At the current, two interpretations are proposed, i.e., rough sets as classification among positive, negative and boundary elements, and rough sets as approximations of a set by means of elementary sets derived from a given relation [12]. The purpose of the application can be distinct depending on the interpretation of rough sets. Since the definition of rough sets is different by the interpretation, the application methods can also be different.

For the treatment of continuous attribute values, fuzzy rough sets would be useful since it might produce fuzzy if-then rules from the given information table and fuzzy if-then rules might be capable of interpolative reasoning. In order to produce fuzzy if-then rules for interpolative reasoning, we should apply the second interpretation of rough sets, i.e., rough sets as approximations of a set by means of elementary sets since by the first interpretation we will not obtain rules for all elements but only for positive and negative elements. While the traditional fuzzy rough sets are based on the first interpretation, the new fuzzy rough sets proposed by Inuiguchi and Tanino [12] are based on the second interpretation.

In this paper, we discuss the treatment of continuous attribute values by fuzzy rough sets proposed by Inuiguchi and Tanino [12]. For the first step of the research, we discuss the approximation of a continuous function by means of fuzzy rough sets when some data about the function is given. In the next section, we describe fuzzy rough sets briefly. Then we propose an interpolative reasoning method derived from fuzzy rough sets in Section 3. We will show the correspondence between fuzzy rough sets and fuzzy if-then rules. A specific construction of fuzzy rules from a given information table (data) is discussed in Section 4. Some properties of the specific method are shown. A simple example of approximation of a function by means of fuzzy rough sets is given. In Section 5, modifications are proposed for getting a better approximation and for minimizing the number of rules. Concluding remarks are described in Section 6.

2 Fuzzy Rough Sets

Fuzzy rough sets are originally defined by Dubois and Prade [10] (independently, Nakamura [13] also defined fuzzy rough sets in a different way). In this definition, the lower and upper approximations are defined by necessity and possibility measures. The fuzzy rough sets are defined by pairs of lower and upper approximations. This definition is valid only when a fuzzy similarity relation is given. Recently, Inuiguchi and Tanino [12] defined a new type of fuzzy rough sets based on certainty qualifications. This type of fuzzy rough sets can be defined even if a family of normal fuzzy sets is given. Moreover the new fuzzy rough sets provide better lower and upper approximations.

In the original fuzzy rough sets, membership function values of lower and upper approximations show the degrees to what extent elements are included in positive and possible regions of the given set, respectively. On the other hand, in the new fuzzy rough sets, lower and upper approximations are best approximations, i.e., largest and smallest fuzzy sets which take the same necessity degrees as a given fuzzy set takes. Thus the new fuzzy rough sets will be useful for approximate reasoning such as interpolative reasoning while the original fuzzy sets will be useful for robust reasoning or for reasoning all possible conclusions. For the approximation of a continuous function implicit in a given information table, interpolation is necessary. We use the new fuzzy rough sets introduced below.

Let $\mathcal{F} = \{F_1, F_2, \ldots, F_p\}$ be a family of fuzzy sets F_i such that there exists $\omega \in \Omega$ satisfying $\mu_{F_i}(\omega) = 1$, where μ_{F_i} is a membership function of F_i and Ω is a universal set. Then, given a fuzzy set $A \subseteq \Omega$, lower and upper approximations of A are defined as

Table 1: Properties of $\mathcal{F}_\Box(A)$ and $\mathcal{F}^\Diamond(A)$

(i)	$\mathcal{F}_\Box(A) \subseteq X \subseteq \mathcal{F}_1^*(A)$
(ii)	$\mathcal{F}_\Box(\emptyset) = \emptyset, \mathcal{F}^\Diamond(\Omega) = \Omega$
(iii)	$\mathcal{F}_\Box(A \cap B) \subseteq \mathcal{F}_\Box(A) \cap \mathcal{F}_\Box(B), \quad \mathcal{F}^\Diamond(A \cup B) \supseteq \mathcal{F}^\Diamond(A) \cup \mathcal{F}^\Diamond(B)$
(iv)	$A \subseteq B$ implies $\mathcal{F}_\Box(A) \subseteq \mathcal{F}_\Box(B), \quad A \subseteq B$ implies $\mathcal{F}^\Diamond(A) \subseteq \mathcal{F}^\Diamond(B)$
(v)	$\mathcal{F}_\Box(A \cup B) \supseteq \mathcal{F}_\Box(A) \cup \mathcal{F}_\Box(B), \quad \mathcal{F}^\Diamond(A \cap B) \subseteq \mathcal{F}^\Diamond(A) \cap \mathcal{F}^\Diamond(B)$
(vi)	$\mathcal{F}_\Box(\Omega - X) = \Omega - \mathcal{F}^\Diamond(A), \quad \mathcal{F}^\Diamond(\Omega - X) = \Omega - \mathcal{F}_\Box(A)$
(vii)	$\mathcal{F}_\Box(\mathcal{F}_\Box(A)) = \mathcal{F}_\Box(A), \quad \mathcal{F}_\Box(A) \subseteq \mathcal{F}^\Diamond(\mathcal{F}_\Box(A)) \subseteq \mathcal{F}^\Diamond(A),$ $\mathcal{F}^\Diamond(\mathcal{F}^\Diamond(A)) = \mathcal{F}^\Diamond(A), \quad \mathcal{F}^\Diamond(A) \supseteq \mathcal{F}_\Box(\mathcal{F}^\Diamond(A)) \supseteq \mathcal{F}_\Box(A)$

follows (see Inuiguchi and Tanino [12]):

$$\mu_{\mathcal{F}_\Box(A)}(x) = \max_{i=1,2,\ldots,p} \xi[I](\mu_{F_i}(x), N_{F_i}(A)), \tag{1}$$

$$\mu_{\mathcal{F}^\Diamond(A)}(x) = \min_{i=1,2,\ldots,p} n(\xi[I](\mu_{F_i}(x), N_{F_i}(\Omega - A))), \tag{2}$$

where $I : [0,1] \times [0,1] \to [0,1]$ is an implication function satisfying

(I1) $I(1,0) = 0$ and $I(0,0) = I(0,1) = I(1,1) = 1$,
(I2) $I(c,b) \leq I(a,d)$ if $0 \leq a \leq c \leq 1$ and $0 \leq b \leq d \leq 1$,
(I3) I is upper semi-continuous.

$n : [0,1] \to [0,1]$ is a strong negation, i.e., a strictly decreasing function such that $n(0) = 1$ and $n(n(a)) = a$. $\xi[I]$, N_{F_i} and $\Omega - A$ are defined by

$$\xi[I](a,b) = \inf_{0 \leq h \leq 1} \{h \mid I(a,h) \geq b\}, \tag{3}$$

$$N_{F_i}(A) = \inf_{\omega \in \Omega} I(\mu_{F_i}(\omega), \mu_A(\omega)), \tag{4}$$

$$\mu_{\Omega - A}(\omega) = n(\mu_A(\omega)). \tag{5}$$

A fuzzy rough set is defined by a pair of the lower and upper approximations.

The lower approximation $\mathcal{F}_\Box(A)$ and the upper approximation $\mathcal{F}^\Diamond(A)$ have properties listed in Table 1.

3 Inference on Function Value from Information Table

We assume an information table is given. Let $X = \{x_1, x_2, \ldots, x_m, y\}$ be a set of attributes. We divide the attributes into two groups: an attribute y composes a group and the remaining attributes compose the other group. Thus the information table is regarded as a decision table with criteria $x_1, x_2, \ldots, x_m$ and a decision attribute y. Moreover we assume that attribute values of x_i, $i = 1, 2, \ldots, m$ and y are continuous, i.e., real numbers. The structure of the decision table is shown in Table 2. For the sake of simplicity, $x_i(\omega)$ and $y(\omega)$ denote the x_i- and y-values of an object ω.

Let V_{x_i} and V_y be sets of possible attribute values of x_i and y, respectively. Let $\boldsymbol{x}(\omega) = (x_1(\omega), x_2(\omega), \ldots, x_m(\omega))^T$ and $\bar{x}^j = (\bar{x}_1^j, \bar{x}_2^j, \ldots, \bar{x}_m^j)^T$. To apply fuzzy rough sets described in the previous section, we should define a family of fuzzy sets $\mathcal{F} = \{F_1, F_2, \ldots, F_p\}$ on $V_{x_1} \times V_{x_2} \times \cdots \times V_{x_m}$ and fuzzy sets Y_i on V_y, $i = 1, 2, \ldots, q$.

Table 2: The structure of the decision table

object	x_1	x_2	$\cdots$	x_m	y
ω_1	$\bar{x}_1^1$	$\bar{x}_2^1$	$\cdots$	$\bar{x}_m^1$	$\bar{y}^1$
ω_2	$\bar{x}_1^2$	$\bar{x}_2^2$	$\cdots$	$\bar{x}_m^2$	$\bar{y}^2$
$\vdots$	$\vdots$	$\vdots$	$\cdots$	$\vdots$	$\vdots$
ω_n	$\bar{x}_1^n$	$\bar{x}_2^n$	$\cdots$	$\bar{x}_m^n$	$\bar{y}^n$

The interpolative reasoning from Table 2 will be based on the following rules:

$$\text{if } x(\omega) \text{ is near } \bar{x}^j \text{ then } y(\omega) \text{ is near } \bar{y}^j, \ j = 1, 2, \ldots, n. \tag{6}$$

We appropriately define a fuzzy set F_j so as to represent 'near $\bar{x}^j$', $j = 1, 2, \ldots, n$. Similarly, we appropriately define a fuzzy set Y_j so as to represent 'near $\bar{y}^j$', $j = 1, 2, \ldots, n$.

Let $\hat{F}_j$ and $\hat{Y}_j$ be fuzzy sets of objects on $\Omega = \{\omega_1, \omega_2, \ldots, \omega_n\}$ defined by

$$\mu_{\hat{F}_j}(\omega) = \mu_{F_j}(x(\omega)), \tag{7}$$

$$\mu_{\hat{Y}_j}(\omega) = \mu_{Y_j}(y(\omega)). \tag{8}$$

Then we obtain a lower approximation $\mathcal{F}_\square(\hat{Y}_j)$ for each $\hat{Y}_j$ as

$$\mu_{\mathcal{F}_\square(\hat{Y}_j)}(\omega) = \max_{i=1,2,\ldots,n} \xi[I]\left(\mu_{\hat{F}_i}(\omega), N_{\hat{F}_i}(\hat{Y}_j)\right). \tag{9}$$

However to obtain $\mathcal{F}_\square(\hat{Y}_j)$ we should calculate $N_{\hat{F}_i}(\hat{Y}_j)$, $i = 1, 2, \ldots, n$. This will be computationally expensive and we expect $N_{\hat{F}_i}(\hat{Y}_j) = 0$ for $i \neq j$ in many cases. From this point of view, we use the following simple lower approximation $\underline{\mathcal{F}}(\hat{Y}_j)$:

$$\mu_{\underline{\mathcal{F}}(\hat{Y}_j)}(x) = \xi[I]\left(\mu_{\hat{F}_j}(\omega), N_{\hat{F}_j}(\hat{Y}_j)\right). \tag{10}$$

Obviously, we obtain

$$\underline{\mathcal{F}}(\hat{Y}_j) \subseteq \mathcal{F}_\square(\hat{Y}_j) \subseteq \hat{Y}_j. \tag{11}$$

Note that each $\underline{\mathcal{F}}(\hat{Y}_j)$ corresponds to a rule in (6).

Using the lower approximation $\underline{\mathcal{F}}(\hat{Y}_j)$, for any $x(\omega) = x^*$, we know that $y(\omega)$ has the membership degree to Y_j more than $\mu_{\underline{\mathcal{F}}(\hat{Y}_j)}(\omega)$. Namely, we have

$$y(\omega) \in [Y_j]_{\mu_{\underline{\mathcal{F}}(\hat{Y}_j)}(\omega)}, \ j = 1, 2, \ldots, n, \tag{12}$$

where $[Y_j]_h$ is an h-level set, i.e.,

$$[Y_j]_h = \{y \in \mathbf{R} \mid \mu_{Y_j}(y) \geq h\}. \tag{13}$$

By (12), we may estimate $y(\omega)$ as

$$y(\omega) \in Y(\omega) = \bigcap_{j=1,2,\ldots,n} [Y_j]_{\mu_{\underline{\mathcal{F}}(\hat{Y}_j)}(\omega)}. \tag{14}$$

In the way described above, we can infer the range of $y(\omega)$ as $Y(\omega)$ defined in (14). However, there is no guarantee that $Y(\omega) \neq \emptyset$. It is said that rules (6) are consistent at $x(\omega)$ if and only if $Y(\omega) \neq \emptyset$. Moreover, it is said that rules (6) are globally consistent if and only if rules (6) are consistent at any $\omega \in \Omega$. Note that even if rules (6) are globally consistent, the estimation of $y(\omega)$ based on rules (6) is not always precise, nor exact. Namely we may have $|Y(\omega)| > 1$ and $y(\omega) \notin Y(\omega)$.

4 A Specific Method

In the previous section, we described an inference mechanism based on fuzzy rough sets. To actualize this inference, we should determine I, F_i and Y_i, $i = 1, 2, \ldots, n$, concretely. This section is devoted to this topic.

First of all, we use an arbitrary implication function I satisfying

(I4) there exists $\hat{h} \in (0, 1]$ such that $I(a, b) \geq \hat{h}$ if and only if $a \leq b$.

Then we have

$$\xi[I](a, \hat{h}) = a. \tag{15}$$

Gödel, Łukasiewicz, Goguyen implications and more generally, R-implication functions [14] defined by t-norms satisfy (I4) with $\hat{h} = 1$. Moreover, implication functions proposed by Inuiguchi and Tanino [14] satisfy (I4) with $\hat{h} = 0.5$.

We define F_i by

$$\mu_{F_i}(\boldsymbol{x}^*) = \eta_i(\psi(\boldsymbol{x}^*)), \quad \boldsymbol{x}^* \in V_{x_1} \times V_{x_2} \times \cdots \times V_{x_m}, \tag{16}$$

where $\eta_i : [0, +\infty) \to [0, 1]$ is a quasi-concave function such that $\eta(\psi(\bar{\boldsymbol{x}}^i)) = 1$ and $\lim_{r \to +\infty} \eta_i(r) = \lim_{r \to -\infty} \eta_i(r) = 0$. $\psi : \mathbf{R}^m \to \mathbf{R}$ is a continuous scalarizing function such that $\psi(\boldsymbol{0}) = 0$ and $\psi(\boldsymbol{r}_1) \geq \psi(\boldsymbol{r}_2)$ if $\boldsymbol{r}_1 \geq \boldsymbol{r}_2$. For example, we may define $\eta_i(r) = \max(1 - |r|/d, 0)$ and $\psi(\boldsymbol{x}) = \sum_{i=1}^m x_i$, where d is a positive number showing the degree of tolerance.

Consider points (y_i^j, μ_i^j), $i = 1, 2, \ldots, n$, $j = 1, 2, \ldots, n$ defined by

$$y_i^j = y^j, \qquad \mu_i^j = \xi[I](\mu_{\hat{F}_i}(\omega_j), h^i), \tag{17}$$

where $h^i \in (0, 1]$ is a predetermined value. By definition, $\mu_i^i = 1$ and $\mu_i^j < 1$, $j \neq i$. We define ν_i^j, $i = 1, 2, \ldots, n$, $j = 1, 2, \ldots, n$ by

$$\nu_i^j = \begin{cases} 0, & \text{if } j = i, \\ \dfrac{1 - \mu_i^j}{y_i^i - y_i^j}, & \text{otherwise.} \end{cases} \tag{18}$$

For each $i \in \{1, 2, \ldots, n\}$, we define

$$J_i^- = \{j \in \{1, 2, \ldots, n\} \mid \mu_i^j > 0 \text{ and } \not\exists k; \nu_i^k < \nu_i^j \text{ and } y_i^k \leq y_i^j\}, \tag{19}$$
$$J_i^+ = \{j \in \{1, 2, \ldots, n\} \mid \mu_i^j > 0 \text{ and } \not\exists k; \nu_i^k < \nu_i^j \text{ and } y_i^k \geq y_i^j\}. \tag{20}$$

We also define μ_i^- and μ_i^+ by

$$\mu_i^- = \begin{cases} \min\{\mu_i^j \mid j \in J_i^-\}, & \text{if } J_i^- \neq \emptyset, \\ 1, & \text{if } J_i^- = \emptyset, \end{cases} \tag{21}$$

$$\mu_i^+ = \begin{cases} \min\{\mu_i^j \mid j \in J_i^+\}, & \text{if } J_i^+ \neq \emptyset, \\ 1, & \text{if } J_i^+ = \emptyset, \end{cases} \tag{22}$$

Renumber elements of a set $\{y_i^j \mid j \in J_i^- \cup J_i^+\}$ so that we have $y_i^{j_i(1)} < y_i^{j_i(2)} < \cdots < y_i^{j_i(s_i)}$. We may have $\{j_i(k) \mid k = 1, 2, \ldots, s_i\} \subset J_i^- \cup J_i^+$ because we may have $y_i^{k_1} = y_i^{k_2}$ for $k_1 \neq k_2$ and $k_1, k_2 \in J_i^- \cup J_i^+$. We have $\mu_i^{k_1} = \mu_i^{k_2}$ if $y_i^{k_1} = y_i^{k_2}$. Then we define Y_i by the following piecewise linear membership function:

$$\mu_{Y_i}(y^*) = \begin{cases} \mu_i^- - \epsilon, & \text{if } y^* < y_i^{j_i(1)}, \\ \dfrac{(y^* - y_i^{j(1)})\mu_i^{j(2)} + (y_i^{j(2)} - y^*)\mu_i^{j(1)}}{y_i^{j(2)} - y_i^{j(1)}}, & \text{if } y_i^{j(1)} \leq y^* \leq y_i^{j(2)}, \\ \cdots\cdots\cdots \\ \dfrac{(y^* - y_i^{j(s_i-1)})\mu_i^{j(s_i)}}{y_i^{j(s_i)} - y_i^{j(s_i-1)}} + \dfrac{(y_i^{j(s_i)} - y^*)\mu_i^{j(s_i-1)}}{y_i^{j(s_i)} - y_i^{j(s_i-1)}}, \\ \qquad\qquad\qquad\qquad\qquad\text{if } y_i^{j(s_i-1)} \leq y^* \leq y_i^{j(s_i)}, \\ \mu_i^+ - \epsilon, & \text{if } y^* > y_i^{j_i(s_i)}. \end{cases} \tag{23}$$

where $\epsilon > 0$ is theoretically an infinitesimal nonstandard number, i.e., a number greater than zero yet smaller than any positive real number. However, in practice, we set ϵ as a very small positive real number.

Under the setting above, we have $N_{\hat{F}_i}(\hat{Y}_i) = h^i$, $i = 1, 2, \ldots, n$. Under certain conditions, we guarantee the global consistency as shown in the following theorem.

Theorem 1. If I satisfies (I4), F_i is defined by (16) with a convex function η_i and Y_i is defined by (23) with $h^i = \hat{h}$, $i = 1, 2, \ldots, n$, then rules (6) are globally consistent.

The next theorems will be used for getting a minimal number of rules which are exact at any $\boldsymbol{x}(\omega_i)$, $i = 1, 2, \ldots, n$.

Theorem 2. If there exist $i, j \in \{1, 2, \ldots, n\}$ such that $k \in J_i^+ \cap J_j^-$, then $Y(\omega_k) = \{y^k\}$.

Theorem 3. If $|J_i^+| > 1$ and $|J_j^-| > 1$ for $i, j \in \{1, 2, \ldots, n\}$, then $Y(\omega)$ is bounded for any object ω such that $\mu_{F_i}(\boldsymbol{x}(\omega)) \geq \mu_i^+$ and $\mu_{F_j}(\boldsymbol{x}(\omega)) \geq \mu_j^-$.

Example 1. Let us apply the proposed method to a decision table given in Table 3 with two attributes x (criterion) and y (decision). We use x instead of x_1 since we have $m = 1$ in this example. Actually those data are obtained by a function $y = x^3 - 9x$ with randomly generated x in the range $[-7, 7]$. Thus we try to approximate a function $y = x^3 - 9x$ using fuzzy rough sets based on Table 3.

Table 3: A Simple Decision Table

object	x	y
ω_1	5.850633977	147.6110151
ω_2	4.5555777	53.54301755
ω_3	4.036283714	29.43091069
ω_4	2.937410143	-1.09160521
ω_5	0.821639766	-6.840075536
ω_6	-0.694559623	5.915971966
ω_7	-2.956441836	0.767053774
ω_8	-4.575922623	-54.63225069
ω_9	-5.08668439	-85.83453386
ω_{10}	-6.711483266	-241.908754

We define $\eta_i(r) = \max(1 - |r|/5, 0)$ and $\psi(x) = x$. Thus we obtain triangular fuzzy numbers F_i, $i = 1, 2, \ldots, 10$. By (23), we can calculate each Y_i with a piecewise linear membership function.

For the obtained rules, J_i^- and J_i^+ are as follows: $J_1^- = \{1, 2, 3, 4\}$, $J_1^+ = \{1\}$, $J_2^- = \{2, 3, 4, 5\}$, $J_2^+ = \{1, 2\}$, $J_3^- = \{3, 4, 5\}$, $J_3^+ = \{1, 3\}$, $J_4^- = \{4, 5\}$, $J_4^+ = \{1, 4\}$, $J_5^- = \{5\}$, $J_5^+ = \{2, 5\}$, $J_6^- = \{6, 8\}$, $J_6^+ = \{3, 6\}$, $J_7^- = \{7, 10\}$, $J_7^+ = \{6, 7\}$, $J_8^- = \{8, 10\}$, $J_8^+ = \{6, 7, 8\}$, $J_9^- = \{9, 10\}$, $J_9^+ = \{6, 7, 8, 9\}$, $J_{10}^- = \{10\}$ and $J_{10}^+ = \{7, 8, 9, 10\}$.

For $\omega \in [-6.5, 5.5]$, we estimate $y(\omega)$ by (14). The upper and lower bounds of $Y(\omega)$ are depicted as 'estimated upper curve' and 'estimated lower curve' in Figure 1. In Figure 1, the curve $y = x^3 - 9x$ is also depicted. We can see that the obtained rules are globally consistent.

5 Modifications

5.1. Improving the approximation

Even when the assumptions of Theorem 1 is satisfied, the approximation is not very good as shown in Figure 1. From Theorems 2 and 3, we expect that the approximation will be improved by increasing $|J_i^-|$ and $|J_i^+|$ for $i = 1, 2, \ldots, n$. To increase $|J_i^-|$ and $|J_i^+|$, we may update η_i so as to satisfy $\eta_i(\psi(\bar{x}^j)) = \mu_{Y_i}(\bar{y}^j)$, for all j such that $\mu_{F_i}(\bar{x}^j) > 0$. Because F_i means 'near $\bar{x}^i$', η_i should satisfy $\eta_i(\psi(\bar{x}^j)) \leq \eta_i(\psi(\bar{x}^k))$ for j and k such that $\psi(\bar{x}^j) < \psi(\bar{x}^k) < \psi(\bar{x}^i)$ or $\psi(\bar{x}^j) > \psi(\bar{x}^k) > \psi(\bar{x}^i)$. Taking into account this requirement, η_i can be updated by the following procedure.

(a) Let $\pi : \{1, 2, \ldots, n\} \to \{1, 2, \ldots, n\}$ be a permutation such that $\psi(\bar{x}^{\pi(1)}) \leq \psi(\bar{x}^{\pi(2)}) \leq \cdots \leq \psi(\bar{x}^{\pi(n)})$. Let i^* be an index such that $\pi(i^*) = i$.

(b) Set $k = 1$ and $v_0 = 1$.

(c) If $\eta_i(\psi(\bar{x}^{\pi(i^*+k)})) \leq 0$ then go to (f).

(d) $v_k = \min(v_{k-1}, \mu_{Y_i}(\bar{y}^{\pi(i^*+k)}))$.

(e) Update $k = k + 1$. Return to (c).

(f) Set $k = -1$.

(g) If $\eta_i(\psi(\bar{x}^{\pi(i^*+k)})) \leq 0$ then go to (f).

(h) $v_k = \min(v_{k+1}, \mu_{Y_i}(\bar{y}^{\pi(i^*+k)}))$.

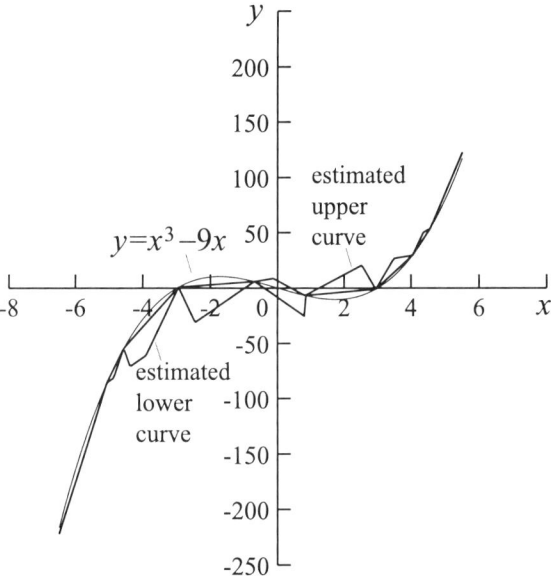

Figure 1: $Y(\omega)$ for $\omega \in [-6.5, 5.5]$

(i) Update $k = k - 1$. Return to (g).
(j) η_i is updated by (24).

$$\eta_i(r) = \begin{cases} v_k - \dfrac{(v_k - v_{k+1})(\psi(\bar{x}^{\pi(i^*+k)}) - r)}{\psi(\bar{x}^{\pi(i^*+k)}) - \psi(\bar{x}^{\pi(i^*+k+1)})} \\ \qquad \text{if } \exists k; r \in (\psi(\bar{x}^{\pi(i^*+k)}), \psi(\bar{x}^{\pi(i^*+k+1)})), \\ v_k, \quad \text{if } \exists k; r = \psi(\bar{x}^{\pi(i^*+k)}), \\ \eta_i(r), \quad \text{otherwise.} \end{cases} \qquad (24)$$

When assumptions of Theorem 1 are satisfied, we have the following theorem.

Theorem 4. If assumptions of Theorem 1 are satisfied then rules (6) are globally consistent with the updated η_i, $i = 1, 2, \ldots, n$.

For rules (6) with modified η_i's, we have the following theorem.

Theorem 5. Let $m = 1$. Assume that x_1^j, $j = 1, 2, \ldots, n$ are all different one another, that each η_i is linear in the range $(0, 1)$ and that assumptions of Theorem 1 are satisfied. Then, for rules (6) with updated η_i's, the following assertion is valid: if the function implicitly in decision table is monotonously increasing or decreasing, then for ω with $x_1(\omega) \in [\min_{i=1,2,\ldots,n} x_1^i, \max_{i=1,2,\ldots,n} x_1^i]$ $Y(\omega) = \{y^L(x_1(\omega))\}$, where

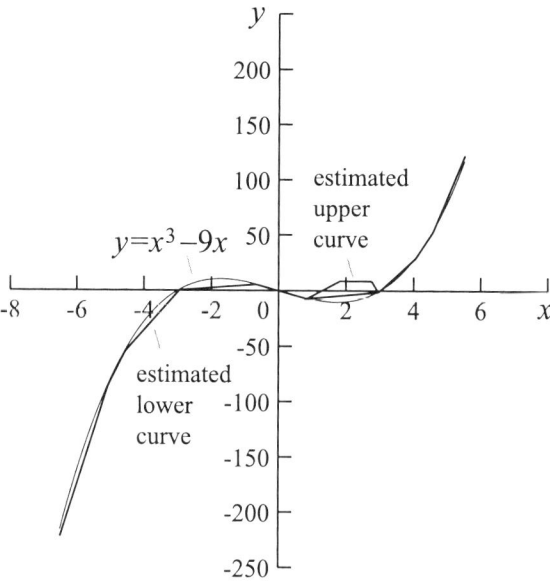

Figure 2: $Y(\omega)$ from the modified rules

$y^L : [\min_{i=1,2,\ldots,n} x_1^i, \max_{i=1,2,\ldots,n} x_1^i] \to \mathbf{R}$ is a piecewise linear function defined by

$$y^L(r) = \begin{cases} \dfrac{(x_1^{i(2)} - r)y^{i(1)}(r - x_1^{i(1)})y^{i(2)}}{x_1^{i(2)} - x_1^{i(1)}}, & \text{if } x_1^{i(1)} \leq r \leq x_1^{i(2)}, \\ \cdots\cdots\cdots \\ \dfrac{(x_1^{i(n)} - r)y^{i(n-1)}(r - x_1^{i(n-1)})y^{i(n)}}{x_1^{i(n)} - x_1^{i(n-1)}}, & \text{if } x_1^{i(n-1)} \leq r \leq x_1^{i(n)}, \end{cases} \quad (25)$$

where $\{x_1^j \mid j = 1, 2, \ldots, n\}$ are renumbered so that we have $x_1^{i(1)} < x_1^{i(2)} < \cdots < x_1^{i(n)}$.

Example 2. Let us apply the modification to the rules obtained in Example 1. The results are depicted in Figure 2. J_i^- and J_i^+ are obtained as follows: $J_1^- = \{1,2,3,4\}$, $J_1^+ = \{1\}$, $J_2^- = \{2,3,4,5\}$, $J_2^+ = \{1,2\}$, $J_3^- = \{3,4,5\}$, $J_3^+ = \{1,2,3\}$, $J_4^- = \{4,5\}$, $J_4^+ = \{1,2,3,4\}$, $J_5^- = \{5\}$, $J_5^+ = \{2,3,4,5,6\}$, $J_6^- = \{5,6,7,8,9\}$, $J_6^+ = \{3,6\}$, $J_7^- = \{7,8,9,10\}$, $J_7^+ = \{6,7\}$, $J_8^- = \{8,9,10\}$, $J_8^+ = \{6,7,8\}$, $J_9^- = \{9,10\}$, $J_9^+ = \{6,7,8,9\}$, $J_{10}^- = \{10\}$ and $J_{10}^+ = \{7,8,9,10\}$. $|J_i^-|$ and $|J_i^+|$ are increased. As is shown in Figure 2, the approximation is improved. Moreover, we can see that the monotonous parts of the function is approximated by linear interpolation.

5.2. Minimizing the number of rules

There is no guarantee that all of the obtained rules are indispensable. The obtained rules may include some superfluous rules. In this subsection, we select indispensable rules from the obtained ones under the restrictions that we can reproduce the given data precisely and we obtain a bounded $Y(\omega)$ for any $\boldsymbol{x}(\omega)$ such that $\psi(\boldsymbol{x}(\omega)) \in \left[\min_{i=1,\ldots,n} \psi(\bar{\boldsymbol{x}}^i), \max_{i=1,\ldots,n} \psi(\bar{\boldsymbol{x}}^i)\right]$.

To do this, Theorems 2 and 3 should be applied. We define the following two kinds of index sets:

$$P(k) = \{i \mid k \in J_i^+\}, \quad M(k) = \{i \mid k \in J_i^-\}. \tag{26}$$

From Theorem 2, if $|P(k)| = 1$ (resp. $|M(k)| = 1$), we cannot reproduce the k-th data unless we select the only element in $P(k)$ (resp. $M(k)$). On the other hand, from Theorem 3, we can observe that even if there exists i, j such that $k \in P(i)$ and $j \in N(j)$ for any $k \in \{1, 2, \ldots, n\}$, we may have unbounded $Y(\omega)$ for some $\boldsymbol{x}(\omega)$ such that $\psi(\boldsymbol{x}(\omega)) \in \left[\min_{i=1,\ldots,n} \psi(\bar{\boldsymbol{x}}^i), \max_{i=1,\ldots,n} \psi(\bar{\boldsymbol{x}}^i)\right]$ if there is no J_i^+ (resp. J_i^-) such that $k \in J_i^+$ (resp. $k \in J_i^-$) and $|J_i^+| > 1$ (resp. $|J_i^-| > 1$) for some $k \in \{1, \ldots, n\}$.

Considering those facts, we can select the rules by the following procedure, where we suppose that the obtained n rules can reproduce the given data precisely and produce a bounded $Y(\omega)$ for any $\boldsymbol{x}(\omega)$ such that $\psi(\boldsymbol{x}(\omega)) \in \left[\min_{i=1,\ldots,n} \psi(\bar{\boldsymbol{x}}^i), \max_{i=1,\ldots,n} \psi(\bar{\boldsymbol{x}}^i)\right]$.

(a) Let $N = \{1, 2, \ldots, n\}$. Calculate $P(k)$ and $M(k)$, $k \in N$. For $k \in N$, calculate
$$l^-(k) = \min_{i \in J_k^-} \psi(\bar{\boldsymbol{x}}^i), \quad q^-(k) = \max_{i \in J_k^-} \psi(\bar{\boldsymbol{x}}^i),$$
$$l^+(k) = \min_{i \in J_k^+} \psi(\bar{\boldsymbol{x}}^i), \quad q^+(k) = \max_{i \in J_k^+} \psi(\bar{\boldsymbol{x}}^i).$$
Let $KP = KN = N$ and $RP = RN = [\min_{i=1,2,\ldots,n} l^-(i), \max_{i=1,2,\ldots,n} q^+(i)]$.

(b) Let $R = \{k \mid P(k) = \{k\} \text{ or } M(k) = \{k\}\}$. Update N by $N = N \backslash R$.

(c) Update KP, KN, RP and RN by
$$KP = KP \backslash R, \quad RP = RP \backslash \bigcup_{k \in R} [l^+(k), q^+(k)],$$
$$KN = KN \backslash R, \quad RN = RN \backslash \bigcup_{k \in R} [l^-(k), q^-(k)].$$

(d) If $KP = KN = \emptyset$ and $RP = RN = \emptyset$ then terminate the algorithm. The members of R are indices of selected rules. Let $IRP = \{i \mid \psi(\bar{\boldsymbol{x}}^i) \in \text{cl}(RP)\}$ and $IRN = \{i \mid \psi(\bar{\boldsymbol{x}}^i) \in \text{cl}(RN)\}$.

(e) For each $k \in N$, calculate
$$Q(k) = \{i \mid \psi(\bar{\boldsymbol{x}}^i) \in \text{cl}(RP \cap [l^+(k), q^+(k)])\},$$
$$L(k) = \{i \mid \psi(\bar{\boldsymbol{x}}^i) \in \text{cl}(RN \cap [l^-(k), q^-(k)])\}.$$

(f) For each $k \in IRP$, calculate $PP(k) = \{j \mid k \in Q(j)\}$. For each $k \in IRN$, calculate $MM(k) = \{j \mid k \in L(j)\}$

(g) Select a set Z with minimum cardinality from $\{P(k) \mid k \in KP\} \cup \{M(k) \mid k \in KP\} \cup \{PP(k) \mid k \in IRP\} \cup \{MM(k) \mid k \in IRN\}$. Select $k \in Z$ lexicographically maximizes $(|KP \cap J_k^+| + |KN \cap J_k^-| + |Q(k)| + |L(k)|, \min(|KP \cap J_k^+|, |KN \cap J_k^-|, |Q(k)|, |L(k)|), |J_k^+| + |J_k^-|, \min(|J_k^+|, |J_k^-|), -k)$. Update R, RP and RN by
$$R = R \cup \{k\}, \quad RP = RP \backslash [l^+(k), q^+(k)] \text{ and } RN = RN \backslash [l^-(k), q^-(k)].$$

Return to (d).

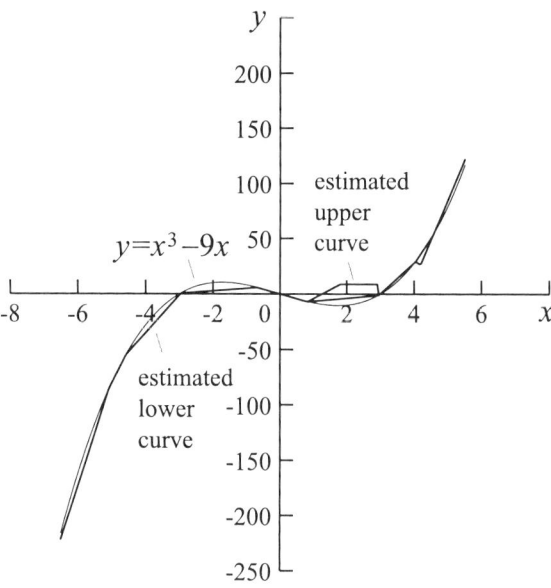

Figure 3: $Y(\omega)$ from the modified rules

Example 3. Applying the algorithm above to the rules obtained in Example 2, rules 1, 3, 5, 6, 8 and 10 are selected. The results of approximation is depicted in Figure 3.

6 Concluding Remarks

In this paper, we showed that we can approximate a continuous function implicit in a given decision table by fuzzy rough sets. This fact shows a possibility of treatment of continuous attributes in decision tables by means of fuzzy rough sets. Moreover, as the first attempt, we proposed a specific method to obtain globally consistent rules. Some properties of the method are examined. Based on the properties, we may design an algorithm to obtain globally consistent rules which estimate values correctly for given data with a minimum number of rules.

The application of fuzzy rough sets to approximation of a continuous function implicit in decision table has just proposed and it is still at a very beginning stage. There are a lot of open problems we should tackle. For example, in the proposed specific method, we implicitly assume that multiple attributes $x_1, x_2, \ldots, x_m$ can be aggregated to one value by a function ψ. This assumption is very strong. We should consider other specific methods introducing a distance function or a similarity measure. Moreover, to obtain a better approximation, we may introduce the change rates of decision attribute values which can be calculated from a given decision table as a decision attribute.

References

[1] Z. Pawlak (1982). Rough Sets. In *Internat. J. Inform. Comput. Sci.*, volume 11, no.5, pages 341-356.

[2] Z. Pawlak (1991). *Rough Sets: Theoretical Aspects of Reasoning about Data*, Kluwer Academic Publishers.

[3] Y. Y. Yao and T. Y. Lin (1996). Generalization of Rough Sets Using Modal Logics. In *Intelligent Automation and Soft Computing*, volume.2, no.2, pages 103-120.

[4] R. Słowinski and D. Vanderpooten (2000). A Generalized Definition of Rough Approximation Based on Similarity. In *IEEE Tran. on Knowledge and Data Engineering*, volume 12, no.2, pages 331-336.

[5] S. Greco, B. Matarazzo and R. Słowinski (1999). Rough Approximation of a Preference Relation by Dominance Relations. In *European Journal of Operational Research*, volume 117, pages 63-83.

[6] S. Greco, B. Matarazzo and R. Słowinski (1999) The Use of Rough Sets and Fuzzy Sets in MCDM. In *T. Gal, T. J. Stewart and T. Hanne (Eds.) Multicriteria Decision Making: Advances in MCDM Models, Algorithms, Theory, and Applications*, pages 14-1-14-59, Kluwer Academic Publishers, Boston.

[7] Y. Y. Yao (1996). Two Views of the Theory of Rough Sets in Finite Universes. In *Int. J. Approximate Reasoning*, volume 15, pages 291-317.

[8] Y. Y. Yao (1998). Relational Interpretations of Neighborhood Operators and Rough Set Approximation Operators. In *Information Sciences*, volume 111, pages 239-259.

[9] M. Inuiguchi and T. Tanino (2001). On Rough Sets under Generalized Equivalence Relations. In *Bulletin of International Rough Set Society: Proceedings of RSTGC 2001*, volume 5, no.1/2, pages 167-171.

[10] D. Dubois and H. Prade (1990). Rough Fuzzy Sets and Fuzzy Rough Sets. In *Int. J. General Systems*, volume 17, pages 191-209.

[11] D. Dubois and H. Prade (1992). Putting Rough Sets and Fuzzy Sets Together. In *R. Słowinski (Ed.) Intelligent Decision Support: Handbook of Applications and Advances of the Rough Sets Theory*, pages 203-232, Kluwer Academic Publishers, Dordrecht.

[12] M. Inuiguchi and T. Tanino (2002). New Fuzzy Rough Sets Based on Certainty Qualification. In *L. Polkowski S.K. Pal and A. Skowron (Eds.) Rough-neuro Computing: a Way to Computing with Words*, Physica-Verlag, Heidelberg.

[13] A. Nakamura (1988). Fuzzy Rough Sets. In *Note on Multiple-Valued Logic in Japan*, volume 9, no.8, pages 1-8.

[14] M. Inuiguchi and T. Tanino (2000). Necessity Measures and Parametric Inclusion Relations of Fuzzy Sets. In *Fundamenta Informaticae*, volume 42, pages 279-302.

Retrieving Information

Query Aggregation Reflecting Domain-knowledge

Troels Andreasen

Roskilde University, P.O. Box 260, 4000 Roskilde, Denmark

troels@ruc.dk

Abstract.

This paper discusses the problem of applying knowledge to guide the evaluation of queries. The objective is to apply common or domain-specific knowledge covering the content of an information base in connection with the evaluation of queries to the base. Two main issues are considered; the derivation of measures of similarity for properties from an ontology in the knowledge base and further the initiation of object similarity partly based on property similarity and partly on knowledge from the complementary knowledge base.

Keywords: Fuzzy Aggregation, Knowledge-based Query-evaluation, Ontology

1 Introduction

The principle of knowledge-guided query evaluation described in this paper is considered in the context of an approach to ontology-based querying, where the employed knowledge base contains domain-specific knowledge comprising a dictionary and an ontology for a given domain.

The general idea with this line of evaluation is to assimilate applicable knowledge during the evaluation to guide and improve this process. We consider two directions for application of knowledge – query transformation and similarity-based relaxation.

A query, which is initially assumed to be posed as a list of words, either unstructured or forming natural language expressions, is transformed into a structured expression based on linguistic knowledge and knowledge about words and concepts of the domain. During the evaluation of the query, measures of similarity for words and concepts are applied to obtain answers that include not only what is strictly reflected by the query, but also what can be considered similar to this.

The approach applies so-called descriptions derived for database objects as well as for queries. A description is an intermediate representation of the "content" derived as described above. In principle querying is performed by transforming into descriptions and evaluating through comparison at the level of these descriptions.

The approach described in this paper is developed as part of the OntoQuery project [4,5] funded by the Danish Research Agency under the Information Technology Program.

2 Descriptions, the knowledge base and queries

This paper is related to the project OntoQuery [4,5] where the issue is querying to an information base that contains unstructured text documents, thus we are concerned with an information retrieval approach. A central aspect of the approach is that descriptions are created as intermediate representations of "content". A description is a set of descriptors describing a text fragment. For a text fragment, e.g. a sentence, a simple form description expresses the content by means of a set of words from the sentence. The approach is however concept rather than word based and the set of words is only a special case.

Descriptions have the general form:
$$D=\{D_1, ..., D_n\}$$
where each descriptor D_i is a set of concepts. A concept is an expression in the OntoLog language [10], where words and concepts can be combined into new concepts by means of semantic relations.

To generate descriptions, text fragments are prepared by a parser that employs the knowledge base. The parser can in principle scale from a simple word recognizer to a complex natural language parser that maps the full meaning content of the sentence into an internal representation. Since the issue here is IR the idea is of course to grab fragments of content rather than representing full meaning and the building stones are concepts. The structure of descriptions should be understood considering this aim.

The approach to description generation in the OntoQuery project is a subject to ongoing development. In the present state descriptions are generated as follows.

A tagger identifies heuristically categories for words. Based on tags and a simple grammar, a parser divides the sentence by framing identified noun phrases (NPs) in the sentence (producing a markup in the sentence identifying beginnings and endings of NP-fragments of the sentence). For each part of the sentence, corresponding to an NP, a descriptor is produced as a set of concepts, that is, a set of expressions in the applied concepts language (OntoLog). A concept produced in this second parse of the sentence have the form of a lemma (a word mapped into lemma form) or a combination of lemmas by semantic relations of the language (such as WRT – with respect to, CBY – caused by, CHR – characterized by, …). Words and concepts that pass through to the description are only those that can be recognized in the knowledge base. Apart from a dictionary of words and morphological forms, the knowledge base includes an ontology relating words and concepts by a number of different relations. Semantic relations combine, as mentioned, words and concepts. The most central relation is concept inclusion (ISA) by which a central lattice of concepts is formed. Further relations as partonomy and association can also be introduced in the ontology.

Take as an example the sentence:
"Hasty postmen sometimes get serious injures from dog bites"

A description in a simple form without taking into account the framing of NP's in the sentence could be:
   (hasty), (postman), (serious), (injure) , (dog), (bite)

With the framing of NP's the descriptors can be gathered giving the description:
   (hasty, postman), (serious, injure) , (dog bite)

When also employing the semantic relations the descriptions can be enriched into for instance the following:
   (postman CHR hasty), (injure CHR serious CBY dog bite)

Descriptions are obviously not unique. The resulting description from parsing a sentence depends on a number of different circumstances. Most important are the actual framing into noun phrases, the point of detail that the second-level (semantic) parser gather concepts and the domain-specific part of the knowledge base (that direct descriptions towards more domain-specific concepts).

This line of description generation is an important aspect of the OntoQuery project as explained in [4, 5, 9, 10].

Queries and objects in the database are preprocessed in a similar manner leading to descriptions, thus querying becomes a matter of comparing descriptions. This approach does not eliminate the possibility to pose queries in the typical (and more convenient) fragment style rather than in a full form natural language style. The fragmentation in the generation of descriptions may well lead to the same description regardless of whether a full form query like:
  "Is it so that hasty dogs sometimes get injures from heavy traffic?"
or a fragment-style query like:
  "hasty dogs, injures from heavy traffic?"
is posed. One possible resulting description could be:
  (dog CHR hasty), (injure CBY traffic CHR heavy)

3 Querying as similarity evaluation

Querying in general is driven by query evaluation, where an answer is produced through comparing the query with objects from the information base. The answer can be considered to be a collection of the objects from the information base that are the most similar to the query according to the means of similarity exploited by the evaluation. The query can be considered to be an indication of the "ideal object". The comparison measures, in principle for each object of the information base, the degree to which it is similar to the "ideal object" indicated by the query.

The issue of query evaluation may thus be seen as a matter of exploiting measures of similarity between information base objects – in a uniform manner information base objects and the query are mapped into object descriptions and the answer is produced through measuring similarity between these object descriptions.

The similarity evaluation view on the query processing applies regardless whether the information base is a conventional database, a base of documents of an information retrieval system or a collection of pages on the web.

While object similarity measures in rare cases may be available directly as independent measures, it is more common to obtain these as derived from property similarity measures. A typical query states compound constraints in the properties of objects. A relational database query combines properties of set membership and attribute values such as 'part where weight=17', where objects that have the property of being member of the relation part and the property of having the value 17 for the attribute weight are queried. A simple information retrieval query lists a number of words as being conjunctive properties of the ideal object. Similar objects to the query 'hairy dogs' are objects that includes both words. Natural language information retrieval can be regarded as retrieval where not only words but also concepts are properties, thus an object embedding the phrase 'the hairy owner of the red dog' is not necessarily among the most similar to the query 'hairy dogs', because the latter is considered to indicate the concept of dogs that are hairy.

In most cases there is an intrinsic derivation of object similarity from property similarity – similar objects are objects with similar properties. A query embedding a number of property constraints describes an ideal object that fulfills these constraints. The answer to the query is the information base objects that have the properties described in the query.

Measuring similarity for crisp querying is in most cases a very simple matter. Similar objects are objects that strictly resemble the properties of the query. If property similarity is expressed as a truth value then the object similarity is derived from aggregating the property similarity values according to the logical connectors combining the properties in the query[1].

The softening of the query process goes in two directions. The property similarity may be relaxed. For instance, when distance measures can be defined on the property domains, the crisp similarity can be generalized to relate as similar values also values that are close according to the defined distance.

The object similarity may be relaxed – that is, the aggregation of the query properties may be softened. As a simple example, in a conjunctive list-of-properties query this could correspond to a preference of most of the properties rather than an insistence on all. However when property similarity is relaxed far more advanced object similarity relaxation possibilities appears.

4   Property similarity

As indicated above the ontology in the knowledge base is assumed to explicate various relations between concepts. We consider below how such relations may contribute to similarity. Apart from hyponymy (concept inclusion) we briefly discuss the relations synonymy, partonomy and association. All of these contribute directly to similarity between concepts. Moreover we consider the semantic relations, used in forming concepts in the ontology. These indirectly contribute to similarity through subsumption.

Hyponymy

Until now the main concern in the OntoQuery project has been the hyponymy or concept inclusion relation. For this relation we should intuitively have strong similarity in the opposite direction of the inclusion (specialization), but also the direction of the inclusion (generalization) must contribute with some degree of similarity.

Take as an example the small fraction of an ontology in figure 1.

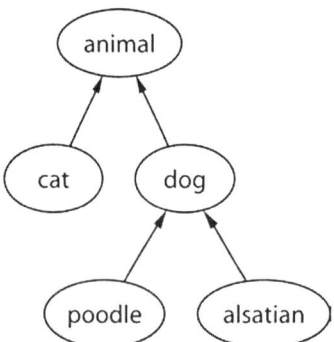

Figure 1: Inclusion relation (ISA) with upwards reading, e.g. dog ISA animal

---

[1] Queries as lists of properties with no explicit logical combination of properties typically indicate the conjunction of the properties listed in the query.

Based on this the term dog could be expanded to for instance the set:
   dog+ = 1/dog + 0.9/poodle + 0.9/alsatian + 0.3/animal
reflecting that more general terms are less similar.

While the hyponymy relation obviously is transitive (e.g. poodle ISA animal), a similarity measure based on this should reflect 'distance' in the relation, where greater distance – longer path in the relation graph – corresponds to smaller similarity. For instance an expansion of the terms animal and poodle into sets of similar values could be:
   poodle + = 1/ poodle + 0.3/ dog + 0.27/alsatian +0.09/animal+ 0.081/cat
   animal+ = 1/animal + 0.9/cat + 0.9/dog + 0.81/poodle + 0.81/alsatian

If we in the knowledge base can distinguish explicitly stated, original references from derived, then we can define a (non-transitive) relation ISA~ as consisting of these (such that ISA becomes the transitive closure of ISA~) and measure similarity from distance (minimal path-length) in the ISA~ graph.
A similarity function sim_{ISA} based on distance in ISA~ dist(X,Y) should have the properties:
   sim: U×U → [0,1], where U is the universe of terms
   sim (X,Y) = 1 only if X=Y
   sim(X,Y) < sim(X,Z) if dist(X,Y) < and dist(X,Z)
Based further on two factors σ and γ expressing costs of specialization and generalization respectively, we can define a simple similarity function as follows.
If there is a path from nodes (terms) X and Y in the hyponymy relation then it has the form
   $P = (P_1, \cdots, P_n)$ with X= P_1 and Y= P_n and either
                           P_i ISA~ P_{i+1} or P_{i+1} ISA~ P_i for each i
Given a path $P = (P_1, \cdots, P_n)$ set s(P) and g(P) to the numbers of specializations and generalizations respectively along the path P thus:
   $s(P) = |\{i | P_i\ ISA \sim\ P_{i+1}\}|$ and $t(P) = |\{i | P_{i+1}\ ISA \sim\ P_i\}|$
If $P^1, \cdots, P^m$ are all paths connecting X and Y then the degree to which Y is similar to X can be defined as
   $sim(X,Y) = \min_{j=1,\ldots,m} \{\sigma^{s(P^j)} \gamma^{t(P^j)}\}$

Notice that the examples given above on expanding to sets of similar values can
If there is no means to determine explicitly stated references in the knowledge base then ISA~ can be derived as the transitive reduction of ISA.

Synonymy
Synonymy obviously implies strong similarity. A synonymy relation is rarely transitive and typically not reflexive. If the knowledge base includes a non-transitive, non-reflexive synonymy relation similarity based on this can be defined using a constant factor:

$sim(X,Y) = \delta$ if Y is a synonym for X

Partonomy

Partonomy, as exemplified in figure 2, is in general difficult to measure in terms of degrees of similarity. Related terms are often not very similar and the most obvious alternatives are either to drop the relation as basis for a similarity measure or to let it contribute to an association relation.

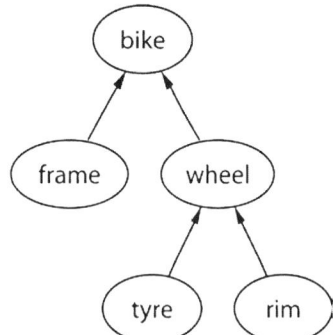

Figure 2: Partonomy relation (Part-of)

Association

Various sources can be used to set up an association relation. Probably among the most reliable and useful as domain-specific association relations, applied in connection with querying, are those variants that are statistically based on a corpus (see for instance [2]). Such a relation has an inherent grading from the statistics and is thus very accurately reflected in graded similarity. If associating statistically by:

$$assoc(X,Y) = \begin{cases} 0 & \|\Omega_X\| = 0 \\ \dfrac{\|\Omega_X \cap \Omega_Y\|}{\|\Omega_X\|} & \|\Omega_X\| > 0 \end{cases}$$

where Ω_X is the number of documents where term X appears, then a similarity measure following this association principle could simply be

$sim(X,Y) = assoc(X,Y)$

thus directly using the association grades as similarities.

Semantic relations

The semantic relations, used in forming concepts in the ontology, indirectly contribute to similarity through subsumption. For instance:

'disease CBY lack WRT vitamin'

is subsumed by – and thus included in – the more general concepts 'disease CBY lack' and 'disease'. Thus while semantic relations does not directly contribute to similarity, they do add to the extension of concept inclusion and thus to similarities based on hyponymy.

Using property similarity

The purpose of similarity measures in connection with querying is of course to look for similar rather than for exactly matching values, that is, to introduce soft rather than crisp evaluation. As indicated through examples above one approach to introduce similar values is to expand crisp values into fuzzy sets including also similar values.

Expansion of this kind, applying similarity based on knowledge in the knowledge base, is a simplification replacing direct reasoning over the knowledge base during query evaluation. The graded similarity is the obvious means to make expansion a useful – by using simple threshold values for similarity the size of the answer can be fully controlled.

5   Object similarity

Given a set of objects in the information base and an object in focus – the query (or the "ideal object" as indicated by the query) – the main exercise is to find those objects in the information base that or most similar to the query object. Object similarity may in rare cases be directly available as for instance from a relevance feedback procedure in Information Retrieval Systems or from other kinds of use, interest or purchase statistics on information base objects. However, more often it is needed to establish object similarity as derived from property similarity. In this case the problem is to choose an aggregation.

In this direction of appropriate object similarity aggregations the class of order weighted averaging (OWA) operators [13] has been shown to be very useful [2, 8]. These operators are especially suitable for aggregating a set of unstructured properties as a list-of words query to an Information Retrieval system. However when structure is introduced to the query expression the OWA principle becomes insufficient. This is the case when queries are mapped into descriptions as explained above where the need for nested aggregation arises.

We describe briefly below OWA and an extension to nested aggregation based on OWA. While the general idea with the line of querying discussed in this paper is to provide an intuitive platform for posing queries we introduce, with the nested aggregation, a very complex specification of parameters for queries. The solution to this problem is to perform a knowledge based modification of query parameters as touched upon in the second subsection below.

Aggregation of property similarities

OWA (Order Weighted Averaging) utilizes an ordering vector $W = [w_1, \cdots, w_n]$. The aggregation of values $a_1, \ldots, a_n$ is

$$F_W(a_1, \cdots, a_n) = \sum_{j=1}^{n} w_j b_j \quad \text{where} \quad \sum_{j=1}^{n} w_j = 1, \quad w_j \in [0,1] \quad \text{and}$$

b_j is the j'th largest $a_1, \cdots, a_n$

Thus with $b_1, \ldots, b_n$ is the (descending) ordering of the values $a_1, \ldots, a_n$. By modifying W we obtain different aggregations, for instance $F_{(1,0,0,\ldots)}$ is max $F_{(1/n,1/n,\ldots)}$ is average and $F_{(0,0,\ldots,1)}$ is min. Querying based on this may proceed from a query on attributes $A_1, \ldots, A_n$ such that the value $A_i(d) \in [0,1]$ is the degree to which the document d satisfies the property A_i The overall valuation of document d is

   Val(d) = $F_W(A_1(d), \ldots, A_n(d))$.

The OWA aggregation principle is very flexible and may further include importance weighting in the form of an n-vector $M=<m_1,\ldots,m_n>$, $m_j \in [0,1]$ giving attribute importances to $A_1,\ldots,A_n$ such that for instance $M=<1,0.8,0.8,\ldots>$ gives more importance to A_1, while importances are not discriminated with $M=<1,1,\ldots>$. Attribute importance may be included as a modification that leads to a new set of order-weights individually for each document d see [14] for details. In addition, the aggregation may be modified by a 'linguistic quantifier', which basically is an increasing function $Q:[0,1]\rightarrow[0,1]$ with $Q(0)=0$ and $Q(1)=1$, such that the order weights are prescribed as:

$$w_j = Q(\frac{j}{n}) - Q(\frac{j-1}{n})$$

Linguistic quantifiers lead to values of W and we can model for instance a quantifier EXISTS by $Q(x)=1$ *for $x>0$*, FOR-ALL by $Q(x)=0$ *for $x<1$*, and SOME by $Q(x)=x$, while one (of many) possibility to introduce MOST is by a power of SOME, e.g. $Q(x)=x^3$.

Thus we have a general query expression:

   $<A_1, ...,A_n:M:Q>$

where $A_1,...,A_n$ are the query attributes, M specifies attribute importance weighting and Q specifies a linguistic quantifier and thereby indicates an order weighting.

In [14], Yager introduces a hierarchical approach to aggregation as a language intended for document retrieval based on order weighted averaging (OWA). The intention is to enable users to better represent their requirements using the language. Basically this is an extension to nested query specification.

Query attributes may be grouped for individual aggregation and the language is orthogonal in the sense that aggregated values may appear as arguments to aggregations. Thus, queries may be viewed as hierarchies.

As an example we could pose a nested query expression:

   <A1(d),
     < A2(d), A3(d),
         < A4(d), A5(d), A6(d):M3:Q3>
            :M2:Q2>,
     :M1:Q1>

where $A_i(d) \in [0,1]$ measures the degree to which attribute A_i conforms to document d, while M_j and Q_j are the importance and quantifier applied in the j'th aggregate.

Rather than requiring the user to specify query at this level of detail the general idea is to perform a knowledge-based transformation of the query, heuristically deducing both nestings of the query attributes and aggregate operators, and inferring what parts of the query are more important as briefly discussed below.

Knowledge-based modification of aggregation

In the OntoQuery project an initial matching principle, which may be considered as based on two-level hierarchies, has been chosen for the first prototype.

As mentioned in section 2 an NLP parsing is performed that heuristically identifies noun phrases (NPs) preparing for descriptions with nestings corresponding to the NPs. For instance the sentence "Is it so that hasty dogs sometimes get injures from heavy traffic?" may lead to the description:

   (hasty, dog), (injure) , (heavy, traffic)

A nested aggregation can then be applied over the groups in the description, where each group is aggregated with individual importance weighting, and quantification. The aggregation parameters are to be derived during the query evaluation process. A simple general principle is that aggregation is restrictive for individual groups and relaxed for the overall query aggregate, corresponding to linguistic quantifiers like 'most' for individual groups and 'some' for the query. This can then be further modified through importance weighting based on domain knowledge, primarily from giving more importance to nouns in general and domain-specific concepts from the ontology in particular.

The manipulation of the example sentence above may in this way result in the nested aggregation expression:

  << hasty, dog : MOST : (½,1)>,
   < injure >,
   < heavy, traffic : MOST : (½,1)>,
   : SOME : (1,1,1)>

Here, importance weighting is only exploited at the level of individual groups, where nouns are given more importance. The restrictive quantification for groups is by MOST and the relaxed quantification for the overall query is by SOME.

The approach to querying is then to derive a knowledge-based nested aggregation expression for the initially posed query and then to evaluate this against the text objects in the information base, calculating degrees of conformity to each sentence description in the base. Based on this, the answer may be given as the most similar objects.

The ontology in the knowledge base comes into play from applying property similarity. The query expression above can be expanded to cover also similar values for argument properties, as for instance from replacing the subexpression

  << hasty, dog : MOST : (½,1)>,

with

  < hasty, <dog, animal : EXISTS : (1, .3)> : MOST : (½,1)>,

The ontology is, as mentioned, also intended to be applied in a conceptual formation leading to concepts rather than words as the smallest units in descriptions. This will lead to modifications that can be introduced in a similar way as expansions to the query.

6   Concluding remarks

We have discussed similarity as derived from property relations in a knowledge base and further introduced an approach to query evaluation, where simple word lists or queries are posed in natural language are transformed into hierarchical expressions over quantified, importance-weighted groups of attributes for order-weighted evaluation. This transformation is also guided by the knowledge base.

Basically, the query evaluation requires only sufficient knowledge to perform a grouping of properties. Improvements to the evaluation may be gained from knowledge about more important properties and from relations that allow deduction of degrees of conformity.

Acknowledgement

The research reported here was partly funded by the Danish Technical research Council and the IT-University of Copenhagen.

References

[1] S. Abney, "Partial parsing via finite-state cascades". Proceedings of the ESSLLI'96 Robust Parsing Workshop, 1996

[2] T. Andreasen, Flexible Database Querying based on Associations of Domain Values. In ISMIS'97, Eight International Symposium on Methodologies for INTELLIGENT SYSTEMS. Charlotte, North Carolina. Springer Verlag, Lecture Notes in Artificial Intelligence, 1997.

[3] T. Andreasen, Query evaluation based on domain-specific ontologies NAFIPS'2001, 20th IFSA / NAFIPS International Conference Fuzziness and Soft Computing, in press, 2001.

[4] T. Andreasen, H. Erdman Thomsen, J. Fischer Nilsson: The OntoQuery project in [6]

[5] T. Andreasen, J. Fischer Nilsson, H. Erdman Thomsen: Ontology-based Querying, in H.L. Larsen et al Flexible Query Answering Systems, Recent Advances, Proceedings of the FQAS'2000 conference

[6] P. Anker Jensen et al. (eds.) Proceedings of the International OntoQuery Workshop "Ontology-based interpretation of NP's", January 17 and 18, 2000, forthcoming, se also www.ontoquery.dk.

[7] E. Brill, "Transformation-based error-driven learning and natural language processing: A case study in part-of-speech tagging". Computational Linguistics, 21(4), pp.543-565, 1995.

[8] H.L. Larsen, T. Andreasen, H. Christiansen: Knowledge Discovery for Flexible Querying. In T. Andreasen et al. (eds.), Proceedings of the International Conference on Flexible Query Answering Systems, 11-15 May 1998, Roskilde, Denmark. Lecture Notes in Artificial Intelligence, Springer Verlag, Berlin. 1998.

[9] B. Nistrup Madsen, B. Sandford Pedersen and H. Erdman Thomsen, Defining Semantic Relations for OntoQuery, in [2].

[10] J. Fischer Nilsson: A Logico-Algebraic Framework for Ontologies, in [2].

[11] P. Paggio: "Parsing in OntoQuery - Experiments with LKB", in [6], 2000.

[12] Pustejovsky, J. (1995). The Generative Lexicon, Cambridge, MA, The MIT Press

[13] R.R. Yager: On order weighted averaging aggregation operators in multicriteria decision making. IEEE Transactions on Systems, Man and Cybernetics, 18(1), pp. 183-190, 1988.

[14] R.R. Yager, A hierarchical document retrieval language Information Retrieval 3, 357-377, 2000

B. Bouchon-Meunier, L. Foulloy and R.R. Yager (Editors)
Intelligent Systems for Information Processing:
From Representation to Applications
© 2003 Elsevier Science B.V. All rights reserved.

Modelling of Fuzzy and Uncertain Spatio-Temporal Information in Databases: A Constraint-based Approach

G. de Tré, R. de Caluwe, A. Hallez and J. Verstraete

Ghent University, Dept. of Telecommunications and Information Processing
Sint-Pietersnieuwstraat 41, B-9000 Ghent, Belgium
{Guy.deTre,Rita.deCaluwe,Axel.Halles,Jorg.Verstraete}@ugent.be

Abstract

A constraint-based generalized object-oriented database model is adapted to manage spatio-temporal information. The presented adaptation is based on the definition of a new data type, which is suited to handle both temporal and spatial information. Generalized constraints are used to describe spatio-temporal data, to enforce integrity rules on databases, to specify the semantics of a database scheme and to impose selection criteria in flexible database querying.

Keywords: Spatio-temporal information modelling, object-oriented database model, (generalized) constraints.

1 Introduction

A constraint can formally be seen as a relationship, which has to be satisfied. With respect to database systems, constraints are considered to be an important and adequate means to define the semantics and the integrity of the data [1, 2, 3, 4, 5]. This is especially true for spatial data and for temporal data. A (spatio-temporal) database instance then belongs to the database in as far as it satisfies all of its defining constraints.

In practice, spatial data usually consist of line segments, and therefore linear arithmetic constraints are particularly appropriate for representing such data [5]. For example, if spatial geographical information is handled, constraints can be used to define the borders of a country, a city, a region, to define a river, a highway, etc. This is illustrated in Figure 1 and in Table 1, which respectively represent a map of France (a real map will be defined by many more constraints, but the basic ideas are the same) and some geometrical descriptions.

Constraints can also be used to impose selection criteria for information retrieval. In this case, each constraint defines an extra condition for the database instances to belong to the result of the retrieval [6, 7]. Every instance belongs to the result in as far as it satisfies all the imposed criteria. For example, if someone wants to retrieve all the young persons who live in Annecy, two constraints can be imposed: one that selects all the young persons and another that selects all persons living in Annecy.

Spatio-temporal information can be fuzzy and/or uncertain [8, 9]. There has been a

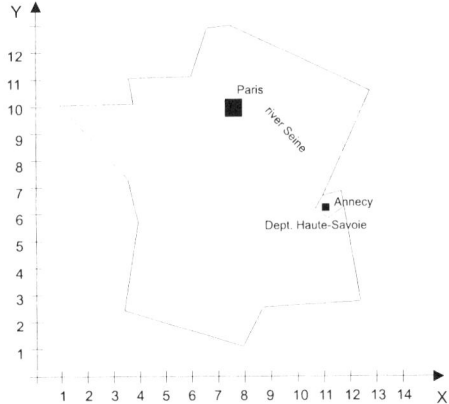

Figure 1: Spatial information: map of France.

Table 1: Geometrical descriptions.

Annecy: $(x \geq 10.9) \wedge (x \leq 11.1) \wedge (y \geq 6) \wedge (y \leq 6.2)$
Seine: $((y \leq 11) \wedge (y + 0.7x = 15.2) \wedge (y \geq 9.6))$ $\vee ((y \leq 9.7) \wedge (y - 0.2x = 8) \wedge (y \geq 9.6))$ $\vee ((y \leq 8) \wedge (y + 1.2x = -20) \wedge (y \geq 9.7))$
Haute-Savoie: $(y - 0.1x \leq 8) \wedge (y + 7x \leq 89.6) \wedge (y - 0.4x \leq 1.5)$ $\wedge (y + 0.2x \geq 8.2) \wedge (y - 4x \geq 36.7)$

considerable amount of research regarding fuzziness in spatio-temporal databases [8, 10, 11, 12]. In this paper an extension of a constraint-based fuzzy object-oriented database model [3] is presented. This extension is based on the introduction of a new data type and on the generalization of linear arithmetic constraints.

In the following section, the main concepts of the fuzzy object-oriented database model are introduced. The modelling of fuzzy spatio-temporal information, by means of generalized linear arithmetic constraints, is discussed in Section 3. Finally, the achieved results and future developments are summarized in the concluding section.

2 Generalized object-oriented database model

The employed fuzzy object-oriented database model [3] has been obtained as a generalization of a crisp object-oriented database model that is consistent with the ODMG de facto standard [13]. The model is build upon a generalized algebraic type system and a generalized constraint system, which are both used for the definition of so-called generalized object schemes and generalized database schemes.

2.1 Type system

To support the definition of types, a (generalized) type system GTS has been built [3]. In order to be consistent with the ODMG data model, the type system supports the generalized definitions of literal types, object types and reference types (which enable to refer to the instances of object types and are used to formalize the binary relationships between the object types in a database scheme).

The semantic definition of a (generalized) type $\tilde{t}$ is based on domains and operators (cf. [14]) and is fully determined by:

- a set of domains $D_{\tilde{t}}$
- a designated domain $dom_{\tilde{t}} \in D_{\tilde{t}}$
- a set of operators $O_{\tilde{t}}$ and
- a set of axioms $A_{\tilde{t}}$

The designated domain $dom_{\tilde{t}}$ is called the domain of the type $\tilde{t}$ and consists of the set of all the possible values for $\tilde{t}$. Every domain value is represented by a fuzzy set, which is defined over the domain of the corresponding ordinary type t. In order to deal with "undefined" values and inapplicability of domain values, a type specific bottom value $\bot_t$, has been added to the domain of every ordinary type t [15]. The set of operators $O_{\tilde{t}}$ contains all the operators that are defined on the domain $dom_{\tilde{t}}$. The set of domains $D_{\tilde{t}}$ consists of all the domains that are involved in the definition of the operators of $O_{\tilde{t}}$, whereas the set of axioms $A_{\tilde{t}}$ contains all the axioms that are involved in the definition of the semantics of $O_{\tilde{t}}$.

The instances of a literal type, an object type and a reference type are respectively called literals, objects and reference instances. Every instance is characterized by its type and a domain value of this type (also called the state of the instance). In accordance with the ODMG de facto standard [13], only objects can have a persistent lifetime. Persistent objects are additionally characterized by a unique object identifier and an optional set of unique object names.

2.2 Constraint system

Constraints are used to enforce integrity rules on databases (e.g. domain rules, referential integrity rules, etc.) and to specify the formal semantics of the database scheme (e.g. the applicability of null values, the definition of keys, etc.). To support the definition of constraints, a (generalized) constraint system GCS has been built. The set of generalized constraint definitions supported by the constraint system can be partitioned into a subset of constraints which can be applied to objects independent of any existing database (e.g. domain constraints) and a subset of "database" constraints which are defined for database objects (e.g. referential integrity constraints) [3, 4].

The semantics of a constraint $\tilde{c}$ are defined by means of a function $\rho_{\tilde{c}}$, which associates with every object $\tilde{o}$ a fuzzy set

$$\{(True, \mu_{True}), (False, \mu_{False}), (\perp_{Boolean}, \mu_{\perp_{Boolean}})\}$$

which represents the extended possibilistic truth value [16] of the proposition

"object $\tilde{o}$ satisfies constraint $\tilde{c}$"

The membership grades μ_{True} and μ_{False} indicate to which degree this proposition is respectively true and false. The membership grade $\mu_{\perp_{Boolean}}$ denotes to which degree the proposition is not applicable, and is used to model those cases where the constraint $\tilde{c}$ is (partially) not applicable to $\tilde{o}$.

2.3 Object schemes

The full semantics of an object are described by an object scheme $\tilde{os}$. This scheme "in fine" completely defines the object, now including the definitions of the constraints that apply to it. Each object scheme is defined by an identifier id, an object type $\tilde{t}$, a "meaning" $\tilde{M}$ and a conjunctive fuzzy set of constraints $\tilde{C}_{\tilde{t}}$, which all have to be applied onto the objects of type $\tilde{t}$ independently of any existing database

$$\tilde{os} = [id, \tilde{t}, \tilde{M}, \tilde{C}_{\tilde{t}}]$$

The "meaning" $\tilde{M}$ is provided to add comments and is usually described in a natural language. The membership grade of a constraint $\tilde{c}$ of $\tilde{C}_{\tilde{t}}$ indicates to which degree $\tilde{c}$ applies to the object type $\tilde{t}$ and represents the relative importance of $\tilde{c}$ within the definition of $\tilde{os}$. A membership grade $\mu_{\tilde{C}_{\tilde{t}}}(\tilde{c}) = 0$ denotes 'not important at all', whereas $\mu_{\tilde{C}_{\tilde{t}}}(\tilde{c}) = 1$ denotes 'fully important'. In order to have an appropriate scaling it is assumed that $\max_{\tilde{c}} \mu_{\tilde{C}_{\tilde{t}}}(\tilde{c}) = 1$.

An instance $\tilde{o}$ of the object type $\tilde{t}$ is defined to be an instance of the object scheme $\tilde{os}$, if it satisfies (with a truth value which differs from $\{(False, 1)\}$) all the constraints of $\tilde{C}_{\tilde{t}}$ and all the constraints of the sets $\tilde{C}_{\widehat{\tilde{t}}}$ of the object schemes, which have been defined for the supertypes $\widehat{\tilde{t}}$ of $\tilde{t}$.

2.4 Database schemes

A database scheme $\tilde{ds}$ describes the full semantics of the objects which are stored in a generalized database and is defined by the quadruple

$$\tilde{ds} = [id, \tilde{D}, \tilde{M}, \tilde{C}_{\tilde{D}}]$$

in which id is the identifier of the database scheme,

$$\tilde{D} = \{\tilde{os}_i | 1 \leq i \leq n, i, n \in \mathbb{N}_0\}$$

is a finite set of object schemes, $\tilde{M}$ is provided to add comments, and $\tilde{C}_{\tilde{D}}$ is a conjunctive fuzzy set of "database" constraints, which imposes extra conditions on the instances of the object schemes of $\tilde{D}$ (e.g. referential constraints between two object schemes). Again, the membership grades denote the relevance of the constraints. Every generalized object scheme in $\tilde{D}$ has a different object type. If an object scheme $\tilde{os} \in \tilde{D}$ is defined for an object type $\tilde{t}$ and $\tilde{t}'$ is a supertype of $\tilde{t}$, or $\tilde{t}'$ is an object type for which a binary relationship with $\tilde{t}$ is defined, then an object scheme $\tilde{os}' \in \tilde{D}$ has to be defined for $\tilde{t}'$.

Every persistent instance $\tilde{o}$ of an object scheme $\tilde{os} \in \tilde{D}$ of a database scheme $\tilde{ds}$ has to satisfy all the constraints of $\tilde{C}_{\tilde{D}}$, with a truth value which differs from $\{(False, 1)\}$.

2.5 Database Model

The generalized database model is finally obtained by extending the formalism with data definition (DDL) and data manipulation operators (DML) [3] (see Figure 2).

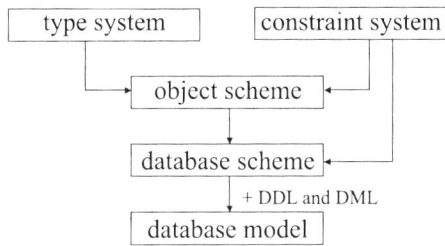

Figure 2: Generalized object-oriented database model: an overview.

3 Modelling of spatio-temporal information

The generalized object-oriented database model presented in the previous section is extended in order to support the modelling of both temporal and spatial information. This is done by adding a new generic literal type *SpaceTime* to the type system. The domain of this new type consists of all the fuzzy sets which are defined over the points of a given geometrical space, which on its turn is defined by a finite number of axes, which all have only one point in common. Each axis either represents a time dimension or a spatial dimension. Generalized linear arithmetic constraints are defined and are used to describe the domain values of the *SpaceTime* type.

In the next subsection the focus is on the modelling of one-dimensional temporal information. In this special case, the literal type *SpaceTime* has to describe a temporal space, which is defined by one time axis. The modelling of spatial information is discussed in Subsection 3.2. The cases of one-dimensional, two-dimensional and n-dimensional spaces are handled. The formal definition of the *SpaceTime* type is given in Subsection 3.3.

3.1 Modelling of temporal information

In order to model temporal information, a new data type *TimeDim* is defined. This data type will not be included directly in the database model, but is necessary for the definition of the type *SpaceTime*.

The domain of *TimeDim* is defined by

$$dom_{TimeDim} = \mathbb{R} \cup \{\perp_{TimeDim}\}$$

where $\mathbb{R}$ denotes the set of real numbers and $\perp_{TimeDim}$ represents an "undefined" domain value.

The considered operators are the binary operators $=, \neq, <, >, \leq, \geq, +, -, *$ and $/$ and a null-ary operator $\perp$, which always results in an "undefined" domain value. When restricted to the set $dom_{TimeDim} \setminus \{\perp_{TimeDim}\}$, all binary operators have the same semantics as their counterparts within $\mathbb{R}^2$. For the bottom value $\perp_{TimeDim}$, the semantics are:

$$\forall\, x \in dom_{TimeDim} : op\,(x, \perp_{TimeDim}) = op\,(\perp_{TimeDim}, x) = \perp_{TimeDim}$$

where "op" is a variable copula whose successive values are respectively the symbols $=, \neq, <, >, \leq, \geq, +, -, *$ and $/$.

The type *TimeDim* can be employed to model time, hereby using the set $\mathbb{R}$ of real numbers as a representation of the continuum of physical time points [8]. However, in prospect of the generalization discussed in Subsections 3.2 and 3.3, the type *SpaceTime* is introduced.

The type *SpaceTime* is structured and consists of a finite number of components. Each component either represents a temporal dimension or a spatial dimension. In this subsection only one (temporal) component is considered, so that the specification of *SpaceTime* is defined as:

$$SpaceTime\ id(id_1 : TimeDim)$$

where id is the identifier of the type and id_1 is the identifier of the component with associated type *TimeDim*.

The domain of type

$$SpaceTime\ id(id_1 : TimeDim)$$

(shortly written as dom_{id}) is defined by

$$dom_{id} = \tilde{\wp}(\{(x)|x \in dom_{TimeDim}\}) \cup \{\perp_{SpaceTime}\}$$

where $\tilde{\wp}(U)$ denotes the set of all fuzzy sets, which can be defined over the universe U and $\perp_{SpaceTime}$ represents an "undefined" domain value.

With the previous definition, every "regular" value of dom_{id} is a fuzzy set, which is defined over the continuum of physical time points. In order to describe the values of dom_{id}, linear arithmetic constraints are generalized. This is done by generalizing the comparison operators $=, \leq$ and $\geq$.

Traditionally, these operators allow to describe crisp subsets of the continuum of physical time points, e.g. $\forall\, t \in dom_{id}$, $x = t$ describes the fuzzy set $\{(t, 1)\}$ which represents the single time point t, $x \leq t$ describes the fuzzy set $\{(x, 1)|x \leq t\}$ which represents the time interval $]-\infty, t]$ and $x \geq t$ describes the fuzzy set $\{(x, 1)|x \geq t\}$ which represents the time interval $[t, +\infty[$.

For the generalization, a normalized fuzzy set $\tilde{V}$ has been associated with each operator. This fuzzy set is defined over the universe of valid distances —the set $\mathbb{R}^*$ of positive real numbers— and the boundary condition $\mu_{\tilde{V}}(0) = 1$ must hold for it.

If $d(x, x')$ denotes the Euclidean distance between the defined elements x and x' of $dom_{TimeDim}$, i.e. $d(x, x') = |x - x'|$, then the membership functions of the fuzzy sets described by the generalized operators $=_{\tilde{V}}, \leq_{\tilde{V}}$ and $\geq_{\tilde{V}}$ are defined as follows:

$\forall\, x, t \in dom_{TimeDim} \setminus \{\perp_{TimeDim}\}:$

- $\mu_{x=_{\tilde{V}}t}((x)) = \mu_{\tilde{V}}(d')$, with $d' = \min\{d(x,x')|x' \in dom_{TimeDim} \wedge x' = t\}$
- $\mu_{x\leq_{\tilde{V}}t}((x)) = \mu_{\tilde{V}}(d')$, with $d' = \min\{d(x,x')|x' \in dom_{TimeDim} \wedge x' \leq t\}$
- $\mu_{x\geq_{\tilde{V}}t}((x)) = \mu_{\tilde{V}}(d')$, with $d' = \min\{d(x,x')|x' \in dom_{TimeDim} \wedge x' \geq t\}$

Figure 3 illustrates the membership functions that result from the application of the generalized comparison operators $\leq_{\tilde{V}}$, $\geq_{\tilde{V}}$ and $=_{\tilde{V}}$ to a given fuzzy set $\tilde{V}$.

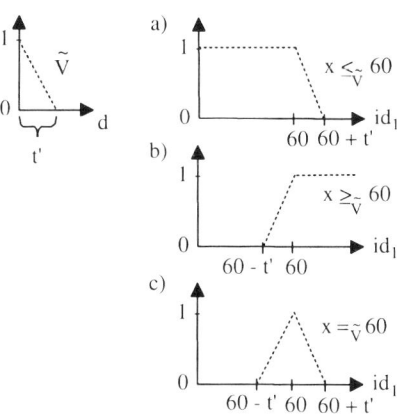

Figure 3: Application of the generalized comparison operators.

Linear arithmetic constraints have been generalized by replacing all regular comparison operators by (adequate) generalized comparison operators and by replacing the regular logical operators $\wedge$, $\vee$ and $\neg$ by their fuzzy counterparts $\tilde{\wedge}$, $\tilde{\vee}$ and $\tilde{\neg}$, which semantics have been defined as follows:

- the impact of the $\tilde{\wedge}$ operator is reflected by applying Zadeh's (standard) intersection operator [17] onto the fuzzy sets described by the arguments of the operator, i.e. with arguments $\tilde{U}$ and $\tilde{V}$, the membership degree of (x), $x \in dom_{TimeDim}$ in the resulting fuzzy set equals

$$\min(\mu_{\tilde{U}}(x), \mu_{\tilde{V}}(x))$$

- the impact of the $\tilde{\vee}$ operator is reflected by applying Zadeh's (standard) union operator [17], i.e. with arguments $\tilde{U}$ and $\tilde{V}$, the membership degree of (x), $x \in dom_{TimeDim}$ in the resulting fuzzy set equals

$$\max(\mu_{\tilde{U}}(x), \mu_{\tilde{V}}(x))$$

- the impact of the $\tilde{\neg}$ operator is reflected by applying Zadeh's (standard) complement operator [17], i.e. with argument $\tilde{U}$, the membership degree of (x), $x \in dom_{TimeDim}$ in the resulting fuzzy set equals

$$1 - \mu_{\tilde{U}}(x)$$

For example, with appropriate fuzzy sets $\tilde{U}$, $\tilde{V}$ and $\tilde{W}$, the fuzzy temporal information "*around* time point 60 or from time point *about* 100 until time point *about* 120" can be described as:

$$(x =_{\tilde{U}} 60) \tilde{\vee} ((x \geq_{\tilde{V}} 100) \tilde{\wedge} (x \leq_{\tilde{W}} 120))$$

3.2 Modelling of spatial information

The data type *SpaceTime* can be adapted to model spatial information. A distinction is made between one-dimensional, two-dimensional and n-dimensional data.

In order to model spatial information a new data type *SpaceDim* is defined. The definition of this type is similar to the definition of the type *TimeDim*, introduced in the previous section: the domain of *SpaceDim* is defined by

$$dom_{SpaceDim} = \mathbb{R} \cup \{\bot_{SpaceDim}\}$$

where $\bot_{SpaceDim}$ represents an "undefined" domain value; furthermore, the same operators $=, \neq, <, >, \leq, \geq, +, -, *, /$ and $\bot$ have been defined. Hereby, the set $\mathbb{R}$ of real numbers is used as a representation of a dimension in a spatial space.

3.2.1 One-dimensional spatial data

The type *SpaceTime* is also suited for the modelling of spatial data. One-dimensional spatial data can be handled by considering one (spatial) component. The specification of *SpaceTime* then becomes:

$$\textit{SpaceTime } id(id_1 : \textit{SpaceDim})$$

where id remains the identifier of the type and id_1 is the identifier of the component with associated type *SpaceDim*. In the one-dimensional case, the modelling of spatial information is then completely analogous to the temporal case discussed in the previous subsection.

3.2.2 Two-dimensional spatial data

Two-dimensional spatial data can be modelled by considering two (spatial) components for the type *SpaceTime*, i.e. by considering the specification:

$$\textit{SpaceTime } id(id_1 : \textit{SpaceDim}, id_2 : \textit{SpaceDim})$$

In this case, the domain dom_{id} is defined by

$$dom_{id} = \tilde{\wp}(\{(x,y) | x, y \in dom_{SpaceDim}\}) \cup \{\bot_{SpaceTime}\}$$

With this definition, each "regular" value of dom_{id} is a fuzzy set, which is defined over the continuum of points in the plane defined by the two spatial axes with identifiers id_1 and id_2.

The generalization of the comparison operators $=, \leq$ and $\geq$ is obtained analogously as in the one-dimensional case. A normalized fuzzy set $\tilde{V}$, which is defined over the universe of valid distances and for which the boundary condition $\mu_{\tilde{V}}(0) = 1$ holds, is associated with each operator.

If $d((x,y),(x',y'))$ denotes the Euclidean distance between the defined elements (x,y) and (x',y') of $dom_{SpaceDim} \times dom_{SpaceDim}$, i.e.

$$d((x,y),(x',y')) = \sqrt{(x-x')^2 + (y-y')^2}$$

then the membership functions of the fuzzy sets described by the generalized operators $=_{\tilde{V}}$, $\leq_{\tilde{V}}$ and $\geq_{\tilde{V}}$ are defined as follows:

$$\forall (x,y) \in (dom_{SpaceDim} \setminus \{\bot_{SpaceDim}\})^2, \forall m, l \in \mathbb{R}:$$

- $\mu_{x+my=_{\tilde{V}}l}((x,y)) = \mu_{\tilde{V}}(d')$, with

$$d' = \min\{d((x,y),(x',y'))|(x',y') \in (dom_{SpaceDim})^2 \wedge x' + my' = l\}$$

- $\mu_{x+my\leq_{\tilde{V}}l}((x,y)) = \mu_{\tilde{V}}(d')$, with

$$d' = \min\{d((x,y),(x',y'))|(x',y') \in (dom_{SpaceDim})^2 \wedge x' + my' \leq l\}$$

- $\mu_{x+my\geq_{\tilde{V}}l}((x,y)) = \mu_{\tilde{V}}(d')$, with

$$d' = \min\{d((x,y),(x',y'))|(x',y') \in (dom_{SpaceDim})^2 \wedge x' + my' \geq l\}$$

For example, with the fuzzy set $\tilde{V}$ of Figure 3 and the fuzzy set $\tilde{W} = \{(0,1)\}$, "the *environment* of Annecy" can be modelled by $(x =_{\tilde{V}} 11)$ $\tilde{\vee}$ $(y =_{\tilde{V}} 6.1)$ and "the *neighborhood* of the Lake of Geneva *in* Haute-Savoie" can be modelled by $(y \geq_{\tilde{V}} 6.7)$ $\tilde{\wedge}$ $(y - x \geq_{\tilde{V}} -3.7)$ $\tilde{\wedge}$ $(y \leq_{\tilde{W}} 6.7)$ $\tilde{\wedge}$ $(y - x \geq_{\tilde{W}} -3.7)$ $\tilde{\wedge}$ $(x \leq_{\tilde{W}} 11.9)$ $\tilde{\wedge}$ $(x \geq_{\tilde{V}} 11)$. Both examples are illustrated in Figure 4 (drawn to scale).

Figure 4: Illustration of the modelling of two-dimensional spatial data.

3.2.3 n-dimensional spatial data

In order to model n-dimensional spatial data, the type *SpaceTime* can be constructed with n spatial components. In this case, the domain dom_{id} is defined by

$$dom_{id} = \tilde{\wp}(\{(x_1, x_2, \ldots, x_n) | x_1, x_2, \ldots, x_n \in dom_{SpaceDim}\}) \cup \{\perp_{SpaceTime}\}$$

The comparison operators $=$, $\leq$ and $\geq$ can be generalized straightforwardly and analogously to previous cases by considering the Euclidean distance

$$d((x_1, x_2, \ldots, x_n), (x'_1, x'_2, \ldots, x'_n)) = \sqrt{(x_1 - x'_1)^2 + (x_2 - x'_2)^2 + \cdots + (x_n - x'_n)^2}$$

and an associated, normalized fuzzy set $\tilde{V}$, which is defined over the universe of valid distances and for which the boundary condition $\mu_{\tilde{V}}(0) = 1$ holds.

3.3 Literal type *SpaceTime*

In general the literal type *SpaceTime* can have both spatial and temporal components. This allows to model spatio-temporal information in its most general form. The type specification then becomes:

$$\text{SpaceTime } id(id_1 : t_1, id_2 : t_2, \ldots, id_n : t_n)$$

where id remains the identifier of the type. Each component $id_i : t_i$, $i = 1, 2, \ldots, n$ represents an axis, that is identified by the identifier id_i and whose nature is denoted by the associated type $t_i \in \{TimeDim, SpaceDim\}$.

The domain of the type *SpaceTime* is defined by

$$dom_{id} = \tilde{\wp}(\{(x_1, x_2, \ldots, x_n) | x_i \in dom_{t_i}, i = 1, 2, \ldots, n\}) \cup \{\perp_{SpaceTime}\}$$

where $\perp_{SpaceTime}$ represents an "undefined" domain value.

Because by definition all the elements of the domain are fuzzy sets, operators have been provided for the handling of fuzzy sets. Among the considered operators are: $\cup$, $\cap$, co, $normalise$, $support$, $core$, $\alpha - cut$ and $\bar{\alpha} - cut$. Each operator preserves its usual semantics. Additionally, a null-ary operator $\perp$, which always results in an "undefined" domain value, is added.

4 Conclusion

A new approach for the handling of spatio-temporal information is presented. The rationale behind this approach is the assumption that linear arithmetic constraints are particularly appropriate for representing such kind of information.

The approach is presented as an extension of a constraint-based fuzzy object-oriented database model, but its application is definitely not restricted to database models. Central to the approach is the introduction of a new generic type *SpaceTime*, which is suited to handle fuzzy multi-dimensional temporal and/or spatial information. The description of domain values of *SpaceTime* by means of so-called generalized linear arithmetic constraints, which have been obtained by generalizing the definition of the comparison operators $=$, $\leq$ and $\geq$, is typical. Future work includes the definition of appropriate data definition and data manipulation operators and the study of appropriate flexible querying techniques.

References

[1] L.A. Zadeh (1996). Fuzzy Logic = Computing with Words. IEEE Transactions on Fuzzy Systems, vol. 2, pp. 103–111.

[2] L.A. Zadeh (1997). Toward a theory of fuzzy information granulation and its centrality in human reasoning and fuzzy logic. Fuzzy Sets and Systems, vol. 90 (2), pp. 111–127.

[3] G. De Tré, R. De Caluwe and B. Van der Cruyssen (2000). A Generalised Object-Oriented Database Model. In: Recent Issues on Fuzzy Databases. G. Bordogna and G. Pasi (eds.). Studies in Fuzziness and Soft Computing, Physica-Verlag, Heidelberg, Germany, pp. 155–182.

[4] G. De Tré and R. De Caluwe (2000). The Application of Generalized Constraints to Object-Oriented Database Models. Mathware and Soft Computing, vol. VII (2–3), pp. 245–255.

[5] G.M. Kuper, L. Libkin and J. Paredaens (eds.) (2000). Constraint Databases. Springer-Verlag, Berlin, Germany.

[6] P.C. Kanellakis, G.M. Kuper and P.Z. Revesz (1995). Constraint Query Languages. Journal of Computer and System Sciences, vol. 51, pp. 26–52.

[7] H. Prade (2001). From flexible queries to case-based evaluation. In: Proc. of ECSQARU-2001 workshop on Management of uncertainty and imprecision in multimedia information systems, Toulouse, France.

[8] R. De Caluwe, G. De Tré, B. Van der Cruyssen, F. Devos and P. Maesfranckx (2000). Time management in Fuzzy and Uncertain Object-Oriented Databases. In: Knowledge Management in Fuzzy Databases. O. Pons, M.A. Vila and J. Kacprzyk (eds.). Studies in Fuzziness and Soft Computing, Physica-Verlag, Heidelberg, Germany, pp. 67–88.

[9] A. Morris (2001). Why Spatial Databases Need Fuzziness. In: Proc. of joint 9th IFSA and 20th NAFIPS International Conference, Vancouver, Canada, pp. 2446–2451.

[10] R. George, F.E. Petry, B.P. Buckles and S. Radharkrishnan (1996). Fuzzy Database Systems – Challenges and Opportunities of a New Era. International Journal of Intelligent Systems, Vol. 11, John Wiley and Sons.

[11] E.L. Usery (1996). A Conceptual Framework an Fuzzy Set Implementation for Geographic Features. In: Geographic Objects with Indeterminate Boundaries. P. Burrough and A. Frank (eds.).GISDATA Series, vol. 2, Taylor and Francis, London, UK, pp. 71–86.

[12] A. Morris, F.E. Petry and M. Cobb (1998). Incorporating Spatial Data into the Fuzzy Object Oriented Data Model. In: Proc. of IPMU98, Paris, France, pp. 604–611.

[13] R.G.G. Cattell et al. (2000). The Object Data Standard: ODMG 3.0. Morgan Kaufmann Publishers Inc., San Francisco, USA.

[14] J. Early (1971). Toward an understanding of data structures. Communications of the ACM, vol. 14 (10), pp. 617–627.

[15] H. Prade and C. Testemale (1984). Generalizing Database Relational Algebra for the Treatment of Incomplete or Uncertain Information and Vague Queries. Information Sciences, vol. 34, pp. 115–143.

[16] G. De Tré and R. De Caluwe (2001). Application of Extended Possibilistic Truth Values in Multimedia Database Systems. In: Proc. of ECSQARU-2001 workshop on Management of uncertainty and imprecision in multimedia information systems, Toulouse, France.

[17] L.A. Zadeh (1965). Fuzzy sets. Information and Control, vol. 8 (3), pp. 338–353.

A general framework for meta-search based on query-weighting and numerical aggregation operators

Mario Gómez, Chema Abásolo

Institut d'Investigació en Intel.ligència Artificial (IIIA-CSIC)
{abasolo,mario}@iiia.csic.es

Abstract

Meta-search engines arise to increase the coverage of single search-engines. Meta-search engines offer a unified way to access multiple search engines, avoiding the user to interact with all the underlying engines; but usually they do not merge results from different sources, neither they calculate a unique ranking for each item retrieved. We present a framework to develop meta-search engines allowing to rank documents retrieved from search-engines that originally do not give any rank; and a way to combine and integrate the results from different engines into a unique collection of documents ranked according to some relevance or utility criteria. Our approach is based on exploiting the filtering capabilities of search-engines and the generalized use of weights and aggregation operators to rank documents.

Keywords: meta-search, query processing, aggregation operators, information integration.

1 Introduction

The vast amount of information available in the Web causes some serious problems to the users when looking for information: it is difficult to find all the relevant information; in addition, the information retrieved is rarely ranked or classified according to the user's utility criteria. Meta-search is one of the most promising approaches to solve the first problem. If the single search-engines store only a portion of all the existing information about some particular domain, various searchers should be queried to increase the search coverage. But in practice, meta-search is under-exploited; existing meta-search engines do not combine results from the different single engines, neither they rank the information retrieved, or the ranking mechanisms are quite poor.

Our approach overcomes these two limitations. We propose a framework to develop meta-search engines that can rank documents retrieved from engines that originally do not perform rankings; and a procedure to combine documents retrieved from different search engines. The originality of our vision is that we exploit the filtering capabilities of the search-engines, thus ranking is achieved with a little computation and storage requirements. We argue that our approach is so flexible that it can be applied in a wide range of search domains, including *semi-structured* and *structured data*. We will demonstrate the feasibility and utility of our ideas by showing some experimental results of an application to search bibliographic references in medicine. The case study uses knowledge about Evidence-Based Medicine (EBM) as the utility criteria to rank the bibliographic references.

2 Meta-search architecture

2.1 Overview of the model

We oriented this work towards a particular kind of meta-search where the queries are described as vectors of elements. The meta-search process is modelled as involving two main processes or phases: expansion and aggregation. Expansion means that a query is transformed into a collection of new queries by applying some transformation to the original query. Expansion is used to enrich or refine queries with different purposes and different meanings. We have considered different kinds of query expansion:

a) At the domain level: using a thesaurus to elaborate semantically equivalent queries, generalise or specialise a query, or enrich a query with specific utility criteria.

b) At the source level: customise a query for multiple search engines, using with different search-modes and filters to get more rich information.

The outcome of query expansion is a new set of queries weighted according to some criteria, like specificity of the search modes in b) or strength of the semantic relationship in a). Queries generated during the expansion phase allow scoring documents although a search engine does not give any ranking by itself. Ranking can be done because the queries are weighted, thus their answers can be considered as inheriting the weight of the query as an assessment of relevance or utility. The answers to all the queries and search engines are combined to obtain a unique set of retrieved items, where repeated items are eliminated, keeping only one instance of each item, with a unique, overall ranking. Ranking synthesis is achieved by applying some aggregation operator, as explained in section 4. Expansion can occur at different levels; therefore aggregation can also be applied at different levels. We will show the weighted query expansion and aggregation procedures including some examples from MELISA [1], a medical literature search agent.

2.2 Assigning weights during query expansion

Query expansion refers to the process of transforming a query in a collection of new queries. The new queries enrich or transform the original query with additional domain information, including semantic and syntactic information. We can also distinguish between domain-based elaboration and source-based elaboration, that is to say, elaboration of queries using domain knowledge and elaboration based on particular characteristics of the information sources.

The key idea is that if queries are weighted, then we can aggregate the results from these queries to obtain a unique ranking just using the weights of the queries as the values to be aggregated. The idea is simple but powerful. Suppose we have a collection of queries with different weights expressing the relative importance of each query with respect to the original query. Then, retrieved –non ranked- items can be assigned a rank equal to the weight of the query they are an answer for. For items already scored by the queried search-engines, a new rank can be calculated by aggregating the given score and the weight of the query. With semantic expansion we can generate queries where some keywords are replaced by synonyms, hyponyms[1] and hyperonyms. This kind of expansion is very useful to increase or decrease the coverage of search; synonyms and hyponyms will increase the recall, while hyperonyms are useful to restrict the search. Resulting queries are weighted according to the strength of the association –correlation- between words. Correlation between words can be obtained by statistical calculations, such as latent semantics analysis [21], or given by an expert.

---

[1] A term α is an hyponym of the term β if the meaning of α subsumes the meaning of β. Then β is an hyperonyms of α.

Example 1: Some possible term-correlations for the term "Guidelines" in the medical domain[2] are given below.

   Practice-guidelines, correlation = 0,8
   Guideline-Adherence, correlation = 0,6
   Clinical-Protocols, correlation = 0,4

Given the query {Levofloxacin, Pneumonia, Guidelines}[3]; three new queries are generated by replacing the term Guidelines by each of their correlated terms, assigning the query with a weight equals to the correlation coefficient. The original query is also included in the result of the expansion and is assigned with a unitary weight.

   Q1 = {levofloxacin, pneumonia, guidelines}; weight=1
   Q2 = {levofloxacin, pneumonia, practice-guidelines}; w=0,8
   Q3 = {levofloxacin, pneumonia, guideline-adherence}; w=0,6
   Q4 = { levofloxacin, pneumonia, clinical-protocols}; w=0,6

The previous examples of query expansion belong to the class of domain-based expansion. In the other side, source-based expansion of queries allows to solve two problems: ranking items according to the strength of the matching, and adapting a query for different, heterogeneous search-engines. We consider a particular class of information, the filters and the search modifiers allowed by the search-engines. We call this procedure *query customisation*, as the queries are customised for each particular search-engine. If the generated queries are weighted taking into account the search modifiers applied by the expansion algorithm, then it is possible to rank items according to the search mode or the type of filter applied, it is not the same to search a keyword by title than by subject or abstract.

Source expansion is very useful to perform more accurate rankings over the retrieved documents, but also generalisation or specialisation can be achieved by using different search modifiers. In addition, weights can depend on the particular search engines, thus allowing to have into account preferences about particular search engines.

Example 2: To better understand the concept of search-modes, let us explain in detail the search modes allowed by the PubMed search-engine:

1. MAJR: search for keywords appearing as major topics (the most important terms describing the subject of a document, belonging to the MeSH thesaurus)
2. MHNOEXP: search terms appearing as MeSH terms (includes major topics) without term explosion[4]
3. MH: search terms appearing as a MeSH term, with term explosion. This mode includes (subsumes) the two previous search modes.
4. TI: search terms appearing in the title
5. TW: search terms appearing in the abstract
6. ALL: include all the former search modes, so it is the less restrictive search mode; in fact, this is the default search mode.

We assign weights to the different search modifiers according to their specificity or relative importance.

---

[2] Vocabulary from the Medical Subject Heading (MeSH), a medical thesaurus from the National Library of Medicine.

[3] We are considering conjunctive keyword-based queries. The meaning of this query is "to find references about guidelines on the use of Levofloxacin in the treatment of pneumonia."

[4] Term explosion refers to the fact that when looking for one MeSH term, it also will retrieve documents matching with hyponyms of that term.

<MAJR, 1>; <MHNOEXP, 0.8>; <MH, 0.6>; <TI, 0.5>; <TW, 0.4>; <ALL, 0.2>

Given a query, we can elaborate new queries using the different search modifiers. If a query has more than one keyword, then different combinations of search-modifiers for the different keywords can be used. For example, we can generate new queries modifying only one keyword per query.

More than one transformation can be applied to the same query, therefore it is necessary to combine the different weights to obtain a unique weight. This is similar to the necessity of a mechanism to propagate weights, that is to say, to obtain a new weight for a query that is already weighted. The way to propagate and combine weights is addressed in sections 2.3 and 3.

Given the query {AIDS, Diagnosis}, a possible expansion algorithm can generate these queries:

   Q1 = {AIDS [ALL], Diagnosis}; weight = 0,2
   Q2= {AIDS [TW], Diagnosis}; w = 0.4
   Q3= {AIDS [TI], Diagnosis}; w = 0.5
   Q2= {AIDS [MH], Diagnosis}; w = 0.6
   Q2= {AIDS [MHNOEXP], Diagnosis}; w = 0.8
   Q6 = {AIDS [MAJR], Diagnosis}; w = 1

Therefore, when aggregation is carried on, the items retrieved for all this queries are combined and the different weights are aggregated to obtain the overall score for each item.

2.3 Propagating and combining weights

We have not discussed yet how to assign weights when two or more transformations are applied to the same query. A similar problem arises when applying a new transformation to a previously elaborated –and weighted- query. Both problems are in fact the same, how to combine or synthesise weights.

Different functions can be used to combine weights. We can consider weights as membership values of queries with respect to the user request, or logical values expressing relevance or utility. Thus, weights can be combined by using numerical aggregation operators or multivalued logical operators.

We will propose a very general rule that will facilitate the specification of the properties these functions must fulfil. This simple rule is that the *weight of one query cannot be increased after applying a transformation*; the meaning is that query transformations move queries further away from the user request. In other words, if we transform a query q with a weight w into a new query q' with a weight w', w' cannot be greater than w. Such class of operators includes –but is not reduced to- the family of *t-norm* operators. Next follows a formal description of our proposal to weight queries during query expansion.

3 Formal description of query-weighting

Definition 1: A q*uery* Q is a vector of non-repeated *query-elements* that can be either keywords or other elements (i.e. search filters).
$Q = \langle k_1..k_n \rangle \qquad \forall i,j : 1 \leq i,j \leq n; k_i \neq k_j$

Definition 2: A *weighted-query* is a pair with a query and a weight in the interval [0, 1]:
$WQ = \langle Q, w \rangle \qquad w \in [0,1]$

Definition 3: A *query-transformation* τ is a relation between two queries and a weight. It is defined as follows:
$\tau(Q_1, Q_2, w) \Leftrightarrow (\exists! k \mid k \in Q_1 \wedge k \notin Q_2) \wedge (\exists! k' \mid k' \in Q_2 \wedge k' \notin Q_1) \wedge \sigma(k, k', w)$

Where k and k' are query elements, and σ is a relation between two elements and a weight in the interval [0,1].

Definition 4: A *chain of query-transformations* T is a relation between two queries and a weight defined recursively as a sequence of multiple *query-transformations*.

$$T(Q_1,Q_2,w) \Leftrightarrow \begin{cases} \tau(Q_1,Q_2,w) \\ \exists Q',\tau \mid \tau(Q',Q_2,w') \wedge T(Q_1,Q',w'') \\ \wedge w = \Theta(w',w'') \end{cases}$$

Where Θ is a *t-norm* operator.

Definition 5: A *query-weighting* function Γ is a function to obtain a weight according to the *chain of query-transformations* between two queries

$$\Gamma(Q_1,Q_2) = \begin{cases} w & \text{iff} \quad T(Q_1,Q_2,w) \\ 0 & \text{otherwise} \end{cases}$$

Where Q_1, Q_2, are queries and T is a *chain of query-transformations*.

Definition 6: A *weighted-query-weighting* function Ω is a function to calculate a weight according to the *chain of query-transformations* between a weighted query and a non-weighted query.

$$\Omega(Q_1,Q_2,w) = \begin{cases} \Theta(w,w') & \text{iff} \quad T(Q_1,Q_2,w') \\ 0 & \text{otherwise} \end{cases}$$

Where Q_1, Q_2, are queries, w is the weight assigned to one of the queries, T is a *chain of query-transformations* and Θ is a *t-norm* operator.

The *query-weighting* and *weighted-query-weighting* functions are used to obtain the weight for the query resulting of applying one or more query-transformations. The former is used when the original query is not weighted, and the last when the original query is already weighted. In fact, both functions can be reduced to a unique function if we consider the non-weighted queries as having a weight equal to 1.

4 Aggregation and fusion

When we assign a weight to a query after applying a transformation, we are expressing the relative importance or representativity of that query with respect to the original one. The meaning of a weight assigned to a query is logically inherited by the documents or items retrieved for that query, thus we can say that the weights associated to the items retrieved represent the membership of those elements to the topic requested by the user. Aggregation operators for numeric values can be used here. See [34] for a review of such operators.

The most widely used operators are the *arithmetic mean* and the *weighted mean*, but there is a big family of aggregation operators, including fuzzy measures.

Example 3: Suppose we have queried PubMed with the four queries on the example 1. Suppose there is a reference that appears in queries 1 and 2, but not in 3 and 4. As PubMed do not rank documents, we assign by default a maximum score of one to both apparitions of the same reference. When doing the aggregation, absences of items are also taken into account; they are scored with zero points, the minimum.

Table 1: Example of aggregation

Item	Weight	Norm-w	Score
a_1	1,0	0,36	1
a_2	0,8	0,29	1
a_3	0,6	0,21	0
a_4	0,4	0,14	0

To aggregate these items we can apply for instance a *weighted-mean* operator:

$$WM(a_1,...,a_n) = \sum w_i a_i$$

$$w_i \in [0,1], \sum w_i = 1$$

Normalizing the weights to fulfil the formal requisites of the weighted-mean operator we obtain the aggregated ranking for that item: $WM(a_1,...,a_4) = (\Sigma w_i\, score) = 0.65$

Fusion refers to the aggregation of items retrieved from different information sources. In our framework, the same procedure used to aggregate items retrieved from a unique search-engine can be used in fusion. In this case, each source can be assigned with a weight expressing the reliability of that source or other kind of "goodness". A query customized for multiple search-engines should take a weight that is a combination of the different weights applied during the different steps of query expansion, including also the goodness of the source.

5 Applications

We have developed two applications where this framework has been evaluated: MELISA and WIM[5].

MELISA[1] is a system to look for medical literature based on the use of ontologies to separate the domain knowledge from the source descriptions.. The notions of query weighting and exploiting the filtering capabilities of search-engines are already used here, but only one search-engine is included. The system demonstrated that the weighting/aggregating framework is an accurate approach to rank documents.[6]. In MELISA the query elaboration applies two kinds of query expansion, one at the domain level and the other at the source level. Query expansion at the domain level is carried over using knowledge categories; each category is a medical topic described as a collection of weighted elements. The elements used to describe medical categories are medical terms, used as keywords, and other search attributes used to look for bibliographic references.[7] Query expansion at the source level is achieved by using the search modes explained in example 2 (section 2.2).

After testing this framework in MELISA, we have modelled the query expansion and aggregation procedures as a library of reusable knowledge components. We have adopted UPML[13] as the modelling language. This approach aims to develop reusable libraries of problem-solving methods [27] based on a clear separation of tasks, problem-solving methods and domain models, plus the use of shared ontologies to facilitate reuse [17]. Such a library of knowledge components for information search has been used to build WIM, a "spin-off" application developed in the IBROW project [5].

WIM is a configurable multi agent system to look for medical literature on Internet. The system is configured at two layers. First a broker agent finds a configuration of agent

---

[5] WIM stands for the Web Information Mediator
[6] From the point of view of the Evidence-Based Medicine.
[7] Publication type is a good example of such an attribute.

capabilities fulfilling the requirements of a problem to be solved (some kind of a search task). Second, a team manager negotiates with available problem solving agents to form a team of agents that will solve the problem using the selected capabilities and knowledge, according to the configuration obtained by the capability broker.

6 Experimental results

We have compared the results of MELISA (using PubMed as the search engine) against those obtained with the PubMed when performing a single query. An expert has proposed us 5 cases to look for medical references based on his everyday work. Every case has translated into a query suitable for both PubMed and MELISA. The results have been scored by two different evaluators, instructing them to score the references according to their degree of relevance for the corresponding query. The experts have evaluated the first 40 references retrieved for each query. An ordinal scale with three levels has been used to score references: 2 for references that satisfy the user's need, 1 if the reference is simply elated, 0 if it is not relevant, and "?" if there is not enough information.

Table 2 Assesment of relevance

Relevance	PubMed	Melisa
2	38%	48%
1	16%	17%
0	13%	9%
?	34%	27%

The utility of the proposed framework to develop and use new ranking criteria has also been tested. Some of the implemented medical categories are specifically designed to represent notions about evidence quality in medical references [1]. Using these categories to enrich queries during the query expansion has demonstrated that the resulting references are well ranked according to the quality of the evidence. The evaluators have been instructed to classify references in three groups: good, medium and poor evidence, plus a group for those references without information about the evidence.

Table 3 Assessment of evidence quality

Evidence	PubMed	Melisa
Good	7%	20%
Medium	2%	3%
Poor	2%	9%
?	56%	69%
In progress	33%	0%

The conclusion of the experimental results are that MELISA has proven their utility of our framework to rank documents according to utility criteria –like evidence quality, as MELISA finds more "good" references than PubMed alone.

7 Related work

A brief review of related areas of research is presented, following with a summary of the main contributions of our work in section 8.

Information retrieval: deals with the problem of retrieving textual documents, where both the query and the documents have a word vector representation [31]. IR is based on matching documents and queries through the use of text analysis and statistical methods Some of the IR techniques could be used in Meta-search to score documents once they are retrieved, but they are in fact two different approaches; while IR is well suited to work with local databases of large textual documents, it requires a lot of computation and space. In the other side, meta-search aims to take benefit of the existing search-engines, avoiding hard computation and space requirements, but rather limited by the speed of the network and the availability of existing search engines.

Search engines: Search engines store large collections of documents indexed by keywords and classified into directories. Search engines are power and very quick tools to find information, but they cannot afford the vast amount of evolving information in Internet. The main research on search-engines is focused on the indexing and retrieval techniques, plus research on information gathering and crawling; hence it is quite different of our approach.

Multiple databases: The term meta-search is mainly used to refer to those search engines that delegate the search task to other search-engines. But some people of the database community are also using this term to describe the activity of accessing and integrating information from multiple databases. Notions such as mediation, information integration and fusion are shared by both the meta-search and the database communities. Nowadays, it is very difficult to distinguish between classical databases and search engines, because databases can be accessed through web-based search engines[8] and vice-versa, search-engines are usually built as an interface to access a database. We think that the main distinction between both approaches is the kind of queries they handle. While our approach works with queries basically consisting of keywords, the database approach deals with more complex query languages, including rule-based languages [32], F-Logic [26], annotated logics [11] and object oriented languages [7] [15], hence the query processing techniques are quite different, see for example [10][20][4]. *Mediators* are responsible of offering a unified way to access heterogeneous information sources, while *wrappers* are responsible of translating queries and results between the mediator representation schema and the particular source schemas [16]. Other examples of ideas inspiring our work are the notions of source descriptions [23], query planning [3] and information fusion. A good example of the cross-fertilisation between heterogeneous databases and web meta-search is found in WebSrcMed [33].

Meta-search engines: There are a moderate number of implemented meta-search engines of general purpose, showing notable differences. From a list of 13 of such meta-search engines [25], we only found complete integration and ranking capabilities in two of them: Debriefing[9] and Metacrawler[10]. Other engines allow to perform fusion, but not ranking, or the ranking is limited to some ordering functions. Some engines allow to refine queries by using logical operators. Other features are the selection of the search-engines to be used for a particular search and other kind of user preferences. Often, the time response is slow and only a subset of the results are really retrieved. See [25] for a review of some meta-search engines.

Information Agents: Information agents is a new field that is becoming more and more interested in web-based services, including all the topics about information search and integration. Some examples of information agents and collaborative information agents can be found in [14][22][24][28][29]. We claim that our framework can be used inside

---

[8] PubMed is a good example, a web-based search-engine (http://www.ncbi.nlm.nih.gov/PubMed/) for the Medline database.
[9] http://www.debriefing.com
[10] http://www.metacrawler.com

information agents, or distributed among a multi agent system, like we have done in the WIM application.

Ontologies: ontologies are shared vocabularies to describe particular worlds or domains, thus it should be very useful when interoperability among heterogeneous components is needed. You can see an overview of the field in [18], some applications in [9] [12] [19]. We are using ontologies extensively, to model the knowledge used during the query expansion, and also to describe our library of components for meta-search that is being used in the WIM application.

8 Conclusions and future work

We claim our framework is able to rank items retrieved from search-engines that originally do not do that. This is not new; the novelty of our approach is that ranking can be achieved with a little computation effort. Our framework is well suited to exploit the search-engines filtering capabilities, while other approaches operates by retrieving all the information about target items, and then applying text analysis techniques similar to those used in IR to rank documents [30]. The advantages of our approach are the following:

- It avoids hard computation requirements. The main computation effort is due to the aggregation and sorting algorithms. The overall time cost of the procedure could be high, because the expansion procedures can generate a lot of queries, thus system performance is very dependent of the network condition. But this is a problem of the particular expansion and querying algorithm, not a limitation of the very general framework we propose here. Massive parallelisation of query execution, query planning and *propose & revise* reformulation are some solutions that can improve the efficiency of such a kind of meta-search engines. Recently we have included a *propose, critique and modify* method in our library of methods for information search and aggregation:.

- It reduces the amount of information retrieved that is needed to calculate rankings; for instance, MELISA only retrieves the Identifier of the documents during the retrieval phase, the rest of the data is retrieved only when needed, that is to say, when that information is of interest for the user. This feature is very useful when the information is retrieved from a slow and unreliable environment like the Internet.

- Several weighting and aggregation functions are allowed, including multivalued and fuzzy logics. Different combinations of such functions can be used together in the same system.

We claim that our approach is so flexible that it can be considered as a general framework to develop meta-search engines. Furthermore, we argue that the query-weighting/aggregation framework is a simple but powerful approach to rank documents, which is suitable to be used for other purposes not addressed here. Some possible applications of this approach include the following:

- Score items according to the credibility or reliability of the sources from which they are retrieved.
- Rank documents according to the user's own utility criteria.
- Combine different criteria and ranking scales to obtain an overall ranking.

Other fruitful ideas to implement meta-search engines have been tested during the application of this framework, mainly the use of ontologies and a clear separation between the domain and the source levels. Having separate models for domain knowledge and information sources is useful to develop scalable systems. The key idea is that we only need a domain model to put queries, plus a collection of source models and mapping schemes between the domain and the source models. Ontologies are a promising approach to separate vocabularies and define mapping schemes between different vocabularies.

The project has evolved towards the development of a problem-solving library where this framework is being tested. Such a library is already implemented using the *product t-norm* to combine weights and the *average, weighted-mean, OWA* and *WOWA* as the aggregation operators. We expect to further parameterise such a library to test and compare different weighting and aggregation functions, but the main goal now is to demonstrate how such a library can be configured to build new applications in new domains, and how to integrate this work in a framework to develop configurable and reusable societies of information agents.

Acknowledgements

The authors would like to thank the Spanish Scientific Research Council for their support. This work has been developed under the SMASH project (TIC96-1038-C04-01) and the IBROW project (IST-1999-190005). Special thanks to Enric Plaza for their proposals and suggestions, and Albert Verdaguer for his assistance in the analysis of the medical domain.

References

[1] Abásolo, J.M.& Gómez, Mario. MELISA. An ontology-based agent for information retrieval in medicine, *Proceedings of the First International Workshop on the Semantic Web*. Lisbon, Portugal. Pp 73-82.

[2] Adali, S. and Emery, R. A uniform framework for integrating knowledge in heterogeneous knowledge systems. In *Proc. Intl. Conference on Data Engineering*, 1995

[3] Ambite, J.L. and Knoblock, C.A. "Flexible and Scalable Cost-Based Query Planning in Mediators: A Transformational Approach," *Artificial Intelligence J.*, vol. 118, no. 1-2 (2000) pp. 115-161.

[4] Arens, Y., Knoblock, C.A. and Hsu, C.N. Query Processing in the SIMS Information Mediator. *Advanced Planning Technology*, editor, Austin Tate, AAAI Press, Menlo Park, CA, 1996.

[5] Benjamins, V.R., Plaza, E., Motta, E., Fensel, D., Studer, R., Wielinga, B., Schreiber, G. and Zdrahal, Z. (1998). An Intelligent Brokering Service for Knowledge-Component Reuse on the World-Wide Web. In *Proceedings of the 11th Banff Knowledge Acquisition for Knowledge-Based System Workshop (KAW 98)*. Banff, Canada, 1998.

[6] Benjamins, V.R. and Fensel, R. Community is Knowledge in (KA)2. *Proceedings of the Eleventh Workshop on Knowledge Acquisition, Modeling and Management. (KAW'98)* Banff, Alberta, Canada, 1998.

[7] Bergamaschi, S., Castano, S. and Vincini, M. Semantic integration of semistructured and structured data sources. *SIGMOD Record*, 28(1):54--59, March 1999

[8] S.Bergamaschi, S.Castano, S.De Capitani di Vimercati, S.Montanari, and M.Vincini. Exploiting schema knowledge for the integration of heterogeneous sources. *Proc. of Convegno Nazionale Sistemi Evoluti per Basi di Dati SEBD98*, june 1998.

[9] Borgo, S, Guarino, N., Masolo, C. and Vetere, G. Using a Large Linguistic Ontology for Internet-based Retrieval of Object-Oriented Components. In *Proceedings of The Ninth Internationsl Conference on Software Engineering and Knowledge Engineering (SEKE '97)*, june 18-20, 1997, Madrid.

[10] Bougamin, L., Fabret, F., Mohan, C. and Valduriez, P. A Dynamic Query Processing Architecture for Data Integration Systems. *Bulletin of the Technical Committee on Data Engineering*, 23-2 (2000), 42-48

[11] Calmet, J., Messing, B., Schu, J., A Novel Approach towards an Integration of Multiple Knowledge Sources. *International Symposium on the Management of Industrial and Cooperate Knowledge ISMICK* (1993).

[12] Fensel, D., Decker, S., Erdmann, M. and Studer, R. Ontobroker: The Very High Idea. In *Proceedings of the 11th International Flairs Conference (FLAIRS-98)*, Sanibal Island, Florid, May 1998

[13] Fensel, D., Benjamins,V.R., Decker, S., Gaspari, M., Groenboom, R., Grosso, W., Musen, M., Motta, Plaza, E., Schreiber, G., Studer, R. & Wielinga, B. The Component Model of UPML in a Nutshell. In WWW *Proceedings of the 1st Working IFIP Conference on Software Architectures (WICSA1)*, San Antonio, Texas, USA, February 1999.

[14] Fowler, J., Nodine, M., Perry, B. and Bargmeyer, B. Agent-based Semantic Interoperability in Infosleuth. *SIGMOD Record* 28, 1999. 11

[15] Garcia-Molina, H. and Hammer, J. Integrating and Accessing Heterogeneous Information Sources in Tsimmis. In *Proc. of ADBIS 97*, St. Petersburg (Russia), 1997.

[16] Genesereth, M., Keller, A. and Duschka, O. Infomaster: An information integration system. In *Proceedings of the ACM SIGMOD International Conference on Management of Data,* Tucson, AZ, 1997.

[17] Gomez Perez, A. and Benjamins, V.R. Applications of Ontologies and Problem-Solving Methods, *AI-Magazine*, 20(1):119-122, 1999.

[18] Guarino, N. (Ed.) *Formal Ontology in Information System.* IOS Press. Amsterdam, The Netherlands, 1998

[19] Guarino, N., Masolo, C. and Vetere, G. OntoSeek: Content-based access to the web. *IEEE Intelligent Systems*, may/june 1999, 70-80

[20] Ives, Z.; Florescu, D.; Friedman, M.; Levy, A.; and Weld, D.. An adaptive query execution engine for data integration. In *Proceedings of the ACM SIGMOD International Conference on Management of Data* (1999)

[21] Landauer, T.K., Foltz. P.W., and Laham, D. An Introduction to Latent Semantic Analysis. *Discourse Processes*, 25 (1998), pp 259-284

[22] Lesser, V. Horling, B, Klassner, F. Raja, A. Wagner, T. and Zhang, S. BIG: A resource-bounded information gathering agent. In *Proceedings of the Fifteenth National Conference on Artificial Intelligence (AAAI-98),* July 1998. See also UMass CS Technical Reports 98-03 and 97-34.

[23] Levy, A. Rajaraman, A. and Ordille, J.. Querying heterogeneous information sources using source descriptions. In *Proceedings of the 22nd International Conference on Very Large Data Bases,* 1996.

[24] Levy, A.Y., Rajaraman, A. and Ordille, J.J. Query-Answering algorithms for Information Agents. In *Proceedings of the Thirteenth National conference on Artificial Intelligence (AAAI-96)*, Portland, OR, 1996

[25] Liu, J. Guide to Meta-Search Engines. 1999 http://www.indiana.edu/~librcsd/search/meta.html

[26] May, M., Himmerder, R., Lausen, G. and Ludscher, B.. A Unified Framework for Wrapping, Mediating and Restructuring Information from the Web. In *International Workshop on the World-Wide Web and Conceptual Modeling (WWWCM)*, number 1727, pp. 307-320, 1999

[27] Motta, E. *Reusable Components for Knowledge Modelling, case studies in parametric-design* IOS Press, Amsterdam, The Netherlands, October 1999.

[28] Oates, T. Nagendra Prasad, M. Lesser, V. and Decker K.S. A distributed problem solving approach to cooperative information gathering. In *AAAI Spring Symposium*, Stanford CA, March 1995.

[29] Reck, C. and B. Koenig-Ries (1997). An Architecture for Transparent Access to Semantically Heterogeneous Information Sources. *1st Int. Workshop on Cooperative Information Agents - LNCS*, Kiel, Germany pp. 260-271.

[30] Selberg, E. and Etzioni, O. The MetaCrawler WWW Search Engine. http://www.cs.washington.edu/research/projects/WebWare1/www/metacrawler/

[31] Sparck Jones, K. Information retrieval and artificial intelligence. *Artificial Intelligence* 114 (1999), 257-281

[32] Subrahmanian, V.S., Adali, S., Brink, A., Emery, R. Lu, J.J. Rajput, A., Rogers, T.J., Ross, R. and Ward, C. HERMES: Heterogeneous Reasoning and Mediator System", submitted for publication, 1997.
http://www.cs.umd.edu/projects/hermes/overview/paper

[33] Vidal, M.E. and Raschid, L. Websrcmed: A mediator for scaling up to multiple web accessible sources (websources). ftp://www.umiacs.umd.edu/pub/mvidal/websrcmed.ps (under review), 1998.

[34] Torra, V. Weighted OWA operators for synthesis of information. Proc. 5th IEEE Intern. Conference on Fuzzy systems, pp 966-971. New Orleans, USA, 1996.

B. Bouchon-Meunier, L. Foulloy and R.R. Yager (Editors)
Intelligent Systems for Information Processing:
From Representation to Applications
© 2003 Elsevier Science B.V. All rights reserved.

On the comparison of aggregates over fuzzy sets

Patrick Bosc, Olivier Pivert
IRISA/ENSSAT
Technopole ANTICIPA BP 447
22305 Lannion Cedex France
bosc|pivert@enssat.fr

Ludovic Liétard
IRISA/IUT
Rue Édouard Branly BP 150
22302 Lannion Cedex France
lietard@iut-lannion.fr

Abstract

This paper deals with flexible queries addressed to regular relational databases where conditions are defined by fuzzy sets. A particular type of fuzzy conditions is investigated, namely where two aggregates applying to fuzzy sets are compared. An example of such a condition is "the maximum salary of *young* employees is lower than the minimum salary of *old* employees". The contribution of the paper is to propose a sound interpretation for such statements in the context of flexible querying, i.e., such that a degree of satisfaction is obtained.

Keywords: Relational databases, Flexible querying, Fuzzy sets, Aggregates.

1 Introduction

This paper considers flexible querying of relational databases where atomic conditions define preferences instead of strict requirements. As a consequence, the set of answers returned to the user is discriminated from the best answers to less satisfactory ones. Many approaches to define flexible querying have been proposed in the last decades and it has been shown that fuzzy set theory provides a unifying framework to define flexible queries [1]. Atomic conditions are defined by fuzzy sets and are called vague or fuzzy conditions (or predicates). An example of a flexible query is: "retrieve *young* and *well-paid* employees working in a *high-budget* department" where *young*, *well-paid* and *high-budget* are vague predicates. The answer is a fuzzy set of employees and the higher the degree of membership of an employee, the more satisfactory he/she is.

Vague conditions can be combined using various operators (generalized conjunctions and disjunctions, linguistic quantifiers [2, 10] for example) and, on this basis, an extension of the SQL language (called SQLf) has been proposed [3]. In SQLf, as well as in SQL, it is possible to consider aggregates which are functions applying to a set of items (such as cardinality, sum, maximum or average). Aggregates can be integrated into SQLf queries as in the following example calling on the aggregate maximum (denoted by max):

    select #firm **from** emp **group by** #firm **having** max(salary) = 'high'.

Here, relation emp whose schema is EMP(#emp, #firm, salary, age, job, education) is assumed to describe employees and this query retrieves firms where the maximum salary is *high* (more precisely, a firm is all the more preferred as it satisfies the fuzzy condition: "the *maximum* salary of its employees is *high*").

This query does not raise particular difficulties since the aggregate applies to a crisp set (of salaries) and, for each firm, the average salary of its employees is computed and matched against the fuzzy condition "high". However, when the items to aggregate are issued from a fuzzy condition, the interpretation is no longer trivial since the referential to which the aggregate applies, becomes fuzzy. It is the case of the query aiming at the retrieval of firms where "the maximum salary of *young* employees is *high*" which could be expressed in SQLf as:

> **select** #firm **from** emp **where** age = 'young'
> **group by** #firm **having** max(salary) = 'high'.

In this case, the aggregate "max" applies to a fuzzy set made of salaries related to *young* employees. In general, such a condition is expressed:

$$\text{agg}(A) \text{ is } C$$

where agg is an aggregate (maximum in the preceding example), A is the set onto which agg applies (fuzzy set of salaries related to *young* employees) and C is a fuzzy condition (*high*).

Another example of a condition where aggregates are involved, is given by the query: "retrieve the firms where the maximum salary of *young* employees is lower than the minimum salary of *old* employees":

> **select** #firm **from** emp E1
> **where** age = 'young' **group by** #firm
> **having** max(salary) < (**select** min(salary) **from** emp E2
>   **where** age = 'old' **and** E1.#firm = E2.#firm).

This query involves the condition "max(A) < min(B)" where A denotes the fuzzy set of salaries of *young* employees of a given firm while B represents the fuzzy set of salaries of its *old* employees. The degree associated with each firm in the result, depends on the satisfaction of this condition. This second type of condition where two aggregates are compared is modeled in the general case by:

$$\text{"agg}_1(A) \ \theta \ \text{agg}_2(B)\text{"}$$

where the sets A and B may be fuzzy and θ is a fuzzy or crisp comparison operator.

The previous approach to interpret statements of type "agg(A) is C" is based on a fuzzy pattern matching process [8, 9]. As a consequence, this interpretation gives two indices, namely a possibility and a necessity degree. Since a unique degree of satisfaction is expected in our approach of flexible querying (SQLf), fuzzy pattern matching cannot be retained. Consequently, we make a new proposal to interpret flexible conditions involving aggregates.

An approach to deal with conditions of type "agg(A) is C" in flexible querying has been suggested recently in [5, 6]. The present paper focuses on conditions of type "agg$_1$(A) θ agg$_2$(B)" where the sets A and B are fuzzy. We limit ourselves to the case where θ is a regular (nonfuzzy) comparison operator and the aggregates are monotonic. The contribution of this work is to propose an interpretation for that kind of statements which is sound and coherent with SQLf, i.e., which returns a grade of satisfaction. The remainder of the paper is structured as follows. In section 2, the context of the paper is detailed and a previous work related to the interpretation of statements of type "agg(A) is C" statements is recalled. It offers a basis for an original approach to the interpretation of conditions of type "$agg_1(A) \theta agg_2(B)$" which is introduced in section 3. Finally, a conclusion summarizes the contribution of the paper and suggests some lines for future research.

2 Context of the work and reminders

2.1 Statement of interest

The present paper is devoted to the interpretation of conditions of the form:

$$agg_1(A) \theta agg_2(B)$$

where θ belongs to $\{\leq, <, \geq, >, =, \neq\}$.

In the remainder of this paper, we limit ourselves to the study of these conditions where θ is "$\leq$" since it can be easily shown that the other formulations can be derived from that case. For instance, "$agg_1(A) < agg_2(B)$" can be expressed as a conjunction :

$$"agg_1(A) \leq agg_2(B)" \textbf{ and } "agg_1(A) \neq agg_2(B)"$$

which, in turn, can be rewritten:

$$"agg_1(A) \leq agg_2(B)" \textbf{ and not } ("agg_1(A) \leq agg_2(B)" \textbf{ and } "agg_2(B) \leq agg_1(A)")$$

which can be simplified, and finally, we have:

$$"agg_1(A) \leq agg_2(B)" \textbf{ and not } "agg_2(B) \leq agg_1(A)"$$

where only statements of the considered type appear.

2.2 About the interpretation of statements of the form "agg(A) is C"

As we will see later, the interpretation of statements of type "$agg_1(A) \theta agg_2(B)$" calls on that of simpler ones of the form "agg(A) is C". It is the reason why we first recall the principle retained for interpreting such statements when A is a fuzzy set and C is a fuzzy condition.

The proposed approach [5, 6] to evaluate "agg(A) is C" is restricted to monotonic fuzzy predicates C on the one hand and monotonic aggregates on the other hand. It covers the aggregates max, min and count, but also sum if the sign of all the values to be added is either positive or negative; of course, the average, the median and the standard deviation do not comply with this requirement and thus are excluded. The limitation to monotonic fuzzy predicates is not a severe limitation in practice since many non monotonic predicates can be decomposed into monotonic predicates ("*around 3*" is the conjunction "*at least 3*" and "*at most 3*").

First, we consider the case where both the aggregate and the predicate are increasing. More precisely, the idea is to start from the following interpretation of the statement "agg(*A*) is *C*" when *A* is a regular set, agg is an increasing aggregate function and *C* is an increasing Boolean condition:

"agg(*A*) is *C*" is true $\Leftrightarrow \exists$ n such that $C(n)$ and agg(*A*) $\geq$ n.

Since *C* is crisp and increasing, it is certain that "agg(*A*) is *C*" is true as soon as agg(*A*) is larger than (or equal to) a value which satisfies *C*.

When *A* becomes a fuzzy set and *C* is a fuzzy condition, the idea is to extend the previous formula, and then, to look for a value n which maximizes (generalization of the existential quantifier) the conjunction of the following two fuzzy conditions: $C(n)$ and agg(*A*) $\geq$ n. Clearly, the hard point is the handling of the second component because the aggregate refers to a fuzzy set. The idea is to consider the α-cuts of the fuzzy set *A* denoted by A_α. Since the aggregate is increasing, if agg(A_α) $\geq$ n it is sure that agg(A_λ) $\geq$ n holds for any level λ in (0, α]. It is possible to show that the truth value of the statement "agg(*A*) is *C*" is given by (see [5, 6] for more details):

$$t(\text{agg}(A) \text{ is } C) = \max_{\alpha \in [0,1]} \min(\alpha, \mu_C(\text{agg}(A_\alpha))) \qquad (1).$$

In this last expression, α takes all values between [0,1] and it implies that the aggregate is defined for each α-level cut. As a consequence, expression (1) cannot be used in case of an aggregate not defined for the empty set (such as maximum or sum) applying to a not normalized fuzzy set *A* (since a not normalized fuzzy has empty α-cuts and vice-versa due to the inclusion of α-cuts). Thus when *A* is not normalized, only aggregates which are defined for the empty set (such as the cardinality) are supported by definition (1). In the following, we assume that either A is normalized or that the aggregate is defined for the empty set.

Then, from a computational point of view, expression (1) can be rewritten in a more convenient manner:

$$t(\text{agg}(A) \text{ is } C) = \max_i \min(\alpha_i, \mu_C(\text{agg}(A_{\alpha_i}))) \qquad (2)$$

where the effective α-level cuts (i.e., the different membership degrees present in *A*) are increasingly rank-ordered: $\alpha_1 < \alpha_2 < ... < \alpha_n$. One may remark that when the aggregate computed on the support of A completely satisfies C (i.e., $\mu_C(\text{agg}(A_{\alpha_1})) = 1$) expression (2) is nothing but a Sugeno fuzzy integral.

Example 1. Let us consider the following fuzzy set of salaries $A = \{1/800 + 0.7/12000 + 0.7/10500 + 0.5/15000 + 0.2/10000 + 0.1/11000\}$ and the statement "max(*A*) is *high*". We get:

$$t(\max(A) \text{ is } high) = \max_i \min(\alpha_i, \mu_{high}(\max(A_{\alpha_i})),$$

where $\alpha_1 = 0.1$, $\alpha_2 = 0.2$, $\alpha_3 = 0.5$, $\alpha_4 = 0.7$ and $\alpha_5 = 1$. Thus t(max(A) is *high*) is:

$$\max(\min(0.1, \mu_{high}(15000)), \min(0.2, \mu_{high}(15000)),$$
$$\min(0.5, \mu_{high}(15000)), \min(0.7, \mu_{high}(12000)),$$
$$\min(1, \mu_{high}(800)))).$$

If $\mu_{high}(800) = 0.2$, $\mu_{high}(12000) = 0.8$ and $\mu_{high}(15000) = 1$, we get:

$$t(\max(A) \text{ is } high) = \max(\min(0.1, 1), \min(0.2, 1), \min(0.5, 1),$$
$$\min(0.7, 0.8), \min(1, 0.2)))$$
$$= 0.7. \blacklozenge$$

Throughout the paper, the aggregate and the condition are assumed to be increasing, but the solution provided for that case can be straightforwardly adapted when at least one of them is decreasing. More precisely (see [5, 6]), expression (1) or (2) can be used when C and the aggregate are monotonic in the same way (both increasing or both decreasing). When the aggregate is increasing (resp. decreasing) while C is decreasing (resp. increasing), the condition "agg(A) is C" can be defined as the negation of the statement "agg(A) is *not C*". In this statement, both the aggregate and *not C* are either both increasing or both decreasing and the evaluation can be performed using expression (1) or (2).

3 Principle of the evaluation of conditions involving two aggregates

Section 3.1 is devoted to the approach proposed in this paper to evaluate a condition of the form "agg$_1$(A) ≤ agg$_2$(B)". Section 3.2 establishes a proof of the given result while section 3.3 provides a final example.

3.1 The approach

Here again, the idea is to start with a definition valid for crisp sets, then to extend it to fuzzy sets. In the case where A and B are crisp, it is possible to express the meaning of the statement "agg$_1$(A) ≤ agg$_2$(B)" using an implication according to the formula:

$$\forall x, [\text{agg}_1(A) \geq x] \Rightarrow [\text{agg}_2(B) \geq x] \qquad (3)$$

where x is used to scan the definition domain of agg$_1$ and agg$_2$.

When A and B are two fuzzy sets, the expression "agg$_1$(A) ≥ x" (resp. "agg$_2$(B) ≥ x") is more or less satisfied. Its degree of truth t(agg$_1$(A) ≥ x) (resp. t(agg$_2$(B) ≥ x)) can be obtained by the approach proposed in section 2 (see 2.2). It is important to recall that only monotonic aggregates can be dealt with. For the sake of simplicity, increasing aggregates are considered and the statement "agg(A) ≥ x" is then evaluated by:

$$t(\text{agg}(A) \geq x) = \max\nolimits_{\alpha \in [0,1]} \min(\alpha, \mu_{\geq x}(\text{agg}(A_\alpha)))$$

which stems from the general case (see formula (1) in section 2). Since "≥ x" is a Boolean predicate, its value of truth is either 1 or 0 and we get:

$$t(agg(A) \geq x) = \max{}_{\alpha \in [0,1] \text{ such that } agg(A_\alpha) \geq x} \alpha \qquad (4).$$

If the universal quantifier in (3) is interpreted as a generalized conjunction, the degree of satisfaction of "$agg_1(A) \leq agg_2(B)$" is given by:

$$\min{}_{x \in D} t(agg_1(A) \geq x) \rightarrow t(agg_2(B) \geq x) \qquad (5)$$

where $\rightarrow$ stands for a fuzzy implication and D is the definition domain of agg_1 and agg_2. Many implications are available [7] (for instance Gödel, Goguen and Kleene-Dienes) and we suggest to choose among R-implications which can be easily interpreted by users in terms of thresholds and penalties [4]. In particular, Lukasiewicz implication (a $\rightarrow_{Lu}$ b = 1 if a $\leq$ b and 1 - a + b otherwise) takes into account the difference (intuitively interpretable as a distance) between the antecedent and the conclusion.

In principle, the computation of expression (5) needs to consider an infinity of x values. However, it is possible to restrict the computation to a finite set of x values, i.e., those which are aggregate values for α-cuts, which leads to:

$$\min{}_{x \in AV} [t(agg_1(A) \geq x) \rightarrow t(agg_2(B) \geq x)] \qquad (6)$$

with AV = $\{agg_1(A_\alpha)$ with α in $(0,1]\} \cup \{agg_2(B_\beta)$ with β in $(0,1]\}$.

3.2 A proof of the validity of the simplified calculus

We demonstrate that x values that do not belong to AV are not useful in the computation of expression (5). Let us recall that a fuzzy implication satisfies the following properties:

a) it is decreasing with respect to the first argument, i.e.,
$\forall (x, y, z) \in [0,1] \times [0,1] \times [0,1]$ such that $x \leq y : (x \rightarrow z) \geq (y \rightarrow z)$,

b) it is increasing with respect to the second argument, i.e.,
$\forall (x, y, z) \in [0,1] \times [0,1] \times [0,1]$ such that $y \leq z : (x \rightarrow y) \leq (x \rightarrow z)$.

In this proof, the n effective α-levels of fuzzy set A (i.e. membership degrees in A) are ranked decreasingly:

$$\alpha_1 > ... > \alpha_n,$$

and due to the montonicity of agg_1, we obviously get:

$$agg_1(A_{\alpha_1}) \leq .... \leq agg_1(A_{\alpha_{j-1}}) \leq agg_1(A_{\alpha_j}) \leq ... \leq agg_1(A_{\alpha_n}).$$

The value for $t(agg_1(A) \geq x)$ depends on the position of x with respect to the aggregate values computed over α-cuts of A. Three cases must be considered, i) when x is smaller then or equal to the smallest aggregate value ($x \leq agg_1(A_{\alpha_1})$), ii) when x is between two aggregate values ($agg_1(A_{\alpha_{j-1}}) \leq x \leq agg_1(A_{\alpha_j})$), iii) when x is larger then (or equal to) the largest aggregate value ($agg_1(A_{\alpha_n}) \leq x$).

Table 1. Values for $t(\text{agg}_1(A) \geq x)$.

	$x \leq \text{agg}_1(A_{\alpha_1})$	$\text{agg}_1(A_{\alpha_{i-1}}) \leq x \leq \text{agg}_1(A_{\alpha_i})$	$\text{agg}_1(A_{\alpha_n}) \leq x$
$t(\text{agg}_1(A) \geq x)$	α_1	α_i	0

Table 2. Values for $t(\text{agg}_2(B) \geq x)$.

	$x \leq \text{agg}_2(B_{\beta_1})$	$\text{agg}_2(B_{\beta_{i-1}}) \leq x \leq \text{agg}_2(A_{\beta_i})$	$\text{agg}_2(A_{\beta_m}) \leq x$
$t(\text{agg}_2(B) \geq x)$	β_1	β_i	0

<u>Case 1</u>: $x \leq \text{agg}_1(A_{\alpha_1})$. In this case, we get:

α	α_1	...	α_{i-1}	α_i	α_{i+1}	...	α_n
$\mu_{\geq x}(\text{agg}_1(A_\alpha))$	1	1	1	1	1	1	1

which gives (using expression (4)):

$t(\text{agg}_1(A) \geq x) = \max_{\alpha \in [0,1] \text{ such that } \text{agg}(A_\alpha) \geq x} \alpha = \alpha_1$.

<u>Case 2</u>: $\text{agg}_1(A_{\alpha_{i-1}}) \leq x \leq \text{agg}_1(A_{\alpha_i})$. In this case, we get:

α	α_1	...	α_{i-1}	α_i	α_{i+1}	...	α_n
$\mu_{\geq x}(\text{agg}_1(A_\alpha))$	0	0	0	1	1	1	1

which gives (using expression (4)):

$t(\text{agg}_1(A) \geq x) = \max_{\alpha \in [0,1] \text{ such that } \text{agg}_1(A_\alpha) \geq x} \alpha = \alpha_i$.

<u>Case 3</u>: $\text{agg}_1(A_{\alpha_n}) \leq x$. In this case, we get:

α	α_1	...	α_{i-1}	α_i	α_{i+1}	...	α_n
$\mu_{\geq x}(\text{agg}_1(A_\alpha))$	0	0	0	0	0	0	0

which gives (using expression (4)):

$t(\text{agg}_1(A) \geq x) = \max_{\alpha \text{ such that } \text{agg}_1(A_\alpha) \geq x} \alpha = 0$.

These results are summarized in Table 1.

Similar cases appear when considering the m effective α-levels of fuzzy set B: $\beta_1 > ... > \beta_m$ and their results are given in Table 2.

When considering a value x out of AV = $\{\text{agg}_1(A_\alpha)$ with α in $(0,1]\} \cup \{\text{agg}_2(B_\beta)$ with β in $(0,1]\}$, 9 cases must be investigated (to situate x values with respect to $\text{agg}_1(A_\alpha)$ and $\text{agg}_2(B_\beta)$). For each of these cases, we demonstrate that x does not play any role in the computation of the final result by showing that one of the two following conditions holds:

a) an element y in AV is such that
$$t(agg_1(A) \geq y) \to t(agg_2(B) \geq y) \leq t(agg_1(A) \geq x) \to t(agg_2(B) \geq x),$$

b) $t(agg_1(A) \geq x) \to t(agg_2(B) \geq x) = 1$.

As the minimum of implication values is delivered by expression (5), value x can be omitted by the computation process.

<u>Case 1</u>: $x < agg_1(A_{\alpha_1})$ and $x < agg_2(B_{\beta_1})$. In that case, from Tables 1 and 2, we get:

$$t(agg_1(A) \geq x) = \alpha_1 \quad \text{and} \quad t(agg_2(B) \geq x) = \beta_1.$$

The contribution of x to the final result is $\alpha_1 \to \beta_1$. Value y from AV defined by $\min(agg_1(A_{\alpha_1}), agg_2(B_{\beta_1}))$ is such that $y \leq agg_1(A_{\alpha_1})$ and $y \leq agg_2(B_{\beta_1})$. From Tables 1 and 2 we get

$$t(agg_1(A) \geq y) = \alpha_1 \quad \text{and} \quad t(agg_2(B) \geq y) = \beta_1.$$

As a consequence we obtain:

$$t(agg_1(A) \geq y) \to t(agg_2(B) \geq y) = t(agg_1(A) \geq x) \to t(agg_2(B) \geq x).$$

<u>Case 2</u>: $x < agg_1(A_{\alpha_1})$ and $agg_2(B_{\beta_{j-1}}) < x < agg_2(B_{\beta_j})$. In that case (see Tables 1 and 2):

$$t(agg_1(A) \geq x) = \alpha_1 \quad \text{and} \quad t(agg_2(B) \geq x) = \beta_j.$$

The contribution of x to the final result is $\alpha_1 \to \beta_j$. Value y from AV defined by $agg_1(A_{\alpha_1})$ is such that (see Table 1):

$$t(agg_1(A) \geq y) = \alpha_1.$$

Since $x < agg_1(A_{\alpha_1}) = y$ and $agg_2(B_{\beta_{j-1}}) < x < agg_2(B_{\beta_j})$, we get:

$$t(agg_2(B) \geq y) \in \{\beta_j, \beta_{j+1}, \ldots, \beta_m\} \text{ and thus } t(agg_2(B) \geq y) \leq \beta_j.$$

Since an implication is increasing with respect to the second argument, we get:

$$(t(agg_1(A) \geq y) \to t(agg_2(B) \geq y)) \leq (t(agg_1(A) \geq x) \to t(agg_2(B) \geq x)).$$

<u>Case 3</u>: $x < agg_1(A_{\alpha_1})$ and $agg_2(B_{\beta_m}) < x$. In that case, from Tables 1 and 2:

$$t(agg_1(A) \geq x) = \alpha_1 \quad \text{and} \quad t(agg_2(B) \geq x) = 0.$$

The value $y = agg_1(A_{\alpha_1})$ from AV is such that (see Table 1):

$$t(agg_1(A) \geq y) = \alpha_1.$$

Since $x < agg_1(A_{\alpha_1}) = y$ and $agg_2(B_{\beta_m}) < x$, we get $agg_2(B_{\beta_m}) < y$ and then (see Table 2):

$$t(agg_2(B) \geq y) = 0.$$

Thus, value y gives the same implication value as x.

<u>Case 4</u>: $agg_1(A_{\alpha_{i-1}}) < x < agg_1(A_{\alpha_i})$ and $x < agg_2(B_{\beta_1})$. In this case, we have (see Tables 1 and 2):

$$t(agg_1(A) \geq x) = \alpha_i \quad \text{and} \quad t(agg_2(B) \geq x) = \beta_1.$$

Consequently, the contribution of value x in the final result is:

$$\alpha_i \to \beta_1.$$

Value $y = agg_1(A_{\alpha_i})$ is such that (see Table 1):

$$t(agg_1(A) \geq y) = \alpha_i.$$

Since β_1 is the maximum β-level and since $t(agg_1(A) \geq y)$ is one of them, we immediatly have $t(agg_1(A) \geq y) \leq \beta_1$. Since an implication is increasing with respect to the second argument we get:

$$(t(agg_1(A) \geq y) \to t(agg_2(B) \geq y)) \leq (t(agg_1(A) \geq x) \to t(agg_2(B) \geq x)).$$

<u>Case 5</u>: $agg_1(A_{\alpha_{i-1}}) < x < agg_1(A_{\alpha_i})$ and $agg_2(B_{\beta_{j-1}}) < x < agg_2(B_{\beta_j})$. From Tables 1 and 2, we get the following result :

$$t(agg_1(A) \geq x) = \alpha_i \quad \text{and} \quad t(agg_2(B) \geq x) = \beta_j.$$

Consequently, the contribution of value x to the final result is :

$$\alpha_i \to \beta_j.$$

The value $y = agg_1(A_{\alpha_i})$ from AV is such that (see Table 1) :

$$t(agg_1(A) \geq y) = \alpha_i.$$

Since $x < agg_1(A_{\alpha_i}) = y$ and $agg_2(B_{\beta_{j-1}}) < x < agg_2(B_{\beta_j})$, we get :

$$t(agg_2(B) \geq y) \in \{\beta_j, \beta_{j+1}, \ldots, \beta_n\} \text{ and thus } t(agg_2(B) \geq y) \leq \beta_j.$$

Since an implication is increasing with respect to the second argument we get:

$$(t(agg_1(A) \geq y) \to t(agg_2(B) \geq y)) \leq (t(agg_1(A) \geq x) \to t(agg_2(B) \geq x)).$$

<u>Case 6</u>: $agg_1(A_{\alpha_{i-1}}) < x < agg_1(A_{\alpha_i})$ and $agg_2(B_{\beta_m}) < x$. From Tables 1 and 2 we get:

$$t(agg_1(A) \geq x) = \alpha_i \quad \text{and} \quad t(agg_2(B) \geq x) = 0.$$

If we consider the value $y = agg_1(A_{\alpha_i})$ from AV, we have (see Table 1):

$$t(agg_1(A) \geq y) = \alpha_i.$$

Since $agg_2(B_{\beta_m}) < x$ and $x < agg_1(A_{\alpha_i}) = y$, we get $agg_2(B_{\beta_m}) < y$ and then (see Table 2):

$$t(agg_2(B) \geq y) = 0.$$

Thus, value y gives the same implication value.

<u>Case 7</u>: $agg_1(A_{\alpha_n}) < x$ and $x < agg_2(B_{\beta_1})$. From Tables 1 and 2 we get:

$$t(agg_1(A) \geq x) = 0 \quad \text{and} \quad t(agg_2(B) \geq x) = \beta_1.$$

The value y from AV is $agg_2(B_{\beta_1})$. Since $agg_1(A_{\alpha_n}) < x$ and $x < agg_2(B_{\beta_1}) = y$ we get $agg_1(A_{\alpha_n}) < y$. From Table 1:

$$t(agg_1(A) \geq y) = 0.$$

Since $y = agg_2(A_{\beta_1})$, we have from Table 2:

$$t(agg_2(B) \geq y) = \beta_1.$$

Thus, value y gives the same implication value as x.

<u>Case 8</u>: $agg_1(A_{\alpha_n}) < x$ and $agg_2(B_{\beta_{j-1}}) < x < agg_2(B_{\beta_j})$. Such a value x can be discarded because $y = agg_2(A_{\beta_j})$ from AV has the same implication value (that case is similar to case 7).

<u>Case 9</u>: $agg_1(A_{\alpha_n}) < x$ and $agg_2(B_{\beta_m}) < x$. From Table 1 and 2, we obtain that $t(agg_1(A) \geq x) = 0$ and $t(agg_2(B) \geq x) = 0$. The contribution of element x is $0 \to 0 = 1$. As the minimum of the implication values is retained, this contribution can be discarded.

3.3 A final example

Let us consider a condition of the form "$max(A) \leq max(B)$" with:

$A = \{1/500 + 1/600 + 0.8/400 + 0.8/550 + 0.6/650 + 0.3/400\}$,

$B = \{1/100 + 0.9/550 + 0.8/800 + 0.1/100\}$.

We have:

$\max(A_1) = 600$, $\max(A_{0.8}) = 600$, $\max(A_{0.6}) = 650$ and $\max(A_{0.3}) = 650$,

and:

$\max(B_1) = 100$, $\max(B_{0.9}) = 550$, $\max(B_{0.8}) = 800$ and $\max(B_{0.1}) = 800$.

The set AV of values to be considered in expression (6) is:

$AV = \{100, 550, 600, 650, 800\}$.

Using Lukasiewicz implication, we get the implication values given by Tables 3 and 4.

Table 3: Aggregate truth values

x	$t(\max(A) \geq x)$ (1)	$t(\max(B) \geq x)$ (2)
100	1	1
550	1	0.9
600	1	0.8
650	0.6	0.8
800	0	0.8

Table 4: Implication values

x	(1) → (2)
100	1
550	0.9
600	0.8
650	1
800	1

Finally, the minimal value of the implication results is taken, which yields 0.8.

4 Conclusion

In this paper, the issue of flexible querying of databases is considered. In such a framework, we advocate the use of fuzzy sets and a given condition leads to a degree of satisfaction. So far, conditions calling on aggregate functions (count, sum, max, min, avg, ...) were restricted to regular sets and this paper tackles the situation where aggregates may apply to fuzzy sets. In order to be coherent with the considered context (i.e, that of a flexible query language such as SQLf), the interpretation of such a condition must be a unique degree of fulfillment.

We have focused on conditions of type "agg$_1$(A) θ agg$_2$(B)" where A and B may be fuzzy, θ is a regular (nonfuzzy) operator belonging to {≤, >, ≥, <, =, ≠} and agg$_1$, agg$_2$ are monotonic aggregates. Such complex conditions are illustrated by the statement "max(salary of *young* employees) ≤ min (salary of *old* employees)". An overall degree of satisfaction is defined as the minimum of implication values issued from simpler conditions, namely conditions of type "agg(A) is C" where the aggregate applies to a fuzzy set A and is matched against a crisp predicate C. These simpler conditions are interpreted by a unique degree (as suggested in [5, 6]) and this interpretation for "agg(A) is C" differs from that based on a fuzzy pattern process where two indices are computed.

In the near future, the next step will be the study of fuzzy comparison operators θ (such as *much larger than*) in expressions of type "agg$_1$(A) θ agg$_2$(B)". It seems that expression (3) used here cannot be the basis for the interpretation of such statements and a new approach must be suggested. Another matter of future research is the design and implementation of algorithms aiming at evaluating conditions involving aggregates computed on fuzzy sets. Experimental measures should be performed in order to assess the extra cost induced by the fuzzy nature of the condition.

References

[1] P. Bosc, O. Pivert (1992), Some Approaches for Relational Databases Flexible Querying. *Journal of Intelligent Information Systems,* volume 1, pages 323-354.

[2] P. Bosc, L. Liétard and O. Pivert (1995), Quantified statements and database fuzzy querying. *Fuzziness in Database Management Systems,* (P. Bosc, J. Kacprzyk eds), Physica-Verlag, pages 275-308.

[3] P. Bosc, O. Pivert (1995), SQLf: A relational database language for fuzzy querying. *IEEE Transactions on Fuzzy Systems,* volume 3, pages 1-17.

[4] P. Bosc (1998), On the primitivity of the division of fuzzy relations. *Soft Computing*, volume 2, pages 35-47.

[5] P. Bosc, L. Liétard, O. Pivert (2001), Aggregate operators in database flexible querying. In *Proceedings of the 10th International Conference on Fuzzy Systems (FUZZ-IEEE'2001),* Melbourne (Australia), December 2001.

[6] P. Bosc, L. Liétard and O. Pivert (2003), Sugeno fuzzy integral as a basis for the interpretation of flexible queries involving monotonic aggregates. *Information Processing and Management,* volume 39, pages 287-306.

[7] D. Dubois, H. Prade (1985), A review of fuzzy set aggregation connectives. *Information Sciences*, volume 36, pages 85-121.

[8] D. Dubois, H. Prade, C. Testemale (1988), Weighted fuzzy pattern matching. *Fuzzy Sets and Systems*, volume 28, pages 315-331.

[9] D. Dubois, H. Prade (1990), Measuring properties of fuzzy sets: a general technique and its use in fuzzy query evaluation, *Fuzzy Sets and Systems*, volume 38, pages 137-152.

[10] J. Kacprzyk, A. Ziolkowski (1986), Databases queries with fuzzy linguistic quantifiers. *IEEE Transactions on Systems, Man and Cybernetics*, volume 16, pages 474-478.

B. Bouchon-Meunier, L. Foulloy and R.R. Yager (Editors)
Intelligent Systems for Information Processing:
From Representation to Applications
© 2003 Elsevier Science B.V. All rights reserved.

Towards an Intelligent Text Categorization for Web Resources: An Implementation

Sławomir Zadrożny[1], Klaudia Ławcewicz[2] and Janusz Kacprzyk[1,2]

[1] Systems Research Institute, Polish Academy of Sciences
ul. Newelska 6, 01-447 Warsaw, Warsaw
[2] Warsaw School of Information Technology
ul. Newelska 6, 01-447 Warsaw, Poland
e-mail: {zadrozny,kacprzyk}@ibspan.waw.pl

Abstract

We propose the concept and implementation of a software system, TCAT (**T**ext **CAT**egorization) system, for an automatic recognition of a topic of an Internet document. In the training mode the user provides the system with a list of topics and sets of documents representing each topic (supervised learning). In the recognition mode the system automatically classifies previously unseen document to a topic category. A simple learning algorithm is devised and implemented. The results of the classification are presented to the user in the form of a set of linguistic terms. Some new measures of correctness of the classification are proposed. The implemented system processes documents in several popular Internet-related formats.

Keywords: automatic classification of documents, Internet, linguistic terms

1 Introduction

The maintenance and processing (notably retrieving) of textual information by means of computerized systems was among the first applications of the computers. The need for such systems has grown essentially along with the popularity of the electronic form of documents implied by the recent widespread and easy access to the Internet. The Internet provides an excellent testbed for methods developed within *information retrieval* (IR). This encompasses various tasks addressed by IR such as fast textual documents retrieval or automatic text categorization. The latter may be understood in several ways. In order to make clearer the task addressed in this paper let us consider the scenarios in which our system may be applicable. The first one is that of a Web Spider: an agent software "traversing" the Web and automatically classifying documents found with the aim of providing us only with the documents of interest for us (i.e., belonging to a prespecified category/categories). The second scenario is that of a "translation agency". In this case, the aim of the system is to automatically assign to interpreters documents sent by customers. The interpreters prefer certain categories of documents and the aim is to match their preferences so as to secure a high efficiency of the whole translation process. In both cases the classification may be done manually. However, it may be not such a good solution as it may seem. Firstly, in particular in the first case, it is unreasonable to expect that all documents are classified by their authors or some other bodies (see, e.g.,

Yahoo). Secondly, the classification provided by the author may be useless for, or inconsistent with the purposes of the document "consumer". Both scenarios require a set of prespecified categories of documents and a training set of documents properly classified along this categories. Thus, we aim at the solution for filtering rather than clustering the documents. In other words, the targeted user is one with a fairly fixed set of categories of interest, who looks to assign new documents to this categories. This may be contrasted with the requirements of a general information retrieval system where a grouping of somehow similar documents is sought in order to make the retrieval more efficient.

The problem concerned in our paper is usually addressed with the help of methods elaborated within the domains of information retrieval and pattern recognition (more specifically, classifier construction). Characteristic features of our approach described in this paper are the assumption of a high dimensional document representation space and the use of fuzzy logic elements both for the classification purposes as well as for the presentation of results obtained to the user. The starting point are our previous experiences with the fuzzy querying of databases [6] as well as the recent advances in the application of soft computing for information retrieval purposes [3].

In Section 2 we briefly review the literature relevant for the text categorization task. Sections 3 and 4 present the general concept of the TCAT system and employed algorithms.

2 Text Categorization Task

Text categorization as discussed here is a typical example of the classification task. More precisely, the process consists of two phases:

- the learning of classification rules (building a classifier) from examples of documents with known class assignment (*supervised learning*),
- the classification of documents unseen earlier using rules derived in Phase I

A human being classifying a document may take into account its usually rich syntactic and semantic structure. In case of an automatic, computerized approach some simplifications as to the representation of the documents are usually done, i.e., only some *features* of the documents are taken into account. The most popular approach consists in treating a document as a sequence of words (*a bag of words*). Many variations are possible as to which words are taken into account – from exactly all words present in the whole collection of documents under consideration up to a limited *controlled vocabulary* (keywords). Typically, a document is preprocessed to remove so called *stopwords*, extract some basic form of words (stemming) etc. Then, each document is represented as a vector. Each component of this vector corresponds to a, possibly normalized, frequency of appearance of a word (keyword) in that document. Sometimes binary vectors are used to represent documents, where 1 and 0 mean that a keyword occurs (no matter how often) or does not occur in the document, respectively.

Having such a numerical representation of the documents one can apply one of numerous classifier construction algorithms including rule-based systems, decision trees, artificial neural networks, etc. (see, e.g., [7]). One of classical algorithms developed in the area of information retrieval is that of Rocchio [12, 5]. The learning phase consists in computing a *centroid* vector for each category of documents. Then, in the classification phase, a document is classified to a category whose centroid is most similar to this document. The similarity may be meant in several ways – in the original Rocchio's approach it corresponds to the Euclidan distance.

The text categorization task exhibits some imprecision. Even a human may be unsure as to a clear cut classification of a document to just one category. Moreover, it is quite natural to consider a degree of belongingness to a category. This becomes even more apparent in case of an automatic classification procedure. We may easily expect that the results of classification may be ambiguous. The fuzzy logic approach has proved to be useful in such a context.

The quality of given text categorization system is assessed based on the error rate. The meaning and usefulness of this approach in the crisp case is quite obvious. However, when we assume a fuzzy response from a categorization system some special considerations are needed.

In the next section we discuss in more detail how the above mentioned elements of a text categorization system have been adopted and implemented within our TCAT system.

3   The Concept of the TCAT System

Our assumptions for the construction of the TCAT (Text CATegorization) system were as follows:

1. a universal, language independent, representation of the documents
2. a simple classifier learning algorithm
3. a provision for the handling of ambiguous classification results
4. an implementation as an Internet (WWW) based service.

The first has been attained by the representation of document as a sequence of 5 or 10 (a parameter of the system) character long strings (tokens or *n*-gram). This gives rise to a very high dimensional representation space and requires a full fledged database management system to be a part of the text categorization system (in our case it is MySQL).

The classifier employed by the TCAT system is of Rocchio type. A simple formula is used to calculate for each token how representative it is for particular categories of documents. Effectively, it yields a kind of a centroid for each category of documents. In the classification phase the document to be classified is divided into tokens (only tokens extracted from training documents are taken into account). Then, the binary vector representing the document is compared with the categories' centroids yielding the degree to which the document belongs to a given category.

Due to a possible ambiguity of classification, its results are presented in the form of a linguistic expression. The degree to which a document belongs to a category is treated as the realization of a linguistic variable. As a result the user obtains a linguistic description of the membership degree of the document to the particular categories.

The TCAT system processes documents in formats typical for the Internet (WWW). The current version of the system is able to classify documents available locally on the user's computer or somewhere in the Internet, as indicated by their URL (*Uniform Resource Locator*).

4 Representation of Documents and a Classification Algorithm

Let $D=\{d_j\}_{j\in[1,M]}$ be a set of training documents and $T=\{t_i\}_{i\in[1,N]}$ be a set of all tokens occurring in them. A document is represented as the vector $d_i=(d_{i1},...,d_{iN})$, where d_{ij} denotes the number of occurrences of token t_j in document d_i.

In the learning phase the TCAT system computes for each token its *membership degree to particular categories*. This degree indicates how characteristic a token is for given category. The following properties are assumed for this indicator:

1) is proportional to the number of occurrences of the token in the documents of a given category,
2) is counter-proportional to the number of occurences of the token in documents of other categories,
3) is biased towards the proportion 1).

We will use the following notation:

SP_k^t - membership degree of token t to category k,

K – the set of all categories considered,

n^t, n_k^t - number of occurrences of token t in all training documents and in the training documents of category k, respectively,

$\Delta_+^t = \dfrac{1}{n^t}$ and $\Delta_-^t = \dfrac{1}{rn^t}$ - some auxiliary coefficients; r is a parameter, in the computational experiments assumed equal 4.

Let w_k^t denote an indicator "favouring" token t proportionally to the number of its occurences in the documents belonging to the category k and "punishing" token t proportionally to the number of its occurrences in documents belonging to other categories:

$$w_k^t = v + \left(n_k^t \cdot \Delta_+^t\right) - \sum_{m \in K\setminus\{k\}} \left(n_m^t \cdot \Delta_-^t\right) \qquad (1)$$

where v is an initial value of the indicator w_k^t (it is a parameter of the method; by default $v = 0.1$).

It is obvious that the indicator w_k^t meets conditions 1) – 3). Formula (1) may be expressed also as:

$$w_k^t = v + \dfrac{1}{r}\left((r+1)\dfrac{n_k^t}{n^t} - 1\right) \qquad (2)$$

what shows its relation to *tf*idf* weighting scheme. It provides also clear interpretation for the parameter r: a term that appears roughly more than r times in the documents belonging to other than k categories gets negative membership degree to the category k. In order to to normalize the value of w_k^t we employ the following transformation yielding the formula for the indicator SP_k^t sought:

$$SP_k^t = \begin{cases} 0 & \text{for } w_k^t \leq 0 \\ -{w_k^t}^3 + {w_k^t}^2 + w_k^t, & \text{for } 0 < w_k^t <= 1 \\ 1 & \text{for } w_k^t > 1 \end{cases} \qquad (3)$$

Thus, both formulas (1) and (2) and (3) taken together provide for the weighting of terms for particular categories as well as excercise a threshold rejecting terms being not-specific for a given category. Obviously, transformations other than (3) may be applicable to this aim and we treat it as a parameter of the method.

Effectively, while computing the SP_k^t indicators for all tokens we obtain a kind of a centroid, C_k, for each category of documents:

$$C_k = (SP_k^{t_1}, \ldots, SP_k^{t_N})$$

In our approach the centroids are computed directly during the analysis of the training documents. Usually, in the literature it is assumed that first the representation of particular documents is obtained. This representation may use various weights for the tokens, e.g., *tf*∗*idf* [14,1]. Then, the centroids are calculated using, e.g., averaging.

In the classification phase the system computes for a document d its *degree of membership to each of the categories k*, sp_k^d. In this phase the document is represented as a set of tokens –i.e., the Boolean (binary) representation is assumed - occuring in it: $d=\{t_i\}$, $t_i \in T$ (i.e., all tokens found in document d, but not present in the training documents, are ignored). The indicator sp_k^d is given by the formula:

$$sp_k^d = \frac{\sum_{t \in d} SP_k^t}{n_d} \qquad (4)$$

where n_d and SP_k^t denote the number of tokens representing document d and the degree of membership of token t to the category k (calculated in the learning phase according to (3)), respectively. Effectively, the formula (4) is a counterpart of the similarity measure between document d and the centroid of category k.

Having calculated for a document d values sp_k^d for each category k, it is natural to classify document d as belonging to that category for which the value of this indicator attains a maximal value. In case the values of this indicator for a few different categories are close, the system is not in a position to propose clear-cut classification. In such a case it seems better to inform the user about all categories to which the document possibly belongs. To this aim system TCAT provides the user with information about the values of indicator sp_k^d for all categories. However, the system does not present to the user raw numerical values of the indicator. Instead, a more human consistent form is employed referring to the concept of a linguistic variable.

A linguistic variable is a variable taking on the values that are not numerical but are linguistic terms. Usually, semantics for this linguistic terms is provided by fuzzy sets defined over the universe under consideration. The linguistic variable is a tuple $(H, T(H), U, G, M)$, where H is a name of the variable, $T(H)$ is a set of its values (linguistic terms); $U = \{u\}$ denotes the universe under consideration [fuzzy sets defined over U provide the interpretation for particular terms beonging to $T(H)$], G is a rule generating values for the linguistic variable [if $T(H)$ is finite then G may be just simple enumeration of the linguistic terms]; M is a semantic rule providing for each value $l \in T(H)$ its meaning $M(l) \subseteq U$. For example, treating *age* as a linguistic variable one may assume: $T("age")=\{"very young", "young", "middle aged", "old", "very old"\}$, $U=[1,100]$, M associates with particular values of $T("age")$ fuzzy numbers defined over the interval $[0,100]$ and intuitively corresponding to individual descriptions of the age. For example, with the term "young" a *trapezoidal fuzzy number* $(0;0;25;35)$ may be associated.

Treating the membership of a document to a category as a linguistic variable we may adopt the following interpretation for the components of the definition of the linguistic variable: the name is "connected with the category k" (H); the universe of discourse is $U = [0,1]$ (i.e., the range of the indicator sp_k^d); as the set of linguistic terms we may assume:

$$T(H) = \{"not", "slightly", "medium", "strongly", "very strongly"\}. \tag{5}$$

The semantic rule M associates with particular linguistic terms fuzzy numbers defined over the interval $[0,1]$. For example, the term "very strongly" may be represented by the trapezoidal fuzzy number $(0.85, 0.95, 1.0, 1.0)$.

Thus, we treat the degrees of membership of a given document to particular categories as realizations of a linguistic variable. We adopt a simple scheme of the choice of a linguistic term to represent the computed value of indicator sp_k^d. Namely, we choose a term such that the computed value of the indicator belongs to it to a maximal degree:

$$SP_k^d = u \quad \rightarrow \quad l = \arg\max_l \mu_{M(l)}(u) \tag{6}$$

i.e., we represent the value u of indicator sp_k^d with a linguistic term l, such that u belongs to a maximal degree to the fuzzy set $M(l)$ being a semantic interpretation of the term l.

Thanks to the definition of indicator SP_k^t a token often occuring in documents of different categories obtains a low value of membership to all categories. Thus, it has virtually no influence on the results of the classification phase.. In turn, a token occuring only in documents of one category obtains a high degree of membership to this category.

Intuitively, an occurence of such a token in a document strongly indicates its connection with a given category. This intuition is formalized by formula (4).

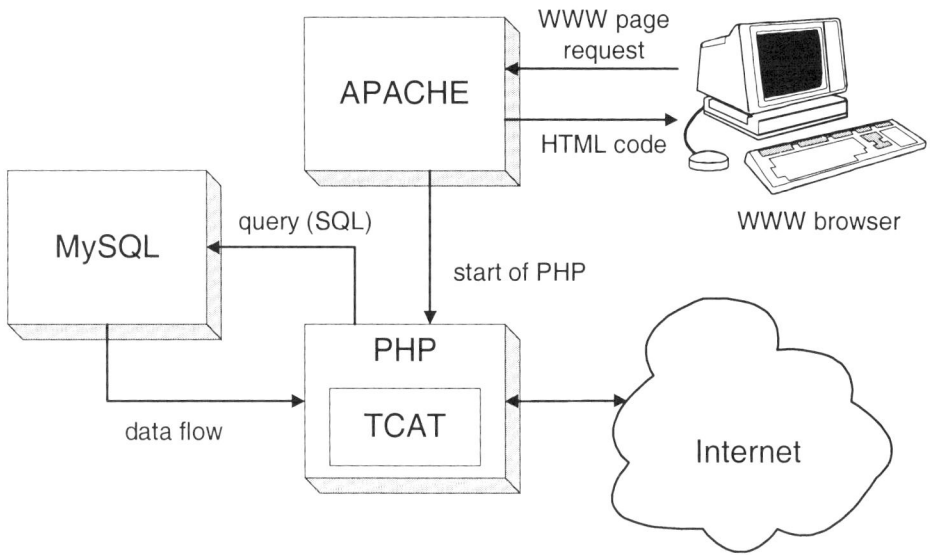

Figure 1: The structure of the TCAT system.

5 Implementation

The TCAT system is a *sensu stricte* Internet based system. It processes primarily documents available in the net. The system itself is implemented as a WWW based application. The system may be perceived in a broader sense as a set of collaborating modules written in PHP, WWW server (Apache) supported by the PHP interpreter, database management system MySQL and a WWW browser. In a narrower sense the TCAT system should be identified with the first of above mentioned components, i.e., a set of PHP modules accomplishing basic functions of the system. In what follows we do not make an explicit distinction as the context usually clearly indicates what meaning of the TCAT system is assumed.

A basic architecture of the TCAT system is shown in Fig. 1. The system operates as follows. First, a user sends a request for the services of the system via a WWW browser. Then, the WWW server (Apache) starts the PHP interpreter. A PHP module – a component of the TCAT system – fetches necessary data from the database and the Internet and generates a HTML code enriched with JavaScript functions. Communication with the MySQL database is carried out using a standard PHP module exposing an interface for accessing this database. The HTML code is interpreted by the browser and the results are displayed as screens forming a user interface of the TCAT system.

The way of the interaction with the system in both phases is simple. The user interface consists of a logically connected set of WWW pages Interfejs containing regular forms.

Processing of the documents goes in a few steps. First, the content of a document is read from either a local disk or directly from the Internet and stored in a buffer. Then, the text read in previous step is preprocessed. All elements of the HTML, DHTML, XHTML, XML, PHP and ASP code are removed. This is accomplished through a simple lexical scanner reading subsequent characters of the string, analyzing them and removing fragments semantically not important. This step is more sophisticated than a mere removal of the whole HTML tags or scriptlets. Their content is also analyzed and the substrings that are possibly important for text categorization are preserved. During the lexical analysis the addresses of the linked documents are extracted. Then, also these documents are processed. A further lexical processing includes the removal of all not alphanumerical characters, and the change of uppercase letters to lowercase. The text of a document preprocessed in such a way is then forwarded to the next stages of the analysis.

The database of the TCAT system is used mainly to store the characteristics of all tokens extracted from the training documents. The most important element of such a characteristic is the degree of membership of the token to particular categories.

6 An Example

The TCAT system has been tested on the set of text documents representing 6 thematic categories. The choice of the testing material was motivated by the popularity of these categories in the resources of the leading Polish Internet portals. The selected categories are:

- nature and ecology,
- economy,
- movies,
- computers and the Internet,
- cars,
- politics, society and law

In the learning phase 20 documents per category have been used. The results of this phase are illustrated in Fig. 2. This figure shows the degrees of membership of selected tokens to the category "politics, society and law".

The testing documents, 10 per category, have been selected from the site of a Polish Internet multimedia encyclopedia (http://wiem.onet.pl/wiem). Each test documents is a description of a certain, relevant for a given category, keyword from the encyclopedia.

While evaluating results of the categorization done by the TCAT system we have to take into account the fuzziness of response of the system. As it is described earlier, the system in the classification phase yields degrees of membership of the document to all (in this case 6) categories. On the other hand, the information about the actual category of test documents is crisp and indicating exactly one category. In order to assess the correctness of fuzzy classification response of the system we have adopted two approaches. In the first approach we accept a fuzzy response as correct if the actual category of the test document belongs to the set of categories for which the system produced the highest membership degree. The second approach poses a stronger requirement: we require that the actual category of the document is pointed out by the system unambigously. Formally it may be denoted as follows:

Approach I ("simple correctness")

$$P1 = 100 * \frac{\sum_{i=1}^{M} \phi(d_i)}{MT}$$

where MT is the cardinality of set $DT=\{d_i\}$ of test documents and

$$\phi(d_i) = \begin{cases} 1 & \text{gdy } SP_{k_*}^{d_i} \geq SP_{k_j}^{d_i} \; \forall j \\ 0 & \text{wpp} \end{cases}$$

and k_* is the actual category of document d_i.

Tabela 1: Accuracy of the classification.

	P1: simple correctness	P2: strong correctness
Category 1 (*nature and ecology*)	90 %	79.5 %
Category 2 (*economy*)	90 %	69.75 %
Category 3 (*movies*)	80 %	79.00 %
Category 4 (*computers and Internet*)	90 %	75.50 %
Category 5 (*cars*)	100 %	66.80 %
Category 6 (*politics, society and law*)	70 %	57.50 %
Total	87 %	71 %

Thus, P1 corresponds to the percentage of the documents correctly classified in the sense of the first approach.

Approach II ("strong correctness")

$$P2 = 100 * \frac{\sum_{i=1}^{MT} p2(d_i)}{MT}$$

$$p2(d) = 100 * \frac{p(d,k_*)}{\sum_j^{|K|} p(d,k_j)}$$

where $p(d, k_j)$ assumes an integer value depending on what linguistic term [see (5)] the system has used to describe the membership degree of document d to category k_j. For the terms "not", "slightly", "medium", "strongly" and "very strongly" the numbers 0, 25, 50, 75 and 100 are used, respectively. As previously, k_* denotes the actual category of document d, and MT – the cardinality of the set of test documents $DT=\{d_i\}$. Thus, $P2$ corresponds to the percentage of the documents correctly classified in the sense of the second approach.

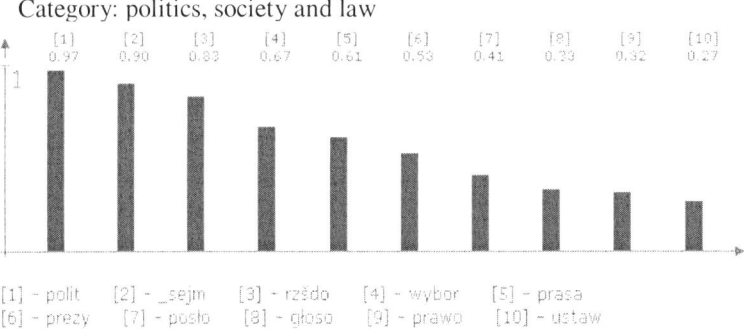

Figure 2: Degrees of membership of some tokens to the selected category

The results of the classification done by the TCAT system as assessed using both approaches are illustrated in Table 1. In most cases the system pointed out the strongest membership of a document to the actual category of the document.

7 Concluding Remarks

We have presented a concept, architecture and implementation of the TCAT system performing an automatic categorization of Internet documents. The results of preliminary experiments with the system have been shown. The original features of our approach include: a human consistent user interface, new measures of accuracy, and a simple, parametrized classifier design.

The proposed approach requires a further research. First of all, the classifier itself has to be thoroughly tested using standard sets of documents and widely adopted testing techniques (e.g., cross-validation). An approach to tuning the systems parameters has to be proposed. Possibly some changes to the very learning algorithm will be introduced. We hope to be able to present more rigorous tests results (for, e.g., Reuters-22173 dataset) during the conference. We plan also to conduct some experiments on the multilingual documents sets.

A further study will also concentrate on the use of well knwn concepts developed within the theory of fuzzy sets. For example, the measure of accuracy (approach II) may be further studied employing the concept of a specificity measure of a fuzzy set proposed by Yager (see [4]). Secondly, the concept of membership of a token to a category directly leads to the interpretation of the category as a fuzzy set defined over the space of tokens. Then, some well known approaches to the measuring of similarity of fuzzy sets may be employed. Finally, the very task of categorization may be redefined taking into account that a document usually belongs, to a varying degree, to different categories. Thus, fuzzy approaches to so-called *multi-class* or *multi-label categorization problem* (see, e.g., [9]) are certainly worth a further study.

References

[1] Aas K., Eikvil L. (1999) Text Categorisation: A Survey. Norwegian Computing Center Rapport 941, June 1999.

[2] Boley D. et al. (1999). Document categorization and query generation on the World Wide Web using WebACE. *AI Review*, vol. 13, N. 5-6, pp. 365-391.

[3] Crestani F., Pasi G. (2000) *Soft Computing in Information Retrieval*. Physica-Verlag, Heidelberg.

[4] Dubois, D. i H. Prade (1987) Properties of measures of information in evidence and possibility theories. *Fuzzy Sets and Systems*, 24, 161-182.

[5] Joachims T. (1997) A probabilistic analysis of the Rocchio algorithm with TFIDF for text categorization. In *Proceddings of ICML-97*.

[6] Kacprzyk J., Zadrożny S. (2001). Computing with words in intelligent database querying: standalone and Internet-based applications. *Information Sciences*, No. 134, pp. 71-109

[7] Kuncheva, L.I. (2000). *Fuzzy Classifier Design*. Physica-Verlag, Heidelberg New York.

[8] Ławcewicz, K.A., Zadrożny S. (2001) System automatycznego rozpoznawania kategorii tematycznych dokumentów internetowych. In J. Studziński, L. Drelichowski, O. Hryniewicz (Eds.): Rozwój i Zastosowania Technologii i Systemów Informatycznych. IBS PAN, Warszawa 2001, pp. 229-238, *in Polish*

[9] McCallum A.K. (1999) Multi-label text classification with a mixture model trained by EM. In *Proceedings of the AAAI'99 Workshop on Text Learning*.

[10] Nigam K., Lafferty J., McCallum A. (1999). Using maximum entropy for text classification. In *Proceedings of the IJCAI-99 Workshop on Machine Learning for Information Filtering*, pp. 61-67.

[11] van Rijsbergen C. J. (1979) *Information Retrieval*. Butterworths, London.

[12] Rocchio J. (1971) Relevance feedback in information retrieval. In "The SMART Retrieval System: Experiments in Automatic Document Processing", Prentice-Hall Inc. pp. 313-323.

[13] Salton G., McGill M. J. (1983). *Introduction to Modern Information Retrieval*. McGraw-Hill, New York.

[14] Salton G, Buckley C. (1988) Term-weighting approaches in automatic text retrieval. *Information Processing and Retrieval.* Vol. 5, 24, pp. 513-523.

[15] Schapire R. E., Singer Y. (2000). BoosTexter: A boosting-based system for text categorization. *Machine Learning*, vol. 39, 2/3, pp. 135-168.

[16] Shanahan J. (2001). Modelling with words: an approach to text categorization. In *Proceedings of the conference FUZZ-IEEE 2001*, vol. 1, Melbourne, Australia, December 2001.

[17] Yang Y. (1999). An Evaluation of Statistical Approaches to Text Categorization. *Information Retrieval*, vol. 1, No. 1 / 2, pp. 69-90.

[18] Yang Y., Liu X. (1999). A re-examination of text categorization methods. In *Proceedings of SIGIR-99, 22nd ACM International Conference on Research and Development in Information Retrieval, Berkeley, 1999.*

[19] Yang Y., Slattery S., Ghant R. (2001) A study of approaches to hypertext categorization. *Journal of Intelligent Information Systems.* Vol. 18, Number 2, March 2002, to appear.

Reasoning

B. Bouchon-Meunier, L. Foulloy and R.R. Yager (Editors)
Intelligent Systems for Information Processing:
From Representation to Applications
© 2003 Elsevier Science B.V. All rights reserved.

Prototype Based Reasoning and Fuzzy Modeling

Ronald R. Yager
Machine Intelligence Institute, Iona College
New Rochelle, NY 10801
yager@panix.com

Abstract
We introduce the methodology of prototype based reasoning and discuss its role as a technology for supplying missing information about some object based on known information about related objects. We show that nearest neighbor based systems and fuzzy rule based models are examples of prototype based reasoning. This perspective allows us to extend the capabilities of fuzzy modeling technology in a number of directions. One such extension discussed here is to suggest a method for fusing multiple fuzzy systems models.

Keywords: fuzzy modeling, fuzzy measures, nearest neighbor principle, information fusion

1. Introduction

An important class of models are those in which we use known information about a collection of objects to provide missing information about some other object of interest. Much of modern information based technologies focus on this problem. Here we consider one framework for modeling this type of inference called **P**rototype **B**ased **R**easoning (PBR).

2. Prototype Based Reasoning Systems

A PBR system consists of a collection of entities called prototypes, $A = \{A_1, A_2, ..., A_n\}$. A common use of PBR is to determine, based on information about the prototypes, the degree to which some non-prototype object has a particular feature. We shall call the feature on which we are focusing the notable feature (NF). It is assumed that for any prototype we know the degree, a_i, to which it has this notable feature, $a_i \in [0, 1]$. An important part of a PBR system is credibility measure μ. Formally $\mu: 2^A \to [0, 1]$ where for any subset $E \subseteq A$, $\mu(E)$ is the degree of credibility associated with any conclusion based upon using the subset E of prototypes. Some natural properties of this credibility measure are: $\mu(\emptyset) = 0$ and if $E \subset F$ then $\mu(E) \leq \mu(F)$, the more prototypes used the more credible the conclusion. We shall assume $\mu(A) = 1$, any conclusion based upon all the prototype is completely credible. This last condition may not be necessary. We note that a set functions having these properties are called fuzzy measures [1, 2]. In PBR we are interested in using the set of prototypes to determine the validity of the notable feature for some target object. In order to accomplish this we must assume the availability of some procedure to determine the similarity (relevance) of a given prototype to the target object. As we would expect the more similar a prototype the more useful it is the determination of the information we are seeking. We shall indicate the degree of relevance of prototype A_i to the object of interest by $\eta_i \in [0, 1]$.

A PBR reasoning system is a type of possibilistic inference engine. Under the prototype based reasoning paradigm we say that the degree to which the target object has the notable feature is equal to the degree to which we can find

a relevant credible subset of prototypes having this feature

That is if we find a relevant credibility set of prototypes that we assume the target has this feature.

In order to formally express this imperative we shall introduce some fuzzy subsets over the power set of A, 2^A. First we note the μ is a fuzzy subset of 2^A, $\mu(E)$ indicating the degree of credibility of the subset E. We next introduce the fuzzy subset Q of 2^A. We define it by $Q(E) = \underset{A_i \in E}{\text{Min}} [a_i]$, it measures the degree to which **all** the elements in E have the feature of interest. Finally we introduce the fuzzy subset R over 2^A in which R(E) indicates the relevancy of the subset E to the object of interest. We shall initially leave the definition of R open but we note that it should depend upon the similarity of the prototypes in E to object of interest: $R(E) = f(\eta_1, ..., \eta_n)$

Using these fuzzy subsets we define the fuzzy set $D = \mu \cap Q \cap R$. D is the fuzzy subset of subsets of prototypes that are credible, similar to the object of interest and having elements that have the property of interest. Using this we calculate

$$\tilde{a} = \text{Max}_{E \subseteq A}[D(E)] = \text{Max}_{E \subseteq A}[\mu(E) \wedge Q(E) \wedge R(E)]$$

it is the degree to which our target object has the feature of interest.

By appropriate selection of μ and R we can generate different manifestations of this PBR paradigm. We shall look at the PBR systems resulting from some different assumptions about the form of μ and R. However before preceding we shall look at some general formulation of these measures. First we shall look as the credibility measure μ.

One important class of credibility measures are cardinality based credibility measures. For these measures $\mu(E) = w_{|E|}$, the credibility of any subset E just depends upon the number of elements in E. Here then we only require we have $w_j \geq w_i$ $j > i$ and $w_0 = 0$ and $w_n = 1$. With cardinality based measures there is no distinction made between the credibilities of the different prototypes, all are assumed to be the same, the credibility just depends upon the number of prototypes considered. Three important special cases of cardinality based credibility measures are worth pointing out. The first is one in which $w_1 = 1$, here we deem any prototype to be completely credible. In this case a conclusion based upon any number of prototypes is assumed completely credible. The second case is the one in which $w_j = 0$, for $j = 1$ to $n - 1$ and $w_n = 1$. Here we any conclusion requires consideration of all the prototypes. A third special case is the one in which subset credibility is directly proportional to the number of prototypes in the subset, $w_i = \frac{i}{n}$, . A general class of cardinality based measures can be defined by using a function $f: [0, 1] \rightarrow [0, 1]$, called a BUM function, having the properties: **1.** $f(0) = 0$, **2.** $f(1) = 0$ and **3.** $f(x) \geq f(y)$ if $x > y$. Using this BUM function we define $w_i = f(i/n)$.

Moving away from the cardinality based credibility measures, which make no distinction between the credibility of individual prototypes, a basic type of credibility measure is one in which we associate with each A_i a value $\alpha_i \in [0, 1]$ and then define $\mu(E) = \frac{1}{T} \sum_{i \in E} \alpha_i$ A generalization of this is one in which we use a BUM function f as defined above and define $\mu(E) = f(\frac{1}{T} \sum_{i \in E} \alpha_i)$.

We now briefly comment on the relevancy function R. One primal example of relevancy function is the one in which $R(E) = 1$ for all E. Essentially, in this case we are not using any information about the similarity of the prototypes to the object of interest. In this case $\tilde{a} = \text{Max}_{E \subseteq A}[Q(E) \wedge \mu(E)]$. This is the fuzzy integral [1, 3].

3. Nearest Neighbor Type PBR Systems

Here we begin to look at some special classes of prototype based reasoning systems. The first class we shall consider are *nearest neighbor based systems* [4-6]. The fundamental feature of these systems is captured by the following imperative characterizing this approach.

Nearest Neighbor imperative:
Don't use any prototype object to determine the notable feature unless you also use all the prototypes that are more similar (more relevant).

As we shall subsequently see, this imperative puts some very interesting restrictions on the form of the relevancy function R. In order to most easily implement this imperative we shall introduce some ideas developed in [7]. Let **P** be a partial ordering over the set **A**. We shall let $A_i >_p A_j$ indicate that A_i is higher in the ordering than A_j and $A_i =_p A_j$ indicate that they are tied in the ordering. We shall call a subset E of **A** a <u>rooted sub-ordering</u> with respect to **P** if for any $A_j \in E$ we also have all $A_k >_p A_j$. Thus E is a rooted sub-ordering if it contains all the elements <u>ahead</u> of its lowest element. It should be emphasized that elements tied with the lowest element may or may not be in the E. Using this definition we can now introduce the nearer neighbor imperative. Let **P** be a partial ordering induced by the relevancy of the prototypes to the object of interest thus if η_j is the relevancy of prototype A_j to the object of interest, then if $\eta_j > \eta_k$ we have $A_j >_p A_k$ and if $\eta_j = \eta_k$ we have $A_j =_p A_k$. The requirement of the nearer neighbor principal can now be introduced by specifying that the relevancy function R be such that $R(E) = 1$ if E is a rooted sub-ordering of **P** and $R(E) = 0$ otherwise. Thus in this case $\tilde{a} = \text{Min}_{E \in RS_p}[\mu(E) \wedge Q(E)]$ where RS_p is the set of rooted sub-orderings of **A** with respect to **P**.

In [7] Yager showed using the properties of $\mu(E)$ and $R(E)$ the following useful result. Let $\hat{P}$ be a linear ordering generated from the partial ordering **P** such that elements that are tied with respect their relevancy value are lexically adjudicated by their notable feature value in descending order, thus if two prototypes are tied in **P** we place the one with the larger a_j value higher in $\hat{P}$. Let $\hat{p}$-index(j) be the index of the prototype that ii in the jth position in the ordering $\hat{P}$. Yager showed [7] that $\tilde{a} = \text{Max}_{j=1 \text{ to } n}[\mu(G_j) \wedge Q(G_j)]$ where $G_j = \{A_{\hat{p}-\text{index}(k)} \mid k = 1 \text{ to } j\}$. We note that G_j consists of the j top elements in the ordering induced by the relevancy, the j nearest neighbors. We further note that $Q(G_j) = \text{Min}_{k=1 \text{ to } j}[a\hat{p}\text{-index}(k)]$, it is the minimal satisfaction among the jth most relevant items.

Special examples of nearest neighbor type prototype based reasoning can be exhibited by different selections of μ. If we consider the case where all prototypes are considered fully credible, $\mu(E) = 1$ for all $E \neq \emptyset$, then $\tilde{a} = \text{Max}_{j=1 \text{ to } n}[Q(G_j)]$. Since $Q(E)$ is a decreasing function, $Q(E) \leq Q(F)$ if $F \subset E$ and since $G_j \subset G_i$ if $i > j$ then for this case of credibility function $\tilde{a} = Q(G_1) = a\hat{p}\text{-index}(1)$. The valuation is the same as that of the single nearest neighbor, this is the simple nearest neighbor rule.

More generally if $\mu(E)$ is a cardinality-based credibility measure such that $\mu(E) = 0$ for $|E| < K$ and $\mu(E) = 1$ for $|E| \geq K$ we see that $\tilde{a} = \text{Min}_{i=1 \text{ to } K}[a\hat{p}\text{-index}(i)]$, the minimal satisfaction by any of the K closest neighbors. This can be seen as a kind of Kth nearest neighbor rule.

In the preceding we just used the ordering over the prototypes. We can consider a modification of this to include the relevancy values in the actual calculation of $\tilde{a}$. Here we let $R(E) = R_1(E) \cap R_2(E)$ where R_1 enforces the nearest neighbor imperative, $R_1(E) = 1$ if $E \neq G_i$ and $R_1(E) = 0$ if $E \neq G_i$. We use R_2 to express information about the actual relevancy values, $R_2(E) = \text{Min}_{j \text{ s.t. } A_j \in E}[\eta_j]$. Combining these we get R such that $R(G_i) = \text{Min}_{j=1 \text{ to } i}[\eta\hat{p}\text{-index}(j)]$ and $R(E) = 0$ for $E \neq G_i$. Using this we have $\tilde{a} = \text{Max}_i[\mu(G_i) \wedge R(G_i) \wedge Q(G_i)]$ with $Q(G_i) = \text{Min}_{k=1 \text{ to } i}[a\hat{p}\text{-index}(k)]$.

If we further assume μ is such that all $\mu(\{A_i\}) = 1$ for all i, $\mu(E) = 1$ for all $E \neq \emptyset$, then $\tilde{a} = \eta\hat{p}\text{-index}(1) \wedge a\hat{p}\text{-index}(1)$, the score of the nearest neighbor "weighted" by its degree of relevance. If we assume $\mu(E) = 0$ for $|E| < K$ and $\mu(E) = 1$ for $|E| \geq K$, then it can be shown that $\tilde{a} = \text{Min}_{j=1 \text{ to } K}[(\eta\hat{p}\text{-index}(j) \wedge a\hat{p}\text{-index}(j)]$. One final case is where $\mu(\{A_i\}) = \alpha_i$ and $\mu(E) = \text{Max}_{i \text{ s.t. } A_i \in E}[\alpha_i]$. Here after some calculations we get

$\tilde{a} = \text{Max}_{i=1 \text{ to } n}[\text{Max}_{j=1 \text{ to } i}[\alpha\hat{p}\text{-index}(j)] \wedge \text{Min}_{j=1 \text{ to } i}[(\eta\hat{p}\text{-index}(j) \wedge a\hat{p}\text{-index}(j))]]$

4. Fuzzy Modeling as Prototype Based Reasoning

We now briefly review the fuzzy systems modeling approach as introduced by Mamdani [8, 9] and Zadeh [10], a more comprehensive discussion can be found in [11, 12]. As we shall subsequently see this provides an example of prototype based reasoning. Our viewing this technology as a case of PBR will enable us to expand its modeling capabilities.

The basic fuzzy systems model consists of a collection of n rules of the form:

If V_1 is B_{1j} **and** ......**and** V_i is B_{ij} **and** .... **and** V_m is B_{mj} **then** W is D_j

Here the V_i are variables taking their values in the spaces X_i and W is a variable taking its value in the space Y. The V_i are called the antecedent (input) variables and W is called the consequent (output) variable. The B_{ij} are fuzzy subsets of X_i and D_j is a fuzzy subset of Y. At times it may be more convenient to write a rule as If V is B_j then W is D_j here

V is a joint variable, $V = (V_1, ..., V_m)$, taking its value in $X = X_1 \times ... \times X_m$ and B_j is a fuzzy of X with $B_j = B_{1j} \times B_{2j} \times ... \times B_{mj}$.

The typical application of fuzzy systems modeling consists of a situation in which we have information about the antecedent variable, $V_i = x_i^*$, $i = 1$ to m, and we are interested in determining the value of W. We shall denote the fuzzy subset of Y corresponding to this output as F. The procedure used to obtain F, called fuzzy inferences, is as follows. First we determine the firing level or relevancy of each rule, as $\eta_j = \text{Min}_{i=1 \text{ to } m}[B_{ij}(x_i^*)]$, here $B_{ij}(x_i^*)$ indicates the membership grade of x_i^* in B_{ij}. We then calculate the effective output of each rule F_j, a fuzzy subset of Y with membership $F_j(y) = D_j(y) \wedge \eta_j$. The next step is the aggregation of the individual rule effective outputs to give us the overall output fuzzy set F, $F(y) = \text{Max}_j[F_j(y)]$.

Often a step of defuzzification is applied to obtain a crisp output [13]. This step is not of concern to us here.

What we shall show in the following is that this fuzzy modeling framework can be viewed as a type of prototype based reasoning with respect to the determination of the membership grades of F, the $F(y)$.

In the PBR framework we shall consider each rule as a prototype. We shall denote the jth rule as A_j, thus $\mathbf{A} = \{A_1, ..., A_n\}$ is our collection of prototypes. In the determination of the membership grade of the element y in the consequent, $F(y)$, we shall associate with each prototype an argument $D_j(y)$, the degree to which the prototype supports the element y. Given an input object, $(x_1^*, ..., x_m^*)$ the relevancy of the prototype A_j is $\eta_j = \underset{i=1 \text{ to } m}{\text{Min}} [B_{ij}(x_i^*)]$.

Using the PBR framework the desired output $F(y) = \text{Max}_{E \subseteq \mathbf{A}}[\mu(E) \wedge Q(E) \wedge R(E)]$. Here R is the relevancy of the subset E of rules, Q(E) is the degree to which all the rules in E support the output value y and $\mu(E)$ is the degree of credibility associated with using the subset E of rules. For the basic fuzzy model we define the relevancy function R as $R(E) = \underset{j \text{ s.t. } A_j \in E}{\text{Min}} [\eta_j]$. The relevancy of a subset of rules is the minimum relevancy of any rule in the subset. Q is defined as $Q(E) = \underset{j \text{ s.t. } A_j \in E}{\text{Min}} [D_j(y)]$. The support for y associated with a subset of rules is the degree to which all the rules support y. Finally in this basic model we assume that $\mu(\{A_i\}) = 1$ for all i, each rule has complete credibility. This of course means that $\mu(E) = 1$ for all $E \neq \emptyset$. Since $\mu(E) = 1$ for all $E \neq \emptyset$ we get $F(y) = \text{Max}_{E \neq \emptyset}[Q(E) \wedge R(E)]$. Since $Q(E) = \underset{j \text{ s.t. } A_j \in E}{\text{Min}} [D_j(y)]$ and $R(E) = \underset{j \text{ s.t. } A_j \in E}{\text{Min}} [\eta_j]$ then for any subset E we have $Q(E) \wedge R(E) \leq Q(\{A_j\}) \wedge R(\{A_j\})$ for $A_j \in E$. From this we get that $F(y) = \text{Max}_{j=1 \text{ to } n}[\eta_j \wedge D_j(y)]$ which is the fuzzy systems model.

Thus we see that with appropriate choice of relevancy (R), credibility (μ) and support (Q) functions the basic fuzzy systems modeling can be viewed as an example of prototype based reasoning.

5. Extending Fuzzy Systems Modeling with Weighted Rules

Viewing the fuzzy system model as a type of PBR allows us to extend capabilities of fuzzy system modeling. As a first extension of the classic fuzzy model we shall look at the situation in which we don't assume all of the individual rules are completely credible we shall assume that $\mu(\{A_i\}) = \alpha_i$ and $\mu(E)$ depends on the prototypes in E.

Here we still have $Q(E) = \underset{j \text{ s.t. } A_j \in E}{\text{Min}} [D_j(y)]$ and $R(E) = \underset{j \text{ s.t. } A_j \in E}{\text{Min}} [\eta_j]$. In this situation using the PBR model we have $F(y) = \text{Max}_{E \subseteq A} [\mu(E) \wedge Q(E) \wedge R(E)]$. We shall find it convenient to denote $d_j = D_j(y)$ and to denote $f_j = d_j \wedge \eta_j$. We shall call f_j the effective value of prototype A_j. Also we shall denote $QR = Q \cap R$. We see that from the current definitions of Q and R we have $QR(E) = Q(E) \wedge R(E) = \underset{j \text{ s.t. } A_j \in E}{\text{Min}} [f_j]$
Using this notation we have $F(y) = \text{Max}_{E \subseteq A} [\mu(E) \wedge QR(E)]$

While we have assumed that the credibilities of the individual prototype rules are α_j we have not indicated any other structure on the credibility measure μ. By imposing particular additional structure on μ we can obtain useful formulations for the determination of the overall consequent value.

As a first case we shall assume that the credibility measure has the properties of a possibility measure [14], $\mu(E) = \underset{j \text{ s.t. } A_j \in E}{\text{Max}} [\alpha_j]$. We can show under this assumption
about the form of the credibility function we obtain as our formulation for consequent fuzzy set a weighted aggregation of effective values of each prototype
$$F(y) = \text{Max}_{j=1 \text{ to } n} [\alpha_j \wedge f_j].$$

More generally we can assume a t-conorm S instead of Max in defining μ. In this case $F(y) = \text{Max}_{i=1 \text{ to } n} [f_{f\text{-index}(i)} \wedge S_{k=1 \text{ to } i} [\alpha_{f\text{-index}(k)}]]$ In the case of the bounded sum, $S(a, b) = (a + b) \wedge 1$, we get

$$F(y) = \text{Max}_{i=1 \text{ to } n} [f_{f\text{-index}(i)} \wedge \sum_{k=1}^{i} \alpha_{f\text{-index}(k)}].$$

To help better understand the relationship of the PBR reasoning to fuzzy modeling we introduce the idea of compounding of fuzzy rules. Let A_1 and A_2 be two fuzzy rules-
A_1: If V_1 is B_2 then W is D_1 and **A_2:** If V_2 is B_2 the W is D_2. Here V_1 and V_2 may be atomic or joint variables, which may or may not be the same. We define a new rule called the compounding of A_1 and A_1 denoted Comp(A_1, A_2) as

If V_1 is B_1 <u>and</u> V_2 is B_2 **then** W is D_1 <u>and</u> D_2

We see compounding A_1 and A_2 defines a new rule that is a conjunction of the antecedents and a conjunction of the consequents of each of the rules being compounded. For a given input x* if the firing levels of A_1 and A_2 are η_1 and η_2 then the firing level of the compound rule is $\eta_1 \wedge \eta_2$, the consequent membership grade is $D_1(y) \wedge D_2(y)$ and the efficient value is $D_1(y) \wedge D_2(y) \wedge \eta_1 \wedge \eta_2$ The extension to the compounding of any number of rules is straight forward.

Using the definitions of Q(E) and R(E) we can see a relationship between the process of compounding and the PBR reasoning mechanism introduced. Let E be a subset of **A** then compounding the rules in E generates a rule, Comp(E), whose antecedent is the conjunction of all the antecedents of the rules in E and whose consequent is the conjunction of all the consequents of the rules in E. If for a given input x* the firing level of the ith rule is η_i then the firing level of the rule Comp(E) is Q(E), the consequent of rule Comp(E) is Q(E) and the effective value of E, is Q(E) ∧ R(E) = QR(E). Thus we see that in the PBR model when using the form $F(y) = Max_{E \subseteq A}[\mu(E) \wedge Q(E) \wedge R(E)]$ we are considering the compounding of the all rules with $\mu(E)$ representing the credibility of the rule Comp(E).

6. Partitioned Fuzzy Rule Sets

We consider another formulation for the credibility function. Let $A = \{A_i, ..., A_n\}$ be our collection of fuzzy rules and assume we partition these into q disjoint classes M_j, $\cup_j M_j = A$ and $M_j \cap M_i = \emptyset$. Using this we first define the degree of inclusion of M_j in E, INC[M_j/E], as follows: INC[M_j/E] = 1 if $E \cap M_j \neq \emptyset$ and INC[M_j/E] = 0 if $E \cap M_j = \emptyset$. Using this indicate the meta–cardinality of a subset E with respect to the partition as M-C(E) = $\sum_{j=1}^{q}$ INC[M_j|E]. Using this we can define a credibility function μ on A: $\mu(E) = w_{M-C(E)}$, where the w_j are collection of weights such that: $w_0 = 0$, $w_q = 1$ and $w_i \geq w_j$ for i > j. Here we see that $\mu(E)$ is related to how many of the different classes of rules are contained in E.

We now shall use the formulation for PBR, $F(y) = Max_{E \subseteq A}[\mu(E) \wedge Q(E) \wedge R(E)]$ where R(E) = Min[$\eta_j \in E$], Q(E) = Min[$d_j \in E$] and $\mu(E)$ is as just defined. We note that if $E' \subset E$ then $Q(E') \geq Q(E)$ and $R(E') \geq R(E)$. We also note that for this μ if E contains multiple elements from one class and, if G is a subset obtained from E by removing all the elements of a class except for one, then $\mu(E) = \mu(G)$. In this situation since $G \subseteq E$ then $Q(G) \geq Q(E)$ and $R(G) \geq R(E)$, this implies $Q(G) \wedge \mu(G) \wedge R(G) \geq Q(E) \wedge \mu(E) \wedge R(E)$.

We shall let H denote the collection Minimal of subsets of A,. a subset of A is contained H if it contains at most one element from each class. Using this notation we see that $F(y) = Max_{G \in H}[\mu(G) \wedge Q(G) \wedge R(G)]$. We shall further use the notation $f_k = \eta_k \wedge d_k$ and recall $Q(G) \wedge R(G) = QR(G) = Min[f_k \in G]$. Using this $F(y) = Max_{G \in H}[\mu(G) \wedge QR(G)]$. Let A_{m_j} be the prototype in M_j having the largest value for f_k. We call A_{m_j} M_j's most effective prototype. Let H^* be the subspace of H consisting of the collection of subsets of A where each subset in H^* contains at most one element from any M_j and that element is always the most effective prototype, A_{m_j}. From the preceding we see that $F(y) = Max_{G \in H^*}[\mu(G) \wedge QR(G)]$. Since $\mu(G) = w_k$ if M-C(G) = k then $F(y) = Max_{k=1 \text{ to } q}[w_k \wedge \underset{G \in H^*, |G|=k}{Max}[QR(G)]$. Let fm-index(k) be the index of the class having the kth largest value for f_{m_j}. Using this notation we can show that $F(y) = Max_{k=1 \text{ to } l}[w_k \wedge c_k]$ where $c_k = f_{m_{\text{fm-index}(k)}}$

The procedure just described for obtaining $F(y)$ can be simply expressed. For each prototype we obtain its firing level for the given input, η_i, and we calculate $f_i = \eta_i \wedge D_i(y)$. For each class M_j, we determine the largest f_i for any $A_i \in M_j$, we denote this f_{m_j}. We then order these f_{m_j} values such that c_k is the kth largest of these and then we calculate $F(y) = Max_{k=1 \text{ to } l}[w_k \wedge c_k]$.

In the following we define a situation which illustrates the preceding structure. Assume we have a model in which we have q antecedent variables, V_j, each with domain X_j and let W be our consequent variable taking its value in Y. Assume we have a collection prototype fuzzy rules of the form *If V_j is B_{ij} then W is F_{ij}*. We shall denote a rule of this type as A_{ij}, the ith rule involving the jth variable. Thus here each rule expresses information about the consequent just based on one variable. Consider now a partitioning of these rules by the antecedent variable, for all i we have $A_{ij} \in M_j$. Here then we see that w_k will be the credibility associated with an inference using k different variables.

We shall now consider an extension of the preceding. Here we still assume a partitioning of the prototypes but allow for different credibility for each of the prototypes. In the preceding illustration this may correspond to a situation in which we attribute more credibility to some variables over different ranges. For example rules over a range of values in which little experience may be stated with less credibility. With a partitioning of the prototypes into q disjoint subsets, M_j, and with A_{ij} being the ith rule in the jth partition and we let α_{ij} be the credibility of A_{ij}, $\mu(\{A_{ij}\}) = \alpha_{ij}$. We let N_j be the number of rules in partition j. We shall also assume for each partition there exists at least one element in each M_j with credibility equal one.

We generalize the idea $INC(M_j|E)$ to $DINC[M_j|E]$, the *degree* that M_j is included in E. We define this as $DINC[M_j|E] = Max_{A_{ik} \in M_j}[\alpha_{ik} \wedge E(A_{ik})]$, maximal credibility of any rule in E. Now we now define $\mu(E)$. In the early case we used $\mu(E) = w_{M-C(E)}$, where M-C(E) was the number of different classes included in E. Here we must generalize this idea. We shall suggest some ways to generalize this. One way to generalize this just using ordinal operations is as follows. Let DC–index(i) be the index of the partition with the ith largest value for $DINC[M_j|E]$ using this we define $\mu(E) = Max_{k=1 \text{ to } q}[DINC(M_{DC\text{-index}(k)}/E) \wedge w_k]$. Here the w_k are a set of weights as defined above.

Another approach is to calculate $\widetilde{DC} = \sum_{j=1}^{q} DINC[M_j|E]$ and then let $Int(\widetilde{DC})$ be the integer portion of $\widetilde{DC}$ and $\Delta = \widetilde{DC} - Int(\widetilde{DC})$ then define

$$\mu(E) = w_{Int(\widetilde{DC})} + \Delta(w_{Int(\widetilde{DC})+1} - w_{Int(\widetilde{DC})})$$

To help better understand the application of the PBR reasoning to fuzzy modeling we introduce the idea of compounding of fuzzy rules. Let A_1 and A_2 be two fuzzy rules-
A_1: If V_1 is B_2 then W is D_1 and **A_2:** If V_2 is B_2 the W is D_2. Here V_1 and V_2 may be atomic or joint variables, which may or may not be the same. We define a new

rule denoted Comp(A_1, A_2) as

$$\text{If } V_1 \text{ is } B_1 \text{ and } V_2 \text{ is } B_2 \text{ then } W \text{ is } D_1 \text{ and } D_2$$

We see compounding A_1 and A_2 defines a new rule that is a conjunction of the antecedents and a conjunction of the consequents of each of the rules being compounded. For a given input x* if the firing levels of A_1 and A_2 are η_1 and η_2 then the firing level of the compound rule is $\eta_1 \wedge \eta_2$, the consequent membership grade is $D_1(y) \wedge D_2(y)$ and the efficient value is $D_1(y) \wedge D_2(y) \wedge \eta_1 \wedge \eta_2$ The extension to the compounding of any number of rules is straight forward.

Using the definitions of Q(E) and R(E) we can see a relationship between the process of compounding and the PBR reasoning mechanism introduced. Let E be a subset of A then compounding the rules in E generates a rule, Comp(E), whose antecedent is the conjunction of all the antecedents of the rules in E and whose consequent is the conjunction of all the consequents of the rules in E. If for a given input x* the firing level of the ith rule is η_i then the firing level of the rule Comp(E) is Q(E), the consequent of rule Comp(E) is Q(E) and the effective value of E, is Q(E) $\wedge$ R(E) = QR(E). Thus we see that in the PBR model when using the form $F(y) = \text{Max}_{E \subseteq A}[\mu(E) \wedge Q(E) \wedge R(E)]$ we are considering the compounding of the all rules with $\mu(E)$ representing the credibility of the rule Comp(E).

7. Fuzzy Systems Models Using a Nearest Neighbor Principle

Now we consider the application of a nearest neighbor principle to fuzzy systems modeling. We assume a collection of fuzzy rules $A = \{A_1, ..., A_n\}$ with consequent variable W taking its value on the space Y. We let the fuzzy subset corresponding to the output of this PBR system, $F(y) = \text{Max}_{E \subseteq A} [\mu(E) \wedge R(E) \wedge Q(E)]$. As in the preceding we shall assume $Q(E) = \text{Min}[d_j \in E]$.

We let η_j be the firing level of the jth rule and let η-index(i) be the index of the ith strongest firing rule.. The nearest neighbor principle says if a > b don't use prototype $A_{\eta\text{-index}(a)}$ without using prototype $A_{\eta\text{-index}(b)}$. Let $G_i = \{A_{\eta\text{-index}(k)} | k = 1 \text{ to } i\}$, it is the set of the i most relevant rules. Under the nearest neighbor principle, $R(G_i) = 1$ and R(E) = 0 for E $\neq G_i$. Using this principle we have

$$F(y) = \text{Max}_{i=1 \text{ to } n}[\mu(G_i) \wedge Q(G_i)]$$
$$F(y) = \text{Max}_{i=1 \text{ to } n}[\mu(G_i) \wedge \text{Min}_{k=1 \text{ to } i}[d_{\eta\text{-index}(k)}])]$$

Let us now focus on the function μ, the credibility function. If μ is, as it was in the basic case, defined as $\mu(E) = 1$ for all E then

$$F(y) = \text{Max}_{i=1 \text{ to } n}[Q(G_i)] = d_{\eta\text{-index}(1)},$$

F(y) is the consequent value of the strongest fired rule, the rule whose antecedent is the nearest neighbor of the input. We note this is different from the usual case of fuzzy modeling in which $F(y) = \text{Max}_i[d_i \wedge \eta_i] = \text{Max}_i[f_i]$.

If we change μ such that $\mu(E) = 0$ if $|E| < K$ and $\mu(E) = 1$ if $|E| \geq K$ then

$$F(y) = \text{Min}_{i = 1 \text{ to } K} [d_{\eta\text{-index}(i)}]$$

It is the minimal membership grade among the K rules that are nearest neighbors.

8. Combining Multiple Fuzzy Rule Based Models

One issue of interest in the use of fuzzy systems modeling is the fusion of multiple fuzzy systems models with the same consequent variable. The PBR approach provides a framework for combining multiple fuzzy models. Assume we have a fuzzy systems model consisting of n_1 rule of the form $\mathbf{R_i}$: If V_1 is B_i then W is D_i, where $i = 1$ to n_1. Here the antecedent V_1 can be an atomic or a joint variable on X and the output W is a variable taking its values in the space Y. Assume we have another fuzzy system model consisting of n_2 rules of the form $\mathbf{R_i}$: If V_2 is B_i if W is D_i here we let $i = n_1 + 1$ to $n_1 + n_2$. Here again V_2 can be an atomic variable or a joint variable which may or may not be the same as V_1. The consequent variable is the same

In combining these two fuzzy models we desire that output values supported by both rules be stronger than those only supported by one rule. Here we shall let F indicate the fuzzy subset resulting from the application of these two rules with input x^*. We use the notation $\eta_i = B_i(x^*)$, the firing level of ith rule, and let $d_i = D_i(y)$ the membership of y in D_i.

We let $\mathbf{A} = \{R_1, ..., R_{n_1}, R_{n_1+1}, ..., R_{n_1+n_2}\}$. We now apply PBR to determine F(y). In particular $F(y) = \text{Max}_{E \subseteq \mathbf{A}}[Q(E) \wedge R(E) \wedge \mu(E)]$ where $Q(E) = \text{Min}[\eta_j \in E]$ and $R(E) = \text{Min}[d_j \in E]$ We see that $F(y) = \text{Max}_{E \subseteq \mathbf{A}}[\text{Min}[f_j \in E) \wedge \mu(E)]$ where $f_j = d_j \wedge \eta_j$

We now consider the form of the credibility measure μ which we define over the space $\mathbf{A}$. Here we want to give a degree of credibility of α_1 to those inferences just based on the first fuzzy rule base, we give a degree of credibility of α_2 for any output just based on the second fuzzy rule base and we give complete credibility of one to those based on both models. In order to capture this we must appropriately define μ. With E denoting a subset of $\mathbf{A}$, a collection of rules, we define $\mu(E)$ as follows:

$\mu(E) = \alpha_1$ if all $R_j \in E$ are s.t. $j \in [1, n_1]$

$\mu(E) = \alpha_2$ if all $R_j \in E$ are s.t. $j \in (n_1+1, n_1, + n_2)$

$\mu(E) = 1$ if there exists at least one $R_j \in E$ s.t. $j \in [1, n_1]$ and one R_j s.t. $j \in [n_1+1, n_1+n_2]$.

We note that if $E_1 \subseteq E_2$ then $\text{Min}[f_j \in E_1] \geq \text{Min}[f_j \in E_2]$. Using this we see that for the assumed form of credibility function μ we get

$$F(y) = (\alpha_1 \wedge \underset{j=1 \text{ to } n_1}{\text{Max}} [f_j]) \vee (\alpha_2 \wedge \underset{j=n_1+1 \text{ to } n_1+n_2}{\text{Max}} [f_j]) \vee (\underset{j=1 \text{ to } n_1}{\text{Max}} [f_j] \wedge \underset{j=n_1+1 \text{ to } n_1+n_2}{\text{Max}} [f_j])$$

Let F_1 and F_2 denote the respective outputs of the individual models for the input x^*. It is easily seen that the membership grade of y in the output of the first model $F_1(y) = \underset{j=1 \text{ to } n_1}{\text{Max}} [f_j]$, and that the membership grade of y in the output of the second $F_2(y) = \underset{j=n_1+1 \text{ to } n_1+n_2}{\text{Max}} [f_j]$. Using this we can express combined rule membership grade for y in a more enlightening form as

$$F(y) = (\alpha_1 \wedge F_1(y)) \vee (\alpha_2 \wedge F_2(y)) \vee (F_1(y) \wedge F_2(y))$$

Here we see $F(y)$ is a weighted combination of the outputs from the individual fuzzy models

What has effectively happened in the PBR approach is that we have created a combined rule base. This rule base consists of the union of the rules making up the two models plus an set additional set of rules each consisting of a compounding of two rules, one from each of the models. Thus for each R_i in the first rule base and each R_j in the second rule base we get an additional rule

$$\text{Comp}(R_i, R_j) = \text{If } V_1 \text{ is } A_i \text{ and } V_2 \text{ is } A_j \text{ the } W \text{ is } B_i \cap B_j.$$

We assign a credibility α_1 to those rules coming from the first model, a credibility α_2 to those rules coming from the second model and a credibility of one to each of the compound rule.

We note from the above formulation for $F(y)$ that if $\alpha_1 = \alpha_1 = 1$ then our output is simply the union of the two models, $F(y) = F_1(y) \vee F_2(y)$. If $\alpha_1 = \alpha_2 = 0$ then our output is the intersection of the two models, $F(y) = F_1(y) \wedge F_2(y)$.

We can extend this technique for combining fuzzy models. Here we shall assume that we have q models and use the prototype based reasoning approach to obtain $F(y)$, the combined output. Initially we shall assume all the models are equally valid. For $j = 1$ to q we shall let w_j be the credibility we associate with an output based on j of the models. Here we have $w_j \geq w_k$ if $j > k$ and assume $w_q = 1$. For a given input x^* we let $F_j(y)$ be the output value obtained from the jth model. We let index(k) be the index of kth largest of the $F_j(y)$. Using this we obtain the combined output of these q models as $F(y) = \text{Max}_{k=1 \text{ to } q}[F_{index(k)}(y) \wedge w_k]$

A further generalization can be made which allows different credibilities to the different models Let $T = \{S_1, ..., S_q\}$, where S_j indicates the jth fuzzy model. Let $\mu: 2^T \rightarrow [0, 1]$ be a such that $\mu(E)$ indicates the degree of credibility associated with a solution determined by the collection of rules bases in E. He we have $\mu(T) = 1$, $\mu(\emptyset) = 0$ and $\mu(E_1) \geq \mu(E_2)$ if $E_2 \subset E_1$. Again let $F_j(y)$ be the output obtained from the jth model and index(k) be the index of kth largest of the $F_j(y)$. Let $H_i = \{S_{index(k)} | k = 1 \text{ to } i\}$, it is the set of the i models with the largest membership grade for y in the output. Using this we get $F(y) = \text{Max}_{k=1 \text{ to } q}[F_{index(k)}(y) \wedge \mu(H_i)]$.

9. Conclusion

We introduced the general framework for prototype based reasoning. We explained its use as a technology for supplying missing information about some target object based on known information about related prototype objects. We looked at some types of prototype based reasoning systems. First we considered nearest neighbor based systems. We then looked at fuzzy rule based models and viewed these as prototype based reasoning. This perspective allowed us to extend the capabilities fuzzy modeling technology on a number of directions. We finally we used the tools provided here to suggest a method for fusing multiple fuzzy systems models.

10. References

[1]. Sugeno, M., "Fuzzy measures and fuzzy integrals: a survey," in Fuzzy Automata and Decision Process, Gupta, M.M., Saridis, G.N. & Gaines, B.R. (eds.), Amsterdam: North-Holland Pub, 89-102, 1977.

[2]. Murofushi, T. and Sugeno, M., "Fuzzy measures and fuzzy integrals," in Fuzzy Measures and Integrals, edited by Grabisch, M., Murofushi, T. and Sugeno, M., Physica-Verlag: Heidelberg, 3-41, 1999.

[3]. Sugeno, M., "Theory of fuzzy integrals and its application," Doctoral Thesis, Tokyo Institute of Technology, 1974.

[4]. Cover, T. M. and Hart, P. E., "Nearest neighbor pattern classification," IEEE Transactions on Information Theory 13, 21-27, 1967.

[5]. Duda, R. O., Hart, P. E. and Stork, D. G., Pattern Classification, Wiley Interscience: New York, 2001.

[6]. Yager, R. R., "Towards the use of nearest neighbor rules in bioinformatics," Proceedings of the Atlantic Symposium on Computational Biology, Genome Information Systems and Technology, Duke University, Durham, 92-96, 2001.

[7]. Yager, R. R., "The induced fuzzy integral aggregation operator," Technical Report# MII-2111 Machine Intelligence Institute, Iona College, New Rochelle, NY, 2001.

[8]. Mamdani, E. H., "Application of fuzzy algorithms for control of simple dynamic plant," Proc. IEEE 121, 1585-1588, 1974.

[9]. Mamdani, E. H. and Assilian, S., "An experiment in linguistic synthesis with a fuzzy logic controller," Int. J. of Man-Machine Studies 7, 1-13, 1975.

[10]. Zadeh, L., "Outline of a new approach to the analysis of complex systems and decision processes," IEEE Transactions on Systems, Man, and Cybernetics 3, 28-44, 1973.

[11]. Yager, R. R. and Filev, D. P., Essentials of Fuzzy Modeling and Control, John Wiley: New York, 1994.

[12]. Nguyen, H. T. and Sugeno, M., Fuzzy Systems: Modeling and Control, Kluwer Academic Press: Norwell, Ma, 1998.

[13]. Yager, R. R. and Filev, D. P., "On the issue of defuzzification and selection based on a fuzzy set," Fuzzy Sets and Systems 55, 255-272, 1993.

[14]. Dubois, D. and Prade, H., Possibility Theory : An Approach to Computerized Processing of Uncertainty, Plenum Press: New York, 1988.

Gradual handling of contradiction in argumentation frameworks

C. Cayrol   M.C. Lagasquie
IRIT – UPS
118 route de Narbonne
31062 Toulouse Cédex
ccayrol@irit.fr   lagasq@irit.fr

Abstract

Argumentation is based on the exchange and evaluation of interacting arguments. In this paper, we concentrate on the analysis of defeat interactions between arguments, within the argumentation framework of [3]. We propose different principles for a gradual evaluation of arguments, which takes into account the defeaters, the defeaters of defeaters, and so on. Following these principles, two formalizations are presented. Finally, existing approaches are restated in our framework.

Keywords: Argumentation, Interactions, Inconsistency, Nonmonotonic logics

1 Introduction

As shown by [3], argumentation frameworks provide a unifying and powerful tool for the study of many formal systems developed for common-sense reasoning, as well as for giving meaning to logic programs. Argumentation is based on the exchange and evaluation of interacting arguments. It can be applied, among others, in the legal domain, for collective decision support systems or for negotiation support.
Two kinds of evaluation can be distinguished :

- intrinsic evaluation : an argument is considered independently of its interactions with other arguments. This enables to express to what extent an argument increases the confidence in the statement it supports. Such valuation methods can take different forms [5], including a numerical value which can be interpreted in probabilistic terms as in [6], or simply by a preference relation on the set of all arguments as in [7] and [1].
- interaction-based evaluation : an argument is valuated according to its defeaters, to the defeaters of its defeaters (the defenders), ...

Intrinsic evaluation and interaction-based evaluation have often been used separately, according to the considered applications. Some recent works however consider a combination of both approaches (see eg. [1]). Interaction-based approaches usually provide a crisp evaluation : the argument is accepted or not. Such an evaluation can be defined individually (per argument), for example by a labelling process [4], or globally, giving sets of arguments that are all accepted together [3]. The common idea is that an argument is acceptable if it can be argued successfully against defeating arguments. Recently, [2] has proposed a gradual interaction-based evaluation in the specific case of deductive arguments (where an argument is a logical proof). An argument tree represents all the chains of defeaters for the argument at the root of the tree. The (computed) value of this tree represents the relative strength of the root argument.

In this paper, we concentrate on the analysis of interactions between arguments, abstracting from the internal structure of an argument. We therefore ignore intrinsic evaluations. Our purpose is to provide a global evaluation of an argument according to the way this argument is defeated by other arguments and by recursion on these defeaters. An argument will be thus considered as more or less acceptable.

Section 2 introduces abstract argumentation frameworks and their representation as attack graphs. In section 3, we present relevant examples that enable us to draw the main principles underlying a gradual evaluation of interactions. Note that our interest is directed to the ordering relation between the values of the arguments and not to the values themselves. Following these principles, two formalizations are proposed in section 4, and related to other existing approaches in section 5.

2 Notations

We consider the abstract framework introduced in [3]. An *argumentation system* $<\mathcal{A}, \mathcal{R}>$ is a set $\mathcal{A}$ of arguments and a binary relation $\mathcal{R}$ on $\mathcal{A}$ called an *attack relation*: let A_i and $A_j \in \mathcal{A}$, $A_i \mathcal{R} A_j$ means that A_i attacks A_j.

Notations: Let $A \in \mathcal{A}$, The set $\{A_i \in \mathcal{A} | A_i \mathcal{R} A\}$ is denoted by $\mathcal{R}^-(A)$ and the set $\{A_i \in \mathcal{A} | A \mathcal{R} A_i\}$ is denoted by $\mathcal{R}^+(A)$.

$<\mathcal{A}, \mathcal{R}>$ defines a directed graph $\mathcal{G}$ (called the *attack graph*). For example, the system $<\mathcal{A}, \mathcal{R}>$ with $\mathcal{A} = \{A_1, A_2, A_3, A_4\}$ and $\mathcal{R} = \{(A_2, A_3), (A_4, A_3), (A_1, A_2)\}$ defines the following graph $\mathcal{G}$:

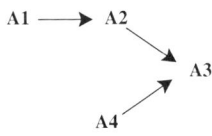

Def 1 *An argument $A \in \mathcal{A}$ such that $\mathcal{R}^-(A) = \emptyset$ is a leaf of the attack graph defined by $<\mathcal{A}, \mathcal{R}>$.*

Def 2 *In the attack graph $\mathcal{G}$, a path from A to B, denoted by $\mathcal{C}(A,B)$[1], is a sequence of arguments $A_1 \ldots A_n$ such that $A = A_1$, $A_1 \mathcal{R} A_2$, ..., $A_{n-1} \mathcal{R} A_n$, $A_n = B$. The length of this path is $n - 1$ (the number of edges that are used in the path) and will be denoted by $l_{\mathcal{C}(A,B)}$.*

We now introduce the notions of direct and indirect attacks and defences. The notions introduced here are inspired by related definitions first introduced in [3] but are not strictly equivalent: in [3]'s work, direct attacks (resp. defences) are also indirect attacks (resp. defences) which is not true in our definitions:

Def 3 *Let $A \in \mathcal{A}$. The direct defeaters of A are the elements of $\mathcal{R}^-(A)$. The direct defenders of A are the direct defeaters of the elements of $\mathcal{R}^-(A)$. The indirect defeaters of A are the elements A_i such that: $\exists \mathcal{C}(A_i, A)$ such that $l_{\mathcal{C}(A_i,A)} = 2k+1$ with $k \geq 1$. The indirect defenders of A are the elements A_i such that: $\exists \mathcal{C}(A_i, A)$ such that $l_{\mathcal{C}(A_i,A)} = 2k$ with $k \geq 2$.*

---

[1] If there are several paths from A to B, they are denoted by $\mathcal{C}(A,B)_1, \mathcal{C}(A,B)_2 \ldots$

Def 4 *Let $A \in \mathcal{A}$, an attack branch (resp. defence branch) for A is an odd (resp. even) length path from a leaf to A in $\mathcal{G}$. When such a branch exists, A is said to be the root of the branch.*

3 Examples

We have identified some specific relevant graphs which allow us to draw some fundamental principles and also problems that will lead to the introduction of two different formalizations of gradual evaluation.

We denote by v an assignment of values to a given ordered set from the arguments of $\mathcal{A}$.

1. case of a single node A:

    The value of A is maximal (the best value) since it suffers no attack.

2. case of a single path of length $n \geq 1$:

   According to the parity of n, we have different results. If n is even, the path $\mathcal{C}(A_n, A_1)$ is an attack branch for A_1 and the A_i such that i odd, $1 < i \leq n-1$, are defenders of A_1 and the other A_i are defeaters of A_1 ; else (if n odd) the path $\mathcal{C}(A_n, A_1)$ is a defence branch for A_1 and the A_i such that i odd, $1 < i \leq n$, are defenders of A_1 and the other A_i are defeaters of A_1.

   So, we have either $v(A_n) \geq v(A_{n-2}) \geq \ldots \geq v(A_2) \geq v(A_1) \geq v(A_3) \geq \ldots \geq v(A_{n-3}) \geq v(A_{n-1})$ (for n even), or $v(A_n) \geq v(A_{n-2}) \geq \ldots \geq v(A_3) \geq v(A_1) \geq v(A_2) \geq \ldots \geq v(A_{n-3}) \geq v(A_{n-1})$ (for n odd).

3. case of a non-linear acyclic attack graph (note that this type of graph is more general than a simple non-trivial tree; for example

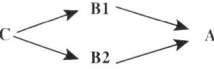

   is a non-linear acyclic attack graph):

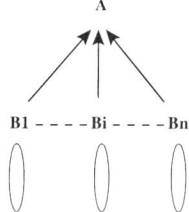

   This example raises the issue of the propagation of the modification of the value of a direct defeater when there are several defeaters.

4. case of an attack graph with cycles:

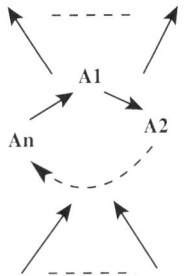

How to take into account the cycles (even or odd) and their lengths ?

4 Principles and formalization

We will consider two types of evaluation:

1. the evaluation of an argument is only a function of the direct defeaters of this argument. Therefore, defenders are taken into account through the defeaters.

2. the evaluation of an argument is a function of the set of all attack and defence branches for this argument.

4.1 Approach 1

Underlying principles   They are 4.
P1: The evaluation is maximal for an argument without defeater and non maximal for an attacked and undefended argument.
P2: The evaluation of an argument is a function of the evaluation of its direct defeaters (the "direct attack").
P3: The evaluation of an argument is a decreasing function of the evaluation of the direct attack.
P4: Each defeater of an argument contributes to the increase of the evaluation of the direct attack for this argument.

Formalization   Let W be a totally ordered set with a minimum element (V_{Min}) and a subset V of W that contains V_{Min} and with a maximum element V_{Max}.

Def 5 *Let $<\mathcal{A}, \mathcal{R}>$ be an argumentation system. An evaluation is a function $v : \mathcal{A} \to V$ such that:*

1. $\forall A \in \mathcal{A}, v(A) \geq V_{Min}$

2. $\forall A \in \mathcal{A}$, if $\mathcal{R}^-(A) = \emptyset$ then $v(A) = V_{Max}$

3. $\forall A \in \mathcal{A}$, if $\mathcal{R}^-(A) = \{A_1, \ldots, A_n\} \neq \emptyset$ then $v(A) = g(h(v(A_1), \ldots, v(A_n)))$

*where $h : V^* \to W$ is such that (V^* denotes the set of all finite sequences of elements of V) $h(x) = x$, $h() = V_{Min}$, $h(x_1, \ldots, x_n, x_{n+1}) \geq h(x_1, \ldots, x_n)$
and $g : W \to V$ is such that $g(V_{Min}) = V_{Max}$, $g(V_{Max}) < V_{Max}$, g is decreasing ($x \leq y \Rightarrow g(x) \geq g(y)$).*

Note: the fact that $h(x_1, \ldots, x_n) \geq max(x_1, \ldots, x_n)$ is a logical consequence of the properties of h.

Properties **Existence of an evaluation** v: see section 5 (property 3) for an example of an evaluation introduced in the framework of [4].
P1 is satisfied since $\forall A \in \mathcal{A}$ if A has no defeater ($\mathcal{R}^-(A) = \emptyset$), then $v(A) = V_{\text{Max}}$ and $g(V_{\text{Max}}) < V_{\text{Max}}$. **P2** is satisfied since if $\mathcal{R}^-(A) = \{A_1, \ldots, A_n\}$, $h(v(A_1), \ldots, v(A_n))$ gives an evaluation of the direct attack of A. **P3** and **P4** are satisfied due to the properties of the functions g and h.

Prop 1 *The function g satisfies for all $n \geq 1$:*

$$g(V_{Max}) \leq g^3(V_{Max}) \leq \ldots \leq g^{2n+1}(V_{Max})$$
$$\leq g^{2n}(V_{Max}) \leq \ldots \leq g^2(V_{Max}) \leq V_{Max}$$

If, moreover, g is strictly decreasing and $g(V_{Max}) > V_{Min}$, the inequalities become strict.

Proof: By induction from $V_{\text{Min}} \leq g(V_{\text{Max}}) < V_{\text{Max}}$ and by applying g twice consecutively. □

Examples Let us consider again some examples from section 3.

Ex 1 $v(A) = V_{\text{Max}}$

Ex 2 $v(A_1) = g^{n-1}(V_{\text{Max}})$
If n is even $v(A_{n-1}) \leq \ldots \leq v(A_3) \leq v(A_1) \leq v(A_2) \leq \ldots \leq v(A_n) = V_{\text{Max}}$
If n is odd $v(A_{n-1}) \leq \ldots \leq v(A_2) \leq v(A_1) \leq v(A_3) \leq \ldots \leq v(A_n) = V_{\text{Max}}$

Ex 3 From the properties of g and h, we conclude that if $v(B_i)$ increases (resp. decreases), then $v(A)$ decreases (resp. increases). Moreover, if a new defeater B_{n+1} is added: $v(A) = g(h(v(B_1), \ldots, v(B_{n+1}))) \leq g(h(v(B_1), \ldots, v(B_n)))$. This example shows that if the number of defeaters is taken into account, then the evaluation of an argument should be able to be strictly lower than $g(V_{\text{Max}})$. This is for example the case if h is strictly decreasing.

Ex 4 The study of the graphs with cycles lays bare the role played by the fixpoints of g, g^2, ... In the case of cycles of length n: if n is odd, all the arguments of the cycle have the same value and this value is a fixpoint of g, if n is even, the values of all the arguments in the cycle are fixpoints of g^n.

4.2 Approach 2

Underlying principles There are some common items with the previous approach.
P1': The evaluation is maximal for an argument without defeater and non maximal for an argument which is attacked (whether it is defended or not).
P2': The evaluation of an argument takes into account all the branches which are rooted in this argument.
P3': The improvement of the defence or the degradation of the attack of an argument leads to an increase of the value of this argument.
P4': The improvement of the attack or the degradation of the defence of an argument leads to a decrease of the value of the argument.
Let us define a *tuple-based evaluation*.

Formalization Let us define recursively the value of an argument A using the following idea: the value must describe the subgraph whose root is A, so we want to memorize in a tuple the length of each branch leading to A. We have two cases:

Rule 1 (*A* does not belong to a cycle) *Let A be an argument which does not belong to a cycle, the value $v(A)$ of A is defined as follows:*

- *if A is a leaf of the attack graph: $v(A) = ()$ (also denoted by $v(A) = (0)$),*
- *else: let $B_1, \ldots, B_n$ be the direct defeaters of A whose respective values are tuples of integers ordered by increasing values $(b_1^1, \ldots, b_{m_1}^1), \ldots, (b_1^n, \ldots, b_{m_n}^n)$, then $v(A) = f_c(b_1^1 + 1, \ldots, b_{m_1}^1 + 1, \ldots, b_1^n + 1, \ldots, b_{m_n}^n + 1)$, where f_c is the function that orders a tuple by increasing values.*

The incrementation of each value in the tuple means that the argument A is more distant from the leaves than its defeaters. Second, the function f_c guarantees that each tuple is ordered.

When cycles occur, we have chosen to memorize only two "branches" (abuse of notation !): the attack branch of minimal length and the defence branch of minimal length. The underlying idea is to consider each cycle as a kind of meta-argument whose value is equal to the tuple $(1, 2)$. The values of the arguments inside the cycle will be defined as the tuple built from the value of the meta-argument and from the values of the arguments that attack the meta-argument *seen from the argument considered*. This leads to the following computation rule:

Rule 2 (*A* belongs to a cycle) *Let A be an argument, A belongs to $C_1 \ldots C_m$ cycles. Let $C_1' \ldots C_n'$ other cycles which are interconnected with an C_i cycle or between them*[2] *(so A does not belong to $C_1' \ldots C_n'$). Let $X^1 \ldots X^p$ arguments which do not belong to the cycles and which are direct defeaters of an element of a cycle. We denote :*

- l_i : *the shortest distance between the cycle C_i' and A ($l_i = Min_{C_j' \in C_i'}(l_{C(C_j', A)})$),*
- $l_{X^i} = l_{C(X^i, A)}$ *(distance between X^i et A),*
- *each argument X^i has the following value : $v(X^i) = (x_1^i, \ldots, x_{k^1}^i)$.*

So, the value $v(A)$ of A is defined as follows :

$$v(A) = f_c(\overbrace{1, 2, \ldots, 1, 2}^{m \text{ times}},$$
$$1 + l_1, 2 + l_1, \ldots, 1 + l_n, 2 + l_n,$$
$$x_1^1 + l_{X^1}, \ldots, x_{k^1}^1 + l_{X^1},$$
$$\ldots,$$
$$x_1^p + l_{X^p}, \ldots, x_{k^p}^p + l_{X_p})$$

Here is a example that summarizes the various cases that can be encountered when cycles are taken into account.

---

[2] 2 cycles are interconnected iff their intersection is not empty.

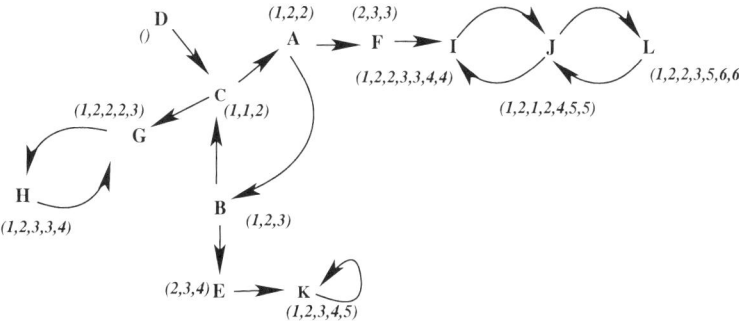

We then consider an algorithm that compares two values (ordered tuples of integers). See algorithm 1 where the following notations are used:

Let X be an argument and $v(X)$ its value. $n_a(X)$ is the number of attack branches of X (0 if X is a leaf, the number of odd integers in the tuple $v(X)$ otherwise). $n_d(X)$ is the number of defence branches of X (∞ if X is a leaf, the number of even integers in the tuple $v(X)$ otherwise). $v_p(X)$ (resp. $v_i(X)$) is the ordered tuple of even (resp. odd) integers in the tuple $v(X)$.

Justification The underlying idea is that an argument A will be better than an argument B if it as a "stronger" defence and a "weaker" attack.

In order to compare two arguments A and B, we start from the numbers $n_a(A)$, $n_a(B)$ on one hand and $n_d(A)$, $n_d(B)$ on the other hand (for a quantitative aspect). We then aggregate these two criteria in the most cautious way (the two criteria have to be in agreement to conclude).

In case of equality of $n_a(A)$, $n_a(B)$ and of $n_d(A)$, $n_d(B)$, we have chosen to take into account the "quality" of the attack and defence branches (again we get two criteria for the comparison), represented by their length. We have chosen to compare each of these criteria lexicographically. Then we cautiously aggregate the results of these two comparisons, as above.

Note that, instead of a lexicographic comparison, one may rely on other mechanisms. Similarly, other types of aggregation may be used.

Properties From the previous definitions, we get the following results:

Prop 1 *The relation defined by the comparison algorithm 1 is a partial order on the set V of all the finite tuples of integers ordered by increasing values and () is the maximum element of V.*

> **Proof:** The fact that the relation is an order relation (reflexive, antisymmetric and transitive) and that () is a maximum element of this relation follows from the definition of the comparison algorithm 1. □

We then define a *per tuple evaluation*:

Def 6 *Let V be the set of all the finite tuples of integers, ordered by increasing values. Let $<\mathcal{A},\mathcal{R}>$ be an argumentation system. The binary relation v on $\mathcal{A} \times V$ such that, for any $A \in \mathcal{A}$, $v(A) \in V$ and $v(A)$ satisfies the computation rules 1 and 2 is called a* per tuple evaluation *of $<\mathcal{A},\mathcal{R}>$.*

Prop 2 *Let v be a per tuple evaluation of $<\mathcal{A}, \mathcal{R}>$, v satisfies the principles* **P1'**, **P2'**, **P3'** *and* **P4'**.

Proof: **P1'** is satisfied due to the use of the computation rules 1 and 2 and from the fact that () is the maximum element of V.

P2' is satisfied due to the computation rules 1 and 2.

P3' and **P4'** are satisfied. It can be checked by examining, one after the other, all the basic possible cases of modification of a value and by comparing these values before and after the modification using the algorithm 1. For principle **P3'**, these different basic cases are removal or lengthening of an attack branch, addition or shortening of a defence branch. And for principle **P4'**, they are removal or lengthening of a defence branch, addition or shortening of an attack branch. Naturally, the change in the length of a branch should not modify the type of the branch itself or else we are no longer in a basic case but in the complex case of a branch removal from a given nature followed by the addition of a branch of another type. □

Algorithm 1: Comparison of two values (tuples)

% **Description of parameters:**   %
% $v(A), v(B)$: 2 tuples   %

begin
  if $((n_a(A) > n_a(B)) \wedge (n_d(A) \leq n_d(B))) \vee ((n_a(A) \geq n_a(B)) \wedge (n_d(A) < n_d(B)))$
  then
  $\quad v(B) > v(A)$ % Case 1 %
  else
  $\quad$ **if** $((n_a(A) < n_a(B)) \wedge (n_d(A) \geq n_d(B))) \vee ((n_a(A) \leq n_a(B)) \wedge (n_d(A) > n_d(B)))$ **then**
  $\quad\quad v(A) > v(B)$ % Case 2 %
  $\quad$ **else**
  $\quad\quad$ **if** $(n_a(A) = n_a(B)) \wedge (n_d(A) = n_d(B))$ **then**
  $\quad\quad\quad$ % lexicographical comparisons %
  $\quad\quad\quad$ % of $v_p(A)$ and $v_p(B)$ and of %
  $\quad\quad\quad$ % $v_i(A)$ and $v_i(B)$ %
  $\quad\quad\quad$ **if** $(v_p(A) \leq_{lex} v_p(B)) \wedge (v_i(A) \geq_{lex} v_i(B))$ **then**
  $\quad\quad\quad\quad v(A) \geq v(B)$ % Case 3 %
  $\quad\quad\quad$ **else**
  $\quad\quad\quad\quad$ **if** $(v_p(A) \geq_{lex} v_p(B)) \wedge (v_i(A) \leq_{lex} v_i(B))$ **then**
  $\quad\quad\quad\quad\quad v(B) > v(A)$ % Case 4 %
  $\quad\quad\quad\quad$ **else**
  $\quad\quad\quad\quad\quad v(B) \not\geq v(A)$ and $v(A) \not\geq v(B)$ % Case 5 %
  $\quad\quad$ **else**
  $\quad\quad\quad v(B) \not\geq v(A)$ and $v(A) \not\geq v(B)$ % Case 6 %
end

Examples On the examples cited in section 3, we get the following results:

Ex 1 $v(A) = ()$ (max of V).

Ex 2 If n is even, $v(A_n) = () > v(A_{n-2}) = (2) > \ldots > v(A_2) = (n-2) > v(A_1) = (n-1) > \ldots > v(A_{n-3}) = (3) > v(A_{n-1}) = (1)$. For n odd,

$$v(A_n) = () > v(A_{n-2}) = (2) > \ldots > v(A_1) = (n-1) > v(A_2) = (n-2) > \ldots > v(A_{n-3}) = (3) > v(A_{n-1}) = (1).$$

Ex 3 If B_i is a defence branch for A and if B_i disappears, then $v(A)$ will decrease. If B_i is an attack branch for A whose length increases then $v(A)$ will increase (see principles **P3'** and **P4'**). In the case of the addition of a new defeater B_{n+1}, if B_{n+1} is on a defence branch for A then $v(A)$ will increase, and if B_{n+1} is on an attack branch for A, $v(A)$ will decrease (see principles **P3'** and **P4'**).

Ex 4 Generally, the value of the arguments in the cycle is $(1, 2) < ()$.

4.3 Comparison of these two approaches

Even if these two approaches rely on different underlying principles, they may yield the same results in some cases. This is the case for example when the attack graph reduces to a unique branch (see example 2). Differences appear as soon as, for a given argument, it is possible to get both attack and defence branches (see example 3 and the following one).

On this example, B has two direct defeaters (C_2 and C_1) and B' has only one direct defeater (C'). So, with the approach 1, B' is better than B.
But, there are also two branches leading to B (one defence branch and one attack branch) and only one attack branch leading to B'. So, with the approach 2, B is better than B' (since there is at least one defence for B and no defence for B'). In this case, the direct defeater C_1 has lost its negatif status of defeater since it became a "carrier of defence" for B !

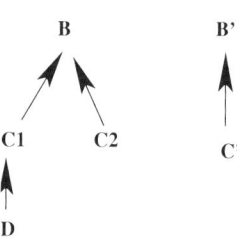

We can synthesize the resultats about these different evaluations on the following table:

Approach 2

arguments having only attack branches	$\preceq$	arguments having attack branches and defence branches	$\preceq$	arguments having only defence branches	$\preceq$	arguments never attacked

Approach 1

arguments having several unattacked direct defeaters	$\preceq$	arguments having only one unattacked direct defeater	$\preceq$	arguments having only one attacked direct defeater (possibly defended)	$\preceq$	arguments never attacked
arguments having several attacked direct defeaters (possibly defended)						

5 Comparison with existing proposals

5.1 The approach of [3]

The argumentation schema introduced in [3] enables the definition of sets of acceptable arguments, called extensions, along with different possible semantics. Under the preferred extension semantics, an extension is an inclusion-maximal conflict-free set which defends all its elements. An argument is therefore accepted iff it belongs to a preferred extension. One may therefore consider that for each semantics, there is a corresponding type of binary interaction-based evaluation and, under a given semantics, a binary evaluation corresponds to each extension.

5.2 The approach of [4]

In the approach of [4], a labelling of a set of arguments assigns a status (accepted, rejected, undecided) to each argument using labels from the set $\{+, -, ?\}$. $+$ (resp. $-$, $?$) represents the "accepted" (resp. "rejected", "undecided") status. Intuitively, an argument labelled with $?$ is both supported and weakened.

Def 7 *Let $<\mathcal{A}, \mathcal{R}>$ be an argumentation system. A complete labelling of $<\mathcal{A}, \mathcal{R}>$ is a function $Lab : \mathcal{A} \to \{+, ?, -\}$ such that:*

1. *If $Lab(A) \in \{?, -\}$ then $\exists B \in \mathcal{R}^-(A)$ such that $Lab(B) \in \{+, ?\}$*
2. *If $Lab(A) \in \{+, ?\}$ then $\forall B \in \mathcal{R}^-(A) \cup \mathcal{R}^+(A)$, $Lab(B) \in \{?, -\}$*

The underlying intuition is that an argument can only be weakened (label $-$ or $?$) if one of its direct defeaters is supported (condition 1); an argument can get a support only if all its direct defeaters are weakened and an argument which is supported (label $+$ or $?$) weakens the arguments it attacks (condition 2). So:

- If A has no defeater $Lab(A) = +$.
- If $Lab(A) = ?$ then $\exists B \in \mathcal{R}^-(A)$ such that $Lab(B) = ?$.
- If $(\forall B \in \mathcal{R}^-(A), Lab(B) = -)$ then $Lab(A) = +$.
- If $Lab(A) = +$ then $\forall B \in \mathcal{R}^-(A) \cup \mathcal{R}^+(A), Lab(B) = -$.

Every argumentation system can be completely labelled. The associated semantics is that S is an acceptable set of arguments iff there exists a complete labelling Lab of $<\mathcal{A}, \mathcal{R}>$ such that $S = \{A | Lab(A) = +\}$.

Other types of labellings are introduced in [4] among which the so-called "rooted labelling" which induces a corresponding "rooted" semantics. The idea is to only reject arguments attacked by accepted arguments: an attack by an "undecided" argument is not rooted since an "undecided" defeater may become rejected.

Def 8 *The complete labelling Lab is rooted iff $\forall A \in \mathcal{A}$, if $Lab(A) = -$ then $\exists B \in \mathcal{R}^-(A)$ such that $Lab(B) = +$.*

The rooted semantics allows to clarify the links between all the other semantics introduced in [4] and some semantics introduced in [3].

On the examples of section 3 We get the following results:

Ex 1 $v(A) = +$

Ex 2 For n even, $v(A_n) = v(A_{n-2}) = \ldots = v(A_2) = +$ and $v(A_{n-1}) = v(A_{n-3}) = \ldots = v(A_1) = -$. For n odd, $v(A_n) = v(A_{n-2}) = \ldots = v(A_1) = +$ and $v(A_{n-1}) = v(A_{n-3}) = \ldots = v(A_2) = -$

More generally, we proved:

- In the directed acyclic graph case, every argument is labelled by $+$ or $-$. An argument A is accepted iff it is (indirectly) defended against each of its direct defeaters (i.e., iff each of its direct defeaters is rejected).
- In the case of a graph with cycle: the arguments that appear on an odd length cycle are necessarily labelled by "undecided". If Lab is a rooted labelling:
  - A is accepted ($+$) iff all its direct defeaters are rejected ($-$).
  - A is rejected iff one of its direct defeaters is accepted.
  - A is undecided iff one of its direct defeaters is undecided and none of its direct defeaters is accepted.

Prop 3 (Link with the approach 1) *Every rooted labelling of $<\mathcal{A},\mathcal{R}>$ can be defined by an evaluation v such that: $V = W = \{-, ?, +\}$ with $- < ? < +$, V_{Min} denoted by $-$ and V_{Max} denoted by $+$.*

Proof: g is defined by $g(-) = +, g(+) = -, g(?) = ?$ and h is the function max. □

5.3 The aggregation function of [2]

[2] introduces a logical framework for "deductive" arguments. The approach can be characterized as follows. An argument is structured as a pair $\langle support, conclusion \rangle$, where *support* is a consistent set of formulae that enables to prove the formula *conclusion*. The attack relation considered here is strict and cycles are not allowed. The notion of a "tree of arguments" allows a concise and exhaustive representation of defeaters and defenders of a given argument, root of the tree. A function, called a "categorizer", allows to assign a value to a tree of arguments. This value represents the relative strength of an argument (root of the tree) given all its defeaters and defenders. A function, called an "accumulator", allows to synthesize the values assigned to all the argument trees whose root is an argument for (resp. against) a given conclusion.
The phase of categorization therefore corresponds to an interaction-based evaluation. [2] introduces the following function Cat:

- if $\mathcal{R}^-(A) = \emptyset$, then $Cat(A) = 1$
- if $\mathcal{R}^-(A) \neq \emptyset$ with $\mathcal{R}^-(A) = \{A_1, \ldots, A_n\}$, $Cat(A) = \frac{1}{1+Cat(A_1)+\ldots+Cat(A_n)}$

Intuitively, the larger the number of direct defeaters of an argument, the lower its value. The larger the number of defenders of an argument, the larger its value.

On the examples of section 3 We get the following results:

Ex 1 $v(A) = 1$

Ex 2 We get the same results as in section 4.1

Prop 4 (Link with the approach 1) *The categorization function of [2] is the restriction to acyclic graphs of an evaluation function as introduced in section 4.1.*

> **Proof:** Let us consider $V = [0,1]$, $W = [0, \infty[$, $V_{\text{Min}} = 0$ and $V_{\text{Max}} = 1$. We define $g : W \to V$ as $g(x) = \frac{1}{1+x}$ and h by $h(\{x_1, \ldots, x_n\}) = x_1 + \ldots + x_n$. □

The evaluation used in the above proof can be used for any attack graph (acyclic or not). The function g defined has a single fixpoint $\alpha = (\sqrt{5} - 1)/2$. This is the inverse of the golden ratio. This is also the unique fixpoint of g^2. Consequently, every argument in a cycle will be assigned this value.

6 Conclusion

This paper gives a first analysis of gradual handling of defeat interaction between arguments. For that purpose, we have chosen the abstract argumentation framework of Dung, represented by an attack graph. Considering some relevant examples, we have drawn principles for a gradual evaluation, and proposed two different formal models. One of these models generalizes a previous proposal for handling interaction between deductive arguments. An extension of this work is to take into account intrinsic values of the arguments in order to define combined evaluation models.

Acknowledgements

Thanks to Thomas SCHIEX and to Philippe BESNARD for their help.

References

[1] Leila Amgoud and Claudette Cayrol. On the acceptability of arguments in preference-based argumentation. In G. F. Cooper and S. Moral, editors, *Proc. of the 14th Uncertainty in Artificial Intelligence*, pages 1–7, Madison, Wisconsin, 1998. Morgan-Kaufmann.

[2] Philippe Besnard and Anthony Hunter. A logic-based theory of deductive arguments. *Artificial Intelligence*, 128 (1-2):203–235, 2001.

[3] Phan Minh Dung. On the acceptability of arguments and its fundamental role in nonmonotonic reasoning, logic programming and n-person games. *Artificial Intelligence*, 77:321–357, 1995.

[4] Hadassa Jakobovits and Dick Vermeir. Robust semantics for argumentation frameworks. *Journal of logic and computation*, 9(2):215–261, 1999.

[5] Paul Krause, Simon Ambler, Morten Elvang, and John Fox. A logic of argumentation for reasoning under uncertainty. *Computational Intelligence*, 11 (1):113–131, 1995.

[6] Simon Parsons. Normative argumentation and qualitative probability. *LNAI*, 1244:466–480, 1997.

[7] Henry Prakken and Giovanni Sartor. Argument-based extended logic programming with defeasible priorities. *Journal of Applied Non-Classical Logics*, 7:25–75, 1997.

Coherent Conditional Probability as a Tool for Default Reasoning

Giulianella Coletti

Dip. di Matematica e Informatica, Univ. di Perugia

Via Vanvitelli 1 - 06100 Perugia (Italy)

coletti@dipmat.unipg.it

Romano Scozzafava

Dip. Metodi e Modelli Matematici, Università "La Sapienza"

Via Scarpa 16 - 00161 Roma (Italy)

romscozz@dmmm.uniroma1.it

Barbara Vantaggi

Dip. Metodi e Modelli Matematici, Università "La Sapienza"

Via Scarpa 16 - 00161 Roma (Italy)

vantaggi@dmmm.uniroma1.it

Abstract

In this paper we face the problem of representing default rules by means of a suitable coherent conditional probability, defined on a family of conditional events. An event is singled-out (in our approach) by a *proposition*, that is a statement that can be either *true* or *false*; a conditional event is consequently defined by means of two propositions and is a multi-valued entity related (in this context) to a conditional probability. We stress the simplicity of our approach (Occam's razor...!), with respect to other well-known methodologies.

Keywords: conditional probability, coherence, default reasoning.

1 Introduction

The concept of conditional event (as dealt with in this paper) plays a central role for the probabilistic reasoning. We generalize the idea of de Finetti of looking at a conditional event $E|H$, with $H \neq \emptyset$ (the *impossible* event), as a three–valued logical entity (*true* when both E and H are true, *false* when H is true and E is false, "undetermined" when H is false) by letting the third value *suitably depend on the given ordered pair* (E, H) and not being just an undetermined *common value* for all pairs: it turns out (as explained in detail in [4]) that this function can be seen as a measure of the degree of belief in the conditional event $E|H$, which under "natural" conditions reduces to the conditional probability $P(E|H)$, in its most general sense related to the concept of *coherence*, and satisfying the classic axioms as given by de Finetti [7], Rényi [15], Krauss [12], Dubins

[8] (see Section 2). Notice that our concept of conditional event differs from that adopted, e.g., by [1], [2], [11].

Among the peculiarities (which entail a large flexibility in the management of any kind of uncertainty) of this concept of *coherent* conditional probability versus the usual one, we recall the following ones:

- due to its *direct* assignment as a whole, the knowledge (or the assessment) of the "joint" and "marginal" unconditional probabilities $P(E \wedge H)$ and $P(H)$ is not required;
- the *conditioning* event H (which *must* be a *possible* one) may have *zero probability*, but in the assignment of $P(E|H)$ we are driven by *coherence*, contrary to what is done in those treatments where the relevant conditional probability is given an *arbitrary* value in the case of a conditioning event of zero probability;
- a suitable interpretation of its extreme values 0 and 1 for situations which are different, respectively, from the trivial ones $E \wedge H = \emptyset$ and $H \subseteq E$, leads to a "natural" treatment of the *default reasoning*.

In this paper we deal with the latter aspect.

2 Coherent Conditional Probability

Given a set $\mathcal{C} = \mathcal{G} \times \mathcal{B}^o$ of conditional events $E|H$ such that $\mathcal{G}$ is a Boolean algebra and $\mathcal{B} \subseteq \mathcal{G}$ is closed with respect to (finite) logical sums, with $\mathcal{B}^o = \mathcal{B} \setminus \{\emptyset\}$, the classic *axioms for a conditional probability* read as follows:

(i) $P(H|H) = 1$, for every $H \in \mathcal{B}^o$,

(ii) $P(\cdot|H)$ is a (finitely additive) probability on $\mathcal{G}$ for any given $H \in \mathcal{B}^o$,

(iii) $P\bigl((E \wedge A)|H\bigr) = P(E|H) \cdot P\bigl(A|(E \wedge H)\bigr)$, for any $A, E \in \mathcal{G}$, $H, E \wedge H \in \mathcal{B}^o$.

Conditional probability P has been defined on $\mathcal{G} \times \mathcal{B}^o$; however it is possible, through the concept of *coherence*, to handle also those situations where we need to assess P on an *arbitrary* set $\mathcal{C}$ of conditional events.

Definition 1 - The assessment $P(\cdot|\cdot)$ on $\mathcal{C}$ is *coherent* if there exists $\mathcal{C}' \supset \mathcal{C}$, with $\mathcal{C}' = \mathcal{G} \times \mathcal{B}^o$, such that $P(\cdot|\cdot)$ can be extended from $\mathcal{C}$ to $\mathcal{C}'$ as a *conditional probability*. •

A characterization of coherence is given (see, e.g., [3]) by the following

Theorem 1 - Let $\mathcal{C}$ be an arbitrary finite family of conditional events $E_1|H_1, ..., E_n|H_n$ and $\mathcal{A}_o$ denote the set of atoms A_r generated by the unconditional events $E_1, H_1, ..., E_n, H_n$. For a real function P on $\mathcal{C}$ the following two statements are equivalent:

(i) P is a *coherent* conditional probability on $\mathcal{C}$;

(ii) there exists (at least) a *class* of probabilities $\{P_0, P_1, \ldots P_k\}$, each probability P_α being defined on a suitable subset $\mathcal{A}_\alpha \subseteq \mathcal{A}_o$, such that for any $E_i|H_i \in \mathcal{C}$ there is a unique P_α with

$$\sum_{\substack{r \\ A_r \subseteq H_i}} P_\alpha(A_r) > 0,$$

(1)
$$P(E_i|H_i) = \frac{\sum_{\substack{r \\ A_r \subseteq E_i \wedge H_i}} P_\alpha(A_r)}{\sum_{\substack{r \\ A_r \subseteq H_i}} P_\alpha(A_r)} \; ;$$

moreover $\mathcal{A}_{\alpha'} \subset \mathcal{A}_{\alpha''}$ for $\alpha' > \alpha''$ and $P_{\alpha''}(A_r) = 0$ if $A_r \in \mathcal{A}_{\alpha'}$. •

According to Theorem 1, a coherent conditional probability gives rise to a suitable class $\{P_o, P_1, \ldots P_k\}$ of "unconditional" probabilities.

Where do the above classes of probabilities come from? Since P is coherent on $\mathcal{C}$, there exists an extension P^* on $\mathcal{G} \times \mathcal{B}^o$, where $\mathcal{G}$ is the algebra generated by the set $\mathcal{A}_o$ of atoms and $\mathcal{B}$ the additive class generated by $H_1, \ldots, H_n$: then, putting $\mathcal{F} = \{\Omega, \emptyset\}$, the restriction of P^* to $\mathcal{A}_o \times \mathcal{F}^o$ satisfies (1) with $\alpha = 0$ for any $E_i|H_i$ such that $P_o(H_i) > 0$. The subset $\mathcal{A}_1 \subset \mathcal{A}_o$ contains only the atoms $A_r \subseteq H_o^1$, where H_o^1 is the union of H_i's with $P_o(H_i) = 0$ (and so on): we proved (see, e.g., [3], [4]) that, starting from a coherent assessment $P(E_i|H_i)$ on $\mathcal{C}$, a relevant family $\mathcal{P} = \{P_\alpha\}$ can be suitably defined that allows a representation such as (1). Every value $P(E_i|H_i)$ constitutes a constraint in the construction of the probabilities P_α ($\alpha = 0, 1, \ldots$); in fact, given the set $\mathcal{A}_o$ of atoms generated by $E_1, \ldots, E_n, H_1, \ldots, H_n$, and its subsets $\mathcal{A}_\alpha$ (such that $P_\beta(A_r) = 0$ for any $\beta < \alpha$, with $A_r \in \mathcal{A}_\alpha$) each P_α must satisfy the following system (S_α) with unknowns $P_\alpha(A_r) \geq 0$, $A_r \in \mathcal{A}_\alpha$,

$$\begin{cases} \sum_{\substack{r \\ A_r \subseteq E_i H_i}} P_\alpha(A_r) = P(E_i|H_i) \sum_{\substack{r \\ A_r \subseteq H_i}} P_\alpha(A_r), & \left[\text{if } P_{\alpha-1}(H_i) = 0\right] \\ \sum_{\substack{r \\ A_r \subseteq H_0^\alpha}} P_\alpha(A_r) = 1 \end{cases}$$

where $P_{-1}(H_i) = 0$ for all H_i's, and H_o^α denotes, for $\alpha \geq 0$, the union of the H_i's such that $P_{\alpha-1}(H_i) = 0$; so, in particular, $H_o^o = H_o = H_1 \vee \ldots \vee H_n$.

Any class $\{P_\alpha\}$ singled-out by the condition *(ii)* is said *to agree* with the conditional probability P. Notice that in general there are infinite classes of probabilities $\{P_\alpha\}$; in particular we have *only one agreeing class* in the case that $\mathcal{C} = \mathcal{G} \times \mathcal{G}^o$, where $\mathcal{G}$ is a Boolean algebra.

A coherent assessment P, defined on a set $\mathcal{C}$ of conditional events, can be extended in a natural way to all the conditional events $E|H$ such that $E \wedge H$ is an element of the algebra $\mathcal{G}$ spanned by the (unconditional) events E_i, H_i, $i = 1, 2, \ldots, n$ taken from the elements of $\mathcal{C}$, and H is an element of the additive class spanned by the H_i's. Obviously, this extension is not unique, since there is no uniqueness in the choice of the class $\{P_\alpha\}$ related to condition *(ii)* of Theorem 1.

In general, we have the following result (see, e.g., [3])

Theorem 2 - If $\mathcal{C}$ is a given family of conditional events and P a corresponding assessment, then there exists a (possibly not unique) coherent extension of P to an arbitrary family $\mathcal{K}$ of conditional events, with $\mathcal{K} \supseteq \mathcal{C}$, *if and only if P is coherent on $\mathcal{C}$*. •

Notice that if P is coherent on a family $\mathcal{C}$, it is coherent also on $\mathcal{E} \subseteq \mathcal{C}$.

3 Zero-layers

Given a class $\mathcal{P} = \{P_\alpha\}$, agreeing with a conditional probability, it *naturally induces* the *zero-layer* $\circ(H)$ of an event H, defined as

$$\circ(H) = \alpha \quad \text{if } P_\alpha(H) > 0,$$

and the zero-layer of a conditional event $E|H$ as

$$\circ(E|H) = \circ(E \wedge H) - \circ(H).$$

Obviously, for the certain event Ω and for any event E with positive probability, we have $\circ(\Omega) = \circ(E) = 0$ (so that, if the class contains only an *everywhere positive* probability P_o, there is only one (trivial) zero-layer, *i.e.* $\alpha = 0$), while we put $\circ(\emptyset) = +\infty$. Clearly,

$$\circ(A \vee B) = \min\{\circ(A), \circ(B)\}.$$

Moreover, notice that $P(E|H) > 0$ if and only if $\circ(E \wedge H) = \circ(H)$, *i.e.* $\circ(E|H) = 0$.

On the other hand, Spohn (see, for example, [19]) considers degrees of plausibility defined via a *ranking* function, that is a map κ that assigns to each *possible* proposition a natural number (its *rank*) such that

(a) either $\kappa(A) = 0$ or $\kappa(A^c) = 0$, or both;

(b) $\kappa(A \vee B) = \min\{\kappa(A), \kappa(B)\}$;

(c) for all $A \wedge B \neq \emptyset$, the conditional rank of B given A is
$\kappa(B|A) = \kappa(A \wedge B) - \kappa(A)$.

Ranks represent degrees of "disbelief". For example, A is *not* disbelieved iff $\kappa(A) = 0$, and it is disbelieved iff $\kappa(A) > 0$. Ranking functions are seen by Spohn as a tool to manage *plain belief* and *belief revision*, since he maintains that probability is inadequate for this purpose. In our framework this claim can be challenged (see [6]), since our tools for belief revision are *coherent conditional probabilities* and the **ensuing** concept of *zero-layers*: it is easy to check that zero-layers have the same formal properties of ranking functions.

4 Coherent Probability and Default Reasoning

We recall that in [6] we showed that a sensible use of events whose probability is 0 (or 1) can be a more general tool in revising beliefs when new information comes to the fore, so that we have been able to challenge the claim contained in [18] that probability is inadequate for revising plain belief. Moreover, as recalled in Section 1, we may deal with the extreme value $P(E|H) = 1$ also for situations which are different from the trivial one $H \subseteq E$.

The aim of this Section is to handle, by means of a *coherent* conditional probability, some aspects of *default reasoning* (see, e.g., [14], [16]): as it is well-known, a default rule is a sort of weak implication.

First of all, we discuss briefly some aspects of the classic example of Tweety. The usual *logical implication* (denoted by $\subseteq$) can be anyway useful to express that a penguin ($\mathcal{P}$) is *certainly* a bird ($\mathcal{B}$), i.e. $\mathcal{P} \subseteq \mathcal{B}$, so that $P(\mathcal{B}|\mathcal{P}) = 1$; moreover we know that Tweety ($\mathcal{T}$) is a penguin, and also this fact can be represented by $P(\mathcal{P}|\mathcal{T}) = 1$.

But we can express as well the statement "a penguin *usually* does not fly" (we denote by $\mathcal{F}^c$ the contrary of $\mathcal{F}$, the latter symbol denoting "flying") by writing $P(\mathcal{F}^c|\mathcal{P}) = 1$.

Then the question "can Tweety fly?" can be faced through an assessment of the conditional probability $P(\mathcal{F}|\mathcal{T})$, which must be coherent with the already assessed ones: by

Theorem 1, it can be shown that *any value* $p \in [0, 1]$ is a coherent value for $P(\mathcal{F}|\mathcal{T})$, so that no conclusion can be reached – *from the given premises* – on Tweety's ability of flying. In other words, interpreting an equality such as $P(E|H) = 1$ like a *default rule* (denoted by $\longmapsto$), which in particular (when $H \subseteq E$) reduces to the usual implication, we have shown its *nontransitivity*: in fact we have

$$\mathcal{T} \longmapsto \mathcal{P} \quad \text{and} \quad \mathcal{P} \longmapsto \mathcal{F}^c,$$

but it *does not* necessarily follow that the default rule $\mathcal{T} \longmapsto \mathcal{F}^c$ (even if we *might* have that $P(\mathcal{F}^c|\mathcal{T}) = 1$, *i.e.* that "Tweety usually does not fly").

Definition 2 - Given a *coherent* conditional probability P on a family $\mathcal{C}$ of conditional events, a *default rule*, denoted by $H \longmapsto E$, is any conditional event $E|H \in \mathcal{C}$ such that $P(E|H) = 1$. •

Clearly, any logical implication $A \subseteq B$ (and so also any equality $A = B$) between events can be seen as a (trivial) default rule.

By resorting to the systems (S_α) to check the coherence of the assessment $P(E|H) = 1$ (which implies, for the relevant zero-layer, $\circ(E|H) = 0$), a simple computation gives $P_o(E^c \wedge H) = 0$ (notice that the class $\{P_\alpha\}$ has in this case only one element P_o). It follows $\circ(E^c|H) = 1$, so that

$$\circ(E^c|H) > \circ(E|H).$$

Obviously, also the converse is true, that is the validity of the latter relation is equivalent to the assessment $P(E|H) = 1$.

In terms of Spohn's ranking functions (we recall – *and underline* – that our zero-layers are – so to say – "incorporated" into a coherent conditional probability, so that **we do not need** an "autonomous" definition of ranking!) we could say, when $P(E|H) = 1$, that the disbelief in $E^c|H$ is greater than that in $E|H$. This conclusion **must not** be read as $P(E|H) > P(E^c|H)$!

Given a *set* $\Delta \subseteq \mathcal{C}$ *of default rules* $H_i \longmapsto E_i$, with $i = 1, ..., n$, we need to check its *consistency*, that is the coherence of the "global" assessment P on $\mathcal{C}$ such that $P(E_i|H_i) = 1$ on Δ. We stress that, even if our definition involves a conditional probability, the condition given in the following theorem refers *only* to logical (in the sense of *Boolean* logic) relations.

Theorem 3 - Let P be a *coherent* conditional probability on a family $\mathcal{C}$ of conditional events. Then a set $\Delta \subseteq \mathcal{C}$ of default rules

$$H_i \longmapsto E_i, \quad i = 1, 2, ..., n,$$

represented by the assessment

$$P(E_i|H_i) = 1, \quad i = 1, 2, ..., n,$$

is consistent (on Δ), i.e. the latter assessment is coherent, if for every subfamily of Δ

$$\{H_{i_1} \longmapsto E_{i_1}, \ldots, H_{i_s} \longmapsto E_{i_s}\}_{s=1,2,...,n}$$

we have

$$\bigvee_{k=1}^{s} (E_{i_k} \wedge H_{i_k}) \not\subseteq \bigvee_{k=1}^{s} (E_{i_k}^c \wedge H_{i_k}). \tag{2}$$

Conversely, if Δ is consistent on $\mathcal{C}$, then (2) holds.

Proof - We prove that, assuming the above logical relations (2), the assessment $P(E_i|H_i) = 1$ (for $i = 1, 2, ..., n$) on Δ is coherent. We resort to the characterization Theorem 2: to begin with, put, for $i = 1, 2, ..., n$, $P(E_i|H_i) = 1$ in the system (S_o); the unconditional probability P_o can be obtained by putting $P_o(A_r) = 0$ for all atoms $A_r \subseteq \bigvee_{j=1}^{n} \left(E_j^c \wedge H_j \right)$.

Then for any atom $A_k \subseteq E_i \wedge H_i$ and not contained in $\bigvee_{j=1}^{n} \left(E_j^c \wedge H_j \right)$ – notice that (2) ensures that there is such an atom A_k, since $\bigvee_{j=1}^{n}(E_j \wedge H_j) \not\subseteq \bigvee_{j=1}^{n}(E_j^c \wedge H_j)$ – we may put $P_o(A_k) > 0$ in such a way that these numbers sum up to 1.

This clearly gives a solution of the first system (S_o). If, for some i, we have that $E_i \wedge H_i \subseteq \bigvee_{j=1}^{n} \left(E_j^c \wedge H_j \right)$, then $P_o(E_i \wedge H_i) = 0$. Then we consider the second system (which refers to all H_i such that $P_o(H_i) = 0$), proceeding as above to construct the probability P_1; and so on. Condition (2) ensures that at each step we can give positive probability P_α to (at least) one of the remaining atoms.

Conversely, consider the (coherent) assignment P on $\mathcal{C}$ such that $P(E_i|H_i) = 1$ (for $i = 1, ..., n$) on Δ. Then, for any index $j \in \{1, 2, ..., n\}$ there exists a probability P_α such that $P_\alpha(E_j \wedge H_j) > 0$ while one has $P_\alpha(E_j^c \wedge H_j) = 0$. Notice that the restriction of P to some conditional events $E_{i_1}|H_{i_1}, ..., E_{i_s}|H_{i_s}$ of Δ is coherent as well.

Let P_o be the first element of an agreeing class, and i_k an index such that $P_o(H_{i_k}) > 0$: then $P_o(E_{i_k} \wedge H_{i_k}) > 0$ and $P_o(E_{i_k}^c \wedge H_{i_k}) = 0$.

Suppose $E_{i_k} \wedge H_{i_k} \subseteq \bigvee_{k=1}^{s}(E_{i_k}^c \wedge H_{i_k})$: then $P_o(E_{i_k} \wedge H_{i_k}) = 0$. This contradiction shows that condition (2) holds. •

Definition 3 - A set $\Delta \subseteq \mathcal{C}$ of default rules *entails* the default rule $H \to E$ if the *only* coherent value for $P(E|H)$ is 1. •

In other words, the rule $H \to E$ is entailed by Δ (or by a subset of Δ) if every possible *extension* of the probability assessment P on $\mathcal{C}$ such that $P(E_i|H_i) = 1$ (for $i = 1, ..., n$) on Δ, assigns the value 1 also to $P(E|H)$.

Going back to the previous example of Tweety, its possible ability (or inability) of flying can be expressed by saying that the default rule $\tau \longmapsto \varphi$ (or $\tau \longmapsto \varphi^c$) *is not entailed* by the premises (the given set Δ).

5 Default logic

Several formalisms for default logic have been studied in the relevant literature with the aim of discussing the minimal conditions that an entailment should satisfy. In our framework this "inferential" process is ruled by the following

Theorem 4 - Given a set Δ of consistent default rules, we have

(Reflexivity)
Δ entails $A \longmapsto A$ for any $A \neq \emptyset$

(Left logical equivalence)
$(A = B)$, $(A \longmapsto C) \in \Delta$ entails $B \longmapsto C$

(Right weakening)
$(A \subseteq B)$, $(C \longmapsto A) \in \Delta$ entails $C \longmapsto B$

(Cut)
$(A \wedge B \longmapsto C)$, $(A \longmapsto B) \in \Delta$ entails $A \longmapsto C$

(Cautious Monotonicity)
$(A \longmapsto B)$, $(A \longmapsto C) \in \Delta$ entails $A \wedge B \longmapsto C$

(Equivalence)
$(A \longmapsto B)$, $(B \longmapsto A)$, $(A \longmapsto C) \in \Delta$ entails $B \longmapsto C$

(And)
$(A \longmapsto B)$, $(A \longmapsto C) \in \Delta$ entails $A \longmapsto B \wedge C$

(Or)
$(A \longmapsto C)$, $(B \longmapsto C) \in \Delta$ entails $A \vee B \longmapsto C$.

Proof - Reflexivity amounts to $P(A|A) = 1$ for every possible event.

Left Logical Equivalence and *Right weakening* trivially follow from elementary properties of conditional probability.

Cut: from $P(C|A \wedge B) = P(B|A) = 1$ it follows that $P(C|A) = 1$, since

$$P(C|A) = P(C|A \wedge B)P(B|A) + P(C|A \wedge B^c)P(B^c|A) = P(C|A \wedge B)P(B|A).$$

Cautious Monotonicity: since $1 = P(B|A) = P(C|A)$, we have that

$$1 = P(C|A) = P(C|A \wedge B)P(B|A) + P(C|A \wedge B^c)P(B^c|A) = P(C|A \wedge B)P(B|A),$$

hence $P(C|A \wedge B) = 1$.

Equivalence: since at least one conditioning event must have positive probability, it follows that A, B, C have positive probability; moreover,

$$P(A \wedge C) = P(A) = P(A \wedge B) = P(B),$$

which implies $P(A \wedge B \wedge C) = P(A) = P(B)$, so $P(C|B) = 1$.

And: since

$$1 \geq P(B \vee C|A) = P(B|A) + P(C|A) - P(B \wedge C|A) = 2 - P(B \wedge C|A),$$

it follows $P(B \wedge C|A) = 1$.

Or: since

$$P(C|A \vee B) =$$
$$= P(C|A)P(A|A \vee B) + P(C|B)P(B|A \vee B) - P(C|A \wedge B)P(A \wedge B|A \vee B) =$$
$$= P(A|A \vee B) + P(B|A \vee B) - P(C|A \wedge B)P(A \wedge B|A \vee B) \geq 1,$$

we get $P(C|A \vee B) = 1$. •

Let us now discuss some "unpleasant" properties, that in fact do not necessarily hold also in our framework:

(Monotonicity)
$(A \subseteq B)$, $(B \longmapsto C) \in \Delta$ entails $A \longmapsto C$

(Transitivity)
$(A \longmapsto B)$, $(B \longmapsto C) \in \Delta$ entails $A \longmapsto C$

(Contraposition)
$(A \longmapsto B) \in \Delta$ entails $B^c \longmapsto A^c$

The previous example about Tweety shows that *Transitivity* can fail.

In the same example, if we add the evaluation $P(\mathcal{F}|\mathcal{B}) = 1$ to the initial ones, the assessment is still coherent (even if $P(\mathcal{F}|\mathcal{P}) = 0$ and $\mathcal{P} \subseteq \mathcal{B}$), so also *Monotonicity* can fail.

Now, consider the conditional probability P defined as follows

$$P(B|A) = 1, \; P(A^c|B^c) = \frac{1}{4} \;;$$

it is easy to check that it is coherent, and so *Contraposition* can fail.

Many authors (cfr., e.g., [13]) claim (and we agree) that the previous unpleasant properties should be replaced by others, that we express below in our own notation and interpretation: we show that these properties hold in our framework. Since a widespread consensus among their "right" formulation is lacking, we will denote them as cs–(Negation Rationality), cs–(Disjunctive Rationality), cs–(Rational Monotonicity), where "cs" stands for "in a coherent setting".

Notice that, given a default rule $H \longmapsto E$, to say $(H \longmapsto E) \notin \Delta$ means that the conditional event $E|H$ belongs to the set $\mathcal{C} \setminus \Delta$.

cs–(Negation Rationality)
If $(A \wedge C \longmapsto B)$, $(A \wedge C^c \longmapsto B) \notin \Delta$ then Δ does not entail $(A \longmapsto B)$.

Proof - Since we have $(A \wedge C \longmapsto B)$ and $(A \wedge C^c \longmapsto B)$ do not belong to Δ, that is $P(B|A \wedge C) < 1$ and $P(B|A \wedge C^c) < 1$, then

$$P(B|A) = P(B|A \wedge C)P(C|A) + P(B|A \wedge C^c)P(C^c|A) < 1\,.$$

cs–(Disjunctive Rationality)
If $(A \longmapsto C)$, $(B \longmapsto C) \notin \Delta$ then Δ does not entail $(A \vee B \longmapsto C)$.

Proof - Starting from the equalities

$$P(C|A \vee B) = P(C|A)P(A|A \vee B) + P(C|A^c \wedge B)P(A^c \wedge B|A \vee B) =$$

$$= P(C|B)P(B|A \vee B) + P(C|A \wedge B^c)P(A \wedge B^c|A \vee B),$$

since $P(C|A) < 1$ and $P(C|B) < 1$, assuming $P(C|A \vee B) = 1$ would imply (by the first equality) $P(A|A \vee B) = 0$ and (by the second one) $P(B|A \vee B) = 0$ (contradiction).

cs–(Rational Monotonicity)
If $(A \wedge B \longmapsto C)$, $(A \longmapsto B^c) \notin \Delta$ then Δ does not entail $(A \longmapsto C)$.

Proof - If it were $P(C|A) = 1$, i.e.

$$1 = P(C|A \wedge B)P(B|A) + P(C|A \wedge B^c)P(B^c|A),$$

we would get either

$$P(C|A \wedge B) = P(C|A \wedge B^c) = 1$$

or one of the following

$$P(C|A \wedge B) = P(B|A) = 1,$$
$$P(C|A \wedge B^c) = P(B^c|A) = 1$$

(contradiction).

6 Conclusions

We want to stress the simplicity of our direct approach to default logic with respect to other well–known methodologies, such as those, e.g., by Adams [1], Benferhat, Dubois and Prade [2], Gilio [9], Goldszmidt and Pearl [10], Lehmann and Magidor [13], Schaub [17], etc. (to show this simplicity is the main aim of the paper). Our methodology, based on the concept of *coherent conditional probability* (whose peculiarities have been briefly recalled in Section 1), encompasses many existing theories and our framework is clearly and rigorously settled: conditional events $E|H$ are **not** 3-valued entities whose third value is looked on as "undetermined" when H is false, but they have been defined instead (see, e.g., [4]) in a way which entails "automatically" (so-to-say) the axioms of conditional probability, which are those *ruling coherence*. In other words ... "tout se tient". For a complete account, a relevant reference is the recent book [5].

Moreover, we deem it interesting that our results are – more or less – not in contrast with those contained in the aforementioned papers. A brief comparison is now in order.

The concept of consistency is usually based on that of *quasi conjunction*: we do not refer to this notion, since it is a particular conditional event (and our concept of conditional event is different from theirs); also what is called *verifiability* of a conditional event $E|H$ (that is $E \wedge H \neq \emptyset$) is a too weak notion – except in the case $H = \Omega$ – to express properly the relevant semantics.

In Adams' framework it is required to probability to be *proper* (i.e., positive) on the given events, but (since the domain of a probability P is an algebra) we need to extend P from the given events to other events (by the way, coherence is nothing but complying with this need). In particular, these "new" events may have zero probability: it follows, according to Adams' definition of *conditional* probability (that is put equal to 1 when the conditioning event has zero probability), that we can easily get *incoherent* assessments. Consider in fact the following example.

Given two (logically independent) events H_1 and H_2, put

$$E_1 = H_1 \wedge H_2, \quad E_2 = H_1^c \wedge H_2.$$

For any $\epsilon > 0$, consider the assessment

$$P(E_1|H_1) = 1, \, P(E_2|H_2) = 1 - \epsilon,$$

so that $\{E_1|H_1, E_2|H_2\}$ is consistent according to Adams, as can be easily checked giving the atoms the probabilities

$$P(H_1 \wedge H_2) = \epsilon, \; P(H_1 \wedge H_2^c) = 0,$$
$$P(H_1^c \wedge H_2^c) = 0, \; P(H_1^c \wedge H_2) = 1 - \epsilon,$$

(notice that the assessment is proper). But for any event $A \subset H_1 \wedge H_2^c$ we can extend P, according to his definition of conditional probability, as

$$P(A|H_1 \wedge H_2^c) = P(A^c|H_1 \wedge H_2^c) = 1,$$

which is **not** coherent.

A coherence–based approach to default reasoning (but in the framework of "imprecise" probability propagation), is that in [9], even if we claim (besides the utmost simplicity of our definitions and results) important semantic and syntactic differences.

For example, our concept of *entailment* is certainly different, as shown by the following simple example.

Consider two (logically independent) events H_1 and H_2, and put

$$E_1 = H_1 \wedge H_2, \; E_2 = H_1^c \wedge H_2,$$
$$E_3 = H_1^c \wedge H_2^c, \; E = E_2, \; H = H_3 = \Omega.$$

Given α, with $0 < \alpha < 1$, the assessment

$$P(E_1|H_1) = P(E_2|H_2) = 1, \; P(E_3|H_3) = \alpha$$

on $\mathcal{C} = \{E_1|H_1, E_2|H_2, E_3|H_3\}$ is coherent; the relevant probabilities of the atoms are

$$P(H_1 \wedge H_2) = P(H_1 \wedge H_2^c) = 0,$$
$$P(H_1^c \wedge H_2^c) = \alpha, \; P(H_1^c \wedge H_2) = 1 - \alpha,$$

so that the set Δ of default rules corresponding to $\{E_1|H_1, E_2|H_2\}$ is consistent.

Does Δ entail $E|H$? A simple check shows that the only coherent assessment for this conditional event is $P(E|H) = 1 - \alpha$. Then the answer is NO, since we require (in the definition of entailment) that 1 is (the only) coherent extension.

On the contrary, according to the characterization of entailment given in [9] – that is: Δ (our notation) entails $E|H$ iff $P(E^c|H) = 1$ is not coherent – the answer to the previous question is YES, since the only coherent value of this conditional probability is $P(E^c|H) = \alpha$ (see the above computation).

The System Z proposed in [10] is based on the concept of ranking function (introduced by Spohn). At the end of Section 3 and in Section 4 it has been discussed that our concept of *zero–layer* encompasses that of ranking function; moreover it is already part (so-to-say) of the coherent conditional probability structure, so that it does not need an autonomous definition. Notice that our concept of default, defined by the assessment $P(E|H) = 1$, coincides *formally* with that given in System Z through ranking functions, since (as we have discussed in Section 4) it is equivalent to the relation $\circ(E^c|H) > \circ(E|H)$ between zero–layers.

References

[1] E. Adams (1975). *The Logic of Conditionals*, Dordrecht, Reidel.

[2] S. Benferhat, D. Dubois, H. Prade (1997). Nonmonotonic Reasoning, Conditional Objects and Possibility Theory. *Artificial Intelligence* **92**: 259–276.

[3] G. Coletti, R. Scozzafava (1996). Characterization of Coherent Conditional Probabilities as a Tool for their Assessment and Extension. *International Journal of Uncertainty, Fuzziness and Knowledge-Based System* **4**: 103–127.

[4] G. Coletti, R. Scozzafava (1999). Conditioning and Inference in Intelligent Systems. *Soft Computing* **3**: 118–130.

[5] G. Coletti, R. Scozzafava (2002). Probabilistic Logic in a Coherent Setting, *Trends in Logic, n.15*, Dordrecht, Kluwer.

[6] G. Coletti, R. Scozzafava, B. Vantaggi (2001). Probabilistic Reasoning as a General Unifying Tool. In: Benferhat, S. and Besnard, P. (eds.), *Lectures Notes AI n. 2143 (ECSQUARU 2001)*, pp. 120–131.

[7] B. de Finetti (1949). Sull'impostazione assiomatica del calcolo delle probabilità. *Annali Univ. Trieste* **19**: 3–55. (Engl. transl.: Ch.5 in: Probability, Induction, Statistics, Wiley, London, 1972).

[8] L.E. Dubins (1975). Finitely Additive Conditional Probabilities, Conglomerability and Disintegration. *Annals of Probability* **3**: 89–99.

[9] A. Gilio (2000). Precise propagation of upper and lower probability bounds in system P. In: *Proc. 8th Int. Workshop on Non-monotonic Reasoning, "Uncertainty Frameworks in Non-Monotonic Reasoning"*, Breckenridge, USA.

[10] M. Goldszmidt, J. Pearl (1996). Qualitative probability for default reasoning, belief revision and causal modeling. *Artificial Intelligence* **84**: 57–112.

[11] I.R. Goodman, H.T. Nguyen (1988). Conditional objects and the modeling of uncertainties. In: Gupta M. and Yamakawa T. (eds.), *"Fuzzy Computing"*, Amsterdam, North Holland, pp. 119–138.

[12] P.H. Krauss (1968). Representation of Conditional Probability Measures on Boolean Algebras. *Acta Math. Acad. Scient. Hungar.* **19**: 229–241.

[13] D. Lehmann, M. Magidor (1992). What does a conditional knowledge base entail? *Artificial Intelligence* **55**: 1–60.

[14] R. Reiter (1980). A Logic for Default Reasoning. *Artificial Intelligence* **13**: 81–132.

[15] A. Rényi (1956). On Conditional Probability Spaces Generated by a Dimensionally Ordered Set of Measures. *Theory of Probability and its Applications* **1**: 61–71.

[16] S.J. Russel, P. Norvig (1995). *Artificial Intelligence. A Modern Approach*. Prentice-Hall, New Jersey.

[17] T. Schaub (1998). The Family of Default Logics. In: Gabbay, D.M. and Smets, P. (eds.), *Handbook of Defeasible Reasoning and Uncertainty Management Systems*, Vol. 2, pp. 77–133.

[18] P.P. Shenoy (1991). On Spohn's Rule for Revision of Beliefs. *International Journal of Approximate Reasoning* **5**: 149–181.

[19] W. Spohn (1994). On the Properties of Conditional Independence. In: Humphreys, P., Suppes, P. (eds), *Scientific Philosopher* **1**: *Probability and Probabilistic Causality*, Kluwer, Dordrecht, pp. 173–194.

B. Bouchon-Meunier, L. Foulloy and R.R. Yager (Editors)
Intelligent Systems for Information Processing:
From Representation to Applications
© 2003 Elsevier Science B.V. All rights reserved.

Detecting Conflict-Free Assumption-Based Knowledge Bases

Rolf Haenni

University of Konstanz, Center for Junior Research Fellows

D-78457 Konstanz, Germany

`rolf.haenni@uni-konstanz.de`

Abstract

This paper presents a simple method that reduces the problem of detecting conflict-free assumption-based knowledge bases to the problem of testing satisfiability.

Keywords: Assumption-Based Reasoning, Argumentation, Abduction, Conflicts, Satisfiability

1 Introduction

Logic-based argumentative and abductive reasoning are closely related tasks of drawing inferences under uncertainty. In both cases, knowledge is expressed in a logic-based or constraint-based language. The complete description of the available knowledge is called *knowledge base*. A particular subset of variables called *assumptions* or *assumables* is used to describe uncertain or unknown circumstances such as possible failure states of system components, different ways of interpreting statements or evidence, unknown outcomes of events, situations of unreliable sensors, witnesses, or information sources, and many more. Because both abduction and argumentation are based on this simple concept, we use *assumption-based reasoning* as a general term that covers both disciplines.[1]

The goal of argumentation is to derive from the given knowledge base *arguments* in favor and *counter-arguments* against certain *hypotheses* about the future or unknown world [1, 3, 4, 7, 12, 14, 15, 17, 18]. Intuitively, every argument provides a possible proof of the hypothesis, while counter-arguments prove the contrary of the hypothesis. Abduction is very similar, as its goal is to derive from the given knowledge base possible *explanations* for some observations [19, 21, 22, 23, 24]. A particular application of abduction is the problem of finding diagnoses for systems with an abnormal behavior [8, 16, 20, 25].

A common feature of argumentation and abduction is that both arguments and explanations are terms (conjunctions of literals) containing assumptions only. These terms are usually assumed to be consistent with the given knowledge base. In order to guarantee consistency, it is important to know the set of inconsistent terms that are in conflict with the knowledge base. As a consequence, prior to computing arguments or explanations, it

---

[1] In Poole's abductive framework [21], assumptions are called *hypotheses*. As a consequence, Poole speaks about *hypothetical reasoning* rather than assumption-based reasoning.

is usually necessary to compute the set of all such conflicts. However, the set of conflicts is sometimes empty. In such cases, in order to avoid unnecessary computations, it would be advantageous to detect conflict-free assumption-based knowledge bases in advance.

This paper presents a simple method that reduces the problem of detecting conflict-free assumption-based knowledge bases to the problem of testing satisfiability (SAT). Of course, since SAT is a well-known member of the class of computationally intractable NP-complete problems [5, 10], it is unlikely to find a SAT algorithm that has a fast worst-case time behavior. However, there is a wide range of clever SAT algorithms that can rapidly solve many SAT problems of practical interest. As we do not further address SAT in this paper, we refer to the literature and especially to [11] for a comprehensive overview of existing techniques.

2 Argumentative and Abductive Reasoning

Let A and P be two distinct sets of propositions and $\mathcal{L}_{A \cup P}$ the corresponding propositional language.[2] The propositions in A are the assumptions and $\xi \in \mathcal{L}_{A \cup P}$ denotes the knowledge base. Often, ξ is given by a conjunctively interpreted set $\Sigma = \{\xi_1, \ldots, \xi_r\}$ of sentences $\xi_i \in \mathcal{L}_{A \cup P}$ or, more specifically, as a set $\Sigma = \{\gamma_1, \ldots, \gamma_r\}$ of clauses $\gamma_i \in \mathcal{C}_{A \cup P}$, where $\mathcal{C}_{A \cup P}$ denotes the set of all (proper) clauses over $A \cup P$ (including the empty clause $\bot$).

A *term* is a conjunction of non-repeating literals. We use $\mathcal{T}_A$ to denote the set of all terms consisting of assumptions only (including the empty term $\top$). Furthermore, $N_A = \{0,1\}^{|A|}$ denotes the set of all possible configurations relative to A. The elements $s \in N_A$ are called *scenarios* and represent possible states of the unknown or future world.[3] Note that every term $\tau \in \mathcal{T}_A$ has a corresponding set $M_A(\tau) \subseteq N_A$ of possible scenarios called *models* of τ for which τ evaluates to 1. Both argumentation and abduction are based on the idea that one particular scenario $\hat{s} \in N_A$ is the *true* scenario.

Of course, it is assumed that the true scenario is not in conflict with the given knowledge base. If $s \in N_A$ is an arbitrary scenario and $\xi_{\leftarrow s}$ the formula obtained from ξ by instantiating all the assumptions according to their values in s, then

$$C_A(\xi) = \{s \in N_A : \xi_{\leftarrow s} \models \bot\} \qquad (1)$$

denotes the set of *conflicting* scenarios of ξ relative to A. Sometimes, the elements of $C_A(\xi)$ are also called *contradictory* or *inconsistent* scenarios. Since the set $C_A(\xi)$ is sometimes intractably large, an appropriate representation is needed. One possibility is to consider the set

$$C(\xi) = \{\tau \in \mathcal{T}_A : M_A(\tau) \subseteq C_A(\xi)\} = \{\tau \in \mathcal{T}_A : \tau \wedge \xi \models \bot\} \qquad (2)$$

of *conflicting* terms whose models are all conflicting scenarios. Note that $C(\xi)$ is an *upward-closed* set. This means that $\tau \in C(\xi)$ implies $\tau' \in C(\xi)$ for all longer terms $\tau' \in \mathcal{T}_A$ with $\tau' \supseteq \tau$. Furthermore, a term $\tau \in C(\xi)$ is called *minimal* with respect to $C(\xi)$, if $C(\xi)$ contains no shorter term $\tau' \subset \tau$. The corresponding set of minimal conflicting terms

$$\mu C(\xi) = \{\tau \in C(\xi) : \neg \exists \tau' \in C(\xi), \tau' \subset \tau\} \qquad (3)$$

---
[2] To simplify matters, we restrict our discussion in this paper to propositional logic.
[3] Note that in Poole's abductive framework [21], terms $\tau \in \mathcal{T}_A$ are called scenarios.

is obtained from $C(\xi)$ by dropping all non-minimal terms. The elements of $\mu C(\xi)$ are also called *minimal conflicts* of ξ. Note that

$$C_A(\xi) = \cup\{M_A(\tau) : \tau \in C(\xi)\} = \cup\{M_A(\tau) : \tau \in \mu C(\xi)\}. \tag{4}$$

Thus, in order to exclude conflicting scenarios, it is sufficient to know the minimal conflicts. However, deriving $\mu C(\xi)$ from ξ may be very expensive, even in cases where $\mu C(\xi)$ is small or even empty.

3 Computing Conflicts

The problem of computing minimal conflicts is closely related to the problems of computing *prime implicates* (or *prime implicants*). Conflicts are conjunctions $\tau \in \mathcal{T}_A$ for which $\tau \wedge \xi \models \bot$ holds. This condition can be rewritten as $\xi \models \neg\tau$. Conflicts are therefore negations of implicates of ξ which are in $\mathcal{C}_A$. In other words, if $\gamma \in \mathcal{C}_A$ is an implicate of ξ, then $\neg\gamma \in \mathcal{T}_A$ is a conflict of ξ. Furthermore, if $\gamma \in \mathcal{C}_A$ is a prime implicate of ξ, then $\neg\gamma$ is a minimal conflict. We use $PI(\xi)$ to denote the set of all prime implicates of ξ. If $\neg\Psi$ is the set of conjunctions obtained from a set of clauses Ψ by negating the corresponding clauses, then we can write

$$\mu C(\xi) = \neg(PI(\xi) \cap \mathcal{C}_A). \tag{5}$$

Since computing prime implicates is known to be NP-complete in general, the above approach is only feasible when ξ is relatively small. However, when A is small enough, many prime implicates of ξ are not in $\mathcal{C}_A$. Such irrelevant prime implicates can be avoided by the method described in [15, 12, 13]. It is assumed that ξ is given as a set $\Sigma \subseteq \mathcal{C}_{A\cup P}$ of clauses over $A \cup P$. The procedure is based on two operations

$$Cons_Q(\Sigma) = Cons_{x_1} \circ \cdots \circ Cons_{x_q}(\Sigma), \tag{6}$$
$$Elim_Q(\Sigma) = Elim_{x_1} \circ \cdots \circ Elim_{x_q}(\Sigma), \tag{7}$$

where $Q = \{x_1, \ldots, x_q\} \subseteq A \cup P$ is subset of propositions appearing in Σ. Both operations repeatedly apply more specific operations $Cons_x(\Sigma)$ and $Elim_x(\Sigma)$, respectively, where x is a proposition in Q. Let Σ_x denote the set of clauses of Σ containing x as a positive literal, $\Sigma_{\bar{x}}$ the set of clauses containing x as a negative literal, and $\Sigma_{\dot{x}}$ the set of clauses not containing x. Of course, we have $\Sigma = \Sigma_x \cup \Sigma_{\bar{x}} \cup \Sigma_{\dot{x}}$. Furthermore, if

$$\rho(\Sigma_x, \Sigma_{\bar{x}}) = \{\vartheta_1 \vee \vartheta_2 : x \vee \vartheta_1 \in \Sigma_x, \neg x \vee \vartheta_2 \in \Sigma_{\bar{x}}\} \tag{8}$$

denotes the set of all resolvents of Σ relative to x, then the two basic operations are defined by

$$Cons_x(\Sigma) = \mu(\Sigma \cup \rho(\Sigma_x, \Sigma_{\bar{x}})), \quad Elim_x(\Sigma) = \mu(\Sigma_{\dot{x}} \cup \rho(\Sigma_x, \Sigma_{\bar{x}})), \tag{9}$$

where the μ-operator means dropping non-minimal clauses. Thus, $Cons_Q(\Sigma)$ computes all the resolvents (consequences) of Σ relative to the propositions in Q and adds them to Σ. Note that if Q contains all the proposition in Σ, then $Cons_Q(\Sigma) = PI(\Sigma)$. In contrast, $Elim_Q(\Sigma)$ eliminates all the propositions in Q from Σ and returns a new set of clauses whose set of models corresponds to the projection of the original set of models to

$(A \cup P) \setminus Q$. Elimination is sometimes called *forgetting* and is known to be NP-complete [6].

The set of the minimal conflicts can then be computed in two different ways by

$$\mu C(\xi) = \neg Cons_A(Elim_P(\Sigma)) = \neg Elim_P(Cons_A(\Sigma)). \tag{10}$$

In most practical applications, computing the consequences relative to the propositions in A is trivial. In contrast, the elimination of the propositions in P is usually more difficult and becomes even infeasible as soon as Σ has a certain size. Note that from a theoretical point of view, the order in which the propositions in P are eliminated is irrelevant [15], whereas from a practical point of view, it critically influences the efficiency of the procedure. The elimination process is a particular instance of Shenoy's fusion algorithm [26, 27] as well as of Dechter's bucket elimination procedure [9].

Today's state of the art among the methods for computing conflicts, arguments, and abductive explanations is a convenient anytime algorithm that can be interrupted at any time returning the solution found so far [13]. The quality of the approximation increases monotonically when more computational resources are available. The method is based on cost functions [12] and returns lower and upper bounds.

4 Detecting Non-Conflicting Knowledge Bases

Let ξ be an arbitrary assumption-based knowledge base. Of course, we can use the procedure of the previous section to find out whether there are conflicts or not. However, running the complete elimination procedure may possibly be very expensive, even in cases where $\mu C(\xi) = \emptyset$. In the following, we will describe a method that simplifies the detection of conflict-free knowledge bases.

Let $\gamma \in \mathcal{C}_{A \cup P}$ be an arbitrary clause over $A \cup P$. Without loss of generality, it is always possible to split γ into sub-clauses $\gamma_A \in \mathcal{C}_A$ and $\gamma_P \in \mathcal{C}_P$ with $\gamma = \gamma_A \vee \gamma_P$. As we will see below, only the sub-clauses γ_P are relevant for detecting conflict-free knowledge bases. Thus, if $\Sigma \subseteq \mathcal{C}_{A \cup P}$ is an arbitrary set of clauses over $A \cup P$, then

$$\Sigma_P = \{\gamma_P : \gamma \in \Sigma\} \subseteq \mathcal{C}_P \tag{11}$$

denotes a new set of clauses obtained by dropping all assumptions. As an effect of this, Σ_P may contain many clauses that are subsumed by others. A corresponding minimal set $\mu \Sigma_P$ is obtained from Σ_P by removing all subsumed clauses. Of course, Σ_P and $\mu \Sigma_P$ are logically equivalent, but $\mu \Sigma_P$ is often considerably smaller.

Theorem 1 *Let $\xi \in \mathcal{L}_{A \cup P}$ be a knowledge base given as a set of clauses $\Sigma \subseteq \mathcal{C}_{A \cup P}$. If $\Sigma_P \subseteq \mathcal{C}_P$ is the set of clauses obtained from Σ as defined above, then $\Sigma_P \not\models \bot$ (or $\mu \Sigma_P \not\models \bot$) implies $C(\xi) = \emptyset$ (and $\mu C(\xi) = \emptyset$).*

Proof: Let $\xi_P = \wedge \Sigma_P$ be the conjunction of clauses of Σ_P. From $\gamma_P \models \gamma$ for all $\gamma \in \Sigma$ follows $\xi_P \models \xi$. Now, suppose $C(\xi) \neq \emptyset$ and $\tau \in C(\xi)$. This implies $\tau \wedge \xi \models \bot$ and thus $\tau \wedge \xi_P \models \bot$. Since ξ_P contains no assumptions, this is only possible if $\xi_P \models \bot$. Thus, $C(\xi) \neq \emptyset$ implies $\xi_P \models \bot$, and the other way round, $\xi_P \not\models \bot$ implies $C(\xi) = \emptyset$. □

Note that $\Sigma_P \models \bot$ does not necessarily mean that $C_A(\xi) \neq \emptyset$. For example, if $A = \{a\}$ and $P = \{p\}$, then $\xi = (a \vee p) \wedge (\neg a \vee \neg p)$ is a conflict-free knowledge base that implies $\xi_P = p \wedge \neg p$ and thus $\xi_P \models \bot$. The above theorem can thus be used to detect some (but not all) cases of conflict-free assumption-based knowledge bases.

Example 1 *Let $A = \{a_1, \ldots, a_n, b_1, \ldots, b_n\}$ and $P = \{p_1, \ldots, p_n\}$ be the given sets of propositions. Furthermore, suppose that ξ is given as a set clauses*

$$\Sigma = \left\{ \begin{array}{l} a_1 \vee p_1, \; b_1 \vee p_1, \\ a_2 \vee \neg p_1 \vee p_2, \; b_2 \vee \neg p_1 \vee p_2, \\ a_3 \vee \neg p_2 \vee p_3, \; b_3 \vee \neg p_2 \vee p_3 \\ \vdots \qquad\qquad \vdots \\ a_n \vee \neg p_{n-1} \vee p_n, \; b_n \vee \neg p_{n-1} \vee p_n \end{array} \right\}.$$

Obviously, ξ is a conflict-free knowledge base. By running the elimination procedure of the previous section, we observe a worst-case scenario with an exponentially increasing number of clauses during the process. Before eliminating the last proposition, we always get a total number of 2^n clauses for which no further resolutions are possible (independently of the actual elimination ordering). Of course, this is not feasible if n exceeds a certain limit. In contrast, by applying Theorem 1, we get

$$\Sigma_P = \{p_1, \neg p_1 \vee p_2, \neg p_2 \vee p_3, \ldots, \neg p_{n-1} \vee p_n\}$$

for which $\Sigma_P \not\models \bot$ and thus $C_A(\xi) = \emptyset$ is easily proved in linear time.

Even if SAT is NP-complete in general, this example demonstrates how the method of Theorem 1 sometimes tremendously reduces the complexity of detecting conflict-free knowledge bases from $O(2^n)$ to $O(n)$. This is typical in many practical examples from the author's domain of interest [2].

5 Conclusion

Even if the set of minimal conflicts of an assumption-based knowledge base is small or empty, effectively computing the minimal conflicts is not feasible in the worst case. However, we have shown how conflict-free knowledge bases, which are common in many practical applications, can sometimes be detected more easily by testing the satisfiability of a corresponding simplified knowledge base. In certain cases, this reduces tremendously the necessary time of computation, even if SAT is also NP-complete in the worst case.

Acknowledgements

Research supported by (1) Alexander von Humboldt Foundation, (2) German Federal Ministry of Education and Research, (3) German Program for the Investment in the Future.

References

[1] L. Amgoud and C. Cayrol. A reasoning model based on the production of acceptable arguments. *Annals of Mathematics and Artificial Intelligence*, 34(1-3):197–215, 2002.

[2] B. Anrig, R. Bissig, R. Haenni, J. Kohlas, and N. Lehmann. Probabilistic argumentation systems: Introduction to assumption-based modeling with ABEL. Technical Report 99-1, Institute of Informatics, University of Fribourg, 1999.

[3] Ph. Besnard and A. Hunter. Towards a logic-based theory of argumentation. In *Proceedings of the National Conference on Artificial Intelligence (AAAI'2000)*, pages 411–416. MIT Press, 2000.

[4] Ph. Besnard and A. Hunter. A logic-based theory of deductive arguments. *Artificial Intelligence*, 128:203–235, 2001.

[5] S. A. Cook. The complexity of theorem-proving procedures. In *Proceedings of the 3rd Annual ACM Symposium on Theory of Computing*, pages 151–158, 1971.

[6] A. Darwiche and P. Marquis. A knowledge compilation map. *Journal of Artificial Intelligence Research*, 17:229–264, 2002.

[7] J. de Kleer. An assumption-based TMS. *Artificial Intelligence*, 28:127–162, 1986.

[8] J. de Kleer, A. K. Mackworth, and R. Reiter. Characterizing diagnosis and systems. *Artificial Intelligence*, 56:197–222, 1992.

[9] R. Dechter. Bucket elimination: a unifying framework for reasoning. *Artificial Intelligence*, 113(1–2):41–85, 1999.

[10] M. Garey and D. Johnson. *A Guide to the Theory of NP-Completeness*. W. H. Freeman and Company, 1979.

[11] J. Gu, P. W. Purdom, J. Franco, and B. W. Wah. Algorithms for the satisfiability (SAT) problem: a survey. In D. Du, J. Gu, and P. M. Pardalos, editors, *Satisfiability Problem: Theory and Applications*, volume 35 of *DIMACS Series in Discrete Mathematics and Theoretical Computer Science*, pages 19–152. American Mathematical Society, 1997.

[12] R. Haenni. Cost-bounded argumentation. *International Journal of Approximate Reasoning*, 26(2):101–127, 2001.

[13] R. Haenni. A query-driven anytime algorithm for argumentative and abductive reasoning. In D. Bustard, W. Liu, and R. Sterrit, editors, *Soft-Ware 2002, 1st International Conference on Computing in an Imperfect World*, LNCS 2311, pages 114–127. Springer-Verlag, 2002.

[14] R. Haenni, B. Anrig, J. Kohlas, and N. Lehmann. A survey on probabilistic argumentation. In *Proceedings of the 6th European Conference on Symbolic and Quantitative Approaches to Reasoning with Uncertainty (ECSQARU'01)*, pages 19–25, Toulouse, France, 2001.

[15] R. Haenni, J. Kohlas, and N. Lehmann. Probabilistic argumentation systems. In J. Kohlas and S. Moral, editors, *Handbook of Defeasible Reasoning and Uncertainty Management Systems, Volume 5: Algorithms for Uncertainty and Defeasible Reasoning*, pages 221–288. Kluwer, Dordrecht, 2000.

[16] J. Kohlas, B. Anrig, R. Haenni, and P. A. Monney. Model-based diagnostics and probabilistic assumption-based reasoning. *Artificial Intelligence*, 104:71–106, 1998.

[17] J. Kohlas and P. A. Monney. *A Mathematical Theory of Hints. An Approach to the Dempster-Shafer Theory of Evidence*, volume 425 of *Lecture Notes in Economics and Mathematical Systems*. Springer-Verlag, 1995.

[18] D. McAllester. Truth maintenance. In R. Smith and T. Mitchell, editors, *Proceedings of the Eighth National Conference on Artificial Intelligence*, volume 2, pages 1109–1116. AAAI Press, 1990.

[19] G. Paul. Approaches to abductive reasoning – an overview. *Artificial Intelligence Review*, 7(2):109–152, 1993.

[20] Y. Peng and J. Reggia. *Aductive Inference Model for Diagnostic Problem Solving*. Springer-Verlag, 1990.

[21] D. Poole. A logical framework for default reasoning. *Artificial Intelligence*, 36(1):27–47, 1988.

[22] D. Poole. Explanation and prediction: An architecture for default and abductive reasoning. *Computational Intelligence*, 5(2):97–110, 1989.

[23] D. Poole. A methodology for using a default and abductive reasoning system. *International Journal of Intelligent Systems*, 5(5):521–548, 1990.

[24] D. Poole. Probabilistic horn abduction and bayesian networks. *Artificial Intelligence*, 64:81–129, 1993.

[25] R. Reiter. A theory of diagnosis from first principles. *Artificial Intelligence*, 32:57–95, 1987.

[26] P. P. Shenoy. Binary join trees. In E. Horvitz and F. Jensen, editors, *Proceedings of the 12th Conference on Uncertainty in Artificial Intelligence (UAI-96)*, pages 492–499. Morgan Kaufmann Publishers, 1996.

[27] P. P. Shenoy. Binary join trees for computing marginals in the shenoy-shafer architecture. *International Journal of Approximate Reasoning*, 17(2–3):239–263, 1997.

Uncertainty

B. Bouchon-Meunier, L. Foulloy and R.R. Yager (Editors)
Intelligent Systems for Information Processing:
From Representation to Applications
© 2003 Elsevier Science B.V. All rights reserved.

Theory of Belief Functions: History and Prospects

A. P. Dempster
Department of Statistics, Harvard University, Cambridge MA 02138 USA.
dempster@fas.harvard.edu

Abstract

The origins and basic elements of the Dempster-Shafer theory of belief functions are explained, including the operations of combination of independent representations of uncertain evidence and propagation among margins of multivariate systems. The theory is described as a tool that scientists can use to formalize subjective uncertainties about objectively formalized unknowns. The future of the theory depends on the development of models that capture common situations and are amenable to modern computational methodologies.

Keywords: Dempster-Shafer theory, subjective probability, propagation and fusion of evidence, independent sources of evidence.

1. Orientation: From Precise Mathematics to Serious Science

My undergraduate studies in Toronto emphasized mathematics, in a department that regularly won the North American Putnam problem-solving competitions. My instincts, however, were and remain to think of mathematics as a tool of science. So when I recognized pure mathematics to be primarily inner directed, I decided that the life of a mathematician, although an important and worthy calling, was not for me, and I switched to probability and statistics for my PhD studies in the Princeton mathematics department of the mid-1950s, where my teachers were S. S. Wilks and J. W. Tukey. Attempting to fashion bridges from mathematics to science has been the prime motivator of my subsequent career. What is science, however? Standard explanations are inadequate to support practice, in my view, because they stress the objective content of science, ignoring and effectively excluding subjective elements obviously present. Mathematics can and should be recognized as a basic support for the subjective side of science, in parallel with its traditional role as provider of models for the objective side. This is a central theme in what follows.

Mathematics aids science through its beauty, clarity, and rigor. Along with this goes a belief among most mathematical scientists that objectivity is a *sine qua non* of science, an attitude that can seem borne out by the triumphs of physics, from mechanics to electro-magnetic theory, and from relativity to quantum theory, all of which are associated with a close symbiosis of detailed mathematical models and spectacularly successful empirical science and engineering. Applied probabilists and mathematical statisticians for the most part tacitly accept this philosophy of objectivity, while seeking with some success to develop parallels to success stories from the physical sciences. In particular, applied statistical scientists for the most part see themselves as applying objective procedures of design and analysis to objective empirical situations, thus obtaining objective results. Having been exposed to all this, and at first accepting the faith with little questioning, I gradually moved to a different philosophy, recognizing obvious roles for subjective elements that complement traditional perceptions of objectivity. The daily work of science involves many choices: among procedures, among patterns of reasoned argument, and among selections of evidence, expert knowledge, and judgment. Much of the reasoning is informal, spontaneous, and unrecorded. A true understanding of specific studies, and thence of the scientific enterprise, is impossible without open recognition of thought processes whose subjective elements are obvious to anyone attuned to examine them.

A fundamental mechanism for controlling subjectivity is to apply the methods of mathematical science to reasoning itself. All reasoning is subjective in the sense that it is the work of human minds. At the same time reasoning is itself amenable to precise mathematical modeling, providing useful tools for the practical scientist. A basic formulation that I find highly relevant for quantifying specific uncertainties combines propositional logic with probability in a way that developed out of my papers of the 1960s, with syntax and semantics broadened in the 1970s by Glenn Shafer. Analyses that adopt this logic with realistic model assumptions are gradually becoming computationally feasible over broad areas of application, often in ways that are close to Bayesian methods, while taking limited but essential steps that extend Bayes to Dempster-Shafer (DS) methods when circumstances appear to warrant. Thus for me Bayes is not a final answer as maintained by many in the active Bayesian school of theoretical and applied statistics.

Most of my statistical colleagues, including many who are deeply and successfully involved in scientific work, pay little heed to formal representations of uncertain reasoning. Their primary concern is with standards of good practice, specifically with choosing and justifying procedures classified under headings such as design, data analysis, estimation, hypothesis testing, and decision-making. While necessary and often admirable, this approach misses much of the story. Procedures should in parallel be seen as embodiments of scientific logic. Thus, for example, formal exploratory data analysis (EDA) tools should be evaluated not only through their mathematical properties, but also *in situ* as directly answering questions that have arisen in the normal course of less formal scientific reasoning. It is a mistake, in my judgment, to mistrust and effectively deny the natural direct interpretation of probabilities. Substitution of theoretical evaluation of procedures through hypothetical long run averages only begs questions about the reasoning behind model choices on which the long run averages are based, and about deeper questions concerning why an unspecified long run has relevance to any specific situation under analysis.

Everyone knows and accepts that science progresses through vigorous and sometimes controversial debates. Going further, a standard faith asserts that truth wins out in the end, once flaws in reasoning as well as gaps in empirical knowledge are all resolved, whence any element of subjectivity is regarded as transient, partly due to error and partly due to ignorance. But if reasoning can have flaws that are important, then reasoning that survives critical analysis is an essential element of findings. Subjective knowing and reasoning are as much part of the scientific process as are claims to have found truth about an objective ambient world. Once this is recognized more openly, the way will be smoothed for scientists to understand and practice the use of tools of formal subjective reasoning, and thence to confidently cross bridges that connect formal logical assertions about idealized small worlds to informal statements of findings about the real objective world that are the bread and butter of normal science. In particular, probability statements about events before the fact or after the fact ("predictively" or "postdictively") will be freed to assume their natural logical interpretations (Dempster, 1998, 2002a).

2. Some History

The earliest history of formal subjective probability (FSP) is shrouded in language that can be interpreted differently through various modern viewpoints. Important references are Hacking (1975) and Shafer (1978). On rereading Shafer (1978) I am impressed by the close correspondence of my outlook with attitudes that he quotes from Part IV of Jakob Bernoulli's posthumous *Ars Conjectandi* (1713) about objective and subjective elements of science and about the fundamentally epistemic nature of formal probabilities. A die thrown by a gambler can only fall with certainty in accord with physical laws that are "necessary" and hence determine an outcome "objectively and in itself", while our knowledge is broadly "contingent" and can only determine an uncertain outcome "subjectively and in relation to us". Bernoulli's much quoted verbal characterization of probability is that a "probability is a degree of certainty and differs from it as a part from a whole". Such probabilities are determined from "arguments" that involve facts about the external world that may be "necessary" or "contingent", together with reasoned "proofs" that also may be "necessary" or "contingent", resulting in three of four cases that yield less than full certainty. Bernoulli classified arguments as "pure" or "mixed", where the former "prove a thing in certain cases in such a way that they prove nothing positively in other cases", while the latter "prove the thing in some cases in such a way that they prove the contrary in the remaining cases". Thus a mixed argument produces complementary probabilities that sum to unity while a pure argument supports only one side of a proposition in a way that presages belief function models, and that Shafer describes as yielding mathematically nonadditive measures. Most relevant to the present paper, Shafer discusses Bernoulli's prescriptions for the combination of independent arguments, an integral part of his calculus of probability, and notes that Lambert (1764) in his *Neues Organon* criticizes Bernoulli's prescriptions and introduces an alternative way to combine arguments that is a special case of the centerpiece combination rule of DS theory. Between Lambert and R. A. Fisher, I am not aware of clearly differentiated instances of DS theory.

Jakob Bernoulli is best known among mathematical statisticians for his derivation of a simple form of what has become known as the law of large numbers. This pathbreaking mathematical work was motivated by his comparably innovative distinction between *a priori* and *a posteriori* scientific arguments, where the former refers to probabilities based solely on a perception of equally likely cases, as in the models for games of chance that had dominated discussions of probability before Bernoulli, while the latter is his at the time novel conception that scientifically useful probabilities often require in addition the input of empirical frequencies. It is interesting that the terms *a priori* and *a posteriori* were later adapted and narrowed to specific technical concepts in Bayesian methodology. Bernoulli did not have the fundamental concept of conditional probability introduced later by Bayes, but instead proposed to use binomial sampling distributions as the basis of statistical inference. Once firmly established by Laplace, the Bayesian paradigm remained dominant in scientific circles throughout the 19th Century (Porter, 1986), while the British-American school of applied statistics rose to prominence in the 20th Century, based upon sampling theory derived by Karl Pearson, 'Student' (W. S. Gosset), and R. A. Fisher.

My sense of relevant history from the middle years of the 20th Century is personal because it has directly motivated my work. When I first encountered probability and statistics, the field was preoccupied with a decision-theoretic frequentist formalism, and the fashion was to base theory on measure-theoretic mathematics derived from Kolmogorov's axioms. Fortunately, I was on the East Coast not the West Coast and my teachers remained firmly in touch with a British tradition of applied statistics whose Fisherian roots had not yet been totally seduced by the siren call of frequentist theory. About the same time, the Bayesian countermovement to the dominant frequentist school started to gain a following, dating most notably from the pioneering book of Savage (1954). Inferential statistics thus became a three-corned battleground for a few years, involving R. A. Fisher who engaged in ceaseless controversy with Jerzy Neyman and his frequentist followers from the 1930s on, and the resurgent neoBayesian school. By the 1950s, however, Fisher was in decline, and his legacy largely vanished from textbooks after his death in 1962, while the other two schools had worn each other out by the 1970s without a resolution, and turned to face the new worlds of statistical science opened up by rapidly evolving information technologies.

Why did Fisher lose his fight? My interpretation is that he was correct to be fundamentally epistemic in his outlook, but tried to tie logic to an objectivist attitude to science, and could never convey a satisfactory understanding of how the marriage worked. I believe that Fisher's fate was unfortunate, because he developed so many genuinely novel and worthwhile inferential tools under the heading of "forms of quantitative inference" (Fisher, 1956) while his antagonists rejected logic altogether and in essence repeated prolonged controversies between Bayesians and frequentists from the 19th Century, as ably reviewed for example by Keynes (1921). The Fisherian initiative that most intrigued me in the 1950s and 1960s was his failed attempt to establish a method that he called the "fiducial argument" (Fisher, 1930) which was essentially an attempt to produce statements of uncertainty about objective unknowns, with precise posterior ("fiducial") probabilities, but without the precise prior probabilities that had long been recognized to be, and indeed remain, a dominant sticking point in the way of scientific applications of Bayes. John Tukey (Dempster 2002b) drew my attention to the central concept of the fiducial argument, namely the "pivotal quantity" that assigns a probability distribution to a margin of the relevant space of unknowns. This distribution became the "bpa" or "mass function" of belief function theory, and led to upper and lower posterior probabilities in the case of discrete sample data, whence the general form of the belief function became clear to me one day in 1965. This general form greatly extends the fiducial argument in that it unifies combination of information from many independent sources, including likelihoods and priors in Bayesian theory, and the familiar union and intersection rules of propositional logic.

3. Elements of DS Theory

My elementary mathematical representation of objective and subjective elements of science is built around the two primitives, of "state space" and "statement space", respectively. The simplest useful example defines the state space to consist of two elements, for convenience labelled here "0" and "1". The term state space is commonly used in engineering sciences to represent the possible "states" of some objective small world under analysis, so "0" might mean "absence of abnormality" while "1" means "presence of abnormality" in a particular slide under a microscope. Evidently, a large amount of implicit understanding is required to make a state space meaningful, such as the source of the slide, the definitions of normal and abnormal in relation to the observational technique, the material being observed, and on and on. The real world is very complex, while formal analysis rests on precise but simplified idealizations. An obvious trend in modern science is to observe, record, and analyze phenomena of escalating complexity, whence formal state spaces are being increasingly challenged to mathematically model hugely complex phenomena, and thence to represent and perform logical analyses that vastly transcend what was contemplated a few short decades earlier. This reality needs to be kept in mind while I introduce my story in elementary terms. In the end I am aimed at many and varied sciences of complex systems.

What can it possibly mean to assert, as I do, that a state space is objective? If you like, this is an "assumption", one of many that are successively wheeled in to make a formal analysis operate. We think that part of an ambient world is captured by a state space, and choose after much review and thinking to base a particular formal analysis on that particular state space, putting aside for the time being urges to revise. Even here, objectivity is of course something of a pretence, designed to create separation between the limited subjectivity required to choose a state space, and the more direct epistemic subjectivity implied by formal statements about which member of the state space is the true member.

Using notation that is easily extended, I represent my simple state space by a list of its elements, written

$$X = (0, 1)$$

where despite the appearance of the printed form, no meaning is attached to the order of the elements. X is abstractly a set with two elements, nothing more. Similarly, the statement space consists of the nonempty subsets of the state space, and will be written

$$S = (\{0\}, \{1\}, \{0,1\})$$

The interpretation of S is that each of its elements corresponds to a statement that some analyst might make about the true state of the small world. The three elements here correspond directly to what is called "three valued logic", namely, the assertions:

"the true state is 0"; "the true state is 1"; "the true state is unknown".

Defenders of the objectivity of science by and large suppress the concept of statement, and make do with the state space alone. I mean this in the sense of formal analysis. The language and notation of elementary propositional logic are foreign territory to most scientists, engineers, and statisticians, including those fluent in advanced mathematical modeling. A scientist will report that his or her slide shows abnormality and sees no need for a formal system of sets and subsets. In a similar way, one almost never encounters in scientific reports talk of disjunctions and conjunctions represented by intersections and unions of subsets of more complex state spaces, although these things are almost universally implied. A corollary is that formal reasoning, whether nonprobabilistic or probabilistic, is effectively sealed off from the everyday practice of science and engineering. Scientific reports are replete with statements about what is true and what is uncertain to varying degrees, but almost all conveyed by informal, often colloquial speech.

Given this circumstance, it is natural for most scientists, engineers, and statisticians to think of probability as something objective. When probabilities are construed as part of "nature", they connect only with the elements of state space X. Thus, in my simple example, it is natural to talk about "the" probability p_1 that the slide shows an abnormal result and "the" probability $p_0 = 1 - p_1$ that the slide shows a normal result. In practice, numerical values for such probabilities might be assigned from past experience with similar slides. Obviously there are important uses for such "frequency" probabilities, but it is mischievous to argue that they are the only kinds of quantities that deserve the name probability. Nor are they distinct from ordinary subjective probabilities. Probabilities directly "approximated" by frequencies are easily interpreted as formal logical inferences, and indeed must be so interpreted, depending upon judgments made in context as well as upon objective facts, in order for them to have the impact intended by their application. Standard practice implicitly leaves subjective interpretation informally understood.

What has all this to do with my topic of "belief functions"? The point is that by associating probability with subjective uncertainty, it becomes natural to assign three probabilities to the elements of S, namely $p_{\{0\}}$, $p_{\{1\}}$ and $p_{\{0,1\}}$, rather than to the elements of the state space X as in objective science. These three probabilities provide a first example of what Shafer (1976) called the "basic probability assignment" or "bpa" of belief function theory, also referred to as the "mass function" and denoted $m(\cdot)$, where the argument is understood to be any element of statement space. The bpa is one of four functions over statement space taking values on the interval (0, 1). Two of these are generally called "belief" and "plausibility" and denoted $Bel(\cdot)$ and $Pl(\cdot)$, where in the simple example

$$Bel(\{0\}) = m(\{0\}),$$
$$Bel(\{1\}) = m(\{1\}),$$
$$Bel(\{0,1\}) = m(\{0,1\}) + m(\{0\}) + m(\{1\}),$$

and

$$Pl(\{0\}) = m(\{0\}) + m(\{0,1\}),$$
$$Pl(\{1\}) = m(\{1\}) + m(\{0,1\}),$$
$$Pl(\{0,1\}) = m(\{0\}) + m(\{1\}) + m(\{0,1\}).$$

Shafer introduced these terms as improvements over my original "lower probability" and "upper probability" (Dempster, 1966, 1967) in order to distinguish our theory from other less articulated theories of bounded probability judgments. To me *Bel* simply means formal subjective probability (FSP) that "must" be assigned to the truth of a statement about an objective state space, while *Pl* concerns FSP that "may" be so assigned. The fourth set function named "commonality" by Shafer is the least interpretable of the four, and is mainly important for a technical role in the calculus of combination of independent pieces of evidence. Such combination or fusion stands at the center of DS theory.

To bypass topological issues it is best to outline the general case using a finite state space X with n elements, whence the statement space S has $2^n - 1$ elements. The oft-repeated standard exposition following Dempster (1967) or Shafer (1976) then defines the four basic set functions and associated operations, with simple examples of various kinds.

4. Two Operators that Drive the Theory.

The *direct sum operator* (also known as the *Dempster rule of combination* or the *product-intersection rule*) takes as input two or more belief functions and outputs a *combined* belief function, all defined over the same state space. The rule is valid only if the inputs are independent. In another language, the combination rule is a technique for *fusing* evidence from independent sources. Independence is critical. Judging independence is an art that must be learned through practice.

The *marginalization* operator inputs a belief function and returns the implied belief function over a partition (also called a margin) of the original state space. The *extension* operator inputs a belief function over a partition and outputs the implied belief function over the full state space. A general term that covers both operators is *propagation*, as in propagation of information through a network.

These operators were presented in abstract form for the general case of a finite state space in Dempster (1967) and Shafer (1976), where the state space may be denoted

$$X = (e_1, e_2, \ldots, e_n)$$

and the associated statement space denoted

$$S = (\{e_1\}, \{e_2\}, \ldots, \{e_1, e_2, \ldots, e_n\}).$$

A belief function model is then defined mathematically by its bpa function $m(\cdot)$ defined over S.

The formal definition of combination starts from a pair of such models with mass functions $m_1(\cdot)$ and $m_2(\cdot)$ over a common statement space. I like to call the combination rule a product-intersection rule because it intersects pairs of elements from two copies of S and assigns the product of the associated bpa values to each such intersection. Since the same intersection may result from many such pair-intersections, it is necessary to sum the corresponding product-probabilities within each distinct pair-intersection, and renormalize to account for empty intersections, thus obtaining the bpa of the combination. I state this in words rather than as the formula, which is available in many places, to emphasize the elements of the formula that must be parsed to understand why the rule is logically compelling.

The combination rule may be studied mathematically, where its most fundamental properties are commutativity and associativity. These imply that any number of input belief functions have a unique combination, whatever the order of successive applications of pairwise combinations. Intuition may be built up by considering first the special case of propositional calculus where each input belief function assigns mass one to an individual statement and the rule then correctly assigns mass one to the intersection statement, as one might expect and require.

Another fundamental special case reproduces the concept of product measure over a product space from standard textbook probability theory. Understanding the latter special case involves subtleties of interpretation that have sometimes led to misunderstandings, and thence criticisms of the rule. For example, suppose that one physician analyzing a slide gives probability 1/4 that it shows abnormalities while with probability 3/4 it is normal. Then a second "independent" physician views the same slide and assigns probabilities 1/3 and 2/3 for the same assertions. The basic rule would then assign probabilities 1/7 and 6/7, while common sense suggests that a judgment of abnormality in the range of 1/4 or 1/3 would be more sensible. In fact, these physicians are not independent in the sense of the theory, because they are accessing the same evidence. Such dependence is commonplace in the Bayesian special case of the theory, where successive sample observations, while independent in the sampling sense with a "fixed" parameter value are marginally dependent in the sense that matters for DS theory because they are dependent on a common unknown parameter value. Independence is a subtle concept, needing careful consideration in all applications of the product-intersection rule, and in particular underlies Bayesian analysis in a way that Bayesian adherents rarely think about, but is uncovered by recognition that the fundamental Bayesian tool of successive conditioning is a special case of the belief function rule (Dempster, 1968). A similar comment applies to routine use of intersections in the propositional calculus. One must always make a judgment that the pieces of evidence being combined do not contaminate each other, and so invalidate applications of the rule.

That said, the application of the rule of combination to ordinary stochastic modeling, as illustrated by many complex stochastic models, such as hidden Markov models, shows that models built this way can provide deep insights into a wide range of scientific phenomena that are conventionally regarded as "random". In a similar way, the combination of many independent elements is the standard way to build up a complex belief function model.

To explain the second operator *propagation* the essential abstract concept is that of a partition of the state space X, namely, a collection of subsets whose intersections are all empty (they are "mutually exclusive") and whose union is X itself (they are "exhaustive"). A partition P of a state space X can also be regarded as a state space representing the same small world, but with less expressive power. That is, statements based on P when reinterpreted as statements about S correspond only to subsets of X that are unions of elements of P. Given a belief function over X with mass function $m_X(\cdot)$ the *marginal* belief function over P is defined by summing collections of $m_X(\cdot)$ to obtain a corresponding element of $m_P(\cdot)$. Specifically, each statement in S projects into a unique statement in P, whence $m_P(\cdot)$ is obtained by summing values of $m_X(\cdot)$ that project into each subset of P. Similarly, given any belief function defined on a partition P, the *extension* to S is defined by assigning the $m_P(\cdot)$ values to the cylinder sets in X that define the partition. A little thought shows that marginalization generally loses information because a marginal belief function does not in general return by extension to the original, while extension neither gains nor loses information. It is interesting that the ordinary theory of probability does not encompass extension, but if an ordinary probability distribution is extended it in fact does have a belief function extension.

The general operator of propagation can now be defined as carrying a belief function on one margin of X to a second margin of X, by first extending to the full space and then marginalizing to the second margin. This operation is fundamental to computations with many useful examples.

5. An Elementary Example

Students of probability theory are typically taught about "product measures" on "product spaces" and then told that these represent "independent" random variables. For example, prior to knowing with certainty the contents of my hypothetical slide I might assign probabilities p_A and p_B to the truth of A or B types, and independently p_0 and p_1 to the truth of normal or abnormal cells, whence by assumption, and the definition of product measure, the four product probabilities are assigned to truth being in corresponding cells of the 2x2 table. The belief function reconstruction of the same story is more detailed. First, the probabilities p_A and p_B are regarded as a belief function on the type margin having probabilities on $\{A\}$ and $\{B\}$ but no probability on $\{A,B\}$, and similarly for the probabilities p_0 and p_1 for the normal/abnormal margin of the 2x2 table. Each of these marginal beliefs is extended to the full table, in one case assigning probabilities to the rows, and in the other case assigning probabilities to the columns. Applying the product-intersection rule over the 2x2 table now produces the standard product measure of ordinary probability theory.

Why should students be burdened with this long way around to a familiar end point? One reason is logical clarity. More important, however, is that belief functions provide a broader range of options for representing uncertainty. Thus on the type margin, I may assign probabilities to $\{A\}$, $\{B\}$ and $\{A,B\}$, and similarly three probabilities may be used for the normal/abnormal model, with the final result being a belief function over the four cells of the 2x2 array with positive mass values on nine of the 15 possible subsets. A common special case might have evidence pointing in special directions, as contemplated by Bernoulli, so that on one margin we might have $P(\{B\}) = 0$ and on the other margin $P(\{0\}) = 0$, in which case the mass function from fusion under independence has only four nonzero values as in the case of ordinary probability, but with genuine belief function uncertainty expressions.

6. What of the Future?

A healthy future for the DS formulation of probabilistic reasoning depends on successful efforts to construct models with credible applications to real situations, and to develop accurate and fast algorithms and software to implement the models. DS theory has demonstrated its logical appeal through gradual development over four decades. The extension of ordinary probability measures from state spaces to statement spaces adds unfamiliarity and complexity, but the foundation of DS methods in the highly developed mathematical theory of probability lends fundamental simplicity, precision, and rigor.

References

[1] Bernoulli, Jakob. (1713). *Ars Conjectandi*. Basel.
[2] Dempster, A. P. (1966). New methods for reasoning towards posterior distributions based on sample data. *Annals of Mathematical Statistics* **37** 355-374.
[3] Dempster, A. P. (1967). Upper and lower probabilities induced by a multivalued mapping. *Annals of Mathematical Statistics* **38** 325-39.
[4] Dempster, A. P. (1968). A generalisation of Bayesian inference (with discussion). *Journal of the Royal Statistical Society B* **30** 205-247.
[5] Dempster, A. P. (1998). Logicist Statistics I. Models and Modeling. *Statistical Science* **13** 248-276.
[6] Dempster, A. P. (2002a). Logicist Statistics II. Inference. Submitted.
[7] Dempster, A. P. (2002b). John W. Tukey as "Philosopher". *Annals of Statistics* **30** 1619-1628.
[8] Fisher, R. A. (1930). Inverse probability. *Proceedings of the Cambridge Philosophical Society* **26** 528-535.
[9] Fisher, R. A. (1956). *Statistical Methods and Scientific Inference*. Oliver and Boyd, Edinburgh. (Slightly revised versions appeared in 1958 and 1960).
[10] Hacking, Ian (1975). The *Emergence of Probability*. Cambridge.
[11] Keynes, J. M. (1921). *A Treatise on Probability*. MacMillan. (Reprinted 1962 Harper Torchbooks)
[12] Lambert, Johann Heinrich (1764). *Neues Organon*. Leipzig.
[13] Porter, Theodore M. (1986). *The Rise of Statistical Thinking 1820-1900*. Princeton.
[14] Savage, L. J. (1954). *The Foundations of Statistics*. Wiley. (Reprinted 1972 Dover).
[15] Shafer, Glenn (1976). *A Mathematical Theory of Evidence*. Princeton.
[16] Shafer, Glenn (1978). Non-additive probabilities in the work of Bernoulli and Lambert. *Archive for History of Exact Science*s **19** 309-370.

Towards another logical interpretation of Theory of Evidence and a new combination rule

Laurence Cholvy
ONERA Centre de Toulouse
2 av. Ed. Belin
31055 Toulouse, France
cholvy@cert.fr

Abstract

Theory of Evidence is a mathematical theory which allows one to reason with uncertainty and which provides a rule for combining uncertain data. It has been shown that Theory of Evidence can be interpreted by logic and probability theory and that the degree of belief of a proposition A can be viewed as the probability that an agent knows (or proves) A. This present paper gives another logical interpretation of masses, when the numbers are rational, in which the degree of belief of a proposition A is interpreted by the proportion, in a given set, of proofs of A. However, this new interpretation does not correctly interpret Dempster's rule. So, a new rule of combination which fits more this interpretation is defined. We show that, in some applications, this new interpretation is adequate. In particular, we study an example that has been used by Zadeh and Smets to criticize normalisation in Dempster's rule. We show that the new interpretation of this example and the new rule lead to acceptable results.

Keywords: Theory of Evidence, logic, combination of uncertain data.

1 Introduction

Theory of Evidence is a mathematical theory defined by Dempster [5] and Shafer [14], which allows one to reason with uncertainty and which defines a rule for combining uncertain data provided by several information sources.

In this theory, the uncertainty is represented by the fact that any proposition of the frame of discernment is associated with a real number, called its mass, which belongs to [0,1] and such that the mass of the contradiction is 0[1], and the sum of all the masses is 1. These numbers intend to represent *a measure of belief committed exactly to the propositions*. Given all the masses, one can define two other numbers: the degree of belief of a proposition, which is also a real number belonging to [0,1] and which represents *the degree of support a body of evidence provides for this proposition*, and the degree of plausibility of a proposition, which is also a real number belonging to [0,1] and which

---
[1] In some extensions of this theory, this constraint is relaxed.

represents *the extend to which one fails to doubt the proposition*. Furthermore, this theory also focuses on the combination of masses through Dempster's rule of combination. This explains why Theory of Evidence is commonly applied in data fusion problems where data are uncertain. For instance, in object identification problems, (i.e, situation assessment [1], candidate assessment [6], decision from MRI images [7], data association [11]) the point is that several sources provide their own beliefs about an observed situation, and the problem is to decide which is the actual situation. For doing so, Theory of Evidence provides different decision strategies like, for instance, Maximum of plausibility strategy.

However, for logicians, Shafer's comments about Theory of Evidence are quite informal. What does it mean that *the mass intends to represent a measure of belief committed exactly to the proposition* ? What does it mean that *the degree of belief represents the degree of support a body of evidence provides for this proposition* ? Finally, what does it mean *the degree of plausibility represents the extend to which someone fails to doubt the proposition* ?

As far as we know, the main logical interpretations of Theory of Evidence have been provided first by Ruspini, then by Pearl and colleagues [12], [10], [13]. They proved that Theory of Evidence can be interpreted by logic and probability theory. Thus, the degree of belief of a proposition A can be viewed as the probability that an agent knows (or proves) A, the mass of a A beeing the probability that the agent knows (or proves) A without knowing (or proving) none of its implicants.

Here, we present a different view of Theory of Evidence, when it is restricted to rational numbers. We interpret the degree of belief of a proposition A as the proportion of proofs of A or, equivalently, the proportion of reasons to believe A. Although it looks close to the previous ones, this logical interpretation is really different. Indeed, even if it correctly interprets the main notions of Theory of Evidence, it fails to correctly interpret Dempster's rule of combination. So we suggest to replace Dempster's rule by a new rule which is more a rule for gathering assignments that a rule for combining assignments.

This paper is organized as follows.

Section 2 presents a new logical view of Theory of Evidence. It is based on the notion of knowledge-sets, introduced in the field of knowledge-bases merging. Section 3 defines a new rule for gathering assignments. It is illustrated in section 4 on an example. Finally, section 5 is devoted to a discussion.

2 A logical interpretation of Theory of Evidence when numbers are rationals

This section presents a new logical interpretation of Theory of Evidence. This interpretation is the one presented in [4], where we also proved some formal equivalences between the Maximum of Plausibility Strategy and one Knowledge-Base Merging operator defined by Konieczny and Pino-Perez [9], [8].

Let us first recall some definitions.

2.1 Preliminaries

Definition 1. A **multi-set** is a set where repeated occurrences of an element may exist. A **knowledge-set** is a multi-set of propositional formulas[2].

---

[2]Konieczny and Pino-Perez define a knowledge-set as a multi-set of sets of formulas but, considering for-

Example 1. If A and B are propositional letters, then $KS = [B, A \vee B, A, A \vee B, A \vee B]$ is a knowledge-set.

Definition 2. Let $KS_1 = [F_1, ..., F_{n_1}]$ and $KS_2 = [F_{n_1+1}, ..., F_p]$ be two knowledge-sets. Their union is: $KS_1 \sqcup KS_2 = [F_1, ..., F_{n_1}, F_{n_1+1}, ..., F_p]$.

Definition 3. Let $KS_1 = [F_1, ..., F_n]$ and $KS_2 = [G_1, ..., G_n]$ two knowledge-sets of the same size. KS_1 and KS_2 are equivalent, noted $KS_1 \leftrightarrow KS_2$, iff there exists a bijection f from KS_1 to KS_2 such that: $\forall i \in \{1...n\} \models F_i \leftrightarrow f(F_i)$

2.2 Main concepts of Theory of Evidence

Theory of Evidence assumes the existence of **a frame of discernment** which is defined as a set Θ of N hypothesis : $\Theta = \{H_1, ..., H_h\}$. Hypothesis correspond to propositions one is dealing with. Their intuitive meaning depend on the context of application. For instance, in an identification problem, the hypothesis H_i will represent the fact "the object to be identified is H_i". The meaning given to hypothesis is out the Theory of Evidence. These hypothesis are supposed to be **exclusive**. This means for instance, that the object to be identified cannot be both H_i and H_j if $i \neq j$. Furthermore, in the initial version of Theory of Evidence (and we will focus on it), the hypothesis are supposed to be **exhaustive**. This means that the object to be identified is H_1 or ... H_h. This assumption is called Closed-World Assumption. Finally, in the Theory of Evidence, propositions are represented by **subsets of** Θ. The set 2^Θ is called Referential of definition.

Let $\Theta = \{H_1, ..., H_h\}$ be a frame of discernment. In the logical formulation, we will say that Θ is a propositional language whose propositional letters are $H_1...H_h$.

As usual, the relation of logical consequences will be denoted by $\models$.

Under the Closed-World Assumption and under the assumption of the hypothesis exhaustivity, we will consider the theory EV whose proper axioms are:

$$(CW) \; H_1 \vee ... \vee H_h$$

$$(EXCL) \; \neg(H_i \wedge H_j) \; if \; i \neq j$$

In EV, the only possible worlds are thus the h worlds $w_1, ...w_h$ where w_i is the world in which only H_i is true.

One can notice that in theory EV, any proposition is equivalent to a positive clause (i.e, a disjunction of positive literals). So here, the Referential of Definition is the set of positive clauses of Θ. This means that any subset of the frame of discernment is logically represented by a positive propositional clause. For instance, the subset $\{H_1, H_2, H_3\}$ of Θ is logically represented by the positive clause $H_1 \vee H_2 \vee H_3$.

The basic notions of the Theory of Evidence are the notion of assignment (or mass function), the notion of belief function, associated with an information source which, by this way, expresses its uncertainty about beliefs and the notion of plausibility function. They are mathematically defined by:

An assignment is a function $m : 2^\Theta \rightarrow [0,1]$ such that:

$$(i) \; m(\emptyset) = 0 \; \text{and} \; (ii) \sum_{A \subseteq \Theta} m(A) = 1$$

---

mulas is enough by assimilating a set of formulas as the conjunction of its formulas

A **belief function** is a function $Bel_m : 2^\Theta \to [0,1]$ which associates any A of to $Bel_m(A)$ with:

$$Bel_m(A) = \sum_{B \subseteq A} m(B)$$

A **plausibility function** is a function $Pl_m : 2^\Theta \to [0,1]$ which associates any A to $Pl_m(A)$ with to $Pl_m(A) = 1 - Bel_m(\overline{A})$, where $\overline{A}$ is the complement of A in Θ.

Shafer gave the following informal interpretation of these numbers: *the number $m(A)$ is understood to be the measure of belief committed exactly to A (.... but) not the total belief that ones commits to A. To obtain the measure of the total belief committed to A, one must add to $m(A)$ the quantities $m(B)$ for all proper subsets B of A. Finally, the degree of plausibility represents the extend to which one fails to doubt the proposition.* So, the problem for us is to give, in logical terms, a meaning to these comments. This is done in the remaining of this section, in the case when the numbers are rational.

The logical representation of an assignment is given by the following definition:

Definition 4. Let m be an assignment defined on Θ by: $m(P_1) = n_1/n$, ..., $m(P_k) = n_k/n$ with $n_1 + ... + n_k = n$. The logical representation of m is the knowledge-set denoted $ks(m)$ defined by: $ks(m) = \{K_1, ..., K_n\}$ with:

$$K_1 = ... = K_{n_1} = \{P_1\}$$
$$K_{n_1+1} = ... = K_{n_1+n_2} = \{P_2\}$$
$$...$$
$$K_{n-n_k+1} = ... = K_n = \{P_k\}$$

Notation. One must notice that, in this definition, the same symbol P_i is used to denote a subset of the Frame of discernment Θ and the propositional clause, of the propositional language associated with Θ, which logically represents it. However, it must be clear that m is defined on subsets while $ks(m)$ is a multi-set of propositional positive clauses.

Example 2. Let $m(A) = 2/3$, $m(A, B) = 1/3$. The logical representation of m is the knowledge-set: $ks(m) = [A, A, A \vee B]$.

Proposition 1. Let m be an assignment. The mass of a proposition P, $m(P)$, is the proportion of formulas K_i in $ks(m)$ which are equivalent, under EV, to P. I.e, the proportion of K_i in $ks(m)$ such that: $EV \models K_i \leftrightarrow P$.

Proposition 2. Let m be an assignment. The degree of belief of a proposition P, $Bel_m(P)$, is the proportion of K_i in $ks(m)$ which, under EV, imply P. I.e, the proportion of K_i in $ks(m)$ such that: $EV \models K_i \to P$.

Example 2 (continued). The three formulas of $ks(m)$ imply $A \vee B$, thus $Bel_m(A \vee B) = 1$. But only two formulas imply A so, $Bel_m(A) = 2/3$.

Proposition 3. Let m be an assignment. The degree of plausibility of a proposition P, $Pl_m(P)$, is the proportion of K_i in $ks(m)$ which do not, under EV, imply $\neg P$. I.e, the proportion of K_i in $ks(m)$ such that: $EV \wedge K_i \wedge P$ is consistent.

Example 2 (continued). All the formulas in $ks(m)$ are consistent (under EV) with A, so $Pl_m(A) = 1$. Only one formula is consistent (under EV) with B, so $Pl_m(B) = 1/3$.

To sum up this section, we can say that *any assignment as defined by Shafer can, if the numbers are rational, be modelled by a knowledge-set as defined previously. Thus,* the mass of a proposition A is the proportion of formulas in this knowledge-set which are equivalent, under EV, to A. The degree of belief of A is the proportion of formulas which imply, under EV, A. The plausibility degree of A is the proportion of formulas which are consistent, under EV, with A.

2.3 Dempster's rule of combination

Given two assignments m_1 and m_2 defined over a frame Θ, Dempster's rule of combination defines a third assignment denoted $m_1 \oplus m_2$ by the following equation:

$$m(A) = \frac{\sum_{A_i \cap B_j = A} m_1(A_i).m_2(B_j)}{N}$$

with

$$N = \sum_{A \neq \emptyset} \sum_{A_i \cap B_j = A} m_1(A_i).m_2(B_j)$$

Obviously, the fraction has a meaning only if $N \neq 0$. This assumption corresponds to the case when the two assignments are not totally in conflict.

In the following, we give the correspondance, in terms of knowledge-sets, of this rule.

Definition 5 Let $KS_1 = \{K_1^1, ..., K_1^n\}$ and $KS_2 = \{K_2^1, ..., K_2^n\}$ be two knowledge sets of the same size. We say that they are in total conflict iff $\forall K_1^i \in KS_1$ $\forall K_2^j \in KS_2$ $EV \cup (K_1^i \wedge K_2^j)$ is inconsistent.

Proposition 4. $ks(m_1)$ and $ks(m_2)$ are in total conflict iff $N = 0$.

The following definition defines an operator, also denoted $\oplus$, which combines two knowledge-sets which are not in total conflict.

Definition 6 Let $KS_1 = \{K_1^1, ..., K_1^n\}$ and $KS_2 = \{K_2^1, ..., K_2^n\}$ be two knowledge-sets of the same size which are not in total conflict. The operator $\oplus$ on knowledge-sets defines a third knowledge-set by:
$KS_1 \oplus KS_2 = [K : \exists K_1^i \in KS_1 \; \exists K_2^j \in KS_2 \; : \; EV \cup \{K_1^i \wedge K_2^j\}$ is consistent and $EV \models (K_1^i \wedge K_2^j) \leftrightarrow K]$

Proposition 5. Let m_1 and m_2 be two assignments which are not in total conflict and let $ks(m_1)$ and $ks(m_2)$ be their logical representations. Then, $ks(m_1 \oplus m_2) \leftrightarrow ks(m_1) \oplus ks(m_2)$.

This proves that the operator $\oplus$ on knowledge-sets corresponds to the logical interpretation of Dempster's rule of combination.

Example 3. Let m_1 and m_2 be two assignments defined by: $m_1(A, B) = 1/2$, $m_1(B) = 1/2$ and $m_2(A) = 2/3$, $m_2(A, B) = 1/3$. Dempster's rule defines the assignment $m_1 \oplus m_2$ by: $m_1 \oplus m_2(A) = 1/2$, $m_1 \oplus m_2(A, B) = 1/4$, $m_1 \oplus m_2(B) = 1/4$. Besides, the logical interpretation of m_1 and m_2 are: $ks(m_1) = [A \vee B, B]$ and $ks(m_2) = [A, A, A \vee B]$. We can then compute $ks(m_1) \oplus ks(m_2)$ and get: $ks(m_1) \oplus ks(m_2) = [A, A, A \vee B, B]$. We can easily check that: $ks(m_1 \oplus m_2) \leftrightarrow ks(m_1) \oplus ks(m_2)$

To sum up this section, we can say that *the assignment $m_1 \oplus m_2$, provided by Dempster's rule on two assignments m_1 and m_2, can be logically interpreted by the knowledge-set denoted $ks(m_1) \oplus ks(m_2)$ previously defined, $ks(m_1)$ and $ks(m_2)$ beeing the knowledge-sets which logically interpret m_1 and m_2.*

2.4 Comments about the logical interpretation of Dempster's rule

It has been previously shown that Dempster's rule leads to an operator for combining two knowledge-sets. Properties of Dempster's rule can thus easily be transferred to this new operator. It is obviously commutative and associative. Its neutral elements is the knowledge-set $KS = [H_1 \vee ... \vee H_h]$. However, it is difficult to give a meaning to that operator, from a knowledge-base point of view. Indeed, let us come back to the intuition underlying the notion of knowledge-set. One must keep in mind that the main characteristic of a knowledge-set is the fact that it is a multi-set (and not a set) of formulas. This can be a way to explicitely count the proofs an agent has made in order to conclude something.

For instance, consider an identification problem in which a witness's accounts is expressed by the knowledge-set $[Peter, Peter]$. It represents the fact that the witness has made two different observations. For instance, it has observed that the person to be identified smokes and is left-handed. Besides the witness knows that $Peter$ is a left-handed smoker. Each observation is used by the witness to deduce that the person to be identified is $Peter$. So the conclusion $Peter$ is deducible twice. This is why the formula $Peter$ appears twice in the knowledge-set.

So, underlying a knowldege-set is an implicit set of observations (see [3]) which is used by the agent to make proofs. Accepting this intuition, we can show that a combined knowledge-set $KS_1 \oplus KS_2$ represents a way to count combined proofs or equivalently to consider conjunctions of observations.

For instance, consider a second witness who is associated with the knowledge-set: $[Peter \vee Paul]$ because he has observed that the person to be identified is a tall person and he knows that both $Peter$ and $Paul$ are tall. Combining the two accounts leads to $[Peter, Peter]$. We can show that it corresponds to the observations: *the person smokes and is tall* and *the person is left-handed and is tall*. But giving a meaning to these conjunctions is not easy. In this example, what can be easily justified is the gathering of all the observations. Gathering the accounts comes to say that the two witnesses have made three observations: two of them leads to conclude that the person is $Peter$ and one of them leads to conclude that it is $Peter$ or $Paul$.

This intuitive idea is developped in the next section.

3 A rule for gathering assignments

In this section, we define a rule for gathering two assignments. This rule is borrowed from the knowledge-base merging community. It is denoted $\sqcup$ and is the reformulation, in Theory of Evidence, of the union of two knowledge-sets as defined in section 2.1.

Definition 7 Let m_1 and m_2 be two assignments defined by: $m_1(P_i) = n_1^i/p_1$ and $m_2(P_i) = n_2^i/p_2$ (where P_i is any subset of the frame of discernment). The rule $\sqcup$ defines a third assignment, denoted $m_1 \sqcup m_2$ by:

$$m_1 \sqcup m_2(P_i) = (n_1^i + n_2^i)/(p_1 + p_2)$$

Example 3 (continued). $m_1 \sqcup m_2$ is defined by: $m_1 \sqcup m_2(A) = 2/5$, $m_1 \sqcup m_2(B) = 1/5$, $m_1 \sqcup m_2(A, B) = 2/5$,

We can prove that $\sqcup$ is commutative and associative. There is no neutral element and no absorbant element. However, $m_1 \sqcup m_2$ is always defined, even if m_1 and m_2 are totally in conflict (i.e, $N = 0$).

Furthermore, it has been shown (see [2]) that for any assignment m_1 and m_2, $m_1 \oplus m_2$ and $m_1 \sqcup m_2$ characterize the same most plausible hypothesis, when there are hypothesis which are most plausible for both assigment. In this case, the same most plausible hypothesis are these ones.

Finally, one must notice that, due to the definition of $\sqcup$, a mass n/p cannot be replaced by the mass n'/p' even if they denote the same rational number. This is illustrated by the following example.

Example 3 (continued). According to $m_1 \sqcup m_2$, the most plausible hypothesis is A. Consider now a third assignment m_1' defined by: $m_1'(A, B) = 3/6$ $m_1'B) = 3/6$. $3/6$ and $1/2$ denote the same number so we could consider that m_1 and m_1' represent the same assignemnt. But we can show that $m_1' \sqcup m_2(A) = 2/9$ $m_1' \sqcup m_2(B) = 3/9$ $m_1' \sqcup m_2(A, B) = 4/9$. Here, B is the most plausible hypothesis.

Obviously, this means that the situation in which the agent has made two observations, one of them leading to $A \vee B$, one of them leading to B is not equivalent to the situation in which the agent has made 6 observations, 3 of them leading to $A \vee B$ and 3 of them leading to B. These two situations are equivalent, from a decision strategy point of view. But they are not when gathering assignments.

4 Application on an example

Zadeh [16] and Smets [15] have critized the normalization introduced in Dempster's rule by showing that, on some example, it leads to questionable results. This example is presented by Smets in these terms: *Suppose a murder case with three suspects:* $\Delta = \{Henry, Tom, Sarah\}$, *and two witnesses. For witness 1, the murderer is not Sarah, it is most probably Henry, but it might also be Tom. Witness 2 holds similar beliefs except for the permutation between Henry and Sarah*

They thus consider the following assignments associated with the two witnesses:

$$m_1(Henry) = 0.99$$
$$m_1(Tom) = 0.01$$
$$m_1(sarah) = 0$$

$$m_2(Henry) = 0$$
$$m_2(Tom) = 0.01$$
$$m_2(sarah) = 0.99$$

Dempster's rule as defined in section 2, leads to the following combined assignment: $m_1 \oplus m_2(Tom) = 1$ which means that Tom is certainly the murder.

Citing Smets again: *Zadeh does not accept this solution as it gives full certainty to a solution (Tom) that is hardly supported at all. In fact, in a totally different situation in which both witnesses migth have been sure that Tom was the murderer, the result of the combination would have been the same.*

Smets adheres to this critics and suggests to define an unormalized rule which leads to the following assignment:

$$m_1 \oplus' m_2(Tom) = 0.0001 \quad \text{and} \quad m_1 \oplus' m_2(\emptyset) = 0.999$$

Citing Smets again: *The unnormalized solution presented within our theory seems much more realistic as it shows Tom to be sligtly supported but $\emptyset$ to be highly supported. (...) the most obvious conclusion in the present situation is that the real murderer must be a fourth person*

In the following, we reformulate this example according to our point of view. We try to show that, in this example, the mass of a proposition can be seen as the proportion of proofs for deducing it without deducing none of its implicants. So using the rule for gathering assignments is more adequate.

According to our interpretation, the assignment m_1 means that the first witness has made 100 observations. Among them, 99 lead him to believe that the murderer is $Henry$ and only 1 leads him to believe that it is Tom. In the same way, the assignment m_2 means that the second witness has made 100 observations. Among them, 99 lead him to believe that the murderer is $Sarah$ and only 1 leads him to believe that it is Tom.

By applying the rule defined in section 3, we get:

$$m_1 \sqcup m_2(Henry) = 99/200$$
$$m_1 \sqcup m_2(Sarah) = 99/200$$
$$m_1 \sqcup m_2(Tom) = 2/200$$

I.e, gathering the two witnesses accounts, one can conclude that there are 99 reasons on 200 to believe that $Henry$ (resp, $Sarah$) is the murderer, and only 2 reasons on 200 to believe that Tom is the murderer.

Obviously, Zadeh and Smets's critics no longer stands: after gathering the two accounts, we still have few beliefs in Tom. And we have the same number of reasons for believing that the murderer is $Henry$ (resp, $Sarah$). Thus, the most plausible hypothesis are $Henry$ and $Sarah$.

5 Discussion

The logical interpretation we gave to masses, degrees of belief, degrees of plausibility, is quite different from the logical interpretations provided by Ruspini and Pearl. Even if it looks close, it has been shown that it is different because it fails to interpret Dempster's rule of combination and suggests to gather assigments instead of combining them.

One can thus argue that this logical interpretation does not interpret Theory of Evidence in a whole. However, in some applications, like the example of section 4, this interpretation looks to be more adequate.

This positive result does not lead us to conclude that this interpretation is the only one valid and that the rule of gathering must always replace Dempster's one. Our feeling is that Theory of Evidence, as defined by Shafer, said nothing formal about the meaning of the masses. Thus, this theory has been used in many applications where the meaning of the masses were completely different. Following Ruspini, we agree on the fact that Theory of Evidence (including Dempster's rule) is adequate when there is an underlying probability measure on observations. Here, we suggest that another interpretation can be given if we only have a set of observations among which we count those which are used to conclude something. And more, following this point of view, Dempster's rule of combination must be replaced by a rule of gathering.

Notice that since this present interpretation rejects Demspter's rule, a question is now: how does it define conditioning ?

Remind that, given an assignment m which represents the beliefs of an agent, conditioning defined a new assignment $m^{|A}$ which is supposed to represent the beliefs of the agent after he learns that the proposition A is true for certain. Shafer suggested to define $m^{|A}$ by combining with Dempster's rule m and the assignment m_A defined by $m_A(A) = 1$. Thus $m^{|A} = m \oplus m_A$.

In the present interpretation, we suggest to solve the problem of learning that an information is certain in a slightly different way. Indeed, we think that the main point, when learning that an information is true for certain, is not to define a new assignment, but to characterize the hypothesis which are the most plausible (if the chosen decision strategy is the Maximum of Plausibility) given the initial beliefs of the agent and given this certain information.

Thus, if, $Maxpl(m, EV)$ are the hypothesis which are the most plausible according to m we claim that the hypothesis which are the most plausible after learning that A is true are $Maxpl(m, EV \wedge A)$.

So, in the present interpretation, integrating certain information does not come to modify the assignment but comes to modify the frame of discernment: not all the hypothesis of the initial frame of discernment are now to be considered: only the hypothesis defined by $EV \wedge A$ must now be considered.

We can prove that this way of implementing conditioning extends Shafer's one. Indeed, if some hypothesis in A are among the most plausible hypothesis according to m, then we get: $Maxpl(m^{|A}, EV) = Maxpl(m, EV \wedge A)$ (i.e, we finally characterize the same most plausible hypothesis than Shafer's conditioning). Obviously, results are different if the premisse is not satisfied, since in this case, our solution gives results while Shafer's one is not defined. However, when another decision strategy is chosen, the comparision remains to be done.

References

[1] A. Appriou. Multi-sensor data fusion in situation assesment processes. In D. Gabbay, R. Kruse, A. Nonengart, and H. Ohlbach, editors, *Lectures Notes in AI 1244: Proc of ECSQARU-FAPR'97*, 1997.

[2] L. Cholvy. Fusion de données pour l'analyse de situations. Technical report, n° 1/04010/DTIM, ONERA, 2000.

[3] L. Cholvy. Applying theory of evidence in multisensor data fusion: a logical interpretation. In *Proceedings of the 3rd International Conference on Information Fusion (FUSION-2000)*, Paris, July 2001.

[4] L. Cholvy. Data merging: Theory of evidence vs knowledge-bases merging operators. In Ph. Besnard, S. Benferhat, editors *Lectures Notes in AI 2143: Proc. of ECSQARU'01*, 2001 September 2001.

[5] A. Dempster. Upper and lower probabilities iduced by a multivalued mapping. *Annals of Mathematical Statistics*, 38:325–339, 1967.

[6] D. Dubois, M. Grabbisch, H. Prade, and Ph. Smets. Assessing the value of a candidate. comparing belief functions and possibility theories. In *Proc. of UAI*, 1999.

[7] L. Gautier, A. Taleb-Ahmed, M. Rombaut, J. Postaire, and H. Leclet. Belief function in low level data fusion: Applications in mri images of vertebra. In *Proc of the 3rd International Conference on Information Fusion, Paris*, 2000.

[8] S. Konieczny and R. Pino-Perez. Merging with integrity constraints. In *Proc. of ESCQARU'99*, 1999.

[9] S. Konieczny and R. Pino-Perez. On the logic of merging. In *Proc. of KR'98*, Trento, 1998.

[10] J. Pearl. *Probabilistic reasoning in intelligent systems: networks of plausible inference*. Morgan Kaufmann Publishers, San Mateo, USA, 1988.

[11] C. Royère, D. Gruyer, and V. Cherfaoui. Data association with believe theory. In *Proc of the 3rd International Conference on Information Fusion, Paris*, 2000.

[12] E. Ruspini. Epistemic logics, probability, and the calculus of evidence. In *Proceedings of IJCAI*, 1987.

[13] E. Ruspini, J. Lowrance, and Th. Strat. Understanding evidential reasoning. *International Journal of Approximate Reasoning*, 6:401–424, 1992.

[14] G. Shafer. *A mathematical theory of evidence*. Princeton University Press, Princeton and London, 1976.

[15] P. Smets. The combination of evidence in the transferable belief model. *IEEE Transactions PAMI*, 12:447–458, 1990.

[16] L. A. Zadeh. A mathematical theory of evidence (book review). *AI magazine*, 5 (3), 1984.

Uncertainty, Type-2 Fuzzy Sets, and Footprints of Uncertainty

Jerry M. Mendel
University of Southern California
Department of Electrical Engineering
3740 McClintock Ave.
Los Angeles, Ca 99089-2564
mendel@sipi.usc.edu

Abstract

Type-2 fuzzy sets let us model and minimize the effects of four kinds of uncertainties that can occur in type-2 fuzzy logic systems. The nature of these uncertainties is captured in the footprint of uncertainty (FOU) of a type-2 membership function. This paper describes the four kinds of uncertainties, defines type-2 fuzzy set terms and provides a small catalog of FOUs. Choosing a FOU is analogous to the well-known starting point of a probability-based design, where one must choose appropriate probability density functions.

Keywords: Uncertainty, type-2 fuzzy sets, footprint of uncertainty, type-2 membership function, type-2 fuzzy logic system

1. Introduction[1]

The original fuzzy logic (FL), founded by Lotfi Zadeh, has been around for more than 37 years, as of the year 2002, and yet it is unable to handle uncertainties. By *handle*, I mean *to model and minimize the effect of*. That the original FL—type-1 FL—cannot do this sounds paradoxical because the word *fuzzy* has the connotation of uncertainty. An expanded FL—type-2 FL—is now able to handle uncertainties because it can model them and minimize their effects. And, if all uncertainties disappear, type-2 FL reduces to type-1 FL in much the same way that if randomness disappears probability reduces to determinism.

Although many applications have been found for type-1 FL, it is its application to *rule-based systems* that has most significantly demonstrated its importance as a powerful design methodology. A rule-based fuzzy logic system (FLS) contains four components —rules, fuzzifier, inference engine, and output processor—that are inter-connected, as shown in Figure 1. This kind of FLS is very widely used in many engineering applications of FL, and is also known as a *fuzzy controller*, *fuzzy system* or *fuzzy model*.

Rules may be provided by experts or can be extracted from numerical data. In either case, they can be expressed as a collection of IF–THEN statements. Fuzzy sets are associated with terms that appear in the antecedents or consequents of such rules, and with the inputs to and output of the FLS. Membership functions (MFs) are used to describe these fuzzy sets; they can be either type-1 or type-2 MFs.

---

[1] Some parts of this paper have been taken from [1] since this material has not appeared in a journal publication.

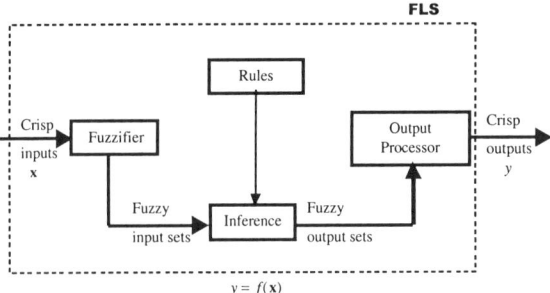

Figure 1: Fuzzy logic system.

A FLS that is described completely in terms of type-1 fuzzy sets is called a *type-1 FLS*, whereas a FLS that is described using at least one type-2 fuzzy set is called a *type-2 FLS*. Type-1 FLSs are unable to directly handle rule uncertainties, because they use type-1 fuzzy sets that are certain. Type-2 FLSs, on the other hand, are very useful in circumstances where it is difficult to determine an exact MF for a fuzzy set; hence, they can be used to handle rule uncertainties, and even measurement uncertainties.

The FLS fuzzifier maps crisp numbers into fuzzy sets. It is needed to activate rules that are in terms of linguistic variables, which have fuzzy sets associated with them. The inputs to the FLS prior to fuzzification may be certain (e.g., perfect measurements) or uncertain (e.g., noisy measurements). The MF for a type-2 fuzzy set lets us handle either kind of measurement.

The inference engine of the FLS maps fuzzy sets into fuzzy sets. It handles the way in which rules are activated and combined. Just as we humans use many different types of inferential procedures to help us understand things or to make decisions, there are many different FL inferential procedures.

In many applications of a FLS, crisp numbers must be obtained at its output. This is accomplished by means of the output processor. The output processor for a type-1 FLS is just a *defuzzifier*; however, the output processor of a type-2 FLS contains two components: the first maps a type-2 fuzzy set into a type-1 fuzzy set—a *type-reduced set*—and the second performs defuzzification on the type-reduced set.

Just as random uncertainties flow through a system and their effects can be evaluated using the mean and variance, linguistic and random uncertainties flow through a type-2 FLS and their effects can be evaluated using the defuzzified output and the type-reduced output of that system. Just as the variance provides a measure of dispersion about the mean, and is often used in confidence intervals, the type-reduced output can be interpreted as providing a measure of dispersion about the defuzzified output and can be thought of as (or related to) a linguistic confidence interval. Just as the variance increases as random uncertainty increases, the type-reduced set also increases as linguistic or random uncertainties increase. So, a type-2 FLS is *analogous* to (but not the same as) a probabilistic system through first and second moments, whereas a type-1 FLS is analogous to a probabilistic system only through the first moment.

Uncertainty comes in many guises and is independent of what kind of FL, or any kind of methodology, one uses to handle it. One of the best sources for general discussions about uncertainty is Klir and Wierman [2]. Regarding the *nature of uncertainty*, they

state (p. 43):
> Three types of uncertainty are now recognized . . . *fuzziness* (or vagueness), which results from the imprecise boundaries of fuzzy sets; *non-specificity* (or imprecision[2]), which is connected with sizes (cardinalities) of relevant sets of alternatives; and *strife* (or discord), which expresses conflicts among the various sets of alternatives.

They divide these three types of uncertainty into two major classes, *fuzziness* and *ambiguity*, where ambiguity ("one to many relationships") includes *nonspecificity* and *strife*.

The following sources of uncertainty can occur for the FLS in Figure 1:
- Uncertainty about the meanings of the words that are used in the rules
- Uncertainty about the consequent that is used in a rule
- Uncertainty about the measurements that activate the FLS
- Uncertainty about the data that are used to tune the parameters of a FLS

Type-1 FL handles **uncertainties about the meanings of words** by using *precise* membership functions that the user believes capture the uncertainty of the words. Once the type-1 MFs have been chosen, all uncertainty about the words disappears, because type-1 MFs are totally precise. Type-2 FL, on the other hand, handles uncertainties about the meanings of words[3] by *modeling* the uncertainties. This is described in Sections 2 and 3. The uncertainty about the meanings of the words that are used in rules seems to be in accord with *fuzziness*, which results from imprecise boundaries of fuzzy sets.

Consequents for rules are either obtained from experts, by means of knowledge mining (engineering), or are extracted directly from data. Because experts don't all agree, a survey of experts will usually lead to a histogram of possibilities for the consequent of a rule. This histogram represents the **uncertainty about the consequent of a rule**, and this kind of uncertainty is different from that associated with the meanings of the words used in the rules. A histogram of consequent possibilities can be handled by a type-2 FLS. Uncertainty about the consequent used in a rule, as established by a histogram of possibilities, seems to be in accord with *strife*, which expresses conflicts among the various sets of alternatives.

Measurements are usually corrupted by noise; hence, they are uncertain. We do not propose to abandon traditional ideas about noisy measurements (i.e., measurement = signal + noise). What we propose to abandon (when appropriate) is the frequently made assumption of a priori knowledge of a probability model (i.e., a probability density function) for either the signal or the noise. Doing this gets around the major shortcoming of a probability-based model, namely the assumed probability model, for which results will be good if the data agree with the model, but may not be so good if the data do not. Uncertain measurements (i.e., randomness in the data) can be handled very naturally within the framework of a FLS; they can be modeled as fuzzy sets (type-1 or type-2); hence, **uncertainty about the measurements that activate the FLS** seems to be in accord with *non-specificity* when nonspecificity is associated with information-based imprecision.

Finally, a FLS contains many design parameters whose values must be set by the

---

[2] This is *information-based imprecision* rather than linguistic imprecision (which is equivalent to fuzziness).
[3] In rules, we distinguish between antecedent words and consequent words; but, there are also connector words (e.g., *and/ or*). There may even be uncertainties associated with them [3], [4]; but, in this paper, we do not focus on such uncertainties.

designer before the FLS is operational. There are many ways to do this, and all make use of a set of data, usually called the *training set*. This set consists of input–output pairs for the FLS, and, if these pairs are measured signals, then usually they are as uncertain as the measurements that excite the FLS. In this case the FLS must be tuned using unreliable data, which is yet another form of uncertainty that can be handled by a type-2 FLS. The **uncertainty about the data that are used to tune the parameters of a FLS** also seems to be in accord with *nonspecificity*.

It would appear that a type-2 FLS is able to directly address all three types of uncertainty—*fuzziness*, *strife*, and *nonspecificity*. Type-2 fuzzy sets are the means to handling all three types of uncertainty totally within the framework of fuzzy set theory. They are able to do this because they directly model uncertainties.

Next, we define type-2 fuzzy sets and some important associated concepts. By doing this we provide a simple collection of mathematically well-defined terms that will let us effectively communicate about type-2 fuzzy sets.

2. Type-2 Fuzzy Sets: Definitions

Imagine blurring the type-1 MF depicted in Figure 2 (a) by shifting the points on the triangle either to the left or to the right and not necessarily by the same amounts, as in Figure 2 (b). Then, at a specific value of x, say x', there no longer is a single value for the MF ($u(x')$); instead, the MF takes on values wherever the vertical line intersects the blur. Those values need not all be weighted the same; hence, we can assign an amplitude distribution to all of those points. Doing this for all $x \in X$, we create a three-dimensional MF—a type-2 MF—that characterizes a type-2 fuzzy set.

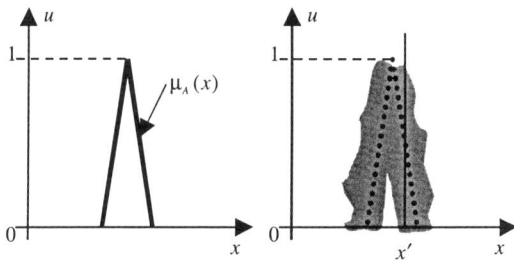

Figure 2: (a) Type-1 MF and (b) blurred type-1 MF

Definition 1: A *type-2 fuzzy set*, denoted $\tilde{A}$, is characterized by a *type-2 MF* $\mu_{\tilde{A}}(x, u)$, where $x \in X$ and $u \in J_x \subseteq [0,1]$, i.e.,

$$\tilde{A} = \left\{ \left((x,u), \mu_{\tilde{A}}(x,u) \right) \mid \forall x \in X, \forall u \in J_x \subseteq [0,1] \right\} \tag{1}$$

in which $0 \leq \mu_{\tilde{A}}(x,u) \leq 1$. $\tilde{A}$ can also be expressed as

$$\tilde{A} = \int_{x \in X} \int_{u \in J_x} \mu_{\tilde{A}}(x, u) / (x, u) \quad J_x \subseteq [0,1] \tag{2}$$

where $\iint$ denotes union over all admissible x and u. For discrete universes of discourse $\int$ is replaced by $\sum$. ∎

In Definition 1, the first restriction that $\forall u \in J_x \subseteq [0,1]$ is consistent with the type-1 constraint that $0 \leq \mu_A(x) \leq 1$, i.e., when uncertainties disappear a type-2 MF must reduce to a type-1 MF, in which case the variable u equals $\mu_A(x)$ and $0 \leq \mu_A(x) \leq 1$. The second restriction that $0 \leq \mu_{\tilde{A}}(x,u) \leq 1$ is consistent with the fact that the amplitudes of a MF should lie between or be equal to 0 and 1.

Definition 2: At each value of x, say $x = x'$, the 2D plane whose axes are u and $\mu_{\tilde{A}}(x', u)$ is called a *vertical slice* of $\mu_{\tilde{A}}(x, u)$. A *secondary MF* is a vertical slice of $\mu_{\tilde{A}}(x, u)$. It is $\mu_{\tilde{A}}(x = x', u)$ for $x' \in X$ and $\forall u \in J_{x'} \subseteq [0,1]$, i.e.,

$$\mu_{\tilde{A}}(x = x', u) \equiv \mu_{\tilde{A}}(x') = \int_{u \in J_{x'} \subseteq [0,1]} f_{x'}(u)/u \tag{3}$$

in which $0 \leq f_{x'}(u) \leq 1$. Because $\forall x' \in X$, we drop the prime notation on $\mu_{\tilde{A}}(x')$, and refer to $\mu_{\tilde{A}}(x)$ as a secondary MF; it is a type-1 fuzzy set, which we also refer to as a *secondary set*. ■

Based on the concept of secondary sets, we can reinterpret a type-2 fuzzy set as the union of all secondary sets, i.e., using (3), *we can re-express $\tilde{A}$ in a vertical-slice manner*, as:

$$\tilde{A} = \{(x, \mu_{\tilde{A}}(x)) \mid \forall x \in X\} \tag{4}$$

or, as

$$\tilde{A} = \int_{x \in X} \mu_{\tilde{A}}(x)/x = \int_{x \in X} \left[\int_{u \in J_x \subseteq [0,1]} f_x(u)/u \right]/x \tag{5}$$

Definition 3: The *domain* of a secondary MF is called the *primary membership* of x. In (5), J_x is the primary membership of x, where $J_x \subseteq [0,1]$ for $\forall x \in X$. ■

Definition 4: The *amplitude* of a secondary MF is called a *secondary grade*. In (5), $f_x(u)$ is a secondary grade; in (1), $\mu_{\tilde{A}}(x', u')$ ($x' \in X$, $u' \in J_{x'}$) is a secondary grade.

■ **Definition 5:** Uncertainty in the primary memberships of a type-2 fuzzy set, $\tilde{A}$, consists of a bounded region that we call the *footprint of uncertainty* (FOU). It is the union of all primary memberships, i.e.,

$$FOU(\tilde{A}) = \bigcup_{x \in X} J_x \quad ■ \tag{6}$$

The shaded region in Figure 2 (b) is the FOU. Other examples of FOUs are given in Section 3. The term *footprint of uncertainty* is very useful, because it not only focuses our attention on the uncertainties inherent in a specific type-2 MF, whose shape is a direct consequence of the nature of these uncertainties, but it also provides a very convenient verbal description of the entire domain of support for all the secondary grades of a type-2 MF. It also lets us depict a type-2 fuzzy set graphically in two-dimensions instead of three dimensions, and in so doing lets us overcome a difficulty about type-2 fuzzy sets—their three-dimensional nature which makes them very difficult to draw. The shaded FOUs imply that there is a distribution that sits on top of it—the new *third dimension* of type-2 fuzzy sets. What that distribution looks like depends on the specific choice made for the secondary grades. When they all equal one, the resulting type-2 fuzzy sets are called *interval type-2 fuzzy sets*. Such sets are the most widely used type-2 fuzzy sets to date.

Definition 6: Consider a family of type-1 MFs $\mu_A(x \mid p_1, p_2, ..., p_v)$ where $p_1, p_2, ..., p_v$ are parameters, some or all of which vary over some range of values, i.e., $p_i \in P_i$ ($i = 1, ..., v$). A *primary MF* is any one of these type-1 MFs, e.g.,

$$\mu_A(x \mid p_1 = p_1, p_2 = p_2, ..., p_v = p_v).$$

For short, we use $\mu_A(x)$ to denote a primary MF. It will be subject to some restrictions on its parameters. The family of all primary MFs creates a FOU. ■

3. Footprints of Uncertainty

The starting point for all applications of type-2 FLSs is *choosing appropriate FOUs for it*. This is presently done in any one of the following ways, by: (1) Analyzing the available data using statistical techniques and examining the variations of the appropriate statistics, (2) Analyzing the natures of the uncertainties by understanding the problem being solved, (3) Collecting surveys about the words that will be used in knowledge-mining questionnaires, or (4) Deciding on the natures of the uncertainties of the measurements that activate a FLS. In this section we provide a catalog of some FOUs that we have found to be useful in applications.

3.1 Triangular Primary MFs

Triangles are perhaps the most popular of all MFs. When there only is uncertainty about the location of the triangle's two base points, then the FOU for a triangular primary MF is the one depicted in Figure 3. Such a FOU can be constructed when people are asked to provide interval end-point information about locations of words on a scale (e.g., "Where on a scale of 0–10 would you locate the end points of an interval associated with the word *some*?"). In Figure 3, the uncertainty interval[4] for point a does not overlap with the uncertainty interval for point b. When the uncertainty interval for point a does overlap with the uncertainty interval for point b, we obtain the FOU depicted in Figure 4. In this case, the entire triangle is the FOU.

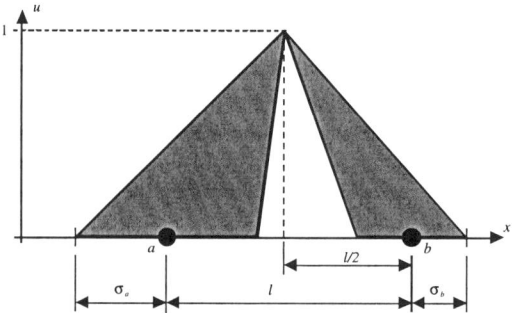

Figure 3: FOU when end-point information is requested. Uncertainty interval for a does not overlap with the uncertainty interval for b.

When there is uncertainty about the location of the triangle's apex as well as its two base points, then the FOU for a triangular primary MF is the one depicted in Figure 5. Such a FOU can be constructed when people are asked to provide center location and interval length about locations of words on a scale (e.g., "Where on a scale of 0–10 would you locate the center of an interval associated with the word *some* and how large would that interval be?"). It is easy to show that the general shape of this FOU is the same when the uncertainty interval for the center location either does or does not overlap with the uncertainty interval for the interval end-points (except for some re-labeling of lengths).

---

[4] In this section and Section 3.2 the uncertainty interval is shown by ± one standard deviation about the mean. How much uncertainty to include in the FOUs in these two sections is somewhat subjective.

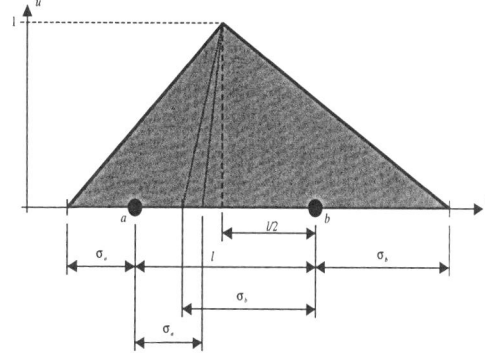

Figure 4: FOU when end-point information is requested. Uncertainty interval for *a* overlaps with the uncertainty interval for *b*.

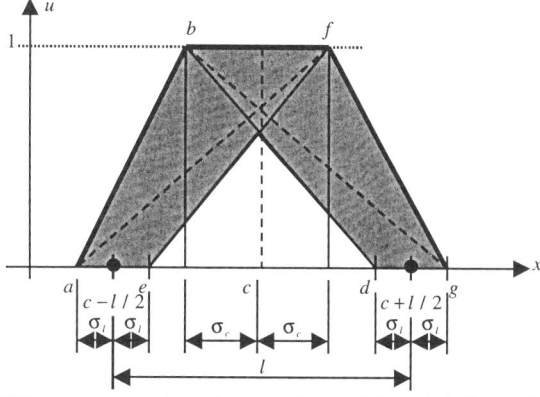

Figure 5: FOU when center location and interval length information is requested.

3.2 Trapezoidal Primary MFs

Another popular primary MF is the trapezoid. Figure 6 depicts the FOU for a trapezoidal primary MF when there is uncertainty about the location of its four defining points, and the amount of uncertainty can be different for each point. When the amount of uncertainty is the same for each point, the FOU is depicted in Figure 7.

3.3 Gaussian Primary MFs

Yet another popular primary MF is the Gaussian. The FOU for a Gaussian primary MF with uncertain mean but certain standard deviation is depicted in Figure 8. We have used this FOU a lot to model antecedents in rule-based forecasters of chaotic time series when only noisy training data are available to tune the parameters of the FLS forecaster.

The FOU for a Gaussian primary MF with uncertain standard deviation but certain mean is depicted in Figure 9. We have used this FOU a lot to model *non-stationary* noisy measurements in rule-based forecasters of chaotic time series.

The FOU for a Gaussian primary MF with uncertain standard deviation and mean is depicted in Figure 10. We are presently using this FOU is a rule-based pattern classification problem for which the measured data is noisy and the features are non-

240

stationary.
Finally, the FOU for a scaled Gaussian primary MF is depicted in Figure 11. The scaling factor is *s* where *s*<1.

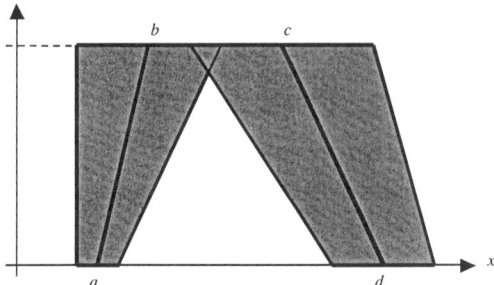

Figure 6: FOU when there is unequal uncertainty about trapezoid's defining points.

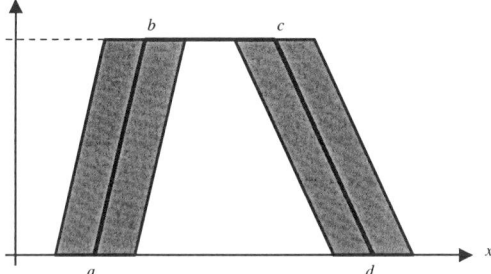

Figure 7: FOU when there is equal uncertainty about trapezoid's defining points.

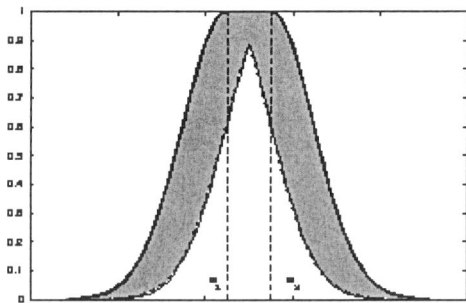

Figure 8: FOU for Gaussian primary MF with uncertain mean.

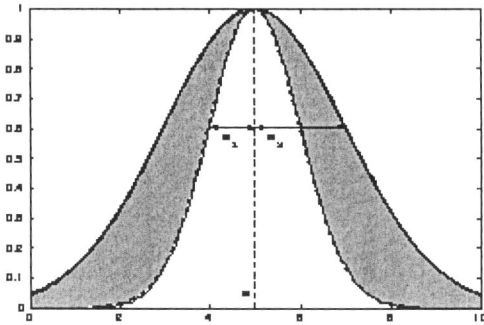

Figure 9: FOU for Gaussian primary MF with uncertain standard deviation.

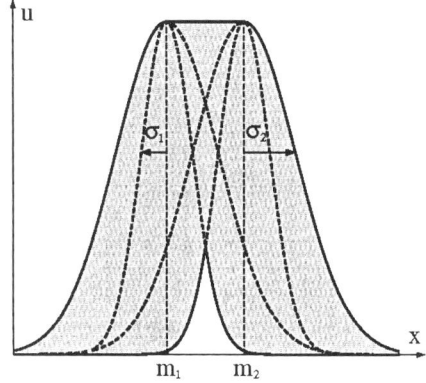

Figure 10: FOU for Gaussian primary MF with uncertain mean and standard deviation.

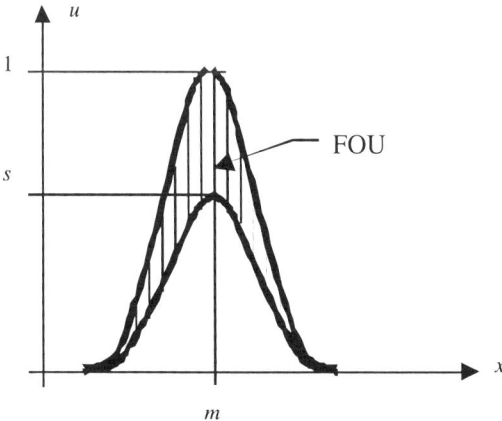

Figure 11: FOU or a scaled Gaussian primary MF.

3.4 Observation

Comparing Figures 5–7 and 10, we see that they are quite similar. Each FOU looks about the same, and if its design parameters are tuned using some training data, our conjecture is that they will look even more the same. We view this as a very healthy development in that it demonstrates that when uncertainties about a primary MF are captured by a FOU then the specific choice for the primary MF is no longer important. This supports the generally known fact that the choice of the shape for a MF is not that critical in a FL controller or FLS, i.e., the designs are robust to that choice.

4. Conclusions

Type-2 fuzzy sets let us model four kinds of uncertainties that can occur in a FLS. The effects of these uncertainties can be minimized by optimizing the MF parameter values of type-2 fuzzy sets. An interval type-2 fuzzy set is completely specified by its FOU. In all of our applications of type-2 FLSs, the very first step in their design is establishing an appropriate FOU. We have provided a small catalog of very useful FOUs in this paper. We have demonstrated that the FOUs of a triangle primary MF that has uncertainties associated with all three of its vertices, a trapezoidal primary MF that has uncertainties associated with all four of its vertices, and a Gaussian primary MF that has uncertainties associated with both its mean and standard deviation all look quite similar. This seems to demonstrate that when uncertainties about a primary MF are captured by a FOU then the specific choice for a primary MF no longer is important.

References

[1] J. M. Mendel (2001). *Uncertain Rule-Based Fuzzy Logic Systems: Introduction and New Directions*. Prentice-Hall, Upper Saddle River, NJ.
[2] G. J. Klir and M. J. Wierman (1998). *Uncertainty-Based Information*. Physica-Verlag, Heidelberg, Germany.
[3] I. Turksen (1986). Interval Valued Fuzzy Sets Based on Normal Forms. In *Fuzzy Sets and Systems*, volume 20, pages 191-210.
[4] R. R. Yager and D. P. Filev (1994). *Essentials of Fuzzy Modeling and Control*. John Wiley, New York.

B. Bouchon-Meunier, L. Foulloy and R.R. Yager (Editors)
Intelligent Systems for Information Processing:
From Representation to Applications
© 2003 Elsevier Science B.V. All rights reserved.

Rough Sets, Bayes' Theorem and Flow Graphs

Zdzisław Pawlak
Institute for Theoretical and Applied Informatics
Polish Academy of Sciences
ul. Bałtycka 5, 44 100 Gliwice, Poland
e-mail: zpw@ii.pw.edu.pl

MOTTO:
"It is a capital mistake to theorize before one has data"
Sherlock Holmes
In: A Scandal in Bohemia

Abstract

Rough set theory is a new approach to vagueness and uncertainty. The theory of rough sets has an overlap with many other theories. Specially interesting is the relationship to fuzzy set theory and the theory of evidence. Recently, it turned out that the theory has very interesting connections with Bayes' theorem. The look on Bayes' theorem offered by rough set theory reveals that any data set (decision table) satisfies total probability theorem and Bayes' theorem. These properties can be used directly to draw conclusions from objective data without referring to subjective prior knowledge and its revision if new evidence is available. Thus the rough set view on Bayes' theorem is rather objective in contrast to subjective "classical" interpretation of the theorem. Besides, it is revealed that Bayes' theorem can be interpreted as a flow *conservation equation* in a flow graph. However the flow graphs considered here are different from those introduced by Ford and Fulkerson. This property gives new perspective for applications of Bayes' theorem.

Thus the paper brings two new interpretation of Bayes' theorem, without referring to its classical probabilistic interpretation: as properties of data tables and properties of flow graphs.

Keywords: roughs sets, Bayes' theorem, decision rules, flow graphs.

1 Introduction

Rough set theory is a new approach to vagueness and uncertainty. Foundation of rough sets can be found in [9]. The theory has found many applications, in particular in data analysis and data mining, offering new look and tools for these domains.

The theory of rough sets has an overlap with many other theories. Specially interesting is the relationship to fuzzy set theory and the theory of evidence. Recently, it turned out that the theory has very interesting connections with Bayes' theorem. This link gives a new look on Bayes' theorem which is significant not only from philosophical point of view but also offers new methods of data analysis.

The look on Bayes' theorem offered by rough set theory reveals that any data set (decision table) satisfies total probability theorem and Bayes' theorem. These properties can be used directly to draw conclusions from objective data without referring to subjective prior knowledge and its revision if new evidence is available. Thus the rough set view on Bayes' theorem is rather objective in contrast to subjective "classical" interpretation of the theorem [6, 7, 8].

It is also interesting that Bayes' theorem can be interpreted as a *flow conservation equation* in a flow graph. However the flow graphs considered here are different from those introduced by Ford and Fulkerson [4].

2 Rough Set Theory - Basic Concepts

In this section we define basic concepts of rough set theory: information system and approximation of sets.

An information system is a data table, whose columns are labeled by attributes, rows are labeled by objects of interest and entries of the table are attribute values.

Formally, by an *information system* we will understand a pair $S = (U, A)$, where U and A, are finite, nonempty sets called the *universe*, and the set of *attributes*, respectively. With every attribute $a \in A$ we associate a set V_a, of its *values*, called the *domain* of a. Any subset B of A determines a binary relation $I(B)$ on U, which will be called an *indiscernibility relation*, and defined as follows: $(x, y) \in I(B)$ if and only if $a(x) = a(y)$ for every $a \in A$, where $a(x)$ denotes the value of attribute a for element x. Obviously $I(B)$ is an equivalence relation. The family of all equivalence classes of $I(B)$, i.e., a partition determined by B, will be denoted by $U/I(B)$, or simply by U/B; an equivalence class of $I(B)$, i.e., block of the partition U/B, containing x will be denoted by $B(x)$.

If (x, y) belongs to $I(B)$ we will say that x and y are B-indiscernible (*indiscernible with respect to B*). Equivalence classes of the relation $I(B)$ (or blocks of the partition U/B) are referred to as *B-elementary sets* or *B-granules*. If we distinguish in an information system two disjoint classes of attributes, called *condition* and *decision attributes*, respectively, then the system will be called a *decision table* and will be denoted by $S = (U, C, D)$, where C and D are disjoint sets of condition and decision attributes, respectively.

Thus the decision table determines decisions which must be taken, when some conditions are satisfied. In other words each row of the decision table specifies a decision rule which determines decisions in terms of conditions.

Observe, that elements of the universe are in the case of decision tables simply labels of decision rules.

Suppose we are given an information system $S = (U, A)$, $X \subseteq U$, and $B \subseteq A$. Our task is to describe the set X in terms of attribute values from B. To this end we define two operations assigning to every $X \subseteq U$ two sets $B_*(X)$ and $B^*(X)$ called the *B-lower* and the *B-upper approximation* of X, respectively, and defined as follows:

$$B_*(X) = \bigcup_{x \in U} \{B(x) : B(x) \subseteq X\},$$

$$B^*(X) = \bigcup_{x \in U} \{B(x) : B(x) \cap X \neq \varnothing\}.$$

Hence, the *B*-lower approximation of a set is the union of all *B*-granules that are included in the set, whereas the *B-upper* approximation of a set is the union of all *B*-granules that have a nonempty intersection with the set. The set

$$BN_B(X) = B^*(X) - B_*(X)$$

will be referred to as the *B-boundary region* of *X*.

If the boundary region of *X* is the empty set, i.e., $BN_B(X) = \emptyset$, then *X* is *crisp* (*exact*) with respect to *B*; in the opposite case, i.e., if $BN_B(X) \neq \emptyset$, *X* is referred to as *rough* (*inexact*) with respect to *B*.

Rough sets can be also defined employing instead of approximations rough membership function, which is defined as follows:

$$\mu_X^B : U \to [0,1]$$

and

$$\mu_X^B(x) = \frac{card(B(x) \cap X)}{card(B(x))},$$

where $X \subseteq U$, $B \subseteq A$ and *card(X)* denotes the cardinality of *X*.

The function measures the degree that *x* belongs to *X* in view of information about *x* expressed by the set of attributes *B*.

The rough membership function has the following properties [9]:

1. $\mu_X^B(x) = 1$ *iff* $x \in B_*(X)$
2. $\mu_X^B(x) = 0$ *iff* $x \in U - B^*(X)$
3. $0 < \mu_X^B(x) < 1$ *iff* $x \in BN_B(X)$
4. $\mu_{U-X}^B(x) = 1 - \mu_X^B(x)$ for any $x \in U$
5. $\mu_{X \cup Y}^B(x) \geq max\left(\mu_X^B(x), \mu_Y^B(x)\right)$ for any $x \in U$
6. $\mu_{X \cap Y}^B(x) \leq min\left(\mu_X^B(x), \mu_Y^B(x)\right)$ for any $x \in U$

Compare these properties to those of fuzzy membership. Obviously rough membership is a generalization of fuzzy membership.

The rough membership function can be used to define approximations and the boundary region of a set, as shown below:

$$B_*(X) = \{x \in U : \mu_X^B(x) = 1\},$$

$$B^*(X) = \{x \in U : \mu_X^B(x) > 0\},$$

$$BN_B(X) = \{x \in U : 0 < \mu_X^B(x) < 1\}.$$

3 Decision Rules

Every decision table describes decisions determined, when some conditions are satisfied. In other words each row of the decision table specifies a decision rule which determines decisions in terms of conditions.

Let us describe decision rules more exactly.

Let $S = (U, C, D)$ be a decision table. Every $x \in U$ determines a sequence $c_1(x),\ldots,c_n(x)$, $d_1(x),\ldots,d_m(x)$ where $\{c_1,\ldots,c_n\} = C$ and $\{d_1,\ldots,d_m\} = D$.

The sequence will be called a *decision rule induced by* x (in S) and denoted by $c_1(x),\ldots,c_n(x)\ d_1(x),\ldots,d_m(x)$, in short $C \rightarrow_x D$.

The number $supp_x(C, D) = card(C(x) \cap D(x))$ will be called the *support* of the decision rule $C \rightarrow_x D$ and the number

$$\sigma_x(C,D) = \frac{supp_x(C,D)}{card(U)},$$

will be referred to as the *strength* of the decision rule $C \rightarrow_x D$. With every decision rule $C \rightarrow_x D$ we associate the *certainty factor* of the decision rule, denoted $cer_x(C, D)$ and defined as follows:

$$cer_x(C,D) = \frac{card(C(x) \cap D(x))}{card(C(x))} = \frac{supp_x(C,D)}{card(C(x))} = \frac{\sigma_x(C,D)}{\sigma_x(C)},$$

where $\sigma_x(C) = \frac{card(C(x))}{card(U)}$.

The certainty factor may be interpreted as conditional probability that y belongs to $D(x)$ given y belongs to $C(x)$, symbolically $\sigma_x(D|C)$. If $cer_x(C, D) = 1$, then $C \rightarrow_x D$ will be called a *certain decision* rule in S; if $0 < cer_x(C, D) < 1$ the decision rule will be referred to as an *uncertain decision* rule in S.

Besides, we will also use a *coverage factor* of the decision rule, denoted $cov_x(C, D)$ and defined as

$$cov_x(C,D) = \frac{card(C(x) \cap D(x))}{card(D(x))} = \frac{supp_x(C,D)}{card(D(x))} = \frac{\sigma_x(C,D)}{\sigma_x(D)},$$

where $\sigma_x(D) = \frac{card(D(x))}{card(U)}$.

Similarly

$$cov_x(C,D) = \sigma_x(C|D).$$

The certainty and coverage factors have been for a long time used in machine learning and data mining, [11, 12] but in fact they have been first introduced in 1913 by Jan Łukasiewicz, in connection with his study of logic and probability [5].

If $C \rightarrow_x D$ is a decision rule then $D \rightarrow_x C$ will be called an *inverse decision rule*. The inverse decision rules can be used to give *explanations* (*reasons*) for decisions.

Let us observe that

$$cer_x(C,D) = \mu_{D(x)}^C(x) \text{ and } cov_x(C,D) = \mu_{C(x)}^D(x).$$

That means that the certainty factor expresses the degree of membership of x to the decision class $D(x)$, given C, whereas the coverage factor expresses the degree of membership of x to condition class $C(x)$, given D.

Decision rules are often represented in the form of "*if...then...*" implications. Thus any decision table can be transformed in a set of "*if...then...*" rules, called a *decision algorithm*. Generation of minimal decision algorithms from decision tables is rather difficult. Many methods for solving this problem have been proposed but we will not discuss this issue in this paper.

4 Properties of Decision Rules

Decision rules have important properties which are discussed below.
Let $C \rightarrow_x D$ be a decision rule in S. Then the following properties are valid:

$$\sum_{y \in C(x)} cer_y(C,D) = 1 \quad (1)$$

$$\sum_{y \in D(x)} cov_y(C,D) = 1 \quad (2)$$

$$\sigma_x(D) = \sum_{y \in C(x)} cer_y(C,D) \cdot \sigma_x(C) = \sum_{y \in C(x)} \sigma_y(C,D) \quad (3)$$

$$\sigma_x(C) = \sum_{y \in D(x)} cov_y(C,D) \cdot \sigma_x(D) = \sum_{y \in D(x)} \sigma_y(C,D) \quad (4)$$

$$cer_x(C,D) = \frac{cov_x(C,D) \cdot \sigma_x(D)}{\sum_{y \in D(x)} cov_y(C,D) \cdot \sigma_x(D)} = \frac{\sigma_x(C,D)}{\sum_{y \in D(x)} \sigma_y(C,D)} = \frac{\sigma_x(C,D)}{\sigma_x(C)} \quad (5)$$

$$cov_x(C,D) = \frac{cer_x(C,D) \cdot \sigma_x(C)}{\sum_{y \in C(x)} cer_y(C,D) \cdot \sigma_x(C)} = \frac{\sigma_x(C,D)}{\sum_{y \in C(x)} \sigma_y(C,D)} = \frac{\sigma_x(C,D)}{\sigma_x(D)} \quad (6)$$

Observe that (3) and (4) refer to the well known *total probability theorem*, whereas (5) and (6) refer to *Bayes' theorem*.

Thus in order to compute the certainty and coverage factors of decision rules according to formulas (5) and (6) it is enough to know the strength (support) of all decision rules only. The strength of decision rules can be computed from data or can be a subjective assessment.

5 Dependences in decision Tables

Next important issue in decision table analysis is the dependency of attributes, particularly dependency of decision attributes on condition attributes [10].

Intuitively speaking the set of decision attributes depends on the set of condition attributes if values of decision attributes are totally (or partially) determined by values of condition attributes.

In other words, this kind of dependency describes which decisions specified in a decision table are to by obeyed if some conditions are satisfied.

In this paper we will introduce another kind of dependences in decision tables, based on some ideas of statistics.

Let $S = (U,C,D)$ be a decision table and let $x \in U$. We say that decisions $d_1(x),\ldots,d_m(x)$ are *independent* on conditions $c_1(x),\ldots,c_n(x)$, where $C = \{c_1,\ldots,c_n\}$, $D = \{d_1,\ldots,d_m\}$, if

$$\sigma_x(C,D) = \sigma_x(C)\sigma_x(D).$$

In other words conditions and decisions in a decision rule $C \to_x D$ are independent if

$$\frac{\sigma_x(C,D)}{\sigma_x(C)} = cer_x(C,D) = \sigma_x(D)$$

or

$$\frac{\sigma_x(C,D)}{\sigma_x(D)} = cov(C,D) = \sigma_x(C).$$

If

$$cer_x(C,D) > \sigma_x(D)$$

or

$$cov_x(C,D) > \sigma_x(C).$$

We say that D *depends positively* on C in a decision rule $C \to_x D$. Similarly if

$$cer_x(C,D) < \sigma_x(D)$$

or

$$cov_x(C,D) < \sigma_x(C).$$

We say that D *depends negatively* on C in a decision rule $C \to_x D$.

Obviously, the relations of independence and positive and negative dependence are symmetric ones.

Again borrowing from statistics the idea of a correlation coefficient we can determine the degree of dependency numerically, defining the *correlation factor*, defined as follows:

$$\eta_x(C,D) = \frac{cer_x(C,D) - \sigma_x(C)}{cer_x(C,D) + \sigma_x(C)} = \frac{cov_x(C,D) - \sigma(D)}{cov_x(C,D) + \sigma_x(D)}.$$

Obviously, $0 \leq \eta_x(C,D) \leq 1$ and if $\eta_x(C,D) = 0$ then C and D are independent, if $\eta_x(C,D) < 0$ then C and D are negatively dependent if $\eta_x(C,D) > 0$, then C and D are positively dependent.

6 Flow Graphs

With every decision table we associate a *flow graph*, i.e., a directed, connected, acyclic graph defined as follows: to every decision rule $C \to_x D$ we assign a *directed branch* x connecting the *input node* $C(x)$ and the *output node* $D(x)$. Strength of the decision rule represents a *throughflow* of the corresponding branch. The throughflow of the graph is governed by formulas (1),...,(6), and can be considered as a *flow conservation equation* similar to that introduced by Ford and Fulkerson [4]. However, let us observe that flow graphs presented in this paper are different from flow networks of Ford and Fulkerson.

Formulas (1) and (2) say that the outflow of an input node or an output node is equal to their inflows. Formula (3) states that the outflow of the output node amounts to the sum of its inflows, whereas formula (4) says that the sum of outflows of the input node equals to its inflow. Finally, formulas (5) and (6) reveal how throughflow in the flow graph is distributed between its inputs and outputs.

It is obvious that the idea of flow graph can be also formulated more generally, independently of decision tables, but we will not consider this issue here.

7 An Example

Let us now illustrate the above ideas by means of a simple example shown in Table 1.

Table 1: Decision table

FACT	DIS.	AGE	SEX	TEST	SUPP.
1	yes	old	man	+	400
2	yes	middle	woman	+	80
3	no	old	man	−	100
4	yes	old	man	−	40
5	no	young	woman	−	220
6	yes	middle	woman	−	60

Attributes **DISEASE**, **AGE** and **SEX** are condition attributes, whereas **TEST** is the decision attribute.

Below a decision algorithm associated with Table 1 is presented.

1. *if (disease, yes) and (age, old) then (test, +)*
2. *if (age, middle) then (test, +)*
3. *if (disease, no) then (test, −)*
4. *if (disease, yes) and (age, old) then (test, −)*
5. *if (age, middle) then (test, −)*

The certainty and coverage factors for the above algorithm are given in Table 2.

Table 2: Certainty and coverage factors

RULE	STRENGTH	CER.	COV.
1	0.44	0.90	0.83
2	0.09	0.56	0.17
3	0.35	1.00	0.77
4	0.04	0.08	0.09
5	0.07	0.44	0.15

Remark. Due to the round-off errors in computations the properties (1)…(6) may not always be satisfied in the table.

The certainty factors of the decision rules lead to the following conclusions:

- 90% ill and old patients have positive test result
- 56% ill and middle aged patients have positive test result
- all healthy patients have negative test result
- 8% ill and old patients have negative test result
- 44% ill and middle aged patients have negative test result

In other words:

- ill and old patients most probably have positive test result (probability = 0.90)
- middle aged patients most probably have positive test result (probability = 0.56)
- healthy patients have certainly negative test result (probability = 1.00)

The inverse decision algorithm is given below:

1'. *if* (*test*, +) *then* (*disease, yes*) *and* (*age, old*)
2'. *if* (*test*, +) *then* (*age, middle*)
3'. *if* (*test*, −) *then* (*disease, no*)
4'. *if* (*test*, −) *then* (*disease, yes*) *and* (*age, old*)
5'. *if* (*test*, −) *then* (*age, middle*)

Employing the inverse decision algorithm and the coverage factors we get the following explanation of test result:

- reasons for positive test results are most probably disease and old age (probability = 0.83)
- reason for negative test result is most probably lack of the disease (probability = 0.77)

The flow graph for the decision algorithm is presented in Fig. 1.

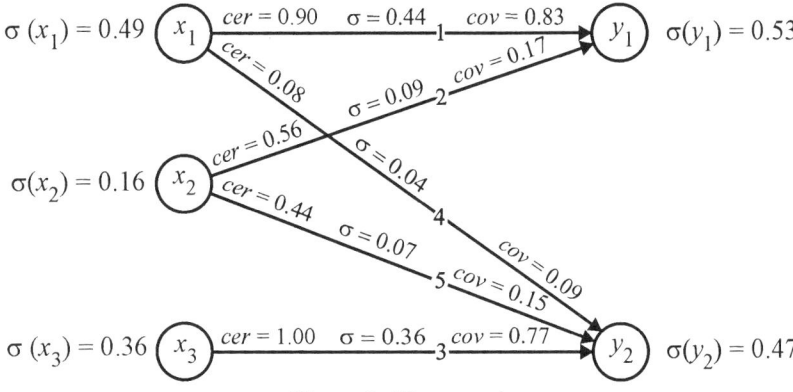

Figure 1: Flow graph

In the flow graph we have $x_1 = \{1, 4\}$, $x_2 = \{2, 6\}$, $x_3 = \{3, 5\}$, $y_1 = \{1, 2\}$ and $y_2 = \{3, 4, 5, 6\}$.

That means that we have in our data base the following groups of patients:

440 – ill and old (x_1)
140 – middle aged (x_2)
320 – healthy (x_3)
and
480 – with positive test result (y_1)
420 – with negative test result (y_2)

Each input node of the flow graph represents a condition of corresponding decision rule, whereas each output node reveals decisions of the rules. The associated numbers can be understood as probabilities (frequencies) of conditions and decisions respectively. Branches of the graph are labeled by strength of associated decision rules.

The flow graph of a decision algorithm shows how probabilities of decisions and conditions are related.

Each node of the graph satisfies equations (1)...(6). Observe, that in order to compute all the conditional and total probabilities it is enough to know the strength of the decision rules only, which makes the computations very easy, and gives also clear insight into the structure of the decision algorithms.

Let us notice that, for example, old age and illness are positively correlated with positive test result, whereas old age and illness are negatively correlated with negative test result.

The corresponding correlation coefficients are 0.26 and – 0.71 respectively.

8 Conclusion

It is clearly seen from the above considerations the difference between Bayesian data analysis and the rough set approach. In the Bayesian inference the data is used to update prior probability (knowledge) into a posterior probability, whereas rough set based Bayesian inference is used to reason directly from data.

The relationship of rough set theory with Bayes' theorem and flow graphs gives new look on Bayesian inference and leads to efficient algorithms for data analysis.

References

[1] J. M. Bernardo, A. F. M. Smith (1994). Bayesian Theory, Wiley Series in Probability and Mathematical Statistics. John Wiley & Sons, Chichester, New York, Brisbane, Toronto, Singapore.

[2] G.E.P. Box, G.C. Tiao (1992). Bayesian Inference in Statistical Analysis. John Wiley and Sons, Inc., New York, Chichester, Brisbane, Toronto, Singapore.

[3] M. Berthold, D.J Hand (1999). Intelligent Data Analysis, an Introduction. Springer-Verlag, Berlin, Heidelberg, New York.

[4] L.R. Ford, D. R. Fulkerson (1962). Flows in Networks. Princeton University Press, Princeton, New Jersey.

[5] J. Łukasiewicz (1970). Die logishen Grundlagen der Wahrscheinilchkeitsrechnung. Kraków (1913). In: L. Borkowski (ed.), Jan Łukasiewicz - Selected Works, North Holland Publishing Company, Amsterdam, London, Polish Scientific Publishers, Warsaw.

[6] Z. Pawlak (2001). New Look on Bayes' Theorem - the Rough Set Outlook. In *Proceeding of International Workshop on Rough Set Theory and Granular Computing RSTGC-2001*, Matsue, Shimane, Japan, May 20-22, S. Hirano, M. Inuiguchi and S. Tsumoto (eds.), Bull. of Int. Rough Set Society vol. 5 no. 1/2 2001, pages 1-8.

[7] Z. Pawlak (2002). In Pursuit of Patterns in Data Reasoning from Data – The Rough Set Way. In: J.J. Alpigini *et al.* (eds.), *Lecture Notes in Artificial Intelligence* 2475, pages 1-9.

[8] Z. Pawlak (2002). Rough Sets, Decision Algorithms and Bayes' Theorem. In *European Journal of Operational Research* 136, pages 181-189.

[9] Z. Pawlak, A. Skowron (1994). Rough Membership Functions. Advances in the Dempster-Shafer Theory of Evidence. R, Yager, M. Fedrizzi, J. Kacprzyk (eds.), John Wiley & Sons, Inc., New York, pages 251-271.

[10] L. Polkowski (2002). Rough Sets – Mathematical Foundations. Physical-Verlag, Springer Verlag Company.

[11] S. Tsumoto, H. Tanaka (1995). Discovery of Functional Components of Proteins Based on PRIMEROSE and Domain Knowledge Hierarchy. In *Proceedings of the Workshop on Rough Sets and Soft Computing* (*RSSC-94*), 1994: Lin, T.Y., and Wildberger, A.M. (eds.), *Soft Computing*, SCS 280-285.

[12] S.K.M.Wong, W. Ziarko(1986). Algorithm for Inductive Learning. In *Bull. Polish Academy of Sciences* 34, 5-6, pages 271-276.

B. Bouchon-Meunier, L. Foulloy and R.R. Yager (Editors)
Intelligent Systems for Information Processing:
From Representation to Applications
© 2003 Elsevier Science B.V. All rights reserved.

Belief Revision as Combinatorial Optimization

Arthur Ramer
School of Computer Science and Engineering
UNSW, Sydney 2052, Australia
ramer@cse.unsw.edu.au

Abstract

AGM framework includes principles for systematic revision of probabilistic beliefs. The key is the principle of minimal change. Expansion P^+ is to be done, by preference, as conditioning, though some form of imaging also gets a nod. Contraction P^-, intended as *inverse conditioning,* is axiomatized but not actually computed.

We demonstrate how both can be achieved through a very natural form of combinatorial optimization. Basic modeling can be effected using standard entropy as the objective function. A reinterpretation of minimality of change leads to plausible and nontrivial formulae for P^-. We next show that stronger results can be obtained using *graph entropies*. We define *graph information distance*, which together give a *GraphMaxEnt* revision rule. It permits modeling of essentially all AGM protocols for P^+ and P^-. More complex patterns, involving conditional probability assertions, can also be realized through this method.

Keywords: AGM model, belief revision, probability revision, graph entropy.

1 Introduction

Our paper deals with reasoning about changes in two types of knowledge—one expressed through logical assertions, the other of quantitative nature (mainly probabilistic). Both domains have their own precise notations; however, when put together, they warrant certain simplifying conventions.

We adopt certain common abbreviations to keep the phrasing of statements and appearance of formulae uncluttered. We write *wrt* for *with respect to* and *iff* for *if and only if*. In summations we do not state the range if it is the entire permissible domain; on occasion we may omit the index of summation.

For a logical assertion A applicable to the elements of the domain X we write A_X for the set of x_i where A holds; thus $x \in A_X$ is the same as $A(x)$.

We deal with probability distributions $P, Q, \ldots$ on a given domain X. We write p_i for $P(x_i)$, q_i for $Q(x_i)$ and so forth. For brevity, in some summations, we have p_i (or q_i) to stand for x_i, eg. we may simplify $\sum_{i:x_i \in A} p_i \log p_i$ to just $\sum_A p_i \log p_i$.[1]

In the later sections we need to refer to various probabilistic entropies. We use the term *entropy* to stand for Shannon entropy $H(P)$ for a finite probability distribution P, without considering any structure on its domain. Its related information distance is *information divergence* $D(P\|Q)$ (also termed *cross-entropy*. Change of the defining formulae gives other entropies, like the Renyi family $H^\alpha(P)$ and $D^\alpha(P\|Q)$.

Graph entropies recognize a double structure - an unordered graph G on the domain X and a probability distribution $P : x_i \mapsto p_i$. They are defined through information-theoretic computations based on ordinary entropies. Thus we write $H(G, P)$ for the graph entropy computed on the basis of Shannon entropy, $H^\alpha(G, P)$ for Renyi and so forth.

2 AGM Probability revisions

Formalizations of belief change have been discussed, in various contexts, since 1970's. Notable specific applications are 'truth maintenance systems' [4] and 'database priorities' [5]. General, abstract protocols were introduced by philosophers Levi [15, 16], Harper [9], and then a series of works by Alchourron, Gardenfors and Makinson [1, 2, 6]. The last one gave the name to the system of postulates for belief revision as the *AGM framework*. Its basic design is founded on a revision scheme addressing needs of the finite propositional knowledge bases. In parallel with the purely logical framework has been proposed a scheme for modifying *beliefs about probability* [6]. Here we consider a finite collection of *possible worlds* X and a probability distribution thereupon. Any proposition A may be held in some subcollection X_A of these worlds, with its probability defined as the sum of probabilities of the worlds where it is held $P(A) := P(X_A)$. A proposition is *accepted* if its probability is 1; it is important to remember that it does not signify a universal acceptance, as there (usually) will be worlds, of probability 0, where the proposition may not hold.

Expansion of the state of beliefs wrt A will mean adjusting the probability distribution to a such P_A^+ that A is *accepted* $P +_A (A) = 1$. In keeping with the overall philosophy of such change, it is postulated that the passage from P to P_A^+ should be effected with a *minimal* change. On a combination of philosophical and logical grounds it is strongly argued that such an expansion should be the *conditioning* wrt A, understood as

$$P_A^+(B) := P(A \wedge B)/P(A), \qquad P(A) > 0$$

and with a pseudo-distribution for the case of $P(A) = 0$. As a probabilistic operation it should be viewed as conditioning wrt the subset A_X; this subset may include some worlds of probability 0, but where A is held. A group of four postulates is given to axiomatize it [6]

P+1 For disjoint A and B ($\vdash \neg(A \wedge B)$), the distribution $P_{A \vee B}^+$ is a suitable convex combination of P_A^+ and P_B^+; taking $\alpha = P(A)/P(A \wedge B)$, $\beta = 1 - \alpha = P(B)/P(A \wedge B)$, one requires $P_{A \vee B}^+ = \alpha P_A^+ + \beta P_B^+$.

P+2 $P_A^+(A) = 1$

---
[1] "Entropy measures information regardless of its content."[3]

P⁺3 If $\vdash A$ then $P_A^+ = P$

P⁺4 If $P(A)$ then $P_A^+ = P_\perp$ - the 'absurd' pseudo-distribution[2]

These axioms do not state that the change from P to P_A^+ should be, in some way, *minimal* though this is clearly a main concern, which is made explicit in the axioms for knowledge base expansions. It is postulated and discussed at length in the context of *contractions* and *revisions* of probability assignments.

It is known [19, 22] that if such minimality is viewed as minimum *information divergence*[3] then conditioning can be derived from just one axiom above—(**P⁺2**), which simply says that P_A^+ is a probability distribution supported on the worlds where A is accepted.

Although Ramer [19] observes that the use of *symmetric divergence* would be as effective, and Williams [22] that *Jeffrey's conditionalization* also obtains, neither notes that use of other entropies would produce exactly the same conditioning and could also be applied to conditioning wrt an arbitrary partition. We discuss these issues in the section below.

The generalization proposed by Jeffrey [10] was to combine conditioning wrt A and its logical complement $\neg A$ as method of revision of probability functions. A revised function P^{+J} should compute probability of an arbitrary assertion B by conditioning wrt the property that $P^{+J}(A) = a$ for some $0 < a < 1$. He was led to the formula

$$P^{+J}(A) = aP(B|A) + (1-a)P(B|\neg A).$$

It is convenient to express it in the language of expansions—Jeffrey conditionalization becomes a linear combination of expansions wrt proposition A and its complement $\neg A$. For the specified $0 < a < 1$

$$P_A^{+J} = aP_A^+ + (1-a)P_{\neg a}^+.$$

This formulation permits formulating *Jeffrey expansions* in context of imaging and other revision methods.

Philosophers proposed alternatives to conditioning when restriction of the domain (expansion of beliefs) is imposed. The principal model is *imaging* [17] and *preservative imaging* [14]. These cannot be obtained through simple entropy; we propose using *graph entropies* [13, 20], applied extensively in communication theory and graph combinatorics. These entropies can be linked to *plausibility* measures, and through those suggest a new distance - *graph information divergence*. Minimization based on this distance appears flexible enough to permit modeling of virtually all probability expansions considered in the literature. We outline this technique in the following section.

While probability expansion can be handled fairly expediently through entropy formalisms, the dual operation of *probability contraction* has been much less clear. The basic idea, propounded by AGM, is to adjust 'minimally' the distribution P so that some logical condition A, for which currently $P(A) = 1$, is no longer a 'sure thing'. Namely, we need to define P_A^- such that $P_A^-(A) < 1$, and such that *its* conditioning wrt A would restore P

$$(P_A^-)_A^+ = P.$$

---

[2] It is defined as assigning probability value 1 to all subsets.
[3] An information distance, based on entropy; not a metric, failing both symmetry and triangle inequalities.

A precise setting is one of $A_X \subset X$ such that $P(x) = 0$ for $x \in X \setminus A_X$. We want to reduce the probability of A to less than one, and distribute the resulting excess among the elements of its complement. We also require that the original distribution, concentrated on A be obtainable by conditioning from the new distribution.

In the original proposal [6] the philosophical considerations are discussed, but nothing is computed. Following a series of contraction postulates, from P^-1 to P^-6, Garderfors still notes (in page 92) *It is, however, by no means clear how such a theory of information can be helpful for constructing contractions ... the sentences that are accepted ... have probability 1. ... a minimal change would would be to assign these sentences probability values that are close to 1. Such an assignment is, however, unrealistic as a model of contraction ...* and further (in page 106) *... neither the traditional conditionalization process nor Jeffrey conditionalization helps us with the problems of how to define contractions and revisions of probability functions.*

We show that they can help very significantly, provided we reinterpret correctly the notion of *minimal change*. We argue that such minimality can take uniform distribution as the reference point. We then minimize the distance from the *most uninformed* prior and impose the *recovery* constraint on the contraction P^-, namely that $(P_A^-)_A^+ = P$. It produces a series of very attractive solutions for the basic model, all relying on Shannon entropy. It is reasonable to posit that an extension to graph entropy would permit modeling of almost any of the main considered proposals. However, a very significant contradistinction arises—as opposed to the conditioning processes—the numerical results depend now on which specific entropy function we use. For example, whether we use Shannon or Renyi entropies we get the same conditional distribution; it applies to equally to the standard case (no special structure on the domain) and to graph entropies. Inverse conditioning (contraction) has different numerical results when the entropy function changes.

This is far from being fully understood. It suggests that conditioning wrt a subset (defined by the logical assertion) is essentially a structural change, while the inverse conditioning is numerical.

3 Conditional assertions

A somewhat controversial issue is the suitability of entropy for conditionalization based itself on conditional premises [7, 21]. A prototypical situation is usually framed as the JB problem (after the film 'Private (Judy) Benjamin'). We show that the 'obvious' answer obtains if the entropy on an *incomplete* graph is used. It appears that similar 'successes' can be generated for majority of like cases. However, it is the ease of creating such solutions that cautions against the automatic use of entropy as the *normative* decision rule. The JB problem is tackled by omitting a specific edge from the complete graph. This can be given a logical basis, but it feels more like an explanation *ex post*, and suggests that the MaxEnt and MinInf are best kept as *descriptive rules*, occasional successes to the contrary notwithstanding [18]. Their prescriptive use would require a supporting logical framework that could decide *ex ante* on choice of the graph of 'information' transfers.

The JB problem is a case in point. The setting consists of four states BH, B2, RH, and R2 (after the 'Blue' and 'Red' armies headquarters and 2nd company areas). Initially JB, totally disoriented, assigns probabilities uniformly to all four states. Receiving a message she needs to reasses the probabilities so that $P(RH|RH \vee R2) = 0.75$. The intuitive answer $(\frac{1}{2}, \frac{1}{2}, \frac{3}{8}, \frac{1}{8})$ preserving *independence* between the 'Red' and 'Blue' states, does not follow from the use of any unstructured entropy. A direct conditional reasoning [7] restores the independence, but cannot be reduced to the basic MaxEnt.

We recover the answer by *removing* the edge $(RH, R2)$ from the interconnection graph. It can be justified as the explication of the fact that states RH and R2 can *confused*. We feel it is best viewed as simply explaining the success of GraphMaxEnt in this instance.

4 Entropies and graph entropies

Given a probability distribution its (Shannon) entropy

$$H(P) = -\sum p_i \log p_i.$$

Its generalization is the Renyi family of α-order entropies

$$H^\alpha(P) = \frac{1}{1-\alpha} \log \sum p_i^\alpha,$$

where $H^0(p) := \lim_{\alpha \to 0} H^\alpha(P) = H(P)$. Related *information divergence* captures the *change* of information entailed by passing from P to Q

$$D(Q\|P) = \sum q_i \frac{q_i}{p_i}.$$

Its Renyi analogs are

$$D^\alpha(Q\|P) = \frac{1}{\alpha-1} \log \sum q_i^{1-\alpha} p_i^\alpha.$$

Many other entropies have been proposed in the literature, eg.[11]. While their initial definitions were often ad hoc, several of them can be obtained systematically from certain generating functions of Dirichlet type, defined for the underlying probability distributions. These functions can also be viewed as Laplace transforms and they appear in the thermodynamic formalism. There are strong reasons to posit that only the entropies which arise from the analytic expansions of these functions are 'reasonable'. We believe that all such reasonable entropies would lead to the same (direct) conditioning formulae, at least on simple domains without a graph structure. Some evidence is given in the next section.

These entropies have an implicit underlying assumption that once the information about $x_i \in X$ is transmitted there is no further confusion about the identity of that x_i. (Such a transmission or choice from X is obviously subject to a probabilistic chance.) Allowing for such confusion should *lower* the entropy—indistinguishable elements could be, in a sense, transmitted together. A formal model [12, 13] recognizes graph G on the vertices $\{x_i\}$, where an edge (x_i, x_j) is formed whenever these two vertices *cannot* be confused. Thus the standard entropy corresponds to the complete graph K_n, $n = |X|$. Contrariwise, a fully confusable arrangement consists of n isolated vertices, with the presumed entropy 0.

Given the distribution P on X, the definition of $H(G, P)$ requires considering probability distributions on the collection $\mathcal{I}$ of the *maximal independent* sets of vertices.[4] Denoting

---
[4]*Independent* means that no two vertices form an edge.

$\mathcal{I} = \{Y \subseteq X, Y\text{- max ind}\}$, we first need a *joint* probability distribution W on $\mathcal{I} \times X$, such that

- $S(Y, x) = 0$ if $x \notin Y$
- S projected onto X is precisely P

Let R be its projection onto $\mathcal{I}$. We put

$$H(G, P) = \min_S (H(P) + H(R) - H(S)).$$

Although the expression may seem convoluted, it is actually quite easy to work with; in particular, there is a simple algorithm finding the minimizing S and computing the entropy. There is an equivalent definition due to Simonyi [20], based on the notion of *vertex packing polytope*. This is less suitable for computations, but better for generalizations. However, it can be recast into a very useful formula using the notion of plausibility. With notation as above, we first consider arbitrary probability distribution R defined on $\mathcal{I}$ and put

$$\mathrm{Pl}^{(R)}(x) = \sum_{Y:x \in Y} R(Y).$$

We have a fairly easy result

$$H(G, P) = \min_R - \sum p_i \log \mathrm{Pl}^{(R)}(x_i).$$

It holds that R that minimizes the expression above is the same distribution as in Korner and Simonyi definitions. We use it to define the *plausibility wrt P* on X

$$\mathrm{Pl}_P(x) = \mathrm{Pl}^{(R)}(x), \quad R = \arg \min H(G, P).$$

It serves to define the *graph information divergence*

$$D(G, Q \| P) = \sum q_i \frac{\mathrm{Pl}_Q(x_i)}{\mathrm{Pl}_P(x_i)}$$

for two distributions Q and P on X, given a (fixed) graph structure G. It is straightforward to offer a similar definition wrt the change of G, but it produces useful results only in restricted cases.

All these entropies have generalized Shannon formula. Variations based on Renyi or other entropies are readily available, along with various limiting and bounding formulae. Graph entropies satisfy additivity [20] wrt vertex substitution, a construction which subsumes additivity and branching properties of the standard entropy. This property seems useful for the future investigations of their uses in belief revision. They also satisfy subadditivity wrt graph complementation (with equality for perfect graphs). Its extreme importance for graph theory suggests relevance for probability revisions.

5 Conditioning and entropy

The simplest expansion problem can be posed as a question about finding $\hat{Q}$ on X where $\hat{Q}(A_X) = 1$ for $A_X \subset X$ - the subset where A holds. This $\hat{Q}$ should be as close as possible to the given P on X. A natural solution would be

$$\arg\min_{Q:Q(A_X)=1} D(Q\|P).$$

The solution is the familiar conditional distribution $P(\cdot|A) : x_i \mapsto p_i/P(A_X)$ if $A(x_i)$, and $x_i \mapsto 0$ if $\neg A(x_i)$. The same result obtains if $D(P\|Q)$ be used instead (hence also for the symmetric distance $D(Q\|P) + D(P\|Q)$). More significantly, use of Renyi entropy (or many others) does not affect the result.

Jeffrey formula intends an expansion where $P_A^+(A) = a$ for some $0 < a < 1$, leaving $P_A^+(\neg A) = 1 - a$, and is defined through

$$P_A^{+J}(x_i) = p_i/a \text{ if } A(x_i),$$
$$P_A^{+J}(x_i) = p_i/(1-a) \text{ otherwise.}$$

It is immediate that

$$\arg\min_{Q:Q(A_X)=a} D(Q\|P)$$

gives this conditionalization. An easy extension is to specify a partition $X = A^{(1)} \cup \ldots \cup A^{(k)}$, a probability assignment on its elements $A^{(i)} \mapsto a_i$, $\sum a_i = 1$ and require that $P^+(A^{(i)}) = a_i$. Such a generalized Jeffrey rule results from a like minimization of information divergence. Again, change of entropy function turns out to be immaterial.

We should also note that we need to use $D(Q\|P)$ and cannot obtain a reasonable answer directly from $H(Q)$. The proximate cause seems to be that we must 'retain' the knowledge of P and then add the fact that $P^+(A) = 1$. Using $H(P)$ would appear to recognize only the latter fact.

6 Inverse conditioning

An attempt to replicate the previous method of direct minimization of a distance between the distributions is bound to fail. The nearest P_A^- which conditionalizes back to P is the very same P. If we insist that P^- must be different, an ϵ-change to $P^-(A) = 1-\epsilon$ would ensue. However, we observe that in case of conditionalization the entropy $H(P^+) < H(P)$, therefore the minimum of $D(Q\|P)$ is somewhat related to minimizing the distance from Q to the most *uninformed* ie. uniform distribution. While this cannot used for deciding on P^+ (as we would loose the knowledge of P), it suggests a useful approach to the P^- problem.

To make the question specific, we assume that P is supported on $A_X \subset X$ and that there are m elements outside A_X. We shall seek distribution $\hat{Q}$, with *maximum* entropy, that conditionalizes back to P. We need to compute

$$\arg\max_{Q:Q_A^+(A)=P} H(Q).$$

The answer has a very attractive form

$$P_A^-(x) = \frac{1}{m + 2^{H(P)}}, \quad x \notin A_X$$
$$P_A^-(A) = \frac{2^{H(P)}}{m + 2^{H(P)}}$$

Noting that $m = 2^{\log m}$, which is the entropy of the uniform distribution on m elements, permits to anticipate the effect of *inverting* Jeffrey conditioning. We first compute $H(P_A^+)$ and $H(P_{\neg A}^+)$. Denoting P^{-J} for the *inverse* Jeffrey rule

$$P^{-J}(A) = \frac{2^{H(P_A^+)}}{2^{H(P_A^+)} + 2^{H(P_{\neg A}^+)}}$$
$$P^{-J}(\neg A) = \frac{2^{H(P_{\neg A}^+)}}{2^{H(P_A^+)} + 2^{H(P_{\neg A}^+)}}$$

The extension to an arbitrary partition is straightforward. Moreover, while 'simple' inverse conditioning assigns to all the elements outside A_X the same probability, one can adopt the inverse Jeffrey rule to recognize some specified proportions. The simplest, albeit somewhat informal method is to view $P(\neg A)$ as having an 'infinitesimal' value, which becomes 0 in actual computations, but permits retaining some meaningful proportions.

The appearance of terms of form $2^{H(P)}$ seems to us significant. Such expressions are closely related to probabilities of *typical events* in coding theory [3] and in thermodynamics, or to the *most likely worlds* in the probabilistic logic [8]. Another significant matter is the dependence of the results on a choice of the entropy. Unlike the expansion, in case of contraction selection of entropy function matters.

7 Graph entropies, imaging and JB problem

As mentioned earlier, alternatives other than conditioning were considered for probability expansions. Conditioning reassigns the weight of probabilities on $(\neg A)_X$ to the elements of A_X in proportion to the latter original weights. One can conceive of the universe where each world has a closest (most similar) counterpart where A is satisfied [17]. One then wants to reassign the weight entirely to that most similar world, rather than spreading it around. More refined is preservative imaging where certain worlds 'communicate' with certain other worlds (for the purpose of weight transfer) but not with all the worlds.

We report in detail only on the simplest case of a graph on three vertices and conditioning onto a set of two vertices (or inverse conditioning from such a set). We denote both the vertices and their probabilities p_1, p_2, p_3. We write the formulae in a form which makes the minimizing plausibilities apparent.

For the complete graph (all three edges) the usual conditioning results - arranging for A so that $A_X = \{p_1, p_2\}$ gives

$$P_A^+ : p_1 \mapsto \frac{p_1}{p_i + p_2}, \quad p_2 \mapsto \frac{p_2}{p_i + p_2}.$$

The inverse conditioning P_A^- admits an interesting presentation (equivalent to one given earlier)

$$P_A^-(p_3) = \frac{1}{1+(\frac{p_1}{p_2})^{p_2}+(\frac{p_2}{p_1})^{p_1}}$$

$$P_A^-(p_1) = \frac{(\frac{p_1}{p_2})^{p_2}}{1+(\frac{p_1}{p_2})^{p_2}+(\frac{p_2}{p_1})^{p_1}}$$

$$P_A^-(p_2) = \frac{(\frac{p_2}{p_1})^{p_1}}{1+(\frac{p_1}{p_2})^{p_2}+(\frac{p_2}{p_1})^{p_1}}$$

For a completely disconnected graph, conditioning is no longer *specific* - any distribution on $\{p_1, p_2\}$ is equally good—all entropies are 0 and, indeed, there is nothing to choose if any two vertices can be confused. The same holds for the inverse conditioning.

For two edges $\{(p_1, p_3), (p_2, p_3)\}$ there are two maximal independent sets $\{p_1, p_2\}$ and $\{p_3\}$, with

$$H(G, P) = -(p_1 + p_2) \log(p_1 + p_2) - p_3 \log p_3.$$

Conditioning onto $\{p_1, p_2\}$ is nonspecific for the same reasons as above. Conditioning onto $\{p_1, p_3\}$ gives *imaging* of p_2 onto p_1 (correctly, as these two are 'confusable').

Inverse conditioning from $\{p_1, p_2\}$ is best viewed as 'unconstrained', thus giving a uniform distribution $(\frac{1}{3}, \frac{1}{3}, \frac{1}{3})$, while starting from $\{p_1, p_3\}$ we should split evenly p_1 for $(\frac{p_1}{2}, \frac{p_1}{2}, p_3)$.

For a single edge (p_1, p_2) there are also two maximal independent sets, but no longer disjoint $\{p_1, p_3\}$ and $\{p_2, p_3\}$. The entropy becomes

$$H(G, P) = -p_1 \log \frac{p_1}{p_1 + p_2} - p_2 \log \frac{p_2}{p_1 + p_2}.$$

Conditioning onto $\{p_1, p_2\}$ is now identical to the standard one, while conditioning onto $\{p_1, p_3\}$ is nonspecific. Inverse conditioning from $\{p_1, p_2\}$ entails no change, giving $(p_1, p_2, 0)$, while from $\{p_1, p_3\}$ it is always $(\frac{1}{2}, \frac{1}{2}, 0)$.

For the JB problem we use a graph G on four vertices BH, B2, RH, and R2, with probabilities $P = (bh, b2, rh, r1)$. It results form the complete graph K_4 on these vertices, by removing the edge $(RH, R2)$. Its entropy is

$$H(G, P) = -bh \log bh - b2 \log b2 - (rh + r1) \log(rh + r1),$$

while the distance from the distribution $Q = (ah, a2, sh, s2)$ is

$$D(G, Q \| P) = ah \log \frac{ah}{bh} + a2 \log \frac{a2}{b2} + (sh + s2) \log \frac{sh + s2}{rh + r2}.$$

For P uniform, the assignment Q satisfying $sh \div s2 = 3 \div 1$ and *closest* to P is necessarily $Q = (\frac{1}{4}, \frac{1}{4}, \frac{3}{8}, \frac{1}{8})$. Other edge structures of G can represent interactions between the 'Blue' and 'Red' states, also supporting two different (undesirable) solutions considered in [21].

We have not computed all the cases of graph entropies based on other than Shannon formulae. However, it seems that the choice of entropies does not affect conditioning, but is material for inverse conditioning.

For graphs on four and five vertices another phenomenon appears—certain coefficients in the graph entropy formulae combine the original probabilities $p_1, p_2, \ldots$ with both positive and negative signs. It implies that the minimizing solutions to conditioning (resp. maximizing for the inverse conditioning) cannot always be obtained by solving Lagrange multipliers equations, but may lie on the boundary of the feasible region.

However, in all the cases we computed, the answers are always simple fractions (rational functions of degree 1) formed from the initial probabilities. We believe it is always the case, and expect the proof to be not too difficult. A different matter might be finding a methodological explanation.

8 Remarks on possibilistic and evidential conditioning

It is known that conditional possibility assignments can be defined on the basis of maximizing the uncertainty measures [18]. In the spirit of the earlier discussion one define *inverse* possibilistic conditioning. The results are quite straightforward - one either 'truncates' a given assignment or leaves is completely intact, and then puts the maximal possibility value of 1 for the 'outside' elements. This pattern is independent of the choice of information measure; moreover, it makes the 'contracted' assignment very similar to the 'expanded' one. A further study is clearly warranted.

A direct use of graph uncertainty measures is not feasible. Possibility assignments have a natural 'built-in' order structure, which needs to be put in a suitable correspondence with any intended graph structure on the underlying domain of discourse.

Similar considerations apply to models based on various forms of Dempster-Shafer theory of evidence. Typical evidence assignments are functions over the complete powersets of the (finite) domain of discourse. Such powersets are boolean algebras and any use of graph entropies should be coordinated with this structure.

References

[1] CE Alchourron, P Gardenfors, D Makinson. *On the logic of theory change: partial meet contractions and revision functions.* J. Symbolic Logic 50(1985), 510–530.

[2] CE Alchourron, D Makinson. *Hierarchies of regulations and their logic.* In R Hilpinen (ed) New Studies in Deontic Logic., 125–148. D Reidel Publ. Co, Dordrecht 1981.

[3] I Csiszar, J Korner. Information Theory: Coding Theorems for Discrete Memoryless Systems. Academic Press, New York 1981.

[4] J Doyle. *A truth maintenance system.* Artificial Intelligence 12(1979), 231–272.

[5] R Fagin, J Ullman, M Vardi. *Updating logical databases.* Adv. Computing Research 3(1986), 1–18.

[6] P Garderfors. Knowledge in Flux. The MIT Press, Cambridge MA 1988.

[7] AJ Grove, JY Halpern. *Probability update: conditioning vs cross-entropy.* Proc. 13th Annual Conf. Uncertainty in Artificial Intelligence, Providence, RI, August 1997.

[8] AJ Grove, JY Halpern, D Koller. *Random worlds and maximum entropy.* J. Artificial Intelligence Research 2(1994), 33–88.

[9] W Harper. *Rational conceptual change.* PSA'76 - Philosophy of Science Asooc. Biennial Meeting, East Lansing, Mich. 1976, 462–494.

[10] RC Jeffrey. The Logic of Decision. McGraw-Hill, New York 1965.

[11] JN Kapur, HK Kesavan. Entropy Optimization Principles with Applications. Academic Press, New York 1992.

[12] J Korner. *Coding of an information source having ambiguous alphabet and the entropy of graphs.* Trans. Sixth Prague Conf. Information Theory. Academia, Prague 1973, pp. 411–425.

[13] J Korner, A Orlitsky. *Zero-error information theory.* IEEE Trans. Information Theory 44(1998), 6:2207–2229.

[14] I Levi. *personal communication to P Gardenfors, in [6].*

[15] I Levi. *Subjunctives, dispositions, and chances.* Synthese 34(1977), 423–455.

[16] I Levi. The Enterprise of Knowledge. The MIT Press, Cambridge, MA 1980.

[17] DK Lewis. *Probabilities of conditionals and conditional probabilities.* Phil. Review 85(1976, 297–315.

[18] A Ramer. *Conditional possibility measures.* Int. J. Cybernetics and Systems 20(1989), 233–247.

[19] A Ramer. *Note on defining conditional probability,* Amer. Math. Monthly, 97(1990), 336–337.

[20] G Simonyi. *Graph entropy: a survey.* DIMACS Series in Discr. Math. and Theor. Comp. Science 20(1995), pp. 399–441.

[21] BC van Fraasen. *Symmetries of personal probability kinematics.* In N Rescher (ed). Scientific Enquiry in Philosophical Perspective. University Press of America, Lanham, MD 1987, pp. 183–223.

[22] PM Williams. *Bayesian conditionalisation and the principle of minimum information.* British J. Phil. Science 31(1980), 131–144.

B. Bouchon-Meunier, L. Foulloy and R.R. Yager (Editors)
Intelligent Systems for Information Processing:
From Representation to Applications
© 2003 Elsevier Science B.V. All rights reserved.

Showing why Measures of Quantified Beliefs are Belief Functions

Philippe Smets
IRIDIA
Université Libre de Bruxelles
50 av. Roosevelt, CP 194-6, 1050 Bruxelles, Belgium
psmets@ulb.ac.be

Abstract

In [8, 10], we present an axiomatic justification for the fact that quantified beliefs should be represented by belief functions. We show that the mathematical function that can represent quantified beliefs should be a Choquet capacity monotone of order 2. In order to show that it must be monotone of order infinite, thus a belief function, we propose several extra rationality requirements. One of them is based on the negation of a belief function, a concept introduced by Dubois and Prade [2]. This concept was essentially abstract, and its applicability was neither established nor illustrated. Here we present an illustrative example of this negation process. This example gives ground to the use of belief functions to represent quantified beliefs.

Keywords: TBM, belief function, belief representation

1 Introduction

The use of any mathematical model to represent quantified beliefs, i.e., weighted opinions, can be supported either by defending convincing definitions with illustrative examples or by producing a set of axioms that justify it. For what concerns the models based on belief functions, examples illustrating the first approach can be found among others in [7, 4, 12, 11] whereas the second approach is developed in [15, 8, 10]).

In any model for quantified beliefs, one considers an agent, the belief holder, called You hereafter, and a finite frame of discernment, denoted Ω. One of the worlds in Ω, denoted ω_0, is the actual world, but, due to Your limited understanding, You cannot state which world is the actual one. All You can express is the strength of Your opinions, called hereafter beliefs, that ω_0 belongs to A for every $A \subseteq \Omega$. We assume that this belief is represented by a pointwise measure defined on 2^Ω, the power set of Ω.

This measure is temporarily denoted by Cr and called a 'credibility function'. So for every $A \subseteq \Omega$, $Cr(A)$ expresses (the strength of) Your belief that the actual world ω_0 belongs to A.

In [8, 10], we produce sets of rationality requirements that should be satisfied by any credibility function and we prove 1) belief functions satisfy them, 2) probability functions, that are special cases of belief functions, are insufficiently expressive to represent degrees of belief and 3) Choquet capacities [1], that are the generalization of belief functions, violate some of the requirements.

During the demonstration, we produce requirements from which we prove 1) the convexity of the set of credibility functions, 2) how credibility functions are adapted by uninformative modifications of the frame of discernment (refinement and coarsening), and 3) how they are revised by conditioning. At that level, we prove that Cr is a Choquet capacity monotone of order 2 [1]. To show that it is monotone of infinite order (i.e., a belief function), we propose several extra requirements. One of them is based on the negation of a belief function, a concept invented by Dubois and Prade [2]. It states: 'the negation of a credibility function is a credibility function'. In that case credibility functions are belief functions.

Unfortunately, this negation concept was only a mathematical property. To be used as a rationality requirement, one must produce at least one practical illustrative and convincing example where the negation is used.

In this paper, we present such an example. Thanks to it, our axiomatic justification presented in [8, 10], is simplified.

2 Credibility functions

Let Cr represent Your belief over Ω, a finite frame of discernment. The only properties of Cr used in this paper are:

1. **Bounded non negativity:** $Cr : 2^\Omega \to [0, 1]$ where $Cr(\emptyset) = 0$ and $Cr(\Omega) \leq 1$

2. **Monotony to inclusion:** $\forall A, B \subseteq \Omega$, if $A \subseteq B$, then $Cr(A) \leq Cr(B)$

3. **Revision:** the revision of Cr by a piece of evidence Ev is represented by a $2^{|\Omega|} \times 2^{|\Omega|}$ matrix H^* which depends on Ev but not on Cr, and the revised credibility function $Cr[Ev]$ is given by the matricial product $Cr[Ev] = H^* \cdot Cr$.

The first requirement is quite strong as it eliminates models based on sets of probability functions [5, 6, 13] or on interval valued probabilities [14]. We accept the closed world assumption ($Cr(\Omega) = 1$) in order to avoid useless discussions. The second requirement is assumed by any model of uncertainty and hardly questionable. The third requirement translates into the belief function framework the transformation achieved by a Markow matrix in classical probability theory. It satisfies quite natural requirements, and could almost be just assumed, what we do here. Proving that it is a necessary property will be presented in a forthcoming paper. That H^* is a stochastic matrix can be deduced when Cr is a belief function. But as far as the purpose of this paper is to prove that Cr is a belief function, we can neither assume nor deduce it.

We present some needed background material (see [12, 11]).

2.1 Notation convention

$Cr^\Omega[Ev](B)$ is the degree of belief given by You that the actual world belongs to B, which is a subset of Ω, Ev is a set of propositions (called the Evidential Corpus) and You accept as true the propositions deduced from those in Ev (Ev covers the classical conditioning event). The domain will be omitted when no confusion can occur.

The term between [and] is what You accept as true. In particular, it can be the conditioning event encountered in probability theory. Note that Your beliefs are based on what You accept as true, not on what is true. There is no necessity that what You accept as true is true, it might perfectly be false. Your beliefs would be 'unjustified', 'inadequate', 'erroneous', but so it is. Rationality is trying to accept only what is true, but this is only an ideal goal, and daily reality is far from that ideal.

By convention when we write $Cr^\Omega[\omega]$ for $\omega \in \Omega$, we mean that You accept as true that $\omega_0 \in \omega$ and do not accept as true that $\omega_0 \in \omega^*$ for any $\omega^* \subset \omega$ (where $\subset$ denotes strict subset). We will say that ω is 'all You accept as true'.

A credibility function can itself be part of the evidential corpus. For example, we write $Cr^\Omega[\omega, Cr^\Theta]$ to mean that You accept as true both that $\omega_0 \in \omega$ and that Your beliefs about Θ is represented by Cr^Θ.

2.2 Doxastic equivalence and consistency

The next definition translates the idea that if two propositions are equivalent given what You know, the credibility functions they induce are equal.

Definition 2.1 *Doxastic equivalence. Let Ev be an 'Evidential Corpus', that is a set of propositions that You accept as true. Two propositions p_1 and p_2 are said to be doxastically equivalent under an Evidential Corpus Ev, what is denoted by $p_1 \equiv_{Ev} p_2$, when $p_1 \wedge Ev$ and $p_2 \wedge Ev$ are logically equivalent.*

We then assume:

Proposition 2.1 *Doxastic consistency. If $p_1 \equiv_{Ev} p_2$, then*

$$Cr^\Omega[Ev](p_1) = Cr^\Omega[Ev](p_2).$$

2.3 The Möbius transform

Let Cr^Ω be a credibility function defined on Ω, its Möbius transform, denoted m^Ω is defined as:

$$m^\Omega(A) = \sum_{B \subseteq A} (-1)^{|A|-|B|} Cr^\Omega(B), \forall A \subseteq \Omega$$

We call m^Ω the Möbius mass function and $m(A)$ a Möbius mass. In particular, the basic belief assignment is the Möbius transform of a belief function. Cr^Ω is a belief function iff $m^\Omega(A) \geq 0$ for all $A \subseteq \Omega$ and $\sum_{A \subseteq \Omega} m^\Omega(A) = 1$.

2.4 Coarsening

Let Ω and Ω^* be two frame of discernments where the elements of Ω^* are the elements of a partition of Ω. For $B \subseteq \Omega$, let $Coars(B)$ denote the smallest subset of Ω^* that contains B. We call Ω^* an uninformative (it means 'just redefining the frame') coarsening of Ω. Given Cr^Ω on Ω, $Cr^{\Omega^*}(A) = Cr^\Omega(A)$ for all $A \subseteq \Omega^*$ (if $A \subseteq \Omega^*$, then $A \subseteq \Omega$). It results from the doxastic consistency requirement. In that case,

$$\forall A \subseteq \Omega^*, m^{\Omega^*}(A) = \sum_{B \subseteq \Omega, Coars(B)=A} m^\Omega(B).$$

2.5 All You accept as true

Let the frame of discernment Ω, and suppose all You accept as true is that $\omega_0 \in \omega$ for an $\omega \subseteq \Omega$. What is the credibility function $Cr^\Omega[\omega]$ induced on Ω under that condition? Requiring $Cr^\Omega[\omega](\omega) = 1$ and $Cr^\Omega[\omega](\overline{\omega}) = 0$ seem natural.

What about You beliefs about $\omega^* \subset \omega$? It seems also natural to require that all strict subsets ω^* of ω receive the same belief. For instance, why should any of them be better supported that it complement relative to ω? The concept of cardinality of the set ω cannot be used, as beliefs would otherwise violate the doxastic consistency requirement (see [10]). Let β be that particular value. The term β cannot be negative as it is among others the belief given to the singletons of ω. If furthermore we require that beliefs can never be smaller than β, then $\beta = 0$ as for $\overline{\omega}$, $Cr^\Omega[\omega](\overline{\omega}) = 0$.

Finally, the beliefs given to any ω^* is equal to the beliefs given to $\omega^* \cap \omega$, as the worlds in ω^* but not in $\omega^* \cap \omega$ belong to $\overline{\omega}$ and are thus accepted by You as impossible given You accept ω as true.

The next proposition express these ideas:

Proposition 2.2 *The credibility function that represents Your beliefs given all You accept as true is $\omega_0 \in \omega$ is given by:*

$$\begin{aligned}[2]Cr^\Omega[\omega](\omega^*) &= 1 & &\text{if } \omega \subseteq \omega^* \\ &= 0 & &\text{if } \omega^* \subset \omega \\ &= Cr^\Omega[\omega](\omega^* \cap \omega) & &\text{otherwise}\end{aligned}$$

The Möbius transform of $Cr^\Omega[\omega]$ is given by:

$$\begin{aligned}[2]m^\Omega[\omega](\omega^*) &= 1 & &\text{if } \omega = \omega^* \\ &= 0 & &\text{otherwise}\end{aligned}$$

2.6 Revision

The next theorem reexpresses the Cr revision requirement under an equivalent but more convenient form. We had assumed that the revision of a credibility function Cr by a piece of evidence Ev can be represented by a matrix H^* such that the revised credibility function $Cr[Ev]$ is equal to $H^* \cdot Cr$.

Theorem 2.1 *Let Cr and Cr^* be two credibility functions defined on Ω. If $Cr^* = H^* \cdot Cr$, there exists a $2^{|\Omega|} \times 2^{|\Omega|}$ matrix H such that $Cr^* = H \cdot m$ where m is the Möbius transforms of Cr. In that case,*

$$Cr^*(A) = \sum_{B \subseteq \Omega} h(A,B)m(B), \forall A \subseteq \Omega$$

where $h(A,B)$ are the elements of the matrix H.

Proof. Let M be the operator (a matrix) that transforms any credibility function on Ω into its Möbius transforms [11]. M is not singular, so M^{-1} exists. $Cr^* = H^* \cdot Cr$ can be rewritten as $Cr^* = H^* \cdot M^{-1} \cdot M \cdot Cr$, thus $Cr^* = H \cdot m$ where $H = H^* \cdot M^{-1}$. The equation is just a rewriting of the matricial equality. □

In order to prove that Cr is a belief function, we must produce an example that shows that if some of the values of m are negative, there exists a H matrix such that some values of Cr^* are negative. To produce such a matrix is mathematically trivial, but the challenge was to find a practical example that leads to such a matrix. This is what we achieve in the next section.

3 The Mischievous Killer

We consider here only the proof that the Möbius mass given to the Ω when $|\Omega| = 3$ may not be negative. The case with 2 was proved in [10]. The real challenge was to go from 2 to 3. So we consider that case first.

Our example is based on a murder scenario, but could as well be rephrased as any diagnostic - detection problem.

3.1 The scenario

Suppose a murder has been committed by a single killer, denoted k_0, and there are three suspects named A, B, and C. We denote by D 'anybody else'. So You know for sure that the killer $k_0 \in \Omega$ with $\Omega = \{A, B, C, D\}$.

You collect a piece of evidence, a cigarette butt which brand, denoted θ_0. The domain for θ_0 is $\Theta = \{a, b, c, d\}$. The butt you observe can only be one of $\{a, b, c\}$. You look at the butt and build a belief Cr^Θ about the actual value θ_0.

The CB piece of evidence. You know the next piece of evidence, denoted CB for cigarette butt,

1. $\theta_0 = a$ iff $k_0 = A$
2. $\theta_0 = b$ iff $k_0 = B$
3. $\theta_0 = c$ iff $k_0 = C$
4. $\theta_0 = d$ iff $k_0 = D$
5. $\theta_0 \in \{a, b, c\}$

Using the doxastic consistency property, Cr^Θ induces a credibility function $Cr^\Omega[CB]$ about k_0 given by:

$$Cr^\Omega[CB](A) = Cr^\Theta(a),$$
$$Cr^\Omega[CB](B) = Cr^\Theta(b),$$
$$Cr^\Omega[CB](C) = Cr^\Theta(c),$$
$$Cr^\Omega[CB](A, B) = Cr^\Theta(a, b),$$
$$Cr^\Omega[CB](A, C) = Cr^\Theta(a, c),$$
$$Cr^\Omega[CB](B, C) = Cr^\Theta(b, c),$$
$$Cr^\Omega[CB](A, B, C) = Cr^\Theta(a, b, c),$$
$$Cr^\Omega[CB](D, \omega) = Cr^\Omega[CB](\omega), \quad \forall \omega \subseteq \{A, B, C\}$$

The FT piece of evidence. Now You learn for sure that if the killer was one of A, B, C, the killer would manage to create a false track. Cigarettes a, b and c can be recognized because, respectively, they have the letters XY, XZ and YZ written on them. The killer will purposely drop a butt that points to the other suspects and surely not to him. For example, if the killer was A, A would have managed to let a butt with XZ or YZ or Z written on it, the last case corresponding to the case where A has erased the missing letter. In the three cases, the butt does not point to A.

What D would do is unknown to You.

Let W be the predicate 'You accept as true that exactly ... is written on the butt', so $W(Z)$ means that 'You accept as true that exactly Z is written on the butt' (and thus nothing more, what means in fact that the butt is either a b or a c butt). This information can be written as:

1. If $k_0 = A$ then $W(XZ) \vee W(YZ) \vee W(Z)$
2. If $k_0 = B$ then $W(XY) \vee W(YZ) \vee W(Y)$
3. If $k_0 = C$ then $W(XY) \vee W(XZ) \vee W(X)$

We write $W()$ to express that nothing is written on the butt. The proposition
$W() \vee W(X) \vee W(Y) \vee W(Z) \vee W(XY) \ldots$
$\vee W(XZ) \vee W(YZ)$
is true. So we have:
$\neg(W(XZ) \vee W(YZ) \vee W(Z)) = \ldots$
$W() \vee W(XY) \vee W(X) \vee W(Y)$
and similarly with the other two consequences.

The three rules can be rewritten as

1. If $W() \vee W(XY) \vee W(X) \vee W(Y)$ then $k_0 \in \{B, C, D\}$
2. If $W() \vee W(XZ) \vee W(X) \vee W(Z)$ then $k_0 \in \{A, C, D\}$
3. If $W() \vee W(YZ) \vee W(Y) \vee W(Z)$ then $k_0 \in \{A, B, D\}$

Let this piece of evidence be denoted by FT (from False Track).

The problem is to build $Cr^{\Omega}[FT]$ and to find out what is the matrix H that transforms $Cr^{\Omega}[CB]$ into $Cr^{\Omega}[FT]$. This revision is achieved by a contraction on CB and an expansion by FT [3]. The whole subtlety of the example comes from the fact that both $Cr^{\Omega}[CB]$ and $Cr^{\Omega}[FT]$ can be deduced from Cr^{Θ} and that Cr^{Θ} is not changed by the pieces of evidence CB or FT. It results from the fact that Your beliefs about the cigarette brand is not affected by what it implies on who is the killer.

3.2 Building the H matrix

Case 1. Suppose You accept as true that the butt is a c butt, thus $W(YZ)$. So the antecedent of the third rule in FT is true and You deduce $k_0 \in \{A, B, D\}$.

In that case, Your belief state is represented by $Cr^{\Theta}(c) = 1$, in which case:

	A	B	C	AB	AC	BC	ABC
A	0	0	0	0	0	0	0
B	0	0	0	0	0	0	0
C	0	0	0	0	0	0	0
A,B	0	0	0	0	0	0	0
A,C	0	0	0	0	0	0	0
B,C	0	0	0	0	0	0	0
A,B,C	0	0	0	0	0	0	0
D	0	0	0	0	0	0	1
A,D	0	0	0	0	0	1	1
B,D	0	0	0	0	1	0	1
C,D	0	0	0	1	0	0	1
A,B,D	0	0	1	0	1	1	1
A,C,D	0	1	0	1	0	1	1
B,C,D	1	0	0	1	1	0	1
A,B,C,D	1	1	1	1	1	1	1

Table 1: The H matrix to transform $Cr^\Omega[CB]$ into $Cr^\Omega[FT]$.

- before revision on FT, Your beliefs were represented by

$$Cr^\Omega[CB, W(YZ)](\omega) = 1 \text{ if } C \in \omega$$
$$= 0 \text{ otherwise}$$

which Möbius transform is given by

$$m^\Omega[CB, W(YZ)](\omega) = 1 \text{ if } \omega = C$$
$$= 0 \text{ otherwise.}$$

- after revision on FT, it becomes

$$Cr^\Omega[FT, W(YZ)](\omega) = 1 \text{ if } \{A, B, D\} \in \omega$$
$$= 0 \text{ otherwise}$$

In that case the coefficients of the $\{C\}$ column of H satisfy

$$h(\omega, \{C\}) = 1 \text{ if } \{A, B, D\} \subseteq \omega$$
$$= 0 \text{ otherwise}$$

The same holds up to a permutation with a and b (see Table 1).

Case 2. Suppose You are sure that Z is written on the butt, but You could not recognize the other letter. So You accept $W(Z)$ as true. It means in fact that You are sure the butt is a b or a c butt and You have no idea to decide if it is a b or if it is a c. So the antecedents of the second and third rules in FT are satisfied and You deduce $k_0 \in \{A,C,D\} \cap \{A,B,D\} = \{A,D\}$

In that case, Your belief state on Θ is represented by $Cr^\Theta(\theta) = 1$ if $\{b,c\} \subseteq \theta$, and 0 otherwise. (see proposition 2.2) in which case:

- before revision on FT, the unique non mull mass of the Möbius transform of Your beliefs $Cr^\Omega[CB,W(Z)]$ was $m^\Omega[CB,W(Z)](B,C) = 1$

- after revision on FT, Your beliefs become: $Cr^\Omega[FT,W(Z)](\omega) = 1$ if $\{A,D\} \subseteq \omega$ and 0 otherwise.

In that case the coefficients of the $\{B,C\}$ column of H satisfy

$$h(\omega, \{B,C\}) = 1 \text{ if } \{A,D\} \subseteq \omega$$
$$= 0 \text{ otherwise.}$$

The same hold up to a permutation with a and b (see Table 1).

Case 3. Suppose You are sure that something is written on the butt, but You could not recognize any letter. It means in fact that You are sure the butt is a a or a b or a c butt and nothing more. So you accept W() as true. So the antecedents of the three rules in FT are satisfied and You deduce $k_0 \in \{A,C,D\} \cap \{A,B,D\} \cap \{B,C,D\} = \{D\}$ This implies that $Cr^\Theta(\theta) = 1$ if $\theta = \Theta$ and 0 otherwise, in which case:

- before revision on FT, the unique non mull mass of the Möbius transform of Your beliefs $Cr^\Omega[CB,W()]$ was $m^\Omega[CB,W()](A,B,C) = 1$

- after revision on FT, Your beliefs become: $Cr^\Omega[FT,W()](\omega) = 1$ if $\{D\} \subseteq \omega$ and 0 otherwise.

In that case the coefficients of the $\{A,B,C\}$ column of H satisfy

$$h(\omega, \{A,B,C\}) = 1 \text{ if } \{D\} \subseteq \omega$$
$$= 0 \text{ otherwise.}$$

The nature of the columns of H for those ω that contain D is not required here as $m^\Omega[CG, m^\Theta](\omega) = 0$ for all ω such that $D \in \omega$.

3.3 General belief on Θ

In general, Your belief on Θ is represented by a credibility function Cr^Θ with m^Θ being its Möbius transform. Under CB, this credibility function on Θ induced a credibility function $Cr^\Omega[CB,Cr^\Theta]$ on Ω. Its Möbius transform is given by: $m^\Omega[CB,Cr^\Theta](\omega) = m^\Theta(\theta)$ where ω contains the same letters as θ, but capitalized, and $m^\Omega[CB,Cr^\Theta](\omega) = 0$ if $D \in \omega$.

Suppose $m^\Theta(\theta) < 0$ for some $\theta \subseteq \Theta$. We consider only two cases: either $\theta = \Theta$ or θ satisfies $|\theta| = |\Theta| - 1$.

Suppose $m^\Theta(\Theta)$ is negative. Then $m^\Omega[CB,Cr^\Theta](\{A,B,C\}) = m^\Theta(\Theta) < 0$. Apply the H transform. That negative mass is allocated to the belief given under FT to

D, and no other mass is given to D, so $Cr^\Omega[FT, Cr^\Theta](D) < 0$ what violates the non negativity of the credibility functions. Hence $m^\Theta(\Theta)$ may not be negative.

Suppose $\theta = \{a, b\}$ and $m^\Theta(\{a, b\})$ is negative. Then $m^\Omega[CB, Cr^\Theta](\{A, B\}) = m^\Theta(\{a, b\}) < 0$. Apply the H transform. We have

$$Cr^\Omega[FT, Cr^\Theta](D) = m^\Omega[CB, Cr^\Theta](\{A, B, C\}) \geq 0$$

as just shown and

$$Cr^\Omega[FT, Cr^\Theta](\{C, D\}) = m^\Omega[CB, Cr^\Theta](\{A, B, C\}) + m^\Omega[CB, Cr^\Theta](\{A, B\}).$$

The negativity of $m^\Omega[CB, Cr^\Theta](\{A, B\})$ implies that

$$Cr^\Omega[FT, Cr^\Theta](\{C, D\}) \leq Cr^\Omega[FT, Cr^\Theta](\{D\})$$

what violates the monotony to inclusion requirement. Hence $m^\Theta(\{a, b\})$ may not be negative.

4 Generalization to any Ω

The previous example can be extended to any set Θ. It is just a matter of rephrasing it accordingly by multiplying the number of cigarette brands and of suspects. So what it proves is that for any Θ, $m^\Theta(\theta) \geq 0$ whenever $|\theta| \geq |\Theta| - 1$.

To prove that credibility functions are belief functions, all we need is to show that if a Möbius mass of the credibility function is negative, we can generate an example where such a credibility function would induce another credibility function where some values are negative.

Suppose a credibility function Cr^Ω with m^Ω its Möbius masses. Suppose that for $\omega \subseteq \Omega$, $m^\Omega(\omega) < 0$. Three cases must be considered. Either $\omega = \Omega$ or $\omega \subset \Omega, \omega \neq \emptyset$ or $\omega = \emptyset$.

Case $\omega = \emptyset$. In that case, $Cr^\Omega(\Omega) > 1$ contrary to the boundness requirement.

Case $\omega = \Omega$. We already know that this case is not acceptable.

Case $\omega \subset \Omega, \omega \neq \emptyset$. Build a coarsening Ω^* of Ω such that the elements of $\overline{\omega}$ are mapped into one singleton of Ω^*, the others being mapped into themselves. Then $m^{\Omega^*}(\omega) = m^\Omega(\omega)$ is negative. Thus we have build a credibility function on Ω^* that allocates a negative mass to a subset of Ω^* such that $|\omega| = |\Omega^*| - 1$. We know that this case is not acceptable.

Therefore the Möbius mass of any credibility function must be non negative, thus credibility functions are belief functions.

4.1 The Negation of a belief function

Dubois and Prade [2] have introduced the concept of the negation of a belief function. The basic belief assignment of the negation of a belief function is obtained by transferring the basic belief mass $m^\Omega(\omega)$ to $m^\Omega(\overline{\omega})$ for all $\omega \subseteq \Omega$. In [9], we have shown that this transformation fits with the idea of a source of evidence that is an absolute liar. Our present example illustrates how to generate such a belief function.

5 Conclusions

We show that if revision of any measure representing quantified beliefs can be represented by a matrix multiplication of the initial beliefs, then the Möbius mass related to the measure must be non negative. This implies that any measure representing quantified beliefs is a belief function. The challenge was to produce an example that produces this effect. Our Mischievous Killer story provides such an example. During this derivation we also illustrate how one can produce the negation of a belief function.

References

[1] CHOQUET, G. Theory of capacities. *Annales de l'Institut Fourier, Université de Grenoble 5* (1953), 131–296.

[2] DUBOIS, D., AND PRADE, H. A set-theoretic view of belief functions: logical operations and approximations by fuzzy sets. *International Journal of General Systems 12* (1986), 193–226.

[3] GÄRDENFORS, P. *Knowledge in Flux: Modelling the Dynamics of Epistemic States.* MIT Press, Cambridge, Mass, 1988.

[4] KOHLAS, J., AND MONNEY, P. A. Theory of evidence: A survey of its mathematical foundations, applications and computations. *ZOR-Mathematical Methods of Operatioanl Research 39* (1994), 35–68.

[5] KYBURG, JR., H. E. *Probability and the Logic of Rational Belief.* Wesleyan Univ. Press, 1961.

[6] LEVI, I. *The Enterprise of Knowledge.* MIT Press, Cambridge, Mass, 1980.

[7] SHAFER, G. *A Mathematical Theory of Evidence.* Princeton Univ. Press. Princeton, NJ, 1976.

[8] SMETS, P. Quantifying beliefs by belief functions: an axiomatic justification. In *Proceedings of the 13th International Joint Conference on Artificial Intelligence* (San Mateo, CA, 1993), Morgan Kaufmann, pp. 598–603.

[9] SMETS, P. The alpha-junctions: combination operators applicable to belief functions. In *Qualitative and quantitative practical reasoning*, D. Gabbay, R. Kruse, A. Nonnengart, and H. Ohlbach, Eds. Springer, 1997, pp. 131–153.

[10] SMETS, P. The normative representation of quantified beliefs by belief functions. *Artificial Intelligence 92* (1997), 229–242.

[11] SMETS, P. The transferable belief model for quantified belief representation. In *Handbook of Defeasible Reasoning and Uncertainty Management Systems* (1998), D. M. Gabbay and P. Smets, Eds., vol. 1, Kluwer, Doordrecht, The Netherlands, pp. 267–301.

[12] SMETS, P., AND KENNES, R. The transferable belief model. *Artificial Intelligence 66* (1994), 191–234.

[13] VOORBRAAK, F. *As Far as I Know: Epistemic Logic and Uncertainty*. Dissertation, Utrecht University, 1993.

[14] WALLEY, P. *Statistical Reasoning with Imprecise Probabilities*. Chapman and Hall, London, 1991.

[15] WONG, S. K. M., YAO, Y. Y., BOLLMANN, P., AND BRGER, H. C. Axiomatization of qualitative belief structure. *IEEE Trans. SMC 21* (1990), 726–734.

Extension of coherent lower previsions to unbounded random variables

Matthias C. M. Troffaes, Gert de Cooman

Universiteit Gent, Onderzoeksgroep SYSTeMS,
Technologiepark – Zwijnaarde 914, 9052 Zwijnaarde, Belgium
{matthias.troffaes,gert.decooman}@rug.ac.be

Abstract

We consider the extension of coherent lower previsions from the set of bounded random variables to a larger set. An *ad hoc* method in the literature consists in approximating an unbounded random variable by a sequence of bounded ones. Its 'extended' lower prevision is then defined as the limit of the sequence of lower previsions of its approximations. We identify the random variables for which this limit does not depend on the details of the approximation, and call them previsible. We thus extend a lower prevision to previsible random variables, and we study the properties of this extension.

Keywords: imprecise probabilities, lower prevision, coherence, unbounded random variable, Dunford integral

1 Introduction

When modelling a system, it often occurs that we do not know all of its aspects, or that we wish to discard certain details in order to simplify the system analysis. These situations give rise to uncertainty, and consequently, we are challenged to find an appropriate system description which takes the uncertainty into account. One particularly successful way of doing so, consists in using a probabilistic description of the uncertainty. Despite its many successes, however, there are quite a number of situations in which this method does not lead to sensible results, simply because there may not be sufficient information available to allow us to select a *single* probability distribution as an appropriate model. There are in the literature a number of uncertainty models that do not assume the uncertainty to be described by a single probability distribution. Among these, Walley's behavioural theory of imprecise probabilities [12] stands out as a very satisfactory choice, certainly from the foundational point of view. It has a clear behavioural interpretation, which leads naturally to a theory of decision making under uncertainty. Moreover, it unifies a large number of other uncertainty models, such as 2-monotone capacities [2], possibility measures [3, 13], comparative and modal probabilities [11], and convex sets of probability measures [8].

One important shortcoming of the existing theory of coherent lower previsions is that it only deals with random variables that are bounded, whereas in engineering, for instance, applications involving unbounded random variables abound. To give only a few examples, the following classes of problems would certainly benefit from an extension of imprecise probability theory able to deal with unbounded random variables: (i) the estimation of unbounded quantities, such as the time to failure of a component in a system [9]; and (ii) optimisation involving an unbounded cost [4]. An intuitive, *ad hoc* way of dealing with an unbounded random variable is to approximate it by a sequence of bounded ones, and to use limit arguments in order to extend notions defined in the context of the bounded random variables to their unbounded counterparts, in the hope that the eventual result will not depend on the exact form of the approximation. Similar types of construction exist in the theory of integration—we shall use them as a source of inspiration.

Our main objectives in this paper are twofold: (i) to construct an extension of coherent lower previsions from bounded random variables to a larger set; and (ii) to study of the properties of this extension in order to motivate that its result can be seen as a coherent lower prevision in its own right. The paper is organised as follows. We give a brief introduction to the theory of imprecise probabilities in Section 2. In Section 3 we introduce the basic concepts of our extension. Important properties are listed in Section 4. Finally, in Section 5 we show that for linear previsions, there is a Dunford integral representation for their extension to unbounded random variables.

2 Imprecise probabilities

We start with a brief introduction to the most important aspects of the existing behavioural theory of imprecise probabilities that are relevant to the problem at hand. More details can be found in [12]. Consider an agent who is uncertain about something, say, the outcome of some experiment. If the set of possible outcomes is Ω, then a *random variable* is a mapping from Ω to $\mathbb{R}$, and it is interpreted as an uncertain reward: if $\omega \in \Omega$ turns out to be the actual outcome of the experiment then the agent receives the amount $X(\omega)$, expressed in units of some linear utility. Bounded random variables are also called *gambles*. They play a very important part in the existing theory. Denote the set of all gambles by $\mathscr{L}(\Omega)$.

The information the agent has about the outcome of the experiment will lead him to accept or reject transactions whose reward depends on this outcome. We can formulate a model for his uncertainty by looking at a specific type of transaction: buying gambles. The agent's *lower prevision* (or supremum acceptable buying price) $\underline{P}(X)$ for a gamble X is the highest price s such that he is disposed to buy the gamble X for any price strictly lower than s. If the agent assesses a supremum acceptable buying price for every gamble X in a subset $\mathcal{K}$ of $\mathscr{L}(\Omega)$, the resulting mapping $\underline{P} : \mathcal{K} \to \mathbb{R}$ is called a *lower prevision*.

It can be argued that $\underline{P}$ must satisfy the following rationality constraint: for all $n \in \mathbb{N}$, all $\lambda_0, \ldots, \lambda_n \geq 0$, and all $X_0, \ldots, X_n \in \mathcal{K}$ we must have that[1]

$$\sup\left[\sum_{i=1}^{n} \lambda_i X_i - \lambda_0 X_0\right] \geq \sum_{i=1}^{n} \lambda_i \underline{P}(X_i) - \lambda_0 \underline{P}(X_0).$$

Here and elsewhere, we denote by $\sup[X]$ the supremum value $\sup_{\omega \in \Omega} X(\omega)$ of the gamble X (and similarly for $\inf[X]$). If the lower prevision $\underline{P}$ satisfies this constraint, we say that it is *coherent*. If $\mathcal{K}$ is a linear space, e.g., when $\mathcal{K} = \mathscr{L}(\Omega)$, then $\underline{P}$ is coherent if and only if

---
[1] For example, take $n = 0$ and $\lambda_0 = 1$, then we find that $\underline{P}(X) \geq \inf[X]$, which means that the agent should be willing to pay at least the lowest possible reward.

(i) $\underline{P}(X) \geq \inf[X]$,

(ii) $\underline{P}(\lambda X) = \lambda \underline{P}(X)$, and

(iii) $\underline{P}(X + Y) \geq \underline{P}(X) + \underline{P}(Y)$.

for all gambles X, Y in $\mathcal{K}$ and $\lambda \geq 0$. This result can be given a simple and natural interpretation: the supremum buying prices should accept a sure gain, they should be independent of the utility scale, and finally, if we are willing to buy X for price s and Y for price t, then we should also be willing to buy $X + Y$ for price $s + t$. The following consequences of coherence will also be used in proofs further on (see [12] for details).

(i) $\overline{P}(X + Y) \leq \overline{P}(X) + \overline{P}(Y)$

(ii) $X \leq Y \implies \underline{P}(X) \leq \underline{P}(Y)$ and $\overline{P}(X) \leq \overline{P}(Y)$

(iii) $|\underline{P}(X) - \underline{P}(Y)| \leq \overline{P}(|X - Y|)$ and $|\overline{P}(X) - \overline{P}(Y)| \leq \overline{P}(|X - Y|)$

It can be shown that if $\underline{P}$ is coherent, there is always a (unique) smallest coherent extension of $\underline{P}$ from its domain $\mathcal{K}$ to $\mathcal{L}(\Omega)$. We call this extension the *natural extension* of $\underline{P}$. It is given by

$$\underline{E}(X) = \sup \left\{ \inf \left[X - \sum_{i=1}^{n} \lambda_i \left[X_i - \underline{P}(X_i) \right] \right] \right\},$$

where $X \in \mathcal{L}(\Omega)$ and the supremum runs over $n \in \mathbb{N}$, $\lambda_1, \ldots, \lambda_n \geq 0$ and $X_1, \ldots, X_n$ in $\mathcal{K}$. This shows that without loss of generality, we may from now on assume that $\underline{P}$ is a coherent lower prevision defined on all of $\mathcal{L}(\Omega)$.

$\overline{P}$ denotes the conjugate *upper prevision* of $\underline{P}$, and is defined by $\overline{P}(X) = -\underline{P}(-X)$ for every $X \in \mathcal{L}(\Omega)$. $\overline{P}(X)$ represents the agent's infimum acceptable selling price for X. The difference $\overline{P}(X) - \underline{P}(X)$ measures the amount of imprecision in the agent's behavioural dispositions toward the gamble X. An *event* A is a subset of Ω. It will be identified with its indicator I_A, which is a gamble.[2] The *lower probability* $\underline{P}(A)$ is then defined as the lower prevision $\underline{P}(I_A)$ of its indicator I_A, and similarly for the upper probability $\overline{P}(A)$.

If it so happens that $\overline{P}(X) = \underline{P}(X)$ for every gamble X, then $\underline{P}$ is called a *linear prevision*, and it is denoted by P. Linear previsions are linear functionals on the linear space $\mathcal{L}(\Omega)$ that are positive and have unit norm ($P(I_\Omega) = 1$). They are the *fair prices* or *previsions* in the sense of de Finetti [6, 7]. The restriction of a linear prevision P to events is a finitely additive probability (also called a probability charge [1]) and $P(X)$ is equal to the expected value of the bounded random variable X with respect to this probability charge (see for instance [1, Theorem 4.7.4]). In this way, any Bayesian model can be considered to be a linear prevision, which is a special kind of lower prevision. The set of all linear previsions on $\mathcal{L}(\Omega)$ is denoted by $\mathcal{P}(\Omega)$.

$\mathcal{M}(\underline{P})$ will denote the set of all linear previsions that dominate $\underline{P}$ point-wise on $\mathcal{L}(\Omega)$: $\mathcal{M}(\underline{P}) = \{Q \in \mathcal{P}(\Omega) : Q \geq \underline{P}\}$. One can show that $\mathcal{M}(\underline{P})$ is a non-empty,

---

[2] I_A is the random variable that takes the value 1 on A and 0 elsewhere.

convex and compact[3] subset of $\mathscr{P}(\Omega)$, and that $\underline{P}$ is the lower envelope of $\mathcal{M}(\underline{P})$, that is,

$$\underline{P}(X) = \min_{Q \in \mathcal{M}(\underline{P})} Q(X),$$

for all $X \in \mathscr{L}(\Omega)$. This equality, and the fact that the lower envelope of any non-empty set of linear previsions is a coherent lower prevision, gives rise to what is called the *Bayesian sensitivity analysis interpretation*, or *Quasi-Bayesian interpretation* of lower previsions: specifying a coherent lower prevision is formally equivalent to specifying a non-empty, convex and compact set of linear previsions (or probability charges).

3 Previsibility

In the previous section, we have seen that within the existing framework it is possible to extend any given coherent lower prevision to the set of all bounded random variables in a natural way. We now investigate whether it can be extended still further to a larger set that includes some unbounded random variables. The first step of our investigation will be the construction of a limit procedure—approximating unbounded random variables by bounded ones—taking the necessary care to ensure that the procedure yields a unique result: we do not want our result to depend on the details of the approximation.

We begin therefore by defining the $\underline{P}$-*norm* of a gamble X by $\|X\|_{\underline{P}} = \overline{P}(|X|)$. Using the coherence of $\underline{P}$, it can be shown that $\|\cdot\|_{\underline{P}}$ is a semi-norm on $\mathscr{L}(\Omega)$. A sequence (X_n) of gambles is called $\underline{P}$-*fundamental*[4] if it is Cauchy with respect to $\|\cdot\|_{\underline{P}}$, i.e., if $\|X_n - X_m\|_{\underline{P}} \to 0$ as $n, m \to \infty$. We say that a sequence of gambles (X_n) converges $\underline{P}$-*hazily*[5] to the random variable X if for every $\epsilon > 0$ we have that

$$\lim_{n \to \infty} \overline{P}(\{\omega \in \Omega : |X(\omega) - X_n(\omega)| > \epsilon\}) = 0.$$

Observe that we do not need to impose any measurability conditions, since $\overline{P}(A)$ is defined for every $A \subseteq \Omega$. The following lemma is the basic result that will guarantee the unicity of the extension introduced in Definition 1.

Lemma 1. *If (X_n) and (Y_n) are $\underline{P}$-fundamental sequences of gambles converging $\underline{P}$-hazily to the same random variable Z, then it holds that the limits $\lim_{n \to \infty} \overline{P}(X_n)$ and $\lim_{n \to \infty} \overline{P}(Y_n)$ exist, are finite real numbers and coincide, and similarily, the limits $\lim_{n \to \infty} \underline{P}(X_n)$ and $\lim_{n \to \infty} \underline{P}(Y_n)$ exist, are finite real numbers and coincide.*

Proof. We first prove that the limits exist and are finite. This follows from the following inequalities, which are consequences of the coherence of $\underline{P}$,

$$|\overline{P}(X_n) - \overline{P}(X_m)| \leq \overline{P}(|X_n - X_m|), \quad |\underline{P}(X_n) - \underline{P}(X_m)| \leq \overline{P}(|X_n - X_m|),$$
$$|\overline{P}(Y_n) - \overline{P}(Y_m)| \leq \overline{P}(|Y_n - Y_m|), \quad |\underline{P}(Y_n) - \underline{P}(Y_m)| \leq \overline{P}(|Y_n - Y_m|).$$

---

[3] We assume in this paper that $\mathscr{P}(\Omega)$ is provided with its topology of point-wise convergence: the relativisation to $\mathscr{P}(\Omega)$ of the weak*-topology on the topological dual $\mathscr{L}(\Omega)^*$, where $\mathscr{L}(\Omega)$ is provided with the supremum norm topology.
[4] Cf. *mean fundamental* in the theory of measures.
[5] Cf. *convergence in measure* in measure theory, and *hazy convergence* [1] in the theory of finitely additive measures.

Since the right hand sides converge to zero, the left hand sides must converge to zero too. This means that $\overline{P}(X_n)$, $\underline{P}(X_n)$, $\overline{P}(Y_n)$ and $\underline{P}(Y_n)$ are Cauchy sequences. By the completeness of $\mathbb{R}$, their limits exist and are finite real numbers.

Next, we prove that $\lim_{n\to\infty} \overline{P}(X_n) = \lim_{n\to\infty} \overline{P}(Y_n)$ and $\lim_{n\to\infty} \underline{P}(X_n) = \lim_{n\to\infty} \underline{P}(Y_n)$. Let $N_n := |X_n - Y_n|$. Again by the coherence of $\underline{P}$, we have that $|\underline{P}(X_n) - \underline{P}(Y_n)| \leq \overline{P}(N_n)$ and $|\overline{P}(X_n) - \overline{P}(Y_n)| \leq \overline{P}(N_n)$. The proof is complete if we can show that $\overline{P}(N_n)$ converges to zero. For every $n \in \mathbb{N}$ and every $A \subseteq \Omega$, define $a_n(A) := \overline{P}(N_n A)$. We must prove that $\lim_{n\to\infty} a_n(\Omega) = 0$.

Every a_n is an element of the function space $\mathbb{R}^{\wp(\Omega)}$. Equip this space with the topology of uniform convergence on $\wp(\Omega)$. Note that by the completeness of $\mathbb{R}$, it follows by that $\mathbb{R}^{\wp(\Omega)}$ is complete with respect to the topology of uniform convergence on $\wp(\Omega)$ (see for instance [10, Section 19.12]). We first claim that a_n converges with respect to the topology of uniform convergence on $\wp(\Omega)$. Indeed, consider $A \subseteq \Omega$, then, using the coherence of $\underline{P}$, we find that

$$|a_n(A) - a_m(A)| = \left|\overline{P}(|X_n - Y_n|A) - \overline{P}(|X_m - Y_m|A)\right|$$
$$\leq \overline{P}\left(\left||X_n - Y_n| - |X_m - Y_m|\right|A\right)$$
$$\leq \overline{P}\left(|(X_n - Y_n) - (X_m - Y_m)|A\right)$$
$$\leq \overline{P}\left(|(X_n - X_m) - (Y_n - Y_m)|\right)$$
$$\leq \overline{P}(|X_n - X_m|) + \overline{P}(|Y_n - Y_m|).$$

Since the right hand side converges to zero independently of A, it follows that a_n is Cauchy with respect to the topology of uniform convergence on $\wp(\Omega)$. By the completeness of $\mathbb{R}^{\wp(\Omega)}$ with respect to the topology of uniform convergence on $\wp(\Omega)$, we find that a_n converges with respect to the topology of uniform convergence on $\wp(\Omega)$.

Uniform convergence implies point-wise convergence, so for every $A \subseteq \Omega$ we can define $a(A) := \lim_{n\to\infty} a_n(A)$. We must prove that $a(\Omega) = 0$. Let $\epsilon > 0$. By the convergence of a_n with respect to the topology of uniform convergence on $\wp(\Omega)$, there is an $M_\epsilon \in \mathbb{N}$ such that for all $A \subseteq \Omega$ and all $n \geq M_\epsilon$

$$|a_n(A) - a(A)| < \epsilon. \tag{1}$$

Define

$$\delta_\epsilon := \begin{cases} \epsilon/\sup N_{M_\epsilon}, & \text{if } \sup N_{M_\epsilon} > 0, \\ 1, & \text{otherwise.} \end{cases}$$

For every $A \subseteq \Omega$, if $\overline{P}(A) < \delta_\epsilon$ then $a_{M_\epsilon}(A) = \overline{P}(N_{M_\epsilon}A) \leq \sup N_{M_\epsilon} \overline{P}(A) < \epsilon$. Since $a(A) \leq |a(A) - a_{M_\epsilon}(A)| + a_{M_\epsilon}(A)$, it holds by (1) that

$$\overline{P}(A) < \delta_\epsilon \implies a(A) < 2\epsilon. \tag{2}$$

Define $B := \{\omega \in \Omega; N_{M_\epsilon}(\omega) \neq 0\}$, then $N_{M_\epsilon} \complement B = 0$. We infer that $\overline{P}(N_{M_\epsilon} \complement B) = a_{M_\epsilon}(\complement B) = 0$. From (1) it follows that $a(\complement B) < \epsilon$. We now prove that $a(\Omega) < 5\epsilon$.

(a) Consider the case that $\overline{P}(B) = 0$. Then $a(B) = \lim_{n \to \infty} \overline{P}(N_n B) = 0$ since $0 \leq \overline{P}(N_n B) \leq \sup N_n \overline{P}(B) = 0$ for every $n \in \mathbb{N}$. By the coherence of $\underline{P}$ it follows that $a(\Omega) \leq a(B) + a(\complement B) < 0 + \epsilon < 5\epsilon$.

(b) Now consider the other case that $\overline{P}(B) > 0$. Since X_n and Y_n converge to Z $\underline{P}$-hazily, it follows easily from coherence of $\underline{P}$ that $N_n = |X_n - Y_n|$ converges $\underline{P}$-hazily to 0. This implies that there is a $K_\epsilon \geq M_\epsilon$ such that for all $n \geq K_\epsilon$

$$\overline{P}(\{\omega \in \Omega; N_n > \epsilon/\overline{P}(B)\}) < \delta_\epsilon. \tag{3}$$

Define $C := \{\omega \in \Omega; N_{K_\epsilon} \leq \epsilon/\overline{P}(B)\}$. By the coherence of $\underline{P}$ we have that $a(\Omega) \leq a(B \cap C) + a(B \cap \complement C) + a(\complement B)$. We now investigate each term.

  (i) By (1) we have that $a(B \cap C) < a_{K_\epsilon}(B \cap C) + \epsilon$, since $K_\epsilon \geq M_\epsilon$. Since $N_{K_\epsilon}(\omega) \leq \epsilon/\overline{P}(B)$ for all $\omega \in C$ and $\overline{P}(B \cap C) \leq \overline{P}(B)$, we have that $a_{K_\epsilon}(B \cap C) = \overline{P}(N_{K_\epsilon}[B \cap C]) \leq (\epsilon/\overline{P}(B))\overline{P}(B \cap C) \leq \epsilon$. We find that $a(B \cap C) < 2\epsilon$.

  (ii) We claim that $a(B \cap \complement C) < 2\epsilon$. By (3) it follows that $\overline{P}(\complement C) < \delta_\epsilon$. The claim is established using $\overline{P}(B \cap \complement C) \leq \overline{P}(\complement C)$ and (2).

  (iii) We already proved that $a(\complement B) < \epsilon$.

In both cases it follows that $a(\Omega) < 5\epsilon$. Since this holds for any $\epsilon > 0$, we find that $a(\Omega) = 0$. $\square$

Observe that the proof given here uses the same techniques as its counterpart in the theory of charges (see for instance the proof of Proposition 4.4.10 in [1]).

Definition 1. A random variable Z is said to be $\underline{P}$-*previsible* if there is a $\underline{P}$-fundamental sequence (X_n) of gambles that converges $\underline{P}$-hazily to Z. We then define $\underline{P}^\times(Z) = \lim_{n \to \infty} \underline{P}(X_n)$, and (X_n) is called a $\underline{P}$-*determining sequence* for Z, or simply a *determining sequence* if there is no ambiguity regarding $\underline{P}$.

By Lemma 1, the limit $\underline{P}^\times(Z)$ is a finite real number, and is independent of the details of the determining sequence (X_n). Moreover, $\underline{P}^\times$ extends $\underline{P}$ in the mathematical sense. This follows simply from the observation that the constant sequence (X_n) defined by $X_n = X$ for each n is a determining sequence for X, whenever X is a gamble.

Proposition 1. $\underline{P}^\times(X) = \underline{P}(X)$ *for every* $X \in \mathscr{L}(\Omega)$.

The set of all $\underline{P}$-previsible random variables will be denoted by $\mathscr{L}_{\underline{P}}^\times(\Omega)$. By Proposition 1 it contains all gambles (bounded random variables). Using the coherence of $\underline{P}$, the following properties of $\mathscr{L}_{\underline{P}}^\times(\Omega)$ can be easily established (denoting the point-wise maximum by $\vee$ and the point-wise minimum by $\wedge$).

Proposition 2. *Let* X *and* $Y \in \mathscr{L}_{\underline{P}}^\times(\Omega)$, *and* $a \in \mathbb{R}$. *Then* $X + Y$, aX, $X \vee Y$, $X \wedge Y$, *and* $|X| \in \mathscr{L}_{\underline{P}}^\times(\Omega)$.

This means that $\mathscr{L}_{\underline{P}}^\times(\Omega)$ is a linear lattice with respect to the point-wise order. In particular, $|Z|$ is $\underline{P}$-previsible if Z is, and therefore we can extend the semi-norm $\|\cdot\|_{\underline{P}}$ introduced above to $\mathscr{L}_{\underline{P}}^\times(\Omega)$ through $\|Z\|_{\underline{P}} = \overline{P}^\times(|Z|)$ for all Z in $\mathscr{L}_{\underline{P}}^\times(\Omega)$. It is not difficult to show that $\|\cdot\|_{\underline{P}}$ is also a semi-norm on $\mathscr{L}^{\times}_{\underline{P}}(\Omega)$, using Proposition 3 below. Moreover, if (Z_n) is a sequence of $\underline{P}$-previsible random variables and $\lim_{n\to\infty}\|Z-Z_n\|_{\underline{P}}=0$ then $\lim_{n\to\infty}\underline{P}^{\times}(Z_n)=\underline{P}^{\times}(Z)$. This shows that topologically indistinguishable random variables are also behaviourally indistinguishable, that is, they have the same extended lower (and upper) prevision.

4 Properties

4.1 Coherence

It turns out that all of the properties of coherent lower previsions listed in [12, Section 2.6.1] extend to the extension $\underline{P}^{\times}$. We mention the three most important ones, which also establish "coherence" of the extension: the supremum buying prices must accept sure gain, they must be independent of the utility scale, and, if we are willing to buy X for price s and Y for price t, then we should certainly be willing to buy $X+Y$ for price $s+t$.

Proposition 3. *Let X and Y in $\mathscr{L}^{\times}_{\underline{P}}(\Omega)$, and let $\lambda \geq 0$. It holds that*

(i) $\underline{P}^{\times}(X) \geq \inf[X]$

(ii) $\underline{P}^{\times}(\lambda X) = \lambda \underline{P}^{\times}(X)$

(iii) $\underline{P}^{\times}(X+Y) \geq \underline{P}^{\times}(X) + \underline{P}^{\times}(Y)$

Proof. The last two properties follow immediately from the coherence of $\underline{P}$ and the fact that (in)equalities are preserved when taking limits whenever both sides of the (in)equality converge. To prove the first property, let (X_n) be a determining sequence for X. Define for every $n \in \mathbb{N}$

$$X'_n(\omega) = \begin{cases} X_n(\omega), & \text{if } X_n(\omega) \geq \inf[X], \\ \inf[X], & \text{otherwise.} \end{cases}$$

Since $|X - X'_n| \leq |X - X_n|$ and $|X'_n - X'_m| \leq |X_n - X_m|$ for every $n, m \in \mathbb{N}$, it follows easily from the coherence of $\underline{P}$ that (X'_n) is also a determining sequence for X. Observe that $X'_n \geq \inf[X]$. From the coherence of $\underline{P}$ we know that $\underline{P}(X'_n) \geq \inf[X]$. Now rely on the fact that this inequality is preserved when taking the limit to conclude that $\underline{P}^{\times}(X) = \lim_{n\to\infty} \underline{P}(X'_n) \geq \inf[X]$. □

4.2 Increasing domain under increasing precision

It turns out that as the precision of a coherent lower prevision increases, its extension will also become more precise, and more random variables become previsible. The proof of this fact is left to the reader as a simple exercise.

Proposition 4. *If $\underline{Q}$ point-wise dominates $\underline{P}$, then $\mathscr{L}^{\times}_{\underline{Q}}(\Omega) \supseteq \mathscr{L}^{\times}_{\underline{P}}(\Omega)$, and $\underline{Q}^{\times}$ point-wise dominates $\underline{P}^{\times}$ on $\mathscr{L}^{\times}_{\underline{P}}(\Omega)$, i.e., $\underline{Q}^{\times}(X) \geq \underline{P}^{\times}(X)$ for all $X \in \mathscr{L}^{\times}_{\underline{P}}(\Omega)$.*

As an example, we consider the case in which the lower prevision $\underline{P}$ describes complete ignorance. In such a case, the supremum price we are willing to buy a gamble X for is given by the lowest possible reward we may expect from X, that is, $\inf_{\omega\in\Omega} X(\omega)$. This lower prevision is called the *vacuous lower prevision*, and we denote it by $\underline{P}_v$. The $\underline{P}_v$-norm is the supremum norm, and $\mathscr{L}^{\times}_{\underline{P}_v}(\Omega) = \mathscr{L}(\Omega)$. Thus the set of all vacuously previsible random variables is exactly the set of bounded random variables.

4.3 Weak*-compactness and a lower envelope theorem

Let $\mathscr{P}_{\underline{P}^{\times}}(\Omega)$ denote the set of all real-valued linear functionals on the linear space $\mathscr{L}_{\underline{P}}^{\times}(\Omega)$ that dominate $\underline{P}^{\times}$ point-wise. These linear functionals have all the properties of a linear prevision—they are linear and positive, and have unit norm.

Theorem 1. $\mathscr{P}_{\underline{P}^{\times}}(\Omega)$ *is weak*-compact.*

Proof. $\mathscr{P}_{\underline{P}^{\times}}(\Omega)$ is a subset of the topological dual $\mathscr{L}_{\underline{P}}^{\times}(\Omega)^*$. The weak*-topology on $\mathscr{L}_{\underline{P}}^{\times}(\Omega)^*$ is the topology of point-wise convergence, and the theorem states that $\mathscr{P}_{\underline{P}^{\times}}(\Omega)$ is compact as a subset of $\mathscr{L}_{\underline{P}}^{\times}(\Omega)^*$ with respect to this topology. Define the set

$$\mathscr{V} = \prod_{X \in \mathscr{L}_{\underline{P}}^{\times}(\Omega)} [\underline{P}^{\times}(X), \overline{P}^{\times}(X)].$$

Members of $\mathscr{V}$ are mappings f on $\mathscr{L}_{\underline{P}}^{\times}(\Omega)$ satisfying $f(X) \in [\underline{P}^{\times}(X), \overline{P}^{\times}(X)]$ for each $X \in \mathscr{L}_{\underline{P}}^{\times}(\Omega)$. In particular, $\mathscr{P}_{\underline{P}^{\times}}(\Omega)$ is a subset of $\mathscr{V}$. We equip $\mathscr{V}$ with the product topology, which is the topology of point-wise convergence.

We see that the relativisation of the weak*-topology on $\mathscr{L}_{\underline{P}}^{\times}(\Omega)^*$ to $\mathscr{P}_{\underline{P}^{\times}}(\Omega)$ is equal to the relativisation of the product topology on $\mathscr{V}$ to $\mathscr{P}_{\underline{P}^{\times}}(\Omega)$, since they are both topologies of point-wise convergence. Since compactness of a set is only determined by the relativisation of the topology to that set, the theorem is established if we can show that $\mathscr{P}_{\underline{P}^{\times}}(\Omega)$ is a compact subset of $\mathscr{V}$ with respect to the product topology on $\mathscr{V}$.

By the ultrafilter principle (i.e., Tychonov's theorem for Hausdorff spaces, see for instance [10, Section 17.22]), $\mathscr{V}$ is compact. Hence, we only need to show that $\mathscr{P}_{\underline{P}^{\times}}(\Omega)$ is a closed subset of $\mathscr{V}$, since any closed subset of a compact space is also compact. By its definition, $\mathscr{P}_{\underline{P}^{\times}}(\Omega)$ is the set of linear mappings in $\mathscr{V}$. Assume that (T_α) is a net in $\mathscr{P}_{\underline{P}^{\times}}(\Omega)$ that converges to $T \in \mathscr{V}$ with respect to the product topology in $\mathscr{V}$. Then T is linear. Indeed, by the point-wise convergence of (T_α) to T we have that

$$T(\lambda X) = \lim_\alpha T_\alpha(\lambda X) = \lim_\alpha \lambda T_\alpha(X) = \lambda \lim_\alpha T_\alpha(X) = \lambda T(X),$$

for any $X \in \mathscr{L}_{\underline{P}}^{\times}(\Omega)$ and any $\lambda \in \mathbb{R}$, and

$$T(X+Y) = \lim_\alpha T_\alpha(X+Y) = \lim_\alpha [T_\alpha(X) + T_\alpha(Y)]$$
$$= \lim_\alpha T_\alpha(X) + \lim_\alpha T_\alpha(Y) = T(X) + T(Y),$$

for any $X, Y \in \mathscr{L}_{\underline{P}}^{\times}(\Omega)$. This shows that $T \in \mathscr{P}_{\underline{P}^{\times}}(\Omega)$, so $\mathscr{P}_{\underline{P}^{\times}}(\Omega)$ must be closed with respect to the topology of point-wise convergence. □

It is not so difficult to establish the following, quite remarkable result (where we denote by $f|_A$ the restriction of a mapping f to the subset A of its domain).

Theorem 2. *There is a canonical one-to-one correspondence between the sets* $\mathcal{M}(\underline{P})$ *and* $\mathscr{P}_{\underline{P}^{\times}}(\Omega)$*, given by*

$$\Phi_{\underline{P}}^{\times}(Q) = Q^{\times}|_{\mathscr{L}_{\underline{P}}^{\times}(\Omega)}, \qquad \Psi_{\underline{P}}^{\times}(R) = R|_{\mathscr{L}(\Omega)},$$

for any $Q \in \mathcal{M}(\underline{P})$ *and any* $R \in \mathscr{P}_{\underline{P}^{\times}}(\Omega)$.

Proof. It is left as an easy exercise to the reader to first verify that $Q^\times|_{\mathscr{L}_{\underline{P}}^\times(\Omega)} \in \mathscr{P}_{\underline{P}^\times}(\Omega)$ and $R|_{\mathscr{L}(\Omega)} \in \mathcal{M}(\underline{P})$ for any $Q \in \mathcal{M}(\underline{P})$ and any $R \in \mathscr{P}_{\underline{P}^\times}(\Omega)$. We still need to show that the mappings $\Phi_{\underline{P}}^\times$ and $\Psi_{\underline{P}}^\times$ are each other's inverses, that is, $\Phi_{\underline{P}}^\times \circ \Psi_{\underline{P}}^\times = 1_{\mathcal{M}(\underline{P})}$ and $\Psi_{\underline{P}}^\times \circ \Phi_{\underline{P}}^\times = 1_{\mathscr{P}_{\underline{P}^\times}(\Omega)}$.

Let $Q \in \mathcal{M}(\underline{P})$. Then, since $Q^\times$ is an extension of Q, it follows immediately that $\Psi_{\underline{P}}^\times \circ \Phi_{\underline{P}}^\times(Q) = \Phi_{\underline{P}}^\times(Q)|_{\mathscr{L}(\Omega)} = Q^\times|_{\mathscr{L}(\Omega)} = Q$. Conversely, let $R \in \mathscr{P}_{\underline{P}^\times}(\Omega)$. We prove that $\Phi_{\underline{P}}^\times \circ \Psi_{\underline{P}}^\times(R) = R$. Define $Q := \Psi_{\underline{P}}^\times(R) = R|_{\mathscr{L}(\Omega)}$. It is easily checked that Q is linear, positive, and has unit norm. It follows that Q is a linear prevision. Moreover, Q dominates $\underline{P}$, and hence, $Q \in \mathcal{M}(\underline{P})$. Let X be a $\underline{P}$-previsible random variable and let (X_n) be a $\underline{P}$-determining sequence for X. Then it is easily established that (X_n) also is a Q-determining sequence for X, using the fact that Q dominates $\underline{P}$. Hence,

$$\Phi_{\underline{P}}^\times(Q)(X) = Q^\times(X) = \lim_{n\to\infty} Q(X_n) = \lim_{n\to\infty} R(X_n).$$

Observe that

$$|R(X) - R(X_n)| = |R(X - X_n)| \leq R(|X - X_n|) \leq \overline{P}^\times(|X - X_n|).$$

But, since $|X_m - X_n|$ is a $\underline{P}$-determining sequence for $|X - X_n|$, it must hold that $\overline{P}^\times(|X - X_n|) = \lim_{m\to\infty} \overline{P}^\times(|X_m - X_n|)$. Hence,

$$\lim_{n\to\infty} |R(X) - R(X_n)| \leq \lim_{n,m\to\infty} \overline{P}^\times(|X_m - X_n|) = 0,$$

from which we infer that $\lim_{n\to\infty} R(X_n) = R(X)$. Thus indeed $\Phi_{\underline{P}}^\times(Q)(X) = R(X)$, for any $\underline{P}$-previsible random variable X. □

Theorem 2 leads to a number of interesting observations. Since $\mathcal{M}(\underline{P})$ is non-empty, so is $\mathscr{P}_{\underline{P}^\times}(\Omega)$. Since $\Phi_{\underline{P}}^\times$ preserves the convex structure, and $\mathcal{M}(\underline{P})$ is convex, we infer that also $\mathscr{P}_{\underline{P}^\times}(\Omega)$ is convex. Finally, Theorem 2 also leads to a very interesting lower envelope theorem for our extension, which shows that the Bayesian sensitivity analysis interpretation still holds for our extension, *in terms of linear previsions on gambles only!*

Theorem 3. *For any $\underline{P}$-previsible random variable X we have that*

$$\underline{P}^\times(X) = \min_{Q \in \mathcal{M}(\underline{P})} Q^\times(X).$$

Proof. Since $\underline{P}^\times$ is concave, and continuous with respect to the $\underline{P}$-norm, it follows from the Hahn-Banach theorem [10, HB17] that $\underline{P}^\times$ is the lower envelope of the continuous affine functions that point-wise dominate $\underline{P}^\times(X)$.

Let $T + a$ be any affine function on $\mathscr{L}_{\underline{P}}^\times(\Omega)$ which point-wise dominates $\underline{P}^\times$, that is, T linear, $a \in \mathbb{R}$ and $T + a \geq \underline{P}^\times$ (we do not assume that $T + a$ is continuous). Then we infer that $T(1) = 1$ and $a \geq 0$, since the inequality $\underline{P}(\alpha) \leq T(\alpha) + a$, which is equivalent to $a \geq \alpha(1 - T(1))$, must be satisfied for all $\alpha \in \mathbb{R}$. Secondly, we infer that T is positive. Indeed, if $X \geq 0$, then $\lambda X \geq 0$ for all $\lambda \geq 0$, and hence, $0 \leq \underline{P}(\lambda X) \leq T(\lambda X) + a$ for all $\lambda \geq 0$, or equivalently, $T(X) \geq -\frac{a}{\lambda}$ for all $\lambda \geq 0$, which can only be satisfied if $T(X) \geq 0$ (recall that we just proved that $a \geq 0$). Finally, T is continuous, since

$|T(X) - T(Y)| \leq \overline{P}^{\times}(|X - Y|)$ for all $X, Y \in \mathscr{L}_{\underline{P}}^{\times}(\Omega)$. Indeed, for all $\lambda > 0$

$$|T(X) - T(Y)| = |T(X - Y)| \leq T(|X - Y|) = \frac{1}{\lambda} T(\lambda |X - Y|)$$
$$\leq \frac{1}{\lambda}[\overline{P}^{\times}(\lambda |X - Y|) + a] = \overline{P}^{\times}(|X - Y|) + \frac{a}{\lambda},$$

which gives the desired inequality. So the affine functions that point-wise dominate $\underline{P}^{\times}$ are continuous, and are exactly of the form $T + a$, with $T \in \mathscr{P}_{\underline{P}^{\times}}(\Omega)$ and $a \in \mathbb{R}$, $a \geq 0$.

This shows that in taking the lower envelope, we may restrict ourselves to the (continuous) *linear* functions (i.e., we may assume that $a = 0$) that are positive, have unit norm and dominate $\underline{P}^{\times}$. But these functions are exactly the elements of $\mathscr{P}_{\underline{P}^{\times}}(\Omega)$. Hence, $\underline{P}^{\times}(X) = \min_{R \in \mathscr{P}_{\underline{P}^{\times}}(\Omega)} R(X)$. Using the one-to-one correspondence of Theorem 2,

$$\underline{P}^{\times}(X) = \min_{Q \in \mathcal{M}(\underline{P})} \Phi_{\underline{P}}^{\times}(Q)(X) = \min_{Q \in \mathcal{M}(\underline{P})} Q^{\times}(X). \quad \square$$

5 The linear case: charges and Dunford integrals

As our extension can be written as the lower envelope of extended linear previsions, it is of particular interest that extended linear previsions can be written as an integral. One possible choice for this integral turns out to be the Dunford integral. Before stating any results, let us first briefly review the theory of *charges*, or finitely additive measures, restricting ourselves to the essentials that are needed to understand the connection with previsibility. In particular, we shall focus on so-called probability charges on the set $\wp(\Omega)$ of all subsets of Ω as we do not need the more general theory of charges, defined on a more general field of subsets. General and detailed accounts can be found in [1] and [5].

A *probability charge* is a real-valued mapping on a field $\mathfrak{F}$ on Ω such that $\mu(\emptyset) = 0$, $\mu(\Omega) = 1$, $\mu(A) \geq 0$ whenever $A \in \mathfrak{F}$, $\mu(A) + \mu(B) = \mu(A \cup B)$ whenever $A, B \in \mathfrak{F}$ and $A \cap B = \emptyset$. From now on, $\mathfrak{F}$ is assumed to be the power set $\wp(\Omega)$. The (Dunford) integral of a simple random variable[6] $X = \sum_{i=1}^n a_i I_{A_i}$ ($n \in \mathbb{N}$, $a_i \in \mathbb{R}$ and $A_i \subseteq \Omega$) is then defined as $D \int X \, d\mu = \sum_{i=1}^n a_i \mu(A_i)$.

A sequence of random variables (X_n) is said to *converge hazily* to a random variable X if for all $\epsilon > 0$ it holds that $\mu(\{\omega \in \Omega \colon |X_n(\omega) - X(\omega)| > \epsilon\}) \to 0$ as $n \to \infty$. X is called *D-integrable* if there is a sequence (X_n) of simple random variables that converges hazily to X, such that moreover $D \int |X_n - X_m| \, d\mu \to 0$ as $n, m \to \infty$. Any such sequence is called a *determining sequence* for X. All determining sequences for X eventually have the same integral, and in this way, one defines the *Dunford integral* of X as the limit of integrals of simple random variables:

$$D \int X \, d\mu = \lim_{n \to \infty} D \int X_n \, d\mu.$$

The Dunford integral has all the properties that we expect from an integral, in particular, it is linear (and therefore finitely additive) and positive.

It is well known that there is a canonical one-to-one correspondence between linear previsions on $\mathscr{L}(\Omega)$ and probability charges on $\wp(\Omega)$: any linear prevision P on $\mathscr{L}(\Omega)$

---

[6]In this paper, a random variable which assumes only a finite number of values is called *simple*.

is uniquely determined through the probability charge μ defined by $\mu(A) = P(I_A)$. The linear prevision $P(X)$ of a gamble X is equal to the Dunford integral of X with respect to μ (see for instance [1, Theorem 4.7.4]).

But it is also well known that the set of all random variables that are D-integrable with respect to μ is much larger than the set of gambles. In fact, it is nothing but the set $\mathscr{L}_P^\times(\Omega)$ of P-previsible random variables, and it is worth nothing that it coincides with the so-called *Lebesgue space* $L_1(\Omega, \wp(\Omega), \mu)$ of those random variables X that are T_1-measurable[7] and whose absolute value $|X|$ is D-integrable. Moreover, our extension of a linear prevision P on $\mathscr{L}(\Omega)$ coincides exactly with the Dunford integral with respect to the probability charge μ. This is the subject of the following theorem.

Theorem 4. *Let P be a linear prevision on $\mathscr{L}(\Omega)$, and let μ be its restriction to the set of events $\wp(\Omega)$. Then the following statements hold.*

(i) *A sequence of random variables (X_n) converges P-hazily to a random variable X if and only if (X_n) converges hazily to X with respect to μ.*

(ii) *A random variable X is P-previsible if and only if it is D-integrable with respect to μ, in which case $P^\times(X) = D \int X \, d\mu$.*

Proof. Compare the definition of P-hazy convergence with the definitions of hazy convergence with respect to μ to establish their equivalence.

Comparing the definitions of P-previsibility and D-integrability with respect to μ, it is immediately clear that D-integrability implies P-previsibility. To see that the converse holds too, assume that (X_n) is a P-determining sequence for X. Observe that (X_n) is a sequence of D-integrable random variables with respect to μ (recall that all gambles are D-integrable). By the definition of P-determining sequence, and the fact that P is uniquely determined by μ through the D-integral, we have that

$$\lim_{n,m \to \infty} D \int |X_n - X_m| \, d\mu, = \lim_{n,m \to \infty} P(|X_n - X_m|) = 0,$$

and also that (X_n) converges hazily to X with respect to μ. From this we may conclude that X is D-integrable, and

$$\lim_{n \to \infty} D \int |X_n - X| \, d\mu = 0.$$

(see [1, Theorem 4.4.20]). Again using the fact that P is uniquely determined by μ through the D-integral, we find that

$$D \int X \, d\mu = \lim_{n \to \infty} D \int X_n \, d\mu = \lim_{n \to \infty} P(X_n) = P^\times(X_n). \qquad \square$$

6 Conclusions

The main message of this paper is that it is possible to define an extension of a coherent lower prevision to a linear space of previsible, not necessarily bounded random variables,

---

[7] A random variable X is called T_1-measurable with respect to μ if there is a sequence of simple random variables converging hazily to X with respect to μ

and that this extension still has properties similar to those of coherent lower previsions. Previsibility coincides with the existing notion of D-integrability when the coherent lower previsions are linear, and an extended lower prevision can be written as the lower envelope of the extensions of the dominating linear previsions of the original.

Acknowledgements

This paper presents research results of project G.0139.01 of the Fund for Scientific Research, Flanders (Belgium), and of the Belgian Programme on Interuniversity Poles of Attraction initiated by the Belgian state, Prime Minister's Office for Science, Technology and Culture. The scientific responsibility rests with the authors.

References

[1] K. Bhaskara Rao, M. Bhaskara Rao, Theory of Charges, a Study of Finitely Additive Measures, Academic Press, London, 1983.

[2] G. Choquet, Theory of capacities, Ann. Inst. Fourier 5 (1953–54) 131–295.

[3] G. de Cooman, Possibility theory I–III, International Journal of General Systems 25 (1997) 291–371.

[4] G. de Cooman, M. C. M. Troffaes, Dynamical programming for deterministic discrete-time systems with uncertain gain, 2003, submitted to the Third International Symposium on Imprecise Probabilities and Their Applications (ISIPTA '03).

[5] N. Dunford, J. T. Schwartz, Linear Operators, John Wiley & Sons, New York, 1957.

[6] B. de Finetti, Sul significato soggettivo della probabilità, Fundamenta Mathematicae 17 (1931) 298–329.

[7] B. de Finetti, Theory of Probability: a Critical Introductory Treatment, Wiley, London, 1975.

[8] I. Levi, The Enterprise of Knowledge. An Essay on Knowledge, Credal Probability, and Chance, MIT Press, Cambridge, 1983.

[9] L. V. Utkin, S. V. Gurov, Imprecise reliability for some new lifetime distribution classes, Journal of Statistical Planning and Inference 105 (1) (2002) 215–232.

[10] E. Schechter, Handbook of Analysis and Its Foundations, Academic Press, San Diego, 1997.

[11] P. Walley, T. L. Fine, Varieties of modal (classificatory) and comparative probability, Synthese 41 (1979) 321–374.

[12] P. Walley, Statistical Reasoning with Imprecise Probabilities, Chapman and Hall, London, 1991.

[13] L. A. Zadeh, Fuzzy sets as a basis for a theory of possibility, Fuzzy Sets Syst. 1 (1978) 3–28.

Learning and Mining

B. Bouchon-Meunier, L. Foulloy and R.R. Yager (Editors)
Intelligent Systems for Information Processing:
From Representation to Applications
© 2003 Elsevier Science B.V. All rights reserved.

Clustering of proximity data using belief functions

Thierry Denœux[1] & Mylène Masson[2]

HEUDIASYC – UMR CNRS 6599
[1]Université de Technologie de Compiègne
[2]Univ. de Picardie Jules Verne
BP 20529 - F-60205 Compiègne cedex
Mylene.Masson@hds.utc.fr

Abstract

A new clustering method for relational data is proposed, based on Evidence theory. In this approach, masses of belief assigned to subsets of classes are used to compute the plausibility that two objects belong to the same class. It is then required that these plausibilities be compatible with the observed dissimilarities between objects. Experiments illustrate the ability of the method to handle noisy or non Euclidean data.

Keywords: Relational clustering, belief functions, outliers, non Euclidean data.

1 Introduction

Whereas evidence theory has been applied to supervised classification problems for a long time (see, e.g., [4]), the work presented in this paper is, to our knowledge, the first incursion of belief functions into the cluster analysis domain. Cluster analysis is concerned with methods for finding groups in data, groups (or classes) being defined as subsets of more or less "similar" objects [11]. The two most frequent data types are *object data*, in which each object is described explicitly by a list of attributes, and *proximity* (or *relational*) data, in which only pairwise similarities, or dissimilarities are given. A quite extensive review of crisp and fuzzy relational clustering models can by found in [1, chapter 3]. These methods can be classified into three broad categories: hierarchical methods, methods based on the decomposition of fuzzy relations, and methods based on the optimization of an objective function. Given n objects to be classified in c classes, methods in the latter category aim at finding a fuzzy partition matrix $U = (u_{ik})$ of size $n \times c$ such that:

$$\sum_{k=1}^{c} u_{ik} = 1 \quad \forall\, i \in \{1, \ldots, n\}$$

and

$$\sum_{i=1}^{n} u_{ik} > 0 \quad \forall\, k \in \{1, \ldots, c\}\,.$$

Each number $u_{ik} \in [0, 1]$ is interpreted as a *degree of membership* of object i to cluster k.

Examples of such methods are the fuzzy non metric (FNM) model [16], the assignment-prototype (AP) model [20] and the relational fuzzy c-means (RFCM) model [9] (a similar approach may be found in [12]). The latter approach was later extended by Hathaway and Bezdek [8] to cope with non-Euclidean dissimilarity data, leading to the non-Euclidean relational fuzzy c-means (NERFCM) model. Finally, robust versions of the FNM and RFCM algorithms were proposed by Davé [3].

In this paper, a novel approach to clustering proximity data is presented, based on Dempster-Shafer (DS) theory of belief functions, also referred to as "Evidence theory". In this approach, the allocation of objects to classes is performed using the concept of basic belief asignment (bba), whereby a "mass of belief" is assigned to each possible subset of classes. Using a suitable noninteractivity assumption, it is possible to compute, for each two objects, the plausibility that they belong to the same class. It is then required that these plausibilities be, in some sense, compatible with the observed pairwise dissimilarities between objects. The rest of this paper is organized as follows. The necessary background on belief functions will be recalled in Section 2. Our method will then be exposed in Section 3, and experimental results will be presented in Section 4. Section 5 will conclude the paper.

2 Evidence theory

Let us consider a variable x taking values in a finite and unordered set Ω. Partial knowledge regarding the actual value taken by x can be represented by a *basic belief assignment* (bba) [18, 19], defined as a function m from 2^Ω to $[0, 1]$, verifying:

$$\sum_{A \subseteq \Omega} m(A) = 1. \qquad (1)$$

The subsets A of Ω such that $m(A) > 0$ are the *focal sets* of m. Each focal set A is a set of possible values for x, and the number $m(A)$ can be interpreted as a fraction of a unit mass of belief, which is allocated to A on the basis of a given evidential corpus. Complete ignorance corresponds to $m(\Omega) = 1$, and perfect knowledge of the value of x is represented by the allocation of the whole mass of belief to a unique singleton of Ω (m is then called a *certain* bba). Another particular case is that where all focal sets of m are singletons: m is then equivalent to a probability function, and is called a *Bayesian* bba.

A bba m such that $m(\emptyset) = 0$ is said to be normal. This condition was originally imposed by Shafer [18], but it may be relaxed if one accepts the *open-world assumption* stating that the set Ω might not be complete, and x might take its value outside Ω [19]. The quantity $m(\emptyset)$ is then interpreted as a mass of belief given to the hypothesis that x might not lie in Ω.

A bba m can be equivalently represented by any of two non additive fuzzy measures: a belief function (BF) bel : $2^\Omega \mapsto [0, 1]$, defined as

$$\mathrm{bel}(A) \triangleq \sum_{\emptyset \neq B \subseteq A} m(B) \quad \forall A \subseteq \Omega, \qquad (2)$$

and a plausibility function pl : $2^\Omega \mapsto [0, 1]$, defined as

$$\mathrm{pl}(A) \triangleq \mathrm{bel}(\Omega) - \mathrm{bel}(\overline{A}) \quad \forall A \subseteq \Omega, \qquad (3)$$

where $\overline{A}$ denotes the complement of A. Whereas bel(A) represents the amount of support given to A, the *potential* amount of support that *could be* given to A is measured by pl(A). Note that both bel and pl boil down to a unique probability measure when m is a Bayesian bba.

Let us now assume that we have two bba's m_1 and m_2 representing distinct items of evidence concerning the value of x. The standard way of combining them is through the conjunctive sum operation $\cap$ defined as:

$$(m_1 \cap m_2)(A) \triangleq \sum_{B \cap C = A} m_1(B) m_2(C) , \qquad (4)$$

for all $A \subseteq \Omega$. The quantity $K = (m_1 \cap m_2)(\emptyset)$ is called the *degree of conflict* between m_1 and m_2. It may be seen as a degree of disagreement between the two information sources. If necessary, the normality condition $m(\emptyset) = 0$ may be recovered by dividing each mass $(m_1 \cap m_2)(A)$ by $1 - K$. The resulting operation is noted $\oplus$ and is called Dempster's rule of combination [18]:

$$(m_1 \oplus m_2)(A) \triangleq \frac{1}{1-K} \sum_{B \cap C = A} m_1(B) m_2(C) . \qquad (5)$$

Consider now a bba m^Ω defined on the Cartesian product $\Omega = \Omega_1 \times \Omega_2$ (from now on, the domain of a bba will be indicated as superscript when necessary). The marginal bba m^{Ω_1} on Ω_1 is defined for all $A \subseteq \Omega_1$ as

$$m^{\Omega_1}(A) \triangleq \sum_{\{B \subseteq \Omega \mid \text{Proj}(B \downarrow \Omega_1) = A\}} m^\Omega(B), \qquad (6)$$

where $\text{Proj}(B \downarrow \Omega_1)$ denotes the projection of B onto Ω_1, defined as

$$\text{Proj}(B \downarrow \Omega_1) \triangleq \{\omega_1 \in \Omega_1 \mid \exists \omega_2 \in \Omega_2, (\omega_1, \omega_2) \in B\} . \qquad (7)$$

The two marginal bba's m^{Ω_1} and m^{Ω_2} are said to be noninteractive iff for all $A \subseteq \Omega_1$ and for all $B \subseteq \Omega_2$

$$m^\Omega(A \times B) = m^{\Omega_1}(A) m^{\Omega_2}(B) . \qquad (8)$$

These definitions can be easily extended to bba's defined over the Cartesian product of n sets $\Omega_1, \ldots, \Omega_n$.

3 The method

3.1 Credal partition of a set of n objects

Let us consider a collection $O = \{o_1, \ldots, o_n\}$ of n objects, and a set $\Omega = \{\omega_1, \ldots, \omega_c\}$ of c classes forming a partition of O. Let us assume that we have only partial knowledge concerning the class membership of each object o_i, and that this knowledge is represented by a bba m_i on the set Ω. We recall that $m_i(\Omega)$ stands for complete ignorance of the class of object i, whereas $m_i(\{\omega_k\}) = 1$ corresponds to full certainty that object i belongs to class k. All other situations correspond to partial knowledge of the class of o_i. For instance, the following bba:

$$m_i(\{\omega_k, \omega_\ell\}) = 0.7$$
$$m_i(\Omega) = 0.3$$

means that we have some belief that object i belongs either to class ω_k or to class ω_ℓ, and the weight of this belief is equal to 0.7.

Let $M = (m_1, \ldots, m_n)$ denote the n-tuple of bba's related to the n objects. We shall call M a *credal partition* of O. Two particular cases are of interest:

- when each m_i is a *certain* bba, then M defines a conventional, crisp partition of Ω; this corresponds to a situation of complete knowledge;

- when each m_i is a *Bayesian* bba, then M specifies a fuzzy partition of Ω, as defined by Bezdek [1].

A credal c-partition (or partition of size c) will be defined as a credal partition $M = (m_1, \ldots, m_n)$ such that, for all $\omega \in \Omega$, we have

$$\text{pl}_i(\{\omega\}) > 0$$

for some $i \in \{1, \ldots, n\}$, pl_i being the plausibility function associated to m_i.

Example 1 Let us consider a collection O of $n = 4$ objects and $c = 3$ classes. A credal partition M of O is given in Table 1. The class of object o_2 is known with certainty, whereas the class of o_4 is completely unknown. The two other cases correspond to situations of partial knowledge. The plausibilities $\text{pl}_i(\{\omega\})$ of each singleton are given in Table 2. Since each class is plausible for at least one object, M is a credal 3-partition of O. Note that the matrix given in Table 2 defines a possibilistic partition as defined in [1].

Table 1: Credal partition of Example 1

F	$m_1(F)$	$m_2(F)$	$m_3(F)$	$m_4(F)$
$\emptyset$	0	0	0	0
$\{\omega_1\}$	0	0	0	0
$\{\omega_2\}$	0	1	0	0
$\{\omega_1, \omega_2\}$	0.7	0	0	0
$\{\omega_3\}$	0	0	0.2	0
$\{\omega_1, \omega_3\}$	0	0	0.5	0
$\{\omega_2, \omega_3\}$	0	0	0	0
Ω	0.3	0	0.3	1

3.2 Compatibility of an evidential partition with a dissimilarity matrix

In this section, we propose a principle that will provide the basis for inferring a credal partition from proximity data.

Table 2: Plausibilities of the singletons for the credal partition of Example 1

i	$\text{pl}_1(\{\omega_i\})$	$\text{pl}_2(\{\omega_i\})$	$\text{pl}_3(\{\omega_i\})$	$\text{pl}_4(\{\omega_i\})$
1	1	0	0.8	1
2	1	1	0.3	1
3	0.3	0	1	1

Without loss of generality, let us assume the available data to consist of a $n \times n$ dissimilarity matrix $D = (d_{ij})$, where $d_{ij} \geq 0$ measures the degree of dissimilarity between objects o_i and o_j. Matrix D will be supposed to be symmetric, with null diagonal elements.

It is reasonable to assume that two similar objects are more likely to be in the same class, than two dissimilar ones. The more similar, the more *plausible* it is that they belong to the same group. To formalize this idea, we need to calculate the plausibility, based on a credal partition, that two objects o_i and o_j are in the same group. This will then allow us to formulate a criterion of compatibility between a dissimililarity matrix D and a credal partition M.

Consider two objects o_i and o_j, and two bba's m_i and m_j quantifying one's beliefs regarding the class of objects i and j. To compute the plausibility that these two objects belong to the same class, we have to place ourselves in the Cartesian product $\Omega^2 = \Omega \times \Omega$, and to consider the joint bba $m_{i \times j}$ on Ω^2 related to the vector variable (y_i, y_j). If m_i and m_j are assumed to be noninteractive, then $m_{i \times j}$ is completely determined by m_i and m_j, and we have $\forall A, B \subseteq \Omega$:

$$m_{i \times j}(A \times B) = m_i(A) m_j(B). \tag{9}$$

In Ω^2, the event "Objects o_i and o_j belong to the same class" corresponds to the following subset of Ω^2:

$$S = \{(\omega_1, \omega_1), (\omega_2, \omega_2), \ldots, (\omega_c, \omega_c)\}$$

Let $\text{pl}_{i \times j}$ be the plausibility function associated to $m_{i \times j}$. We have

$$\begin{aligned}
\text{pl}_{i \times j}(S) &= \sum_{(A \times B) \cap S \neq \emptyset} m_{i \times j}(A \times B) \\
&= \sum_{A \cap B \neq \emptyset} m_i(A) m_j(B) \\
&= 1 - \sum_{A \cap B = \emptyset} m_i(A) m_j(B) \\
&= 1 - K_{ij},
\end{aligned} \tag{10}$$

where K_{ij} is the degree of conflict between m_i and m_j.

Hence, the plausibility that objects o_i and o_j belong to the same class is simply equal to one minus the degree of conflict between the bba's m_i and m_j associated to the two objects. Given any two pairs of objects (o_i, o_j) and $(o_{i'}, o_{j'})$, it is natural to impose the following condition:

$$d_{ij} > d_{i'j'} \Rightarrow \text{pl}_{i \times j}(S) \leq \text{pl}_{i' \times j'}(S) \tag{11}$$

or, equivalently:

$$d_{ij} > d_{i'j'} \Rightarrow K_{ij} \geq K_{i'j'}, \qquad (12)$$

i.e., the more dissimilar the objects, the less plausible it is that they belong to the same class, and the higher the conflict between the bba's. A credal partition M verifying this condition will be said to be *compatible* with D.

3.3 Learning a credal partition from data

To extract a credal partition from dissimilarity data, we need a method that, given a dissimilarity matrix D, generates a credal partition M that is either compatible with D, or at least "almost compatible" (in a sense to be defined).

This problem happens to be quite similar to the one addressed by multidimensional scaling (MDS) methods [2]. The purpose of MDS methods is, given a dissimilarity matrix D, to find a configuration of points in a p-dimensional space, such that the distances between points approximate the dissimilarities. There is a large literature on MDS methods, which are used extensively in sensory data analysis for interpreting subjectively assessed dissimilarities, and more generally in exploratory analysis for visualizing proximity data as well as high dimensional attribute data (in this case, the dissimilarities are computed as distances in the original feature space).

In our problem, each object is represented as a bba, which can be seen as a point in a 2^c-dimensional space. Hence, the concept of "credal partition" parallels that of "configuration" in MDS. The degree of conflict K_{ij} between two bba's m_i and m_j may be seen as a form of "distance" between the representations of objects o_i and o_j. This close connection allows us to transpose MDS algorithms to our problem.

MDS algorithms generally consist in the iterative minimization of a *stress function* measuring the discrepancies between observed dissimilarities and reconstructed distances in the configuration space. The various methods available differ by the choice of the stress function, and the optimization algorithm used. The simplest one is obtained by imposing a linear relationship between "distances" (i.e., degrees of conflict in our case) and dissimilarities, which is referred to as *metric* MDS. The stress function used in our case is:

$$\sigma(M, a, b) \triangleq \frac{\sum_{i<j}(aK_{ij} + b - d_{ij})^2}{\sum_{i<j} d_{ij}^2}, \qquad (13)$$

where a and b are two coefficients, and the denominator is a normalizing constant. This stress function can be minimized iteratively with respect to M, a and b using a gradient-based procedure. Note that this method is invariant under any affine transformation of the dissimilarities.

Remark 1 Each bba m_i must satisfy Eq. (1). Hence, the optimization of σ with respect to M is a constrained optimization problem. However, the contraints vanish if one uses the following parameterization:

$$m_i(A_l) = \frac{\exp(\alpha_{il})}{\sum_{k=1}^{2^c} \exp(\alpha_{ik})}, \qquad (14)$$

where $A_l, l = 1, \ldots, 2^c$ are the subsets of Ω, and the α_{il} for $i = 1, \ldots, n$ and $l = 1, \ldots, 2^c$ are $n2^c$ real parameters.

3.4 Controlling the number of parameters

An important issue is the dimension of the non linear optimization problem to be solved. The number of parameters to be optimized is linear in the number of objects but exponential in the number of clusters. If c is large, the number of free parameters has to be controlled. This can be achieved in two ways:

First, the number of parameters may be drastically decreased by considering only a subclass of bba's with a limited number of focal sets. For example, we may constrain the focal sets to be either Ω, the empty set, or a singleton. In this way, the total number of parameters is reduced to $n(c+2)$, without sacrificing too much of the flexibility of belief functions.

Another very efficient means of reducing the number of free parameters is to add a penalization term to the stress function. This approach does not reduce the number of parameters to be optimized but limits the *effective* number of parameters of the method. It is thus a way to control the complexity of the classification model. In our case, we would like to extract as much information as possible from the data, so that it is reasonable to require the bba's to be as "informative" as possible. The definition of the "quantity of information" contained in a belief function has been the subject of a lot of research in the past few years [14, 13], and it is still, to some extent, an open question. However, several entropy measures have been proposed. The total uncertainty introduced by Pal et al. [15] satisfies natural requirements and has interesting properties. It is defined, for a normal bba m, as:

$$H(m) \triangleq \sum_{A \in \mathcal{F}(m)} m(A) \log_2 \left(\frac{|A|}{m(A)} \right), \qquad (15)$$

where $\mathcal{F}(m)$ denotes the set of focal sets of m. $H(m)$ is minimized when the mass is assigned to few focal sets, with small cardinality (it is proved in [15] that $H(m) = 0$ iff $m(\{\omega\}) = 1$ for some $\omega \in \Omega$).

To apply (15) to a subnormal bba m (i.e., such that $m(\emptyset) > 0$), some normalization has to be performed. Two common normalization procedures are Dempster's normalization (in which the mass given to $\emptyset$ is deleted and all other belief masses are divided by $1 - m(\emptyset)$ [18], and Yager's normalization, in which the mass $m(\emptyset)$ is transferred to Ω [21]. The latter approach has been preferred in our approach, because it allows to penalize subnormal bba's more efficiently. The expression of total uncertainty for a subnormal bba m then becomes:

$$\begin{aligned} H(m) &= \sum_{A \in \mathcal{F}(m) \setminus \{\emptyset\}} m(A) \log_2 \left(\frac{|A|}{m(A)} \right) \\ &+ m(\emptyset) \log_2 \left(\frac{|\Omega|}{m(\emptyset)} \right). \end{aligned} \qquad (16)$$

Finally, the objective function to be minimized is:

$$J(M, a, b) \triangleq \sigma(M, a, b) + \lambda \sum_{i=1}^{n} H(m_i). \qquad (17)$$

3.5 From credal clustering to fuzzy or hard clustering

Although we believe that a lot of information may be gained in analyzing a credal partition, it is always possible to transform it into a fuzzy or hard partition. This conversion is based on the concept of *pignistic* probability [19] defined, for a normalized bba m, by:

$$BetP(A) \triangleq \sum_{\emptyset \neq B \subseteq \Omega} m(B) \frac{|A \cap B|}{|B|} \qquad (18)$$

To obtain a fuzzy partition, one calculates the pignistic probability of each singleton ω_k. In the case where these singletons, Ω and the empty set are the only focal sets of the bba, the expression of the pignistic probabilities is given by:

$$BetP(\{\omega_k\}) = m(\{\omega_k\}) + \frac{m(\Omega) + m(\emptyset)}{c}, \qquad (19)$$

for all $k = 1, c$ (we assume that Yager's normalization is used). A hard partition can then be easily obtained from the values of pignistic probabilities. In this sense, a credal partition may be viewed as a general model of partionning, including fuzzy and hard partitions.

4 Results

4.1 Synthetic dataset

This first example is inspired from a classical dataset [20]. A (13 × 13) dissimilarity matrix was generated by computing the squared Euclidean distances of a two dimensional object dataset represented in figure 1. The 13*th* object, an outlier, is useful to study the robustness of the method. The first object is assumed to be close to all other objets, and is not represented in this figure. The dissimilarity between this point and objects 2 to 12 is arbitrarily set to 1 and to 200 with the 13th object. This object is intended to reflect either noisy, unreliable data, or imprecise evaluations coming from subjective assessments. We compare the results obtained with our method and five classical clustering methods based on relationnal data: Windham's assignment-prototype algorithm (AP) [20], the Fuzzy Non Metric algorithm (FNM) [16], the Relational Fuzzy c-means algorithm (RFCM) [9], and its "Noise" version (NRFCM) [3], and the non-Euclidean RFCM algorithm (NERF) [8]. NRFCM, by using a "noise" cluster, is well-adapted to datasets containing noise and outliers, whereas NERF is intended to cope with non-Euclidean dissimilarities. The task is to find a reasonable 2-partition of object 2 to 12 and to detect the particularity of objects 1 and 13. The figure 2 shows the resulting fuzzy membership functions for the five classical algorithms, and the bba obtained with evidential clustering. Note that only 4 focal elements were considered: $\{\omega_1, \omega_2, \Omega, \emptyset\}$. As could be expected, among the five algorithms, only NRFCM is able to detect the outlier but the method fails with the first object (which is classified in class 2). The evidential clustering method (EVCLUS) provides a clear understanding of the data by allocating an important mass to the empty set for the outlier and to Ω for the first point.

4.2 "Cat cortex" data set

This real data set consists of a matrix of connection strengths between 65 cortical areas of the cat. It was collected by Scannell [17] and used by several authors to test visual-

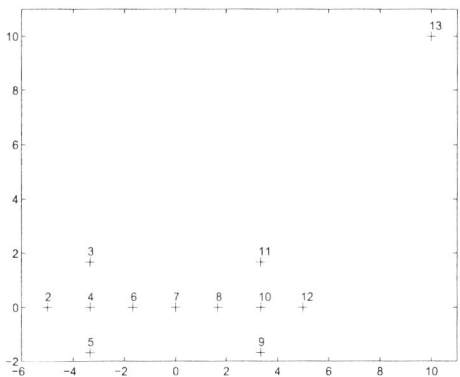

Figure 1: Synthetic dataset.

ization, discrimination or clustering algorithms based on proximity data [6, 7, 10]. The proximity values range from 0 (self-connection), to 4 (absent or unreported connection) with intermediate values : 1 (dense connection), 2 (intermediate connection) and 3 (weak connection). The cortex has been divided into four functional areas: auditory (A), visual (V), somatosensory (S), and frontolimbic (F). The clustering task is to find a four-class partition of the 65 cortical areas, based on the dissimilarity data, which is consistent with the functional regions. Six focal elements were considered for applying the evidential clustering method: 4 singletons $\{\omega_i\}$ ($i = 1, 4$), Ω and $\emptyset$. In order to provide a simple display of the results with EVCLUS, a two dimensional representation of the cortical areas has been obtained from the proximity matrix using a classical MDS algorithm [2]. The classification displayed on figure 3 is done according to the maximum of the pignistic probabilities. The clusters are represented by different symbols and the size of the symbols is proportional to the maximum of the pignistic probabilities. It can be seen that the four functional areas of the cortex are well-recovered. The error rate (only three points among 65 are misclassified), competes honourably with those reported in discrimination studies [6, 7].

5 Conclusion

In this paper we have suggested a new way of classifying relational data based on the theory of evidence. The classification task is performed in a very natural way, by only imposing that, the more two objects are similar, the more likely they belong to the same cluster. The concept of credal partition can be considered as a generalization of a probabilistic or possibilistic partition and offers a very flexible framework to handle noisy, imprecise or non-Euclidean data. Experiments on various datasets, which are not all reported here, have shown the efficiency of this approach.

Acknowledgements

The authors would like to thank Thore Graepel for providing the cat cortex dataset.

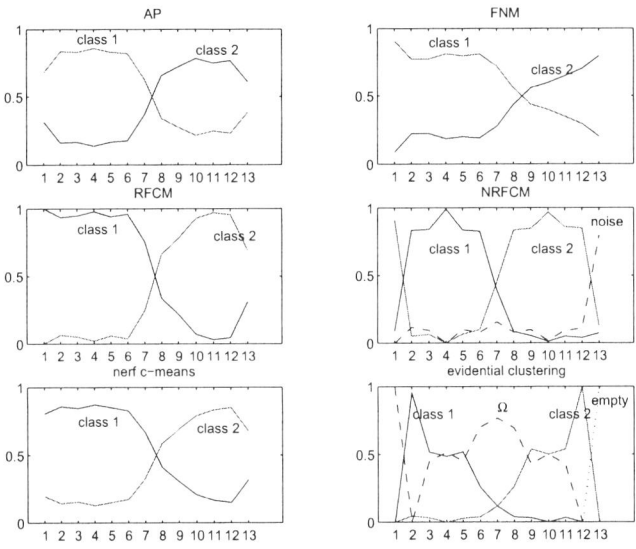

Figure 2: Synthetic dataset. Results of the six algorithms.

References

[1] J. C. Bezdek, J. Keller, R. Krishnapuram, and N. R. Pal. *Fuzzy models and algorithms for pattern recognition and image processing*. Kluwer Academic Publishers, 1999.

[2] I. Borg and P. Groenen. *Modern multidimensional scaling*. Springer, 1997.

[3] R.N. Davé. Clustering relational data containing noise and outliers. In *FUZZ'IEEE 98*, 1411–1416, 1998.

[4] T. Denœux. A k-nearest neighbor classification rule based on Dempster-Shafer theory. *IEEE Trans. on Systems, Man and Cybernetics*, 25(05):804–813, 1995.

[5] D. Dubois and H. Prade. Representation and combination of uncertainty with belief functions and possibility measures. *Computational Intelligence*, 4:244–264, 1988.

[6] T. Graepel, R. Herbrich, P. Bollmann-Sdorra, and K. Obermayer. Classification on pairwise proximity data. in *Advances in Neural Information Processing Systems 11*, MIT Press, Cambridge, MA, 438–444, 1999.

[7] T. Graepel, R. Herbrich, B. Schölkopf, A. Smola, P. Bartlett, K. Robert-Müller, K. Obermayer, and R. Williamson. Classification on Proximity Data with LP–Machines. in *Proceedings of ICANN'99*, 304–309, 1999.

[8] R.J. Hathaway and J.C. Bezdek. Nerf c-means : Non-euclidean relational fuzzy clustering. *Pattern Recognition*, 27:429–437, 1994.

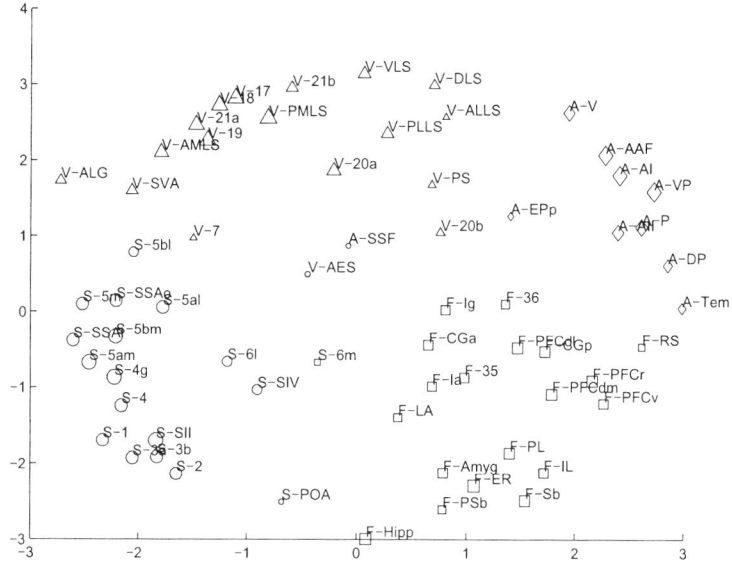

Figure 3: Cat cortex data set. Pignistic probabilities and hard assignments.

[9] R.J. Hathaway, J.W. Davenport, and J.C. Bezdek. Relational duals of the c-means clustering algorithms. *Pattern recognition*, 22(2):205–211, 1989.

[10] T. Hofmann, and J. Buhmann. Multidimensional scaling and data clustering. in *Advances in Neural Information Processing Systems 7*, MIT Press, Cambridge, MA, 459–466, 1995.

[11] A. K. Jain and R. C. Dubes. *Algorithms for clustering data.* Prentice-Hall, Englewood Cliffs, NJ., 1988.

[12] L. Kaufman and P. J. Rousseeuw. *Finding groups in data.* Wiley, New-York, 1990.

[13] G. J. Klir and M. J. Wierman. *Uncertainty-Based Information. Elements of Generalized Information Theory.* Springer-Verlag, 1998.

[14] N. R. Pal, J. C. Bezdek, and R. Hemasinha. Uncertainty measures for evidential reasoning I: A review. *International Journal of Approximate Reasoning*, 7:165–183, 1992.

[15] N. R. Pal, J. C. Bezdek, and R. Hemasinha. Uncertainty measures for evidential reasoning II: New measure of total uncertainty. *International Journal of Approximate Reasoning*, 8:1–16, 1993.

[16] M. Roubens. Pattern classification problems and fuzzy sets. *Fuzzy sets and systems*, 1:239–253, 1978.

[17] J.W. Scannell, C. Blakemore, and M.P. Young. Analysis of connectivity in the cat cerebral cortex. *Journal of Neuroscience*, 15(2):1463–1483, 1995.

[18] G. Shafer. *A mathematical theory of evidence*. Princeton University Press, Princeton, N.J., 1976.

[19] P. Smets and R. Kennes. The Transferable Belief Model. *Artificial Intelligence*, 66:191–243, 1994.

[20] M.P. Windham. Numerical classification of proximity data with assignment measures. *Journal of classification*, 2:157–172, 1985.

[21] R. R. Yager. On the normalization of fuzzy belief structure. *International Journal of Approximate Reasoning*, 14:127–153, 1996.

B. Bouchon-Meunier, L. Foulloy and R.R. Yager (Editors)
Intelligent Systems for Information Processing:
From Representation to Applications
© 2003 Elsevier Science B.V. All rights reserved.

A Hierarchical Linguistic Clustering Algorithm for Prototype Induction

I. González Rodríguez,[*] J. Lawry, J.F. Baldwin

A.I. Group, Dept. of Engineering Mathematics,
University of Bristol, Bristol BS8 1TR (U.K.)
{ines.gonzalez, j.lawry, jim.baldwin}@bristol.ac.uk

Abstract

A clustering algorithm is described which learns fuzzy prototypes to represent data sets and decides the number of prototypes needed. This algorithm is based on a modified hierarchical clustering scheme and incorporates ideas taken from mass assignment theory. It is illustrated using a model classification problem and its potential is shown by its application to a benchmark problem for glass identification.

Keywords: Fuzzy Prototypes, Clustering, Mass Assignment.

1 Introduction

In many of the emerging information technologies, there is a clear need for automated learning from databases. Data mining methods attempt to extract useful general knowledge from the implicit patterns contained in databases. Machine learning approaches learn models of complex systems capable of accurate prediction. Such methods have applications to classification problems, as well as vision, function approximation and control. For example, supermarkets are interested in learning prototypical descriptions of customers with certain purchasing behaviour; these models can then be used to learn descriptions of certain types of customers and to make informed decisions about levels of certain goods to stock, as well as pricing.

It is this need for automated learning that motivates our clustering algorithm, which tries to learn fuzzy prototypes to represent data sets and also decides the number of prototypes needed. Here, prototypes correspond to tuples of fuzzy sets on words over attribute universes and as such represent amalgams of similar objects sharing particular properties. It is this idea of grouping similar object that is central to the model of prototype induction proposed in this paper.

---

[*]Present address: Área de Matemática Aplicada, Universidad Rey Juan Carlos, Avenida Tulipán s/n, 28339 Móstoles (Madrid), Spain, e-mail: igonzalez@escet.urjc.es

2 Basic Mass Assignment Theory

The mass assignment for a fuzzy concept, first introduced by Baldwin [1, 4], can be interpreted as a probability distribution over possible crisp definitions of the concept. We might think of these varying definitions as being provided by a population of voters where each voter is asked to give his or her crisp definition of the concept.

Definition (Mass Assignment) Let f be a fuzzy set on a finite universe Ω such that the range of the membership function of f, χ_f, is $\{y_1, \ldots, y_n\}$ where $y_i > y_{i+1} > 0$. Then, the *mass assignment* of f, denoted m_f, is a probability distribution on 2^Ω satisfying

$$\begin{aligned} m_f(\emptyset) &= 1 - y_1 \\ m_f(F_i) &= y_i - y_{i+1} \text{ for } i = 1, \ldots, n-1 \\ m_f(F_n) &= y_n \end{aligned}$$

where $F_i = \{x \in \Omega \mid \chi_f(x) \geq y_i\}$ for $i = 1, \ldots, n$. $\{F_i\}_{i=1}^n$ are referred to as the *focal elements(sets)* of m_f.

The notion of mass assignment suggests a means of conditioning a variable X relative to a fuzzy constraint 'X is f' to obtain a probability distribution, by redistributing the mass associated with every focal set uniformly to the elements of that set. The probability distribution on X generated in this way is referred to as the *least prejudiced distribution* of f [1].

Definition (Least Prejudiced Distribution) For f a fuzzy subset of a finite universe Ω such that f is normalised, the *least prejudiced distribution of f*, denoted lp_f, is a probability distribution on Ω given by

$$lp_f(x) = \sum_{F_i : x \in F_i} \frac{m_f(F_i)}{|F_i|}$$

The idea of least prejudiced distribution provides us with an alternative definition of the conditional probability of fuzzy events [1].

Definition (Conditional Probability) For f and g fuzzy subsets of a finite universe Ω where g is normalised assuming no prior knowledge we define

$$\Pr(f|g) = \sum_{x \in \Omega} \chi_f(x) lp_g(x)$$

The least prejudiced distribution allows us, in a sense, to convert a fuzzy set into a probability distribution. That is, in the absence of any prior knowledge, we might on being told f naturally infer the distribution lp_f. Now, if we can find a method by which, when presented with a probability distribution, we can infer the fuzzy constraint generating that distribution, we can use fuzzy sets as descriptors of probability distributions.

Theorem Let Pr be a probability distribution on a finite universe Ω taking as a range of values $\{p_1, \ldots, p_n\}$ where $0 \leq p_{i+1} < p_i \leq 1$ and $\sum_{i=1}^{n} p_i = 1$. Then, Pr is the least prejudiced distribution of a fuzzy set f if and only if f has a mass assignment given by

$$m_f(F_i) = y_i - y_{i+1} \text{ for } i = 1, \ldots, n-1$$
$$m_f(F_n) = y_n$$

where

$$F_i = \{x \in \Omega \mid \Pr(x) \geq p_i\}$$
$$y_i = |F_i| p_i + \sum_{j=i+1}^{n} (|F_j| - |F_{j+1}|) p_j$$

Proof (See [2])

It is interesting to note that this transformation algorithm is identical to the bijective transformation method proposed by Dubois and Prade [6], although the motivation here is quite different. A further justification for this transformation can be found in [10].

Definition (Fuzzy Description) For a probability distribution Pr on a finite universe Ω, we refer to the fuzzy set generated from Pr according to the previous theorem as the *fuzzy description* of Pr, denoted $FD(\Pr)$.

3 Fuzzy Prototypes

We now use ideas from mass assignment theory to infer a number of prototypes representing a set of instances and where each prototype corresponds to a grouping of similar points (sometimes called a granule [11]). Unlike many current clustering methods, we do not intend to define single instances as being prototypical. Instead, a prototype is taken to be an amalgam of points represented by a tuple of fuzzy sets on each of the attributes describing an instance.

Definition (Fuzzy Prototype) A *fuzzy prototype* in $\Omega_1 \times \cdots \times \Omega_n$ is a n-tuple of fuzzy sets $\langle f_1, \ldots, f_n \rangle$ where f_i is a fuzzy subset of Ω_i.

In particular, we are interested in fuzzy prototypes generated from a set of elements so that they constitute a description of these elements. In order to determine which vectors should be associated with which prototypes, we need to define a notion of similarity. Furthermore, since data vectors can be viewed as a special case of prototypes, we need to define a similarity relation between fuzzy prototypes. Tversky's statement that similarity "may be better described as a comparison of features rather than as a computation of metric distance between points" [8] inspires the following definition, based on similarity measures between fuzzy sets.

Definition (Prototype Similarity) Let s_i be a similarity measure between fuzzy subsets on Ω_i and Π the class of all prototypes in $\Omega_1 \times \cdots \times \Omega_n$. The *prototype similarity measure* is a function $Sim : \Pi \times \Pi \mapsto [0,1]$ where, given the prototypes $P_1 = \langle f_1, \ldots, f_n \rangle$ and $P_2 = \langle g_1, \ldots, g_n \rangle$ in Π,

$$Sim(P_1, P_2) = \frac{1}{n} \sum_{i=1}^{n} s_i(f_i, g_i)$$

There is wide variety of similarity measures proposed in the literature which are possible candidates for s_i. For example, in [7, p24] there is a list of similarity indices which are a generalisation of the classical set-theory similarity functions. It is from this list that we have chosen a specific similarity measure to obtain the results shown in sections 6 and 7. Given f, g fuzzy sets on a finite universe Ω, we will take the similarity between them to be:

$$\begin{aligned} s(f,g) &= \frac{\sum_{x \in \Omega} \min(\chi_f(x), \chi_g(x))}{\sum_{x \in \Omega} \max(\chi_f(x), \chi_g(x))} \\ &= \frac{|f \cap g|}{|f \cup g|} \end{aligned}$$

Once it is clear what we understand by similar points (or prototypes), we need to define a means of grouping them. We do so by defining prototype addition.

Definition (Prototype Addition) Let $P_1 = \langle f_1, \ldots, f_n \rangle$ and $P_2 = \langle g_1, \ldots, g_n \rangle$ be prototypes in $\Omega_1 \times \cdots \times \Omega_n$ representing a granule of k and c data points respectively. Then, $P_1[+]P_2$ is a prototype in $\{\Omega_1, \ldots, \Omega_n\}$ such that

$$P_1[+]P_2 = \langle FD(r_1), \ldots, FD(r_n) \rangle$$

where r_i is the probability distribution in Ω_i given by

$$\forall x \in \Omega_i \; r_i(x) = \frac{k \, lp_{f_i}(x) + c \, lp_{g_i}(x)}{k + c}$$

4 Linguistic Variables

In the above, we have only considered finite universes, but most real-world problems involve continuous attributes. In this case, we need some way of converting infinite universes into finite ones so that the methods described in section 2 can be applied. Thus, we require some way of partitioning the universes associated with the continuous attributes. Fuzzy sets can be used to divide such universes into information granules, a term defined by Zadeh [11] as a group drawn together by similarity which can be viewed as corresponding to the meaning of words from natural language. Then, a linguistic variable can be defined which is associated with the original continuous attribute and takes as it values the words. Fuzzy sets on words can then be inferred.

Definition (Linguistic Variable) A *linguistic variable* is a quadruple[1]

$$\langle L, T(L), \Omega, M \rangle$$

---

[1] Zadeh [12] originally defined linguistic variable as a quintuple by including syntactic rule according to which new terms (i.e. linguistic values) could be formed by applying quantifiers and hedges to existing words. In this context, however, we shall asume that the term set is predefined and finite

in which L is the name of the variable, $T(L)$ is a finite term set of labels or words (i.e. the linguistic values), Ω is a universe of discourse and M is a semantic rule.

The semantic rule M is defined as a function that associates a normalised fuzzy subset of Ω with each word in $T(L)$. In other words, the fuzzy set $M(w)$ can be viewed as encoding the meaning of w so that for $x \in \Omega$ the membership value $\chi_{M(w)}(x)$ quantifies the suitability or applicability of the word w as a label for the value x. This generates a fuzzy subset of $T(L)$ describing the value of x.

Definition (Linguistic Description of a Value) For $x \in \Omega$, the *linguistic description* of x relative to the linguistic variable L is the fuzzy subset of $T(L)$

$$des_L(x) = \sum_{w \in T(L)} w/\chi_{M(w)}(x)$$

In cases where the linguistic variable is fixed, the subscript L is dropped and we write $des(x)$.

Once we have a fuzzy set on words for a given $x \in \Omega$, $des(x)$, which is a fuzzy subset of a finite universe $T(L)$, the basic mass assignment theory is applicable and it is possible to consider the least prejudiced distribution of $des(x)$, $lp_{des(x)}$, this being a probability distribution in the set of labels $T(L)$.

Now recall that the least prejudiced distribution is defined only for normalised fuzzy sets and, hence, it is desirable that linguistic descriptions as defined above be normalised. This will hold if and only if the semantic function generates a linguistic covering defined as follows.

Definition (Linguistic Covering) A set of fuzzy sets $\{f_i\}_{i=1}^n$ forms a *linguistic covering* of the universe Ω if and only if

$$\forall x \in \Omega \;\; \max_{i=1}^n (\chi_{f_i}(x)) = 1$$

Having described how fuzzy sets can be used as descriptors for probability distributions, we shall introduce a prototype induction algorithm based on these ideas. The prototypes induced belong to an specific type of fuzzy prototype (as described in section 3) where each attribute is described by a fuzzy set on words, with labels provided by the linguistic covering of the attribute's universe. We may think of each of these fuzzy sets as a possibility distribution on the set of linguistic labels describing the attribute's universe. In this way we can evaluate the possibility that a particular label describes an attribute for a certain prototype.

5 A Hierarchical Linguistic Clustering Algorithm

Traditional hierarchical clustering algorithms produce a series of partitions of the data, $P_n, P_{n-1}, \ldots, P_1$. The first, P_n, consists of n single-member clusters, the last P_1 consists of a single group containing all n individuals. At each particular stage, the methods fuse individuals or groups of individuals which are closest (or most similar), it being the chosen definition of 'closeness' that differentiates one method from another. A description of these methods can be found in [5, pp55–90].

However, it is often the case, when hierarchical clustering techniques are used, that what is of interest is not the complete hierarchy obtained by the clustering, but only one or two partitions obtained from it that contain an 'optimal' number of clusters. Therefore, it is necessary to select one of the solutions in the nested sequence of clusterings that comprise the hierarchy. This in itself is a very challenging problem.

Another issue that should be addressed is the computational complexity of exhaustively searching for the pair of most similar elements at each stage of the clustering. This may become an important issue when the number of data points n is 'large', which is usually the case in practical applications.

In our algorithm, we try to overcome these problems by introducing several changes in the scheme described above. First of all, we have already mentioned that each prototype corresponds to a grouping of similar points. Therefore, if two prototypes are not 'similar enough', this implies that the elements in the clusters they represent are not close and should not be merged into a single group. For this reason, we define a similarity threshold $\sigma \in [0,1]$, according to which the grouping of clusters will terminate once the similarity between prototypes falls below σ.

Also, we will introduce a heuristic search for pairs of similar prototypes and we will allow more than one pair to fuse at each level of the clustering. The search for similar elements in one partition P_i will start with the first cluster in P_i, according to some arbitrary ordering, and will go through the elements in this partition to select the most similar one to it. If the similarity between these two clusters is high enough according to the threshold σ, then they should be merged into only one cluster and this new group should be added to the next partition P_{i-1}. If, on the contrary, they are not similar enough, the first cluster alone will be added to the next partition. We repeat this process with the remaining clusters in P_i until all of them have been considered.

This reduces the complexity of the original algorithms in two ways. First of all, by using this heuristic we do not need to compare all the elements to find the most similar pair. Secondly, if we have m clusters, our search will allow us to find up to $m/2$ pairs of similar prototypes as our candidates to merge. Whilst the standard hierarchical algorithms reduce the number of elements from one partition, P_i, to the next one, P_{i-1}, by only one cluster, our search allows us to reduce the number of clusters by up to $m/2$.

Let $S = \{\langle i, \vec{x}(i)\rangle \mid i = 1, \ldots, N\}$, where $\vec{x}(i) \in \Omega_1 \times \cdots \times \Omega_n$, be a data set. For each continuous attribute $j \in \{1, \ldots, n\}$, let us suppose that we have a linguistic covering of the universe Ω_i. We can rewrite the attribute's value for each data point in S, $x_j(i)$ as a fuzzy set of words, namely, its linguistic description $des(x_j(i))$. For the sake of a simpler notation, let us identify $x_j(i)$ with $des(x_j(i))$. Finally, let us suppose that we have Sim the similarity measure between fuzzy objects defined in section 3. Then, having set a threshold $\sigma \in [0,1]$ for the similarity, our linguistic hierarchical clustering algorithm is described as follows:

   C= $\{\vec{x}(i)|\langle i, \vec{x}(i)\rangle \in S\}$
   CHANGED=true
   while CHANGED==true **do**
     NEWC= $\emptyset$
     CHANGED=false
     TO-MERGE= $\{i \mid \langle i, \vec{x}(i)\rangle \in C\}$
     MERGED= $\emptyset$

```
while TO-MERGE≠ ∅ do
    pick i ∈ TO-MERGE
    pick j ∈ TO-MERGE such that
        Sim(x⃗(i), x⃗(j)) =
        max_{k∈TO-MERGE} Sim(x⃗(i), x⃗(k))
    if Sim(x⃗(i), x⃗(j)) > σ then
        add x⃗(i)[+]x⃗(j) to NEWC
        delete {i, j} from TO-MERGE
        add {i, j} to MERGED
        CHANGED=true
    else
        add x⃗(i) to NEWC
        delete {i} from TO-MERGE
        add {i} to MERGED
    end if
end while
C=NEWC
end while
NEWC contains the final clusters for S
```

It may be worth noting that, in the case that new data points are obtained, there is no need to re-run the clustering from the beginning with the enhanced data set in order to update the final set of prototypes. Instead, it is enough to run the clustering algorithm with the union of the old prototypes and the new data points as our new data set.

For supervised learning where the data set is partitioned according to class, $S = \bigcup_{i=1}^{k} S_i$ where S_i is the set of data points in S of class C_i, the hierarchical linguistic clustering algorithm is applied to the subsets of data formed by each of the classes.

Having learnt a set of fuzzy prototypes describing the set S and given any tuple of values $\vec{x} \in \Omega_1 \times \cdots \times \Omega_n$, we can determine the support for it belonging to or being associated with a particular fuzzy prototype $P = \langle f_1, \ldots, f_n \rangle$ as follows:

$$supp(P|\vec{x}) = \prod_{i=1}^{n} \Pr(f_i | des(x_i))$$

In particular, for supervised learning, if we are given a tuple of values $\vec{x} \in \Omega_1 \times \cdots \times \Omega_n$ and we are asked to classify it, we can evaluate the support for each of the prototypes learned. Then, the vector is classified as belonging to the class associated with the prototype with highest support.

6 Application to a Model Classification Problem

To illustrate the above algorithm and its potential, let us consider a toy problem. In this problem, the data set consists of 916 data points in $[-1, 1] \times [-1, 1]$. The data set is divided in two classes, *legal* and *illegal*. If we consider two concentric circles, then the 345 points in the inner circle and the exterior circular crown are labelled as legal; the other 616 points in $[-1, 1] \times [-1, 1]$ are labelled as illegal. A plot of the legal points can be seen in Figure 1.a.

As we have continuous attributes, we use linguistic coverings with 12 trapezoidal fuzzy sets uniformly distributed over each universe. If we set our similarity threshold to $\sigma = 0.5$, the clustering terminates with 36 prototypes to represent the *illegal* points and 31 to represent the *legal* ones. Even though there has been a considerable reduction in the number of clusters from our initial partitions of 616 and 345 data points respectively, we might want to merge more and reduce even further the number of final clusters. If we lower the similarity threshold to $\sigma = 0.4$, we allow clustering at one partition level higher and reduce the number of final prototypes to 17 for each class. Therefore, with only one more intermediate partition, we have approximately halved the number of final clusters for each class. This information is sumarised in Table 1. Of course, we are interested

Table 1: Number of Clustering Stages (NoCS) and Number of Prototypes (NoP) for each class

threshold	NoCS		NoP	
	illegal	legal	illegal	legal
$\sigma = 0.5$	6	5	36	31
$\sigma = 0.4$	7	6	17	17

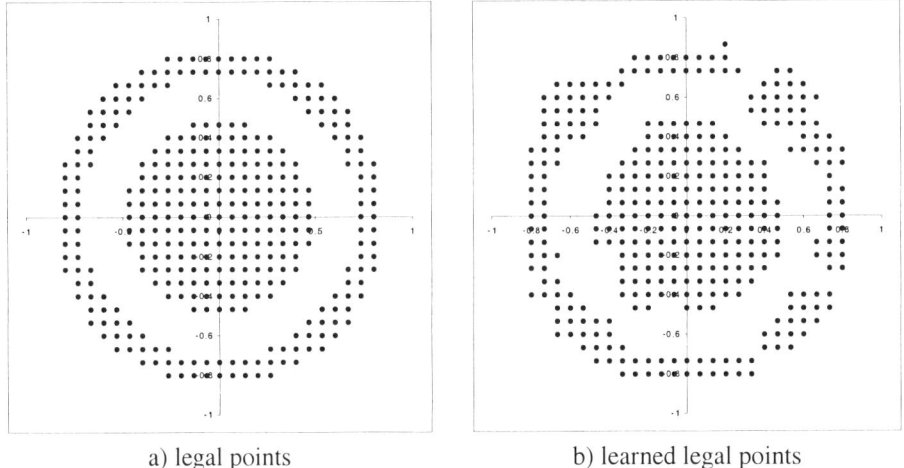

a) legal points   b) learned legal points

Figure 1: Concentric Circles

in knowing how well these prototypes represent our original data. For this purpose, we ran a classifier through the same data and obtained a predictive accuracy of 95.4% for the threshold $\sigma = 0.5$ and of 93% for the threshold $\sigma = 0.4$. Obviously, the reduction in the number of prototypes gives some loss of predictive accuracy, although both results can be considered as good. Figure 1.b is a plot of those points predicted *legal* according to prototypes learned with $\sigma = 0.4$.

7 Application to a Real-World Classification Problem

This database was taken from the UCI machine learning repository [9] and originates from a project carried out by the British Home Office Forensic Science Service Central Research Establishment on the identification of glass fragments found at crime scenes [4]. The study is motivated by the fact that in a criminal investigation the glass found at the scene of the crime can only be used as evidence if it is correctly identified. Glass fragments are divided into 7 possible classes, although the database only contains examples of six (there are no instances of class 4). These are:

1. Building windows—float processed
2. Building windows—non float processed
3. Vehicle windows—float processed
4. Vehicle windows—non float processed
5. Containers
6. Tableware
7. Headlamps

The classification is to be made on the basis of the following 9 attributes, relating to certain chemical properties of the glass (the unit measurement for attributes 2–9 is the weight percent of the corresponding oxide):

1. RI—refractive index
2. Na—sodium
3. Mg—magnesium
4. Al—aluminium
5. Si—silicon
6. K—potasium
7. Ca—calcium
8. Ba—barium
9. Fe—iron

The database, consisting of 214 instances, was split into a training and test set of 107 instances each in such a way that the instances of each class were divided equally between the two sets. A linguistic covering of 5 trapezoidal fuzzy sets was then defined for each attribute where a percentile approach was used to determine the exact position of the fuzzy sets (see Figure 2).

The threshold parameter for the similarity was set at $\sigma = 0.5$ and the linguistic hierarchical clustering was applied to the data. The number of clustering stages (consecutive partitions of the data) used and the number of 'optimal' prototypes for each class can be seen in Table 2.

Figure 3 shows the first attribute, RI, of the 5 prototypes obtained for the first class, *building windows—float processed*.

The accuracy obtained using the prototypes for classification was 98% on the training set and 77.5% on the test set. This compares favourably with other learning algorithms. For instance, a previous mass assignment based prototype induction algorithm [4] gave

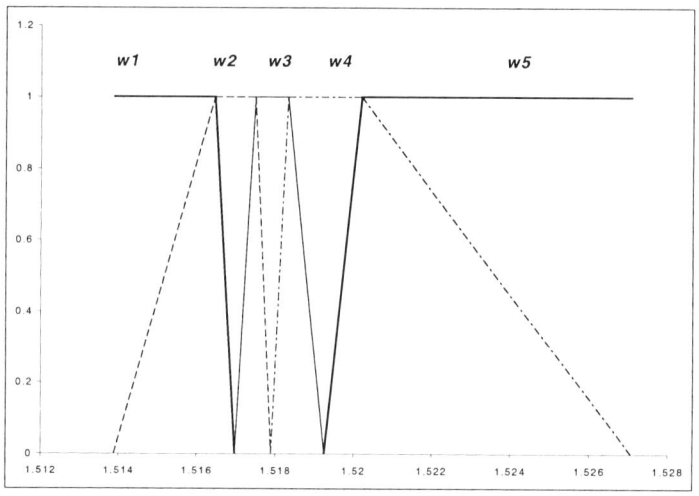

Figure 2: Linguistic covering for attribute 1, RI

Table 2: Number of Clustering Stages (NoCS) and Number of Prototypes (NoP) for each class

CLASS	NoCS	NoP
1	5	5
2	5	4
3	2	4
5	3	2
6	2	1
7	4	5

an accuracy of 71% on the test set. Also, mass assignment ID3 [3] gave an accuracy of 68% on the test set and a neural network with topology 9–6–6 gave 72% on a smaller test set where the network was trained on 50% of the data and validated on 25% and tested on 25% [4].

8 Conclusions

A linguistic hierarchical clustering algorithm has been described for learning fuzzy prototypes to represent a data set as well as the number of prototypes needed. This algorithm incorporates ideas from mass assignment theory and similarity relations between fuzzy objects. The potential of the linguistic hierarchical clustering has been illustrated with both a toy example and a real world problem.

Acknowledgements

This work is partially funded by an E.P.S.R.C. studentship.

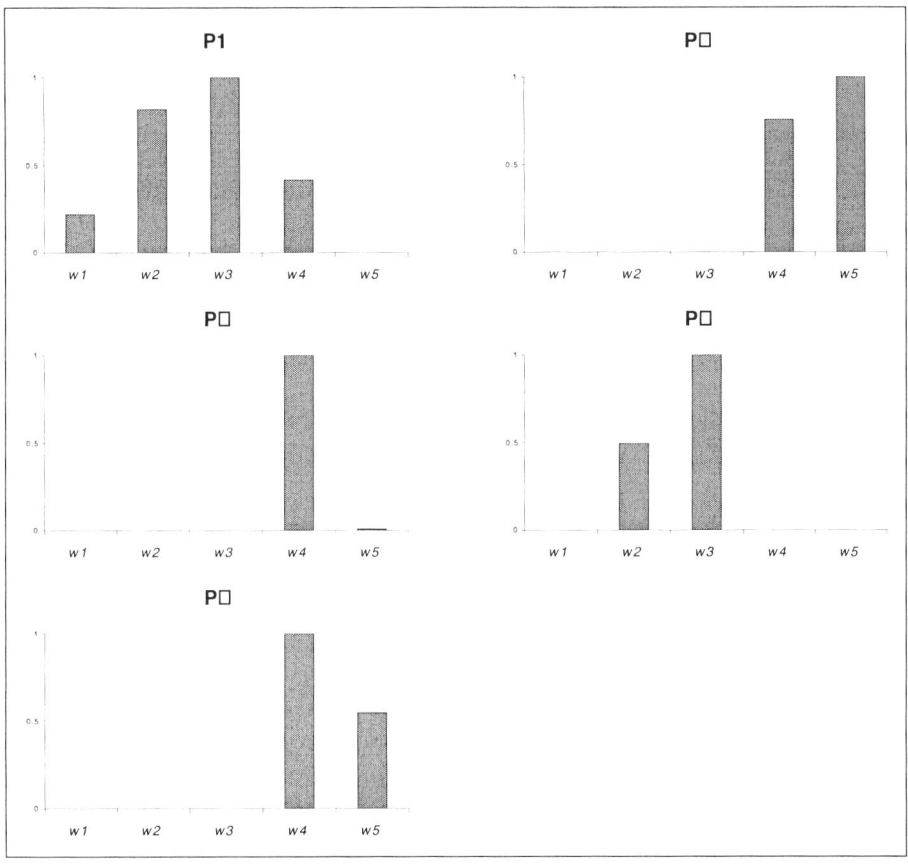

Figure 3: RI in the prototypes for class 1

References

[1] J.F. Baldwin, J. Lawry, T.P. Martin, *Mass Assignment Theory of the Probability of Fuzzy Events*, Fuzzy Sets and Systems, Vol. 83, pp353–367, 1996.

[2] J.F. Baldwin, J. Lawry, T.P. Martin, *The Application of Generalised Fuzzy Rules to Machine Learning and Automated Knowledge Discovery*, International Journal of Uncertainty, Fuzzyness and Knowledge-Based Systems, Vol. 6, No. 5, pp459–487, 1998.

[3] J.F. Baldwin, J. Lawry, T.P. Martin, *Mass Assignment Based Induction of Decision Trees on Words*, Proceedings of IPMU98, Paris, France, 1998.

[4] J.F. Baldwin, J. Lawry, T.P. Martin, *A Mass Assignment Method for Prototype Induction*, International Journal of Intelligent Systems, Vol. 14, No. 10, pp1041–1070, 1999.

[5] B. S, Everitt, *Cluster Analysis*, Edward Arnold (Ed.), third edition, 1993.

[6] D. Dubois, H. Prade, *Unfair Coins and Necessity Measures: Towards a Possibilistic Interpretations of Histograms*, Fuzzy Sets and Systems, Vol, 10, pp15–20, 1979.

[7] D. Dubois, H. Prade, *Fuzzy Sets and Systems: Theory and Applications*, Academic Press, New York, 1980.

[8] A. Tversky, *Features of similarity*, Psychological Review, Vol. 84, No. 4, pp327–353, 1977.

[9] WWW, *UCI Machine Learning Repository*, http://www.ics.uci.edu/~mlearn/MLRepository.html.

[10] K. Yamada, *Probability-Possibility Transformation Based on Evidence Theory*, Proceedings of IFSA-NAFIPS'2001, Vancouver, Canada, 2001.

[11] L. A. Zadeh, *Fuzzy Sets and Information Granularity* in M. Gupta, R. Ragade, R. Yager (Eds), Advances in Fuzzy Set Theory and Applications, pp3–18, North-Holland, Amsterdam, 1979.

[12] L. A. Zadeh, *The concept of a linguistic variable and its applications to approximate reasoning*, Part I: Information Sciences, Vol. 8, pp199–249, 1975; Part II: Information Sciences, Vol. 8, pp301–357; Part III: Information Sciences, Vol. 9, pp43–80, 1976.

A Multiobjective Genetic Algorithm for Feature Selection and Data Base Learning in Fuzzy-Rule Based Classification Systems

O. Cordón[a], F. Herrera[a], M.J. del Jesus[b]
L. Magdalena[c], A.M. Sánchez[d], P. Villar[d] *

[a]Dept. of Computer Science and A.I.; University of Granada,
18071 - Granada, Spain. e-mail: ocordon,herrera@decsai.ugr.es

[b]Dept. of Computer Science; University of Jaén,
23071 - Jaén, Spain. e-mail: mjj@ujaen.es

[c]Dept. Mat. Aplicada; Universidad Politécnica de Madrid
28040 Madrid, Spain. e-mail: llayos@mat.upm.es

[d]Dept. Computer Science; University of Vigo
32004 - Ourense, Spain. e-mail: amlopez,pvillar@uvigo.es

Abstract

In this contribution, we propose a genetic process to select an appropiate set of features in a Fuzzy Rule-Based Classification System (FRBCS) and to automatically learn the whole Data Base definition. An ad-hoc data covering learning method is considered to obtain the Rule Base. The method uses a multiobjective genetic algorithm in order to obtain a good balance between accuracy and interpretability.

Keywords: Fuzzy Rule-Based Classification Systems, Data Base, Learning, Multiobjective Genetic Algorithms

1 Introduction

An FRBCS presents two main components: the Inference System and the Knowledge Base (KB). The KB is composed of the Rule Base (RB) constituted by the collection of fuzzy rules, and of the Data Base (DB), containing the membership functions of the fuzzy partitions associated to the linguistic variables. The composition of the KB of an FRBCS directly depends on the problem being solved. If there is no expert information about the problem under solving, an automatic learning process must be used to derive the KB from examples.

Although, there is a large quantity of RB learning methods proposed in the specialized literature [3, 4, 10, 15], among others, there is not much information about the way to

---

*This research has been supported by CICYT under project TIC2002-04036-C05-01

derive the DB and most of these RB learning methods need of the existence of a previous definition for it. The usual way to proceed involves choosing a number of linguistic terms (granularity) for each linguistic variable, which is normally the same for all of them, and building the fuzzy partition by a uniform partitioning of the variable domain into this number of terms. This operation mode makes the granularity and fuzzy set definitions have a significant influence on the FRBCS performance.

Moreover, high dimensionality problems present a new trouble to obtain FRBCSs with good behaviour: the large number of features, that can originate a RB with a high number of rules, thus presenting a low degree of interpretability and a possible over-fitting. This problem can be tackled from a double perspective:

- Via the compactness and reduction of the rule set, minimising the number of fuzzy rules included in it.

- Via a feature selection process that reduces the number of features used by the FRBCS.

Rule reduction methods have been formulated using different approaches (Neural Networks, clustering techniques, orthogonal transformation methods, similarity measures and Genetic Algorithms). Notice that, for high dimensional problems and problems where a high number of instances is available, it is difficult for the latter approaches to get small rule sets, and therefore the system comprehensibility and interpretability may not be as good as desired. For high dimensionality classification problems, a feature selection process, that determines the most relevant variables before or during the FRBCS inductive learning process, must be considered [2, 19]. It increases the efficiency and accuracy of the learning and classification stages.

Our objective is to develop a genetic process for feature selection and whole DB learning (granularity and membership functions for each variable) to obtain FRBCSs composed of a compact set of comprehensible fuzzy rules with high classification ability. This method uses a multiobjective GA and considers a simple generation method to derive the RB.

To carry out this task, this paper is organised as follows. In Section 2, the FRBCS components will be introduced joint with a brief description of the two main problems tackled by the learning method proposed, feature selection and DB learning. In Section 3 we will expose the characteristics of our proposal for the FRBCS design. The results obtained with Sonar data set will be shown in Section 4. In the last section, some conclusions will be pointed out.

2 Preliminaries

2.1 Fuzzy Rule-Based Classification Systems

An FRBCS is an automatic classification system that uses fuzzy rules as knowledge representation tool. Two different components are distinguished within it:

1. The **KB**, composed of:

- **DB**, which contains the fuzzy set definitions related to the labels used in the fuzzy rules. So, the DB components for every variable are the number of linguistic terms (granularity) and the membership function shape of each term.

- **RB**, comprised by a set of fuzzy rules that in this work are considered to have the following structure:

$$R_k : \text{If } X_1 \text{ is } A_1^k \text{ and } \ldots \text{ and } X_N \text{ is } A_N^k \quad \text{then} \quad Y \text{ is } C_j \text{ with } r^k$$

where $X_1, \ldots, X_N$ are features considered in the problem and $A_1^k, \ldots, A_N^k$ are linguistic labels employed to represent the values of the variables. These kinds of fuzzy rules represent, in the antecedent part, a subspace of the complete search space by means of a linguistic label for each considered variable and, in the consequent part, a class label (C_j) and a certainty degree (r^k). This numerical value indicates the degree of certainty of the classification in that class for the examples belonging to the fuzzy subspace delimited by the antecedent part.

2. The **Fuzzy Reasoning Method (FRM)**, an inference procedure which, combining the information provided by the fuzzy rules related with the example to classify, determines the class to which it belongs to.

The majority of FRBCSs (see [3, 10] among others) use the classical FRM that classifies a new example with the consequent of the fuzzy rule having the highest degree of association. Another family of FRMs that use the information provided by all the rules compatible with the example (or a subset of them) have been developed [3, 5]. In this work, we use two different FRMs, each one belonging to one of the said groups: maximum and normalised sum.

2.2 Feature Selection and DB Learning in FRBCS design

As we mentioned before, our FRBCS learning method generates the KB by selecting an adequate feature set and by learning the appropiate DB components for each selected variable. In this section, we briefly describe these problems jointly solved in our proposal.

2.2.1 Feature Selection Process

The main objective of any feature selection process is to reduce the dimensionality of the problem for the supervised inductive learning process. This fact implies that the feature selection algorithm must determine the best features for its design.

There are two kinds of feature selection algorithms:

- *Filter feature selection algorithms* [17], which remove the irrelevant characteristics without using a learning algorithm (e.g. by means of class separability measures). They are efficient processes but, on the other hand, the feature subsets obtained by them may not be the best ones for a specific learning process because of the exclusion of the heuristic and the bias of the learning process in the selection procedure [16].

- *Wrapper feature selection algorithms* [16, 17]. This kind of feature selection algorithms selects feature subsets by means of the evaluation of each candidate subset with the precision estimation obtained by the learning algorithm. In this form, they obtain feature subsets with the best behaviour in the classifier design. Their problem is their inefficiency since the classifier has to be built for the evaluation of each candidate feature subset.

In our proposal we will use a wrapper feature selection algorithm which utilises the precision estimation provided by an efficient fuzzy rule generation process (Wang and Mendel's fuzzy rule generation process) and a GA as search algorithm. Inside the DB derivation, the granularity learning will provide us an additional way to select features when the number of linguistic labels assigned to a specific variable is only one (we will explain this in a further section).

2.2.2 DB Learning

As previously said, the derivation of the DB highly influences the FRBCS performance. In fact, some studies in Fuzzy Rule-Based Systems have shown that the system performance is much more sensitive to the choice of the semantics in the DB than to the composition of the RB [7]. Some approaches have been proposed to improve the FRBCS behaviour by means of a tuning process once the RB has been derived [4]. However, these tuning processes only adjust the shapes of the membership functions and not the number of linguistic terms in each fuzzy partition, which remains fixed from the begining of the design process.

The methods that try to learn appropiate DB components per variable usually work in collaboration with an RB derivation method. A DB generation process wraps an RB learning one working as follows: each time a DB has been obtained by the DB definition process, the RB generation method is used to derive the rules, and some type of error measure is used to validate the whole KB obtained. The works proposed in [6, 7, 8] use Simulated Annealing and GAs to learn an appropiate DB in a Fuzzy Rule-Based System. The method proposed in [14] considers a GA to design an FRBCS, working in the said way.

3 Genetic Algorithm for Feature Selection and DB Learning

In this section, we propose a new learning approach to automatically generate the KB of an FRBCS composed of two methods with different goals:

- A genetic learning process for the DB that allows us to define:

    - The relevant variables for the classification process (feature selection).

    - The number of labels for each variable (granularity learning).

- The form of each fuzzy membership function in non-uniform fuzzy partitions, using a non-linear scaling function that defines different areas in the variable working range where the FRBCS has a higher or a lower relative sensibility.

- A quick *ad hoc data-driven method* that derives the fuzzy classification rules considering the DB previously obtained. In this work we use the extension of Wang and Mendel's fuzzy rule generation method [21] for classification problems [3], but other efficient generation methods can be considered.

We should note that the granularity learning allows us another way of feature selection: if a variable is assigned only to one label, it has no influence in the RB, so it will not be considered as a relevant variable. A similar double-level feature selection process has been previously considered in genetic learning processes of FRBCSs such as SLAVE [11].

All the components of the DB will be adapted throughout a genetic process. Since it is interesting to reduce the dimensionality of the search space for that process, the use of non-linear scaling functions is conditioned by the necessity of using parameterized functions with a reduced number of parameters. We consider the scaling funtion proposed in [6], that has a single sensibility parameter called a ($a \in \mathbb{R}$). The function used is ($f : [-1, 1] \to [-1, 1]$)

$$f(x) = sign(x) \cdot |x|^a, \quad with \ a > 0$$

The final result is a value in $[-1, 1]$ where the parameter a produces uniform sensibility ($a = 1$), higher sensibility for center values ($a > 1$), or higher sensibility for extreme values ($a < 1$). In this paper, triangular membership functions are considered due to their simplicity. So, the non-linear scaling function will only be applied on the three definition points of the membership function (which is equal to transform the scaling function in a continuous piece-wise linear function), in order to make easier the structure of the generated DB and to simplify the defuzzification process. Figure 1 shows a graphical representation of the three possibilities of fuzzy partition depending on the value of parameter a.

We should note that the previous scaling function is recommended to be used with symmetrical variables since it causes symetrical effects around the center point of the interval. For example, it can not produce higher sensibility in only one of the working range extents. In the method presented in this paper, we add a new parameter (called S) to the non-linear scaling function as described also in [6]. S is a parameter in $\{0, 1\}$ to distinguish between non-linearities with symmetric shape (lower sensibility for middle or for extreme values, Figure 1) and asymmetric shape (lower sensibility for the lowest or for the highest values, Figure 2).

Furthermore, the main purpose of our KB design process is to obtain FRBCSs with good accuracy and high interpretability. Unfortunately, it is not easy to achieve these two objectives at the same time. Normally, FRBCSs with good performance have a high number of selected variables and also a high number of rules, thus presenting a low degree

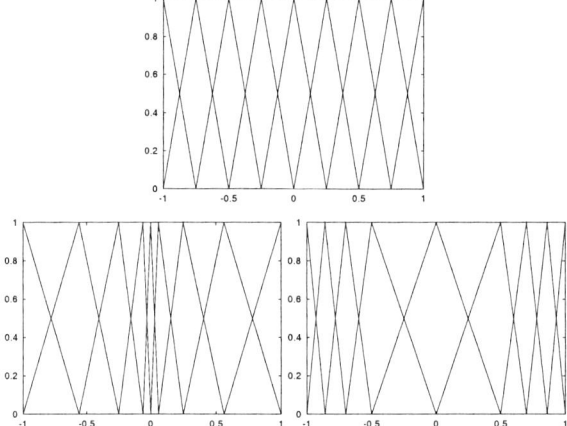

Figure 1: Fuzzy partitions with $a = 1$ (top), $a > 1$ (down left), and $a < 1$ (down right)

of readability. On the other hand, the KB design methods sometimes lead to a certain overfitting to the training data set used for the learning process.

To avoid these problems, our genetic process uses a multiobjective GA with two goals:

- Minimise the classification error percentage over the training data set.

- Design a compact and interpretable KB. This objective is performed by penalising FRBCSs with a large number of selected features and high granularity.

The next subsections describe the main components of the genetic learning process.

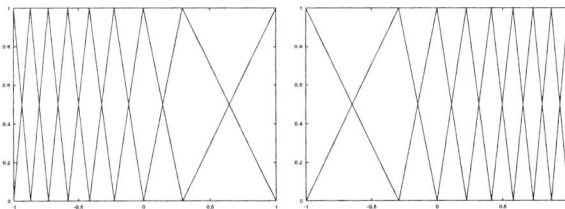

Figure 2: Fuzzy partitions with $S = 1$ (left with $a > 1$ and right with $a < 1$)

3.1 Encoding the DB

The two main DB components of the selected variables are the number of linguistic terms and the membership functions that define their semantics. Therefore, each chromosome will be composed of three parts:

- Relevant variables (C_1): For a classification problem with N variables, the selected features are stored into a binary coded array of length N. In this array, an 1 indicates that the corresponding variable is selected for the FRBCS.

- **Number of labels (C_2):** The number of labels per variable is stored into an integer array of length N. In this contribution, the possible values considered are taken from the set $\{1,\ldots,5\}$.

- **Sensibility parameters (C_3):** An array of lenght $N \times 2$, where the sensibility parameters (a,S) are stored for each variable. In our case, the range considered for the parameter a is the interval $(0,8]$.

If v_i is the bit that represents whether the variable i is selected and l_i is the granularity of variable i, a graphical representation of the chromosome is shown next:

$$C_1 = (v_1,\ldots,v_N) \quad C_2 = (l_1,\ldots,l_N)$$
$$C_3 = (a_1,\ldots,a_N,S_1,\ldots,S_N)$$
$$C = C_1 C_2 C_3$$

3.2 Evaluating the chromosome

There are three steps that must be done to evaluate each chromosome:

1. Generate the fuzzy partitions using the information contained in the chromosome. Obviously, this process is only applied to the selected variables ($v_i = 1$ and $l_i > 1$). First, each variable is linearly mapped from its working range to $[-1,1]$. In a second step, uniform fuzzy partitions for all the variables are created considering the number of labels per variable (l_i). Finally, the non-linear scaling function with its sensibility parameters (a_i, S_i) is applied to the definition points of the membership functions obtained in the previous step, obtaining the whole DB definition.

2. Generate the RB by running a fuzzy rule learning method considering the DB obtained in the previous step.

3. Calculate the two values of the evaluation function:

    - CPE: classification percentage error over the training set.

    - $SV \cdot AL$: with SV being the number of selected variables and AL being the averaged granularity of the selected variables.

3.3 Genetic operators

The initial population is selected considering several groups of chromosomes, each one of them with different percentage for the selected variables (randomly chosen). The remaining values of the chromosome are chosen at random. For the rest of GA components, the following operators are considered.

3.3.1 Selection

We have used the selection mechanism of MOGA [9], which is based on the definition of Pareto-optimality. It is said that a solution dominates another when the former achieves better or equal values than the latter in all but one objective, where the former outperforms the latter. Hence, the pareto is composed of all the non-dominated solutions.

Taking this idea as a base, MOGA assigns the same selection probability to all non-dominated solutions in the current population. The method involves dividing the population into several classes depending on the number of individuals dominating the members of each class. Therefore, the selection scheme of our multiobjective GA involves the following five steps:

1. Each individual is assigned a rank equal to the number of individuals dominating it plus one (chromosomes encoding non-dominated solutions receive rank 1).

2. The population is increasingly sorted according to that rank.

3. Each individual is assigned a selection probability which depends on its ranking in the population, with lower ranking receiving lesser probabilities.

4. The selection probability of each equivalence class (group of chromosomes with the same rank, i.e., which are non-dominated among them) is averaged.

5. The new population is created by following the Baker's stochastic universal sampling [1].

3.3.2 Crossover

Two different crossover operators are considered depending on the two parents' scope:

- *Crossover when both parents have the same selected variables and equal granularity level per variable:* If the two parents have the same values in C_1 and C_2, the genetic search has located a promising space zone that has to be adequatelly exploited. This task is developed by applying the max-min-arithmetical (MMA) crossover operator [13] in the chromosome part based on real-coding scheme (parameters a_i) and obviously by maintaining the parent C_1 and C_2 values in the offspring. Both combinations of parameter S_i are tested and the best two chromosomes are selected.

- *Crossover when the parents encode different selected variables or granularity levels:* This second case highly recommends the use of the information encoded by the parents to explore the search space in order to discover new promising zones. So, a standard crossover operator is applied over the three parts of the chromosome. This operator performs as follows: a crossover point p is randomly generated in C_1 and the two parents are crossed at the p-th variable in all the chromosome parts, thereby producing two meaningful descendents.

3.3.3 Mutation

Three different operators are used, each one of them acting on different chromosome parts:

- *Mutation on C_1 and on the second part of C_3 (parameters S_i)*: As these parts of the chromosome are binary coded, a simple binary mutation is developed, flipping the value of the gene.

- *Mutation on C_2*: The mutation operator selected for the granularity levels is similar to the one proposed by Thrift in [20]. A local modification is developed by changing the number of labels of the variable to the immediately upper or lower value (the decision is made at random). When the value to be changed is the lowest (1) or the highest one, the only possible change is developed.

- *Mutation on first part of C_3 (parameters a_i)*: As this part is based on a real-coding scheme, Michalewicz's non-uniform mutation operator is employed [18].

4 Experimentation

We have applied the learning method to an example base with a high number of features, Sonar data set [12], which has 208 instances of a sonar objective classification problem. Each one of these instances is described by 60 features to discriminate between a sonar output corresponding to a cylindrical metal or an approximately cylindrical rock. The training set contains 104 elements and the test set contains the remaining 104 elements. Table 1 shows the parameter values considered for the experiments developed.

Table 1: Parameter values

Parameter	Value
Granularity values	$\{1,\ldots,5\}$
Population size	100
Crossover probability	0.6
Mutation probability	0.2
Number of generations	$\{100, 500\}$

The best results obtained by our genetic learning process for the two FRMs considered are shown in Table 2. The best results found with the Wang and Mendel's RB generation method considering all the features selected and the same number of labels for each one of them are also shown in the top line of each FRM. The table contains the following columns:

- **FRM**: Fuzzy Reasoning Method used.

- **SV**: Number of selected variables.

- **AL**: Average of the number of labels considered for the selected variables.

- **NR**: The number of rules of the FRBCS.
- **% tra**: Classification percentage error obtained in the training data set.
- **% tst**: Classification percentage error obtained in the test data set.

Table 2: Best results obtained (% error)

FRM	SV	AL	NR	% tra	% tst
	60	3	104	0.9	23.1
	9	3.6	101	0.0	16.3
	8	3.6	91	2.8	13.4
Maximum	6	3.6	91	4.8	15.3
	4	4.2	72	7.7	19.2
	4	3.2	55	15.3	15.3
	3	3.6	42	19.2	19.2
	2	2.5	9	27.8	25.0
	60	3	104	2.8	25.9
	9	4.1	100	0.0	18.2
	7	4.1	92	1.9	19.2
Normalised	6	3.8	93	6.7	13.4
Sum	5	4.0	76	12.5	16.3
	4	3.7	65	15.3	19.2
	3	4.0	25	19.2	20.1
	2	4.0	16	24.1	22.1

As it can be observed, the proposed method achieves a significant reduction in the number of variables selected (about the 90% of the original number of features, or even more in some cases) even with an important increase of the generalization capability (classification rate over the test data set). Besides, many solutions present also a significant decrease in the number of rules, reducing the complexity of the KB. Therefore, our multiobjective GA provides a wide set of solutions that permit an adequate choice depending on the main goal required: good performance or high degree of interpretability.

5 Conclusions

This contribution has proposed a multiobjective genetic process for jointly performing feature selection and DB components learning, which is combined with an efficient fuzzy classification rule generation method to obtain the complete KB for a descriptive FRBCS. Our method achieves an important reduction of the relevant variables selected for the final system and also adapts the fuzzy partition of each variable to the problem being solved. So, we can conclude that the proposed method allows us to significantly enhance the interpretability and accuracy of the FRBCSs generated. We have used a simple RB generation algorithm but another more accurate one can be used, having in mind its run time. Our future work will focus on improving the performance of the multiobjective

GA by using a niching technique or employing a co-evolutive GA and on comparing the results with other feature selection approaches.

References

[1] J. E. Baker. Reducing bias and inefficiency in the selection algorithms. In *Proc. of the Second International Conference on Genetic Algorithms (ICGA'87)*, pages 14–21, Hillsdale, 1987.

[2] J. Casillas, O. Cordón, M. J. del Jesus, and F. Herrera. Genetic feature selection in a fuzzy rule-based classification system learning process for high dimensional problems. *Information Sciences*, 136(1-4):135–157, 2001.

[3] Z. Chi, H. Yan, and T. Pham. *Fuzzy algorithms with applications to image processing and pattern recognition*. World Scientific, 1996.

[4] O. Cordón, M. J. del Jesus, and F. Herrera. Genetic learning of fuzzy rule-based classification systems co-operating with fuzzy reasoning methods. *International Journal of Intelligent Systems*, 13(10/11):1025–1053, 1998.

[5] O. Cordón, M. J. del Jesus, and F. Herrera. A proposal on reasoning methods in fuzzy rule-based classification systems. *International Journal of Approximate Reasoning*, 20(1):21–45, 1999.

[6] O. Cordón, F. Herrera, L. Magdalena, and P. Villar. A genetic learning process for the scaling factors, granularity and contexts of the fuzzy rule-based system data base. *Information Sciences*, 136(1-4):85–107, 2001.

[7] O. Cordón, F. Herrera, and P. Villar. Analysis and guidelines to obtain a good uniform fuzzy partition granularity for fuzzy rule-based systems using simulated annealing. *International Journal of Approximate Reasoning*, 25(3):187–216, 2000.

[8] O. Cordón, F. Herrera, and P. Villar. Generating the knowledge base of a fuzzy rule-based system by the genetic learning of the data base. *IEEE Transactions on Fuzzy Systems*, 9(4):667–675, 2001.

[9] C.M. Fonseca and P.J. Fleming. Genetic algorithms for multiobjective optimization: Formulation, discussion and generalization. In S. Forrest, editor, *Proc. of the Fifth International Conference on Genetic Algorithms (ICGA'93)*, pages 416–423. Morgan Kaufmann, 1993.

[10] A. González and R. Pérez. SLAVE: A genetic learning system based on an iterative approach. *IEEE Transactions on Fuzzy Systems*, 7(2):176–191, 1999.

[11] A. González and R. Pérez. Selection of relevant features in a fuzzy genetic learning algorithm. *IEEE Transctions on Systems, Man and Cybernetics - Part B*, 31, 2001.

[12] R. P. Gorman and T. J. Sejnowski. Analysis of hidden units in a layered network trained to classify sonar targets. *Neural Networks*, 1:75–89, 1988.

[13] F. Herrera, M. Lozano, and J.L. Verdegay. Fuzzy connectives based crossover operators to model genetic algorihtms population diversity. *Fuzzy Sets and Systems*, 92(1):21–30, 1997.

[14] H. Ishibuchi and T. Murata. A genetic-algorithm-based fuzzy partition method for pattern classification problems. In F. Herrera and J.L. Verdegay, editors, *Genetic Algorithms and Soft Computing*, pages 555–578. Physica-Verlag, 1996.

[15] H. Ishibuchi, K. Nozaki, and H. Tanaka. Construction of fuzzy classification systems with rectangular fuzzy rules using genetic algorithms. *Fuzzy Sets and Systems*, 65:237–253, 1994.

[16] R. Kohavi and G.H. John. Wrappers for feature subset selection. *Artificial Intelligence*, 97:273–324, 1997.

[17] H. Liu and H. Motoda. *Feature selection for knowledge discovery and data mining*. Kluwer Academic Publishers, 1998.

[18] Z. Michalewicz. *Genetic Algorithms + Data Structures = Evolution Program*. Springer-Verlag, 1996.

[19] T. Nakashima, T. Morisawa, and H. Ishibuchi. Input selection in fuzzy rule-based classification systems. In *Proc. of the Sixth IEEE International Conference on Fuzzy Systems (FUZZ-IEEE'97)*, pages 1457–1462, 1997.

[20] P. Thrift. Fuzzy logic synthesis with genetic algorithms. In *Proc. Fourth Int. Conference on Genetic Algorithms (ICGA'91)*, pages 509–513, 1991.

[21] L. X. Wang and J. M. Mendel. Generating fuzzy rules by learning from examples. *IEEE Transactions on Systems, Man, and Cybernetics*, 25(2):353–361, 1992.

Mining Implication-Based Fuzzy Association Rules in Databases

Eyke Hüllermeier and Jürgen Beringer
Informatics Institute, University of Marburg
{eyke,beringer}@informatik.uni-marburg.de

Abstract

We introduce a model of fuzzy association rules that employs multiple-valued implication operators in order to represent the relation between the antecedent and the consequent of a rule. For this type of association, adequate quality measures are proposed and some semantic issues are discussed. Moreover, we outline an efficient data mining algorithm for discovering implication-based fuzzy associations in (relational) databases and illustrate this algorithm by means of an example.

Keywords: Data mining, association rules, fuzzy sets, generalized implications.

1 Introduction

Among the techniques that have recently been developed in the field of data mining, so-called *association rules* (or associations for short) have gained considerable attraction [1]. Such rules, syntactically written $A \rightsquigarrow B$, provide a means for representing (apparent) dependencies between attributes in a database. Typically, A and B denote sets of binary attributes, also called features or items. The intended meaning of a (binary) rule $A \rightsquigarrow B$ is that a data record (or transaction) stored in the database that contains the set of items A is likely to contain the items B as well. To illustrate, consider the simple rule $\{eggs, butter\} \rightsquigarrow \{sugar\}$, suggesting that it is likely to find sugar in a transaction (which is here a purchase) which does already contain eggs and butter. Several efficient algorithms have been devised for mining association rules in large databases [2, 10, 13].

Generalizations of (binary) associations of this type have been developed in order to cope with quantitative (e.g. cardinal or ordinal) attributes. Typically, a *quantitative association rule* specifies attribute values by means of intervals, as e.g. in the simple rule "Employees at the age of 30 to 40 have incomes between $50,000 and $70,000." This paper investigates *fuzzy association rules*, which are basically obtained by replacing intervals in quantitative rules by fuzzy sets (fuzzy intervals). The use of fuzzy sets in connection with association rules – as with data mining in general [11] – has recently been motivated by several authors (e.g. [3, 4, 6]). Among other aspects, fuzzy sets avoid an arbitrary determination of crisp boundaries for intervals. Furthermore, fuzzy association rules are very interesting from a knowledge representational point of view: By acting as an interface between a numeric scale and a symbolic scale composed of linguistic terms, fuzzy sets allow for representing the rules discovered in a database in a linguistic and hence comprehensible and user-friendly way. Example: "Middle-aged employees dispose of considerable incomes."

The common approach to fuzzy association rules is to replace set-theoretical by fuzzy set-theoretical operations. The type of rule thus obtained is *conjunction-based*, as will be explained later on. In this paper, we are concerned with an alternative type of fuzzy association rule, namely *implication-based* rules which make use of multiple-valued implication operators for modeling associations. A formal foundation and semantic interpretation of this type of association rule has been presented in [5]. Here, we focus on algorithmic aspects and the practical realization of these results. By way of background, Section 2 reviews some basics of association rules. Implication-based fuzzy association rules and quality measures for rating them are introduced in Section 3. In Section 4, we discuss an algorithm for mining fuzzy association rules. Finally, this algorithm is illustrated by means of an example in Section 5.

2 Association Rules

Consider a set $\mathcal{A} = \{a_1, \ldots, a_m\}$ of items, and let a transaction be a subset $T \subseteq \mathcal{A}$. The intended meaning of the association rule $A \rightharpoonup B$, where $A, B \subseteq \mathcal{A}$, is that a transaction T which contains the items in A is likely to contain the items in B as well.

A rule $A \rightharpoonup B$ is generally rated according to several criteria and is "accepted" if none of them falls below a certain (user-defined) threshold. The most important measures are the following ($D_X \doteq \{T \in D \mid X \subseteq T\}$ denotes the transactions in the database D which contain the items $X \subseteq \mathcal{A}$, and $|D_X| = \text{card}(D_X)$ is its cardinality): A measure of *support* defines the absolute number or the proportion of transactions in D that contain both, A and B:

$$\text{supp}(A \rightharpoonup B) \doteq |D_{A \cup B}| \qquad (1)$$

or $\text{supp}(A \rightharpoonup B) \doteq |D_{A \cup B}|/|D|$. The *confidence* is the proportion of correct applications of the rule:

$$\text{conf}(A \rightharpoonup B) \doteq \frac{|D_{A \cup B}|}{|D_A|}. \qquad (2)$$

These criteria can be complemented by other measures, e.g. a criterion that favors "unexpected" over "expected" associations [14].

In the above setting, a transaction T can be identified with a data record (tuple) $(x_1, \ldots, x_m) \in \{0, 1\}^m$ in a unique way. To this end, one simply identifies each item a_i with a binary variables X_i and lets $x_i = T[X_i] = 1$ if a_i is contained in T and $x_i = 0$ otherwise.

Now, let X and Y be quantitative attributes (such as age or income) with completely ordered domains $\mathfrak{D}_X$ and $\mathfrak{D}_Y$, respectively. Without loss of generality we can assume that $\mathfrak{D}_X, \mathfrak{D}_Y$ are subsets of the real numbers, $\mathfrak{R}$. A quantitative association rule involving the variables X and Y is then of the following form:

$$A \rightharpoonup B: \text{ If } X \in A = [x_1, x_2] \text{ THEN } Y \in B = [y_1, y_2], \qquad (3)$$

where $x_1, x_2 \in \mathfrak{D}_X$ and $y_1, y_2 \in \mathfrak{D}_Y$. This approach can quite simply be generalized to the case where X and Y are multi-dimensional vectors and, hence, A and B hyper-rectangles rather than intervals. Subsequently, we proceed from fixed variables X and Y, and consider the database D as a collection of tuples $(x, y) = (T[X], T[Y])$, i.e. as a projection of the original database.

Note that the quality measures from Section 2 are applicable in the quantitative case as well:[1]

$$\operatorname{supp}(A \twoheadrightarrow B) = \operatorname{card}\{(x, y) \in D \mid x \in A \land y \in B\} \quad (4)$$
$$= |D_{A_{x_1,x_2} \cup B_{y_1,y_2}}|,$$

$$\operatorname{conf}(A \twoheadrightarrow B) = \frac{\operatorname{card}\{(x, y) \in D \mid x \in A \land y \in B\}}{\operatorname{card}\{(x, y) \in D \mid x \in A\}} \quad (5)$$
$$= \frac{|D_{A_{x_1,x_2} \cup B_{y_1,y_2}}|}{|D_{A_{x_1,x_2}}|}.$$

In fact, each interval $[x_1, x_2]$ does again define a binary attribute $A_{x_1,x_2} = \mathbb{I}_{[x_1,x_2]}$. Thus, not only the rating but also the mining of quantitative rules can be reduced to the mining of binary association rules by simply fixing a class of intervals and transforming the numerical data into binary data [9, 14]. Still, finding a useful transformation (binarization) of the data is a non-trivial problem by itself which affects both, the efficiency of subsequently applied mining algorithms and the potential quality of discovered rules. Apart from data transformation methods, clustering techniques can be applied which create intervals and rules at the same time [7, 15].

3 Fuzzy Association Rules

Replacing the sets (intervals) A and B in (3) by fuzzy sets (intervals) leads to fuzzy (quantitative) association rules. Thus, a fuzzy association rule is understood as a rule of the form $A \twoheadrightarrow B$, where A and B are now fuzzy subsets rather than crisp subsets of the domains $\mathfrak{D}_X$ and $\mathfrak{D}_Y$ of variables X and Y, respectively. We shall use the same notation for ordinary sets and fuzzy sets. Moreover, we shall not distinguish between a fuzzy set and its membership function, that is, $A(x)$ denotes the degree of membership of the element x in the fuzzy set A. Recall that an ordinary set A can be considered as a "degenerate" fuzzy set with membership degrees $A(x) = \mathbb{I}_A(x) \in \{0, 1\}$.

The standard approach to generalizing the quality measures for fuzzy association rules is to replace set-theoretic by fuzzy set-theoretic operations. The Cartesian product $A \times B$ of two fuzzy sets A and B is usually defined by the membership function $(x, y) \mapsto A(x) \otimes B(y)$, where $\otimes$ is a t-norm. Moreover, the cardinality of a finite fuzzy set is simply the sum of its membership degrees [8]. Thus, (4) and (5) can be generalized as follows:

$$\operatorname{supp}(A \twoheadrightarrow B) \doteq \sum_{(x,y) \in D} A(x) \otimes B(y), \quad (6)$$

$$\operatorname{conf}(A \twoheadrightarrow B) \doteq \frac{\sum_{(x,y) \in D} A(x) \otimes B(y)}{\sum_{(x,y) \in D} A(x)}. \quad (7)$$

---
[1] Subsequently we focus on *support* and *confidence* measures. The results can be transferred to other measures in a straightforward way.

Note that the support of $A \rightsquigarrow B$ corresponds to the sum of the *individual supports* provided by tuples $(x,y) \in D$:

$$\text{supp}_{[x,y]}(A \rightsquigarrow B) = A(x) \otimes B(y). \tag{8}$$

According to (8), the tuple (x,y) supports $A \rightsquigarrow B$ if both, $x \in A$ and $y \in B$.

In the non-fuzzy case, a tuple (x,y) supports a rule $A \rightsquigarrow B$ iff the conjunction

$$A(x) \wedge B(y) \tag{9}$$

holds true. The measure (8) is a straightforward generalization of this condition, replacing the conjunction by a t-norm. One might object, however, that this approach does not take the implicative nature of a rule into account. And indeed, (6) and (7) entail properties that might appear questionable. For example, consider a rule $A \rightsquigarrow B$, where A is *perfectly associated* with B. By this we mean that $A(x) = B(y)$ for all tuples $(x,y) \in D$. Of course, one would expect this rule to have full confidence: $\text{conf}(A \rightsquigarrow B) = 1$. However, since $\alpha \otimes \alpha < \alpha$ if $\otimes$ is not idempotent (i.e. $\otimes \neq \min$), we usually have

$$\text{conf}(A \rightsquigarrow B) = \frac{\sum_{(x,y) \in D} A(x) \otimes B(y)}{\sum_{(x,y) \in D} A(x)} < 1.$$

3.1 Support

The above problem can be avoided if one takes an alternative specification of the (non-fuzzy) support measure as a point of departure, namely the condition

$$A(x) \wedge (A(x) \Rightarrow B(y)), \tag{10}$$

where $\Rightarrow$ is the standard (material) implication. In fact, (9) and (10) are logically equivalent.

Proceeding from (10) rather than (9), we obtain the following measure of individual support in the fuzzy case:

$$\text{supp}_{[x,y]}(A \rightsquigarrow B) = A(x) \otimes (A(x) \rightsquigarrow B(y)), \tag{11}$$

where $\rightsquigarrow$ is a multiple-valued implication operator. This measure has been proposed recently in [5], with a special justification for letting $\otimes$ the product operator. From a semantic point of view, (11) makes (individual) support conditional on two properties that must be satisfied:

- Correctness: The rule is correct for the tuple (x,y) in a logical sense.

- Non-triviality: The rule is not trivially satisfied for (x,y) in the sense that the premise is not completely false.

Correctness is modeled by the second term in (11), $A(x) \rightsquigarrow B(y)$, non-triviality by the first term, $A(x)$, and these two conditions are combined in a conjunctive way through $\otimes$.

Interestingly enough, (8) and (11) are identical if A and B are intervals, that is $A(x), B(y) \in \{0,1\}$. Thus, the difference between the implication-based and the classical (conjunction-based) approach to association rules disappears in the non-fuzzy case, which might explain why semantic issues have not received much attention as yet. Further, it is interesting to mention that (8) and (11) coincide if the following conditions hold: $\otimes = \min$ in (8), the t-norm $\otimes$ in (11) is continuous and $\leadsto$ is the residuated implication associated with $\otimes$.

Subsequently, we proceed from (11) with product t-norm as an individual support measure. Moreover, the overall support of a rule $A \rightharpoonup B$ is again the sum of the individual supports:

$$\mathsf{supp}(A \rightharpoonup B) \doteq \sum_{(x,y) \in D} \mathsf{supp}_{[x,y]}(A \rightharpoonup B). \tag{12}$$

3.2 Confidence

As in the non-fuzzy case, a measure of confidence of an association rule can be derived from a corresponding measure of support. Interestingly enough, however, the minimal confidence condition, $\mathsf{conf}(A \rightharpoonup B) \geq \Delta$, can be interpreted in different ways, which in turn suggest different generalizations in the fuzzy case.

First, the support of $A \rightharpoonup B$ can be related to the support of A, that is to the support of the default rule $A \rightharpoonup \mathfrak{D}_Y$. Thus, one obtains

$$\mathsf{conf}(A \rightharpoonup B) \doteq \frac{\sum_{(x,y) \in D} \mathsf{supp}_{[x,y]}(A \rightharpoonup B)}{\sum_{(x,y) \in D} A(x)} \tag{13}$$

since $\alpha \leadsto 1 = 1$ for all α and, hence, $\mathsf{supp}_{[x,y]}(A \rightharpoonup \mathfrak{D}_Y) = A(x)$ according to (11). A second possibility is to relate the support of $A \rightharpoonup B$ to the support of $A \rightharpoonup \neg B$. In that case, the minimal confidence condition means that the rule $A \rightharpoonup B$ should be supported much better than $A \rightharpoonup \neg B$:

$$\mathsf{conf}(A \rightharpoonup B) = \frac{\sum_{(x,y) \in D} \mathsf{supp}_{[x,y]}(A \rightharpoonup B)}{\sum_{(x,y) \in D} \mathsf{supp}_{[x,y]}(A \rightharpoonup \neg B)}, \tag{14}$$

where $\neg B(y) = 1 - B(y)$. Interestingly enough, (13) and (14) are again equivalent in the non-fuzzy case, in the sense that $\mathsf{conf}(A \rightharpoonup B) \geq \Delta$ according to (13) iff $\mathsf{conf}(A \rightharpoonup B) \geq \Delta/(1-\Delta)$ according to (14). This is due to the fact that

$$A(x) = \mathsf{supp}_{[x,y]}(A \rightharpoonup \mathfrak{D}_Y) = \mathsf{supp}_{[x,y]}(A \rightharpoonup B) + \mathsf{supp}_{[x,y]}(A \rightharpoonup \neg B)$$

for all (x,y), an equation which is no longer valid in the fuzzy case.

4 Mining Fuzzy Association Rules

We proceed from a set of variables (attributes) $X_1, \ldots, X_N$ (a relational database or a projection thereof) and fixed fuzzy partitions of the domains $\mathfrak{D}_{X_\iota}$ of these variables; a fuzzy partition is simply a collection $\mathcal{F}_\iota = \{A_{\iota,1}, \ldots, A_{\iota,n_\iota}\}$ of fuzzy subsets of $\mathfrak{D}_{X_\iota}$ that cover the domain: For each $x \in \mathfrak{D}_{X_\iota}$ there is a fuzzy set $A_{\iota j} \in \mathcal{F}_\iota$ such that $A_{\iota j}(x) > 0$. Besides, the stronger condition $\sum_{j=1}^{n_\iota} A_{\iota j}(x) = 1$ is often imposed [12]. This, however, is by no means imperative for our approach.

Note that $\mathcal{F}_\iota$ defines a new domain for the variable X_ι on a more abstract level. That is, each fuzzy set $A_{\iota j}$ corresponds to an attribute value and "X_ι is $A_{\iota j}$" means that X_ι

takes the "fuzzy value" A_{ij}. We shall also refer to an A_{ij} as a *fuzzy item* or simply an *item*; an itemset is a collection of attribute values related to *different* variables. (Our current implementation does not support the generation of new items (= fuzzy intervals) by merging adjacent (fuzzy) intervals [14], but it could clearly be extended in this direction.)

The general form a fuzzy association rule is

$$\text{IF} \quad X_1 \text{ is } A_1 \text{ and } \ldots \text{ and } X_n \text{ is } A_n$$
$$\text{THEN} \quad Y_1 \text{ is } B_1 \text{ and } \ldots \text{ and } Y_m \text{ is } B_m,$$

where the involved variables $X_1, \ldots, X_n$ and $Y_1, \ldots, Y_m$ constitute a subset of the variables in the database and none of them occurs more than once (the fuzzy sets A_i and B_j are elements of the corresponding fuzzy partitions). Compactly, this can again be written as

$$\text{If } X = (X_1, \ldots, X_n) \text{ is } A \text{ THEN } Y = (Y_1, \ldots, Y_m) \text{ is } B,$$

where the membership functions of the multi-dimensional fuzzy sets A and B are defined through

$$A : x \mapsto A_1(x_1) \otimes A_2(x_2) \otimes \ldots \otimes A_n(x_n),$$
$$B : y \mapsto B_1(y_1) \otimes B_2(x_2) \otimes \ldots \otimes B_m(y_m).$$

In the non-fuzzy case, the assumption of a fixed partition of a numerical domain, defined by a set of (mutually exclusive and exhaustive) intervals, is often regarded as critical. In the fuzzy case, however, the specification of a fixed partition appears reasonable. In fact, one of the main reasons to employ fuzzy sets is the possibility of expressing relationships between variables in a linguistic form, using terms such as "young" or "high income." Of course, the user or data miner will generally have a concrete idea of such terms, which may depend on the specific context at hand but not on the data. Since it is the user who interprets the association rules, these rules should exactly reflect the meaning he has in mind and, hence, the user himself should characterize each linguistic expression in terms of a fuzzy set. Loosely speaking, it is the user's linguistic concepts rather than the data that defines the fuzzy partition and, hence, the class of rules expressible in the corresponding language. Clearly, one might also think of adapting a fuzzy partition to the data. Then, however, the resulting fuzzy sets might not agree with the idea of the user. Even worse, for some of these sets it might be impossible to find a suitable linguistic term.

Below we propose an extension of the well-known APRIORI algorithm [2] for mining fuzzy association rules. The original APRIORI algorithm generates rules from so-called *frequent itemsets*: One subset of the itemset becomes the premise of the rule and the complement becomes the conclusion. Due to definition (1), the support of any rule derived from a frequent itemset equals the support of the itemset itself. Thus, the problem of finding minimally supported rules reduces to the problem of finding frequent (= minimally supported) itemsets, which constitutes the main part of the APRIORI algorithm. In order to find these itemsets, APRIORI efficiently exploits the following monotonicity property: Each subset of a frequent itemset is a frequent itemset by itself or, stated differently, a set of items having a non-frequent itemset as a subset cannot be a frequent itemset.

In our implication-based approach to association rules, the premise and the conclusion do no longer play symmetric roles. Consequently, the support of an implication-based rule

$A \rightsquigarrow B$ is not determined by the support of the associated itemset $A \cup B$. However, one can take advantage of the fact that the support (11) of $A \rightsquigarrow B$ is *lower-bounded* by the support of the premise A:

$$\mathsf{supp}_{[x,y]}(A \rightsquigarrow B) = A(x) \otimes (A(x) \rightsquigarrow B(y)) \le A(x).$$

Consequently, the premise A of a minimally supported rule $A \rightsquigarrow B$ must be a frequent itemset or, put in a different way, the frequent itemsets constitute a superset of the condition parts of minimally supported association rules. Therefore, it is reasonable to determine the frequent itemsets in a first step. This can be realized by means of the standard APRIORI algorithm with the only difference that an observation $x = (x_1, \ldots, x_n)$ increments the support of an itemset $\{A_1, \ldots, A_n\}$ not by 1 but by

$$\begin{aligned}\mathsf{supp}_x(A) &= A(x) \\ &= A_1(x_1) \otimes A_2(x_2) \otimes \ldots \otimes A_n(x_n).\end{aligned} \quad (15)$$

Now, let $\mathcal{S}$ denote the set of frequent itemsets found by APRIORI. For each $A \in \mathcal{S}$, a *candidate rule* of level one is a rule of the form $A \rightsquigarrow \{B_1\}$, where $B_1 \notin A$. The support and confidence (and maybe other measures) of all candidate rules are computed by scanning the database. Only those rules which do not fall below one of the corresponding thresholds are accepted.

The rules of level one, i.e. with only one attribute in the conclusion part, are perhaps most convenient from a knowledge representational point of view. Yet, rules of higher order can be derived as well. To this end, one can take advantage of the following monotonicity property:

$$\forall B' \subseteq B : \mathsf{supp}(A \rightsquigarrow B) \le \mathsf{supp}(A \rightsquigarrow B').$$

This inequality holds, since an implication $\rightsquigarrow$ is non-decreasing in the second argument and, moreover, $B(y) \le B'(y')$ if the itemset B' is a subset of the itemset B (and y an extension of y'). If the rule $A \rightsquigarrow B$ satisfies the minimum support condition, the same condition hence holds for each rule $A \rightsquigarrow B'$. Moreover, the same relation holds for the confidence since the premise A (hence the denominator in (13)) does not change: $\mathsf{conf}(A \rightsquigarrow B) \le \mathsf{conf}(A \rightsquigarrow B')$. Therefore, the rules $A \rightsquigarrow B$ of level m satisfying the minimal support and minimal confidence condition have the following property: Each rule $A \rightsquigarrow B'$, where B' is derived from B by taking one item away, also has minimal confidence and minimal support. In other words, the candidate rules of level m can be obtained by *joining* the set of rules of level $m-1$ with itself: Two rules $A \rightsquigarrow B$ and $A' \rightsquigarrow B'$ are joined if $A = A'$, B and B' share exactly $m-2$ items, and the items that distinguish them are not related to the same attribute (hence each of B and B' involve $m-1$ variables, and $B \cup B'$ involves m variables). This yields the candidate rule $A \rightsquigarrow B \cup B'$. In a *pruning* step, the candidate rules of level m can be filtered according to the following criterion: A rule $A \rightsquigarrow B$ is deleted if there is an itemset $B' \subseteq B$ of size $m-1$ such that $A \rightsquigarrow B'$ is not an accepted rule of level $m-1$. Finally, the acceptance of the remaining candidate rules is determined by computing their support and confidence.

In summary, our approach to mining implication-based fuzzy association rules consists of the following steps: (1) Apply the APRIORI algorithm (with fuzzy support measure (15)) in order to derive the frequent itemsets. (2) Use the frequent itemsets in order to generate cadidate rules of level one. (3) Filter these rules, using the minimum support and confidence conditions. (4) Derive accepted rules of order m from accepted rules of order $m - 1$ by joining the latter, pruning the rules thus obtained, and computing support and confidence for the remaining candidates.

We refrain from describing implementational details here. Still, concerning implementation and computational complexity let us mention that also the second part of our algorithm is quite similar to the APRIORI algorithm (for quantitative attributes [14]). In fact, each pass first generates candidate rules from lower-level rules and then derives support and confidence measures by scanning the database. The generation of candidates in turn consists of a join and a prune procedure. Roughly, our procedure can hence be seen as a composition of two APRIORI-like algorithms. Experimental results have shown, however, that the execution time is not doubled, which can be explained by the fact that less attributes are considered in the second part (only attributes which are not already part of the antecedent of a rule can appear in the precedent).

Note that usually a kind of post-processing of the accepted association rules is advised, since several of these rules might provide redundant information. For example, consider two accepted rules $\mathcal{A} \rightharpoonup \mathcal{B}$ and $\mathcal{A}' \rightharpoonup \mathcal{B}$ where the first one is more general than the second one in the sense that $\mathcal{A} \subsetneq \mathcal{A}'$. Retaining the second rule has then to be justified, for example through a comparatively high confidence. Thus, it might be pruned if, e.g.,

$$\text{conf}(\mathcal{A}' \rightharpoonup \mathcal{B}) - \text{conf}(\mathcal{A} \rightharpoonup \mathcal{B}) < \varepsilon. \tag{16}$$

An interesting idea is to "tune" the accepted association rules by applying so-called *modifier functions* to their attribute values. In connection with the linguistic modeling of fuzzy concepts, modifiers such as $x \mapsto x^2$ or $x \mapsto \sqrt{x}$ are utilized for depicting the effect of linguistic hedges such as "very" or "almost" [16]. For example, if the fuzzy set A represents the linguistic expression "high income," the modified fuzzy set with membership function $m \circ A = A^2$ models the concept "very high income." Now, suppose a (small) class of such modifier functions to be given. A rule with n items in the condition part and m items in the conclusion part can then be adapted by applying a modifier to each of these items. Thus, k^{n+m} rules can be defined if k modifiers are available. One can then assess these rules by computing their support and confidence and finally replace the original rule by the best modification. Of course, this procedure is practicable only for small k.[2] Apart from that, it is nothing more than an option which might slightly improve the rules which are accepted anyhow.

5 Illustrative Example

To illustrate our approach, we have applied it to the ADULT data set.[3] This data is comprised of 48842 instances (45222 of which remain after removing those containing missing values), where each instance corresponds to a person characterized by 14 attributes, see Table 1. Most attribute names are self explaining, and we refrain from listing the corresponding domains (fnlwgt is the "final weight" of a person as computed by the U.S. Census Bureau according to specific criteria, education-num specifies the duration of education in years). Table 1 also shows the sizes of the fuzzy partitions underlying

---
[2] A reasonable choice is the set of $k = 3$ modifiers $x \mapsto \sqrt{x}, x \mapsto x, x \mapsto x^2$.
[3] Available at http://www.ics.uci.edu/~mlearn/MLRepository.html.

	Attribute Name	Attribute Type	Fuzzy Partition
1	age	continuous	3
2	workclass	discrete (8 values)	8
3	fnlwgt	continuous	3
4	education	discrete (16 values)	3
5	education-num	continuous	3
6	marital-status	discrete (7 values)	7
7	occupation	discrete (14 values)	14
8	relationship	discrete (6 values)	6
9	race	discrete (5 values)	5
10	sex	discrete (2 values)	2
11	capital-gain	continuous	1
12	capital-loss	continuous	1
13	hours-per-week	continuous	3
14	native-country	discrete (41 values)	41

Table 1: Attributes in the ADULT data set.

our data mining procedure (we used standard equi-width partitions with triangular fuzzy sets). If the size of a partition equals the cardinality of the attribute's domain, it simply means that we used the degenerate partition in which each value defines a singleton fuzzy set. To illustrate, Table 2 shows the definition of the fuzzy partition used for the attribute education which is comprised of fuzzy sets for low, middle, and high qualification.

In our experiments, we used the measures (12) and (13) with the product t-norm and the Goguen implication ($\alpha \rightsquigarrow \beta = 1$ if $\alpha \leq \beta$ and β/α otherwise). The support and confidence threshold were set to 1,200 and .7, respectively. Moreover, we used the pruning rule described in the previous section with ε in (16) given by .05. Table 3 shows a list of the ten most confident implication-based rules among those that satisfy the support and confidence conditions (and have not been pruned). As can be seen, the first rules establish a close relationship between the attributes education and education-num. On the one hand, this is hardly astonishing and, hence, hardly interesting. On the other hand, it proves the plausibility of the results. Moreover, it is a nice example for the problem pointed out in Section 3: The duration of the education in years allows for a more or less precise prediction of the level of qualification and vice versa (at least if one defines the corresponding fuzzy partitions in a suitable way, as we have done). This fact is correctly reflected by the implication-based approach but not by the conjunction-based model which assigns, e.g., to the first rule a confidence of only 0.67. More generally, when we compare our result with the rules obtained for the standard approach (also using the product as a t-norm) we find that the respective top 20 rules have only 7 rules in common.

6 Summary

We have introduced an alternative type of fuzzy association rule which is more logic-oriented and makes use of a multiple-valued implication operator in order to connect the condition and conclusion parts of a rule. Consequently, the two parts do no longer play symmetric roles in associations, a property that appears quite reasonable and that avoids some difficulties of the classical (conjunction-based) approach. Interestingly enough, the

Attribute Value	low qual.	middle qual.	high qual.
Bachelors	0	0	0.8
Some-college	0	0.75	0.2
11th	0.2	0.75	0
HS-grad	0	1	0
Prof-school	0	0	1
Assoc-acdm	0	0.25	0.6
Assoc-voc	0	0.5	0.4
9th	0.6	0.25	0
7th-8th	0.8	0	0
12th	0	1	0
Masters	0	0	1
1st-4th	1	0	0
10th	0.4	0.5	0
Doctorate	0	0	1
5th-6th	1	0	0
Preschool	1	0	0

Table 2: Fuzzy partition for the attribute education.

Rule	conf
education-num.low $\rightarrow$ education.low	1.00
education.low $\rightarrow$ education-num.low	1.00
education.high $\rightarrow$ education-num.high	1.00
education.middle $\rightarrow$ education-num.middle	1.00
education-num.middle $\rightarrow$ education.middle	1.00
education-num.high $\rightarrow$ education.high	1.00
age.middle-age, fnlwgt.middle $\rightarrow$ hours-per-week.middle	0.78
age.middle-age, fnlwgt.high fnlwgt $\rightarrow$ hours-per-week.middle	0.78
education.high edu, education-num.middle $\rightarrow$ hours-per-week.middle	0.77
education.middle edu, education-num.high $\rightarrow$ hours-per-week.middle	0.77

Table 3: Ten most confident implication-based association rules.

difference between the implication-based and the classical approach becomes obvious only in the fuzzy case, while both approaches are equivalent in the non-fuzzy case.

Our method for mining fuzzy associations in databases is an extension of the well-known APRIORI algorithm. The basic difference concerns the generation of association rules from frequent itemsets. In our approach, these itemsets are merely used to generate condition parts, not complete rules. Finding accepted rules thus requires an additional step in which the conclusion parts are defined. Still, this can be realized quite efficiently by exploiting a monotonicity property for the support and confidence of implication-based rules, by analogy with the generation of frequent itemsets in APRIORI.

References

[1] R. Agrawal, T. Imielinski, and A. Swami. Mining association rules between sets of items in large databases. ACM SIGMOD, pages 207–216, Washington, D.C., 1993.

[2] R. Agrawal and R. Srikant. Fast algorithms for mining association rules. In *Proceedings of the 20th Conference on VLDB*, Santiago, Chile, 1994.

[3] G. Chen, Q. Wei, and E.E. Kerre. Fuzzy data mining: Discovery of fuzzy generalized association rules. In G. Bordogna and G. Pasi, editors, *Recent Issues on Fuzzy Databases*. Springer-Verlag, 2000.

[4] M. Delgado, D. Sanchez, and M.A. Vila. Acquisition of fuzzy association rules from medical data. In S. Barro and R. Marin, editors, *Fuzzy Logic in Medicine*. Physica Verlag, 2000.

[5] E. Hüllermeier. Implication-based fuzzy association rules. In L. De Raedt and A. Siebes, editors, *Proc.* PKDD–01. Springer-Verlag, 2001.

[6] C. Man Kuok, A. Fu, and M. Hon Wong. Mining fuzzy association rules in databases. *SIGMOD Record*, 27:41–46, 1998.

[7] B. Lent, A. Swami, and J. Widom. Clustering association rules. In *Proceedings* ICDE–97, Birmingham, UK, 1997.

[8] A. De Luca and S. Termini. Entropy of L-fuzzy sets. *Information and Control*, 24:55–73, 1974.

[9] R.J. Miller and Y. Yang. Association rules over interval data. In *Proc.* ACM SIGMOD *International Conference on Management of Data*, pages 452–461, 1997.

[10] J.S. Park, M.S. Chen, and P.S. Yu. An efficient hash-based algorithm for mining association rules. In *Proceedings* ACM SIGMOD *International Conference on Management of Data*, 1995.

[11] W. Pedrycz. Data mining and fuzzy modeling. In *Proc. of the Biennial Conference of the NAFIPS*, pages 263–267, Berkeley, CA, 1996.

[12] E.H. Ruspini. A new approach to clustering. *Information Control*, 15:22–32, 1969.

[13] A. Savasere, E. Omiecinski, and S. Navathe. An efficient algorithm for mining association rules in large databases. In VLDB–95, *Proceedings of the 21th Conference*, Zurich, 1995.

[14] R. Skrikant and R. Agrawal. Mining quantitative association rules in large relational tables. In *Proceedings of the* ACM SIGMOD *International Conference on Management of Data*, pages 1–12, 1996.

[15] Y. Yang and M. Singhal. Fuzzy functional dependencies and fuzzy association rules. In *Data Warehousing and Knowledge Discovery, Proceedings* DAWAK–99, pages 229–240. Springer-Verlag, 1999.

[16] L.A. Zadeh. A fuzzy-set theoretic interpretation of linguistic hedges. *J. Cybernetics*, 2(3):4–32, 1972.

– B. Bouchon-Meunier, L. Foulloy and R.R. Yager (Editors)
Intelligent Systems for Information Processing:
From Representation to Applications
© 2003 Elsevier Science B.V. All rights reserved.

Learning Graphical Models by Extending Optimal Spanning Trees

Christian Borgelt and Rudolf Kruse

Dept. of Knowledge Processing and Language Engineering
Otto-von-Guericke-University of Magdeburg
Universitätsplatz 2, D-39106 Magdeburg, Germany
e-mail: {borgelt,kruse}@iws.cs.uni-magdeburg.de

Abstract

In learning graphical models we often face the problem that a good fit to the data may call for a complex model, while real time requirements for later inferences force us to strive for a simpler one. In this paper we suggest a learning algorithm that tries to achieve a compromise between the goodness of fit of the learned graphical model and the complexity of inferences in it. It is based on the idea to extend an optimal spanning tree in order to improve the fit to the data, while restricting the extension in such a way that the resulting graph has hypertree structure with maximal cliques of at most size 3.

Keywords: Graphical Model, Learning from Data, Optimal Spanning Tree

1 Introduction

In recent years graphical models [Whittaker 1990, Lauritzen 1996]—especially Bayesian networks [Pearl 1988, Jensen 1996] and Markov networks [Lauritzen and Spiegelhalter 1988], but also the more general valuation-based networks [Shafer and Shenoy 1988] and, though to a lesser degree, the newer possibilistic networks [Gebhardt 1997, Borgelt and Kruse 2002]—gained considerable popularity as powerful tools to model dependences in complex domains and thus to make inferences under uncertainty in these domains feasible. Graphical models are based on the idea that under certain conditions a multidimensional (probability or possibility) distribution can be decomposed into (conditional or marginal) distributions on lower dimensional subspaces. This decomposition is represented by a graph, in which each node stands for an attribute and each edge for a direct dependence between two attributes.

The graph representation also supports drawing inferences, because the edges indicate the paths along which evidence has to be transmitted [Jensen 1996, Castillo et al 1997]. However, in order to derive correct and efficient evidence propagation methods, the graphs have to satisfy certain conditions. In general, cycles can pose unpleasant problems, because they make it possible that the same information can travel on different routes to other attributes. In order to avoid erroneous results in this case, the graphs are often transformed into singly connected structures, namely so-called *join* or *junction trees* [Lauritzen and Spiegelhalter 1988, Jensen 1996, Castillo et al 1997].

Since constructing graphical models manually can be tedious and time consuming, a large part of recent research has been devoted to learning them from a dataset of sample cases [Cooper and Herskovits 1992, Heckerman et al 1995, Gebhardt and Kruse 1995, Gebhardt 1997, Borgelt and Kruse 2002]. However, many known learning algorithms do not take into account that the learned graphical model may later be used to draw time-critical inferences and that in this case the time complexity of evidence propagation may have to be restricted, even if this can only be achieved by accepting approximations. The main problem is that during join tree construction edges may have to be added, which can make the graph more complex than is acceptable. In such situations it is desirable that the complexity of the join tree can be controlled at learning time, even at the cost of a less exact representation of the domain under consideration.

To achieve this we suggest an algorithm that constructs a graphical model by extending an optimal spanning tree in such a way that the resulting graph has hypertree structure with maximal cliques of at most size 3.

2 Optimal Spanning Trees

Constructing an optimum weight spanning tree is a special case of methods that learn a graphical model by measuring the strength of marginal dependences between attributes. The idea underlying these heuristic, but often highly successful approaches is the frequently valid assumption that in a graphical model correctly representing the probability or possibility distribution on the domain of interest an attribute is more strongly dependent on adjacent attributes than on attributes that are not directly connected to it. Consequently, it should be possible to find a proper graphical model by selecting edges that connect strongly dependent attributes. Among the methods based on this idea constructing an optimum weight spanning tree is the simplest and best known learning algorithm. It is at the same time the oldest approach, as it was suggested in [Chow and Liu 1968].

In general, the algorithm consists of two components: an evaluation measure, which is used to assess the strength of dependence of two attributes, and a method to construct an optimum weight spanning tree from given edge weights (which are, of course, provided by the evaluation measure). The latter component may be, for example, the well-known Kruskal algorithm [Kruskal 1956] or the Prim algorithm [Prim 1957]. For the former component, i.e., the evaluation measure, there is a variety of measures to choose from. In [Chow and Liu 1968], in which learning probabilistic graphical models was considered, *mutual information* (also called *information gain* or *cross entropy*) was used. It is defined as (A and B are attributes):

$$I_{\mathrm{mut}}(A, B) = H(A) + H(B) - H(AB),$$

where $H(A)$ is the *Shannon entropy* of the probability distribution on A, i.e.,

$$H(A) = - \sum_{a \in \mathrm{dom}(A)} P(a) \log_2 P(a).$$

(Here $P(a)$ is an abbreviation of $P(A = a)$ and denotes the probability that A assumes—as a random variable—the value a. We will adopt such abbreviations throughout this paper.) $H(B)$ and $H(AB)$ are defined analogously. Alternatively, one may use the χ^2 *measure*

$$\chi^2(A, B) = N \sum_{\substack{a \in \mathrm{dom}(A) \\ b \in \mathrm{dom}(B)}} \frac{(P(a)P(b) - P(a,b))^2}{P(a)P(b)},$$

where N is the number of cases in the dataset to learn from (which is often dropped from this measure in applications). Furthermore one may use the *symmetric Gini index* (see, for example, [Zhou and Dillon 1991, Borgelt and Kruse 2002] for a definition), or one of several other *symmetric* evaluation measures (see [Borgelt and Kruse 2002]), where symmetric means that the measure does not change its value if its arguments A and B are exchanged.

While the above measures are designed for learning probabilistic networks, it is clear that the same approach may also be used to learn possibilistic networks: We only have to choose a measure for the possibilistic dependence of two attributes (while the method to construct an optimal spanning tree can be kept). Best known among such measures is the *specificity gain*

$$S_{\mathrm{gain}}(A, B) = \mathrm{nsp}(A) + \mathrm{nsp}(B) - \mathrm{nsp}(AB),$$

where $\mathrm{nsp}(A)$ denotes the U-uncertainty measure of nonspecificity [Klir and Mariano 1987] of the (marginal) possibility distribution π_A on attribute A:

$$\mathrm{nsp}(A) = \int_0^{\sup(\pi_A)} \log_2 |[\pi_A]_\alpha| d\alpha.$$

($[\pi_A]_\alpha$ denotes the α-cut of the possibility distribution, i.e., the set of values that have a degree of possibility of at least α.) $\mathrm{nsp}(B)$ and $\mathrm{nsp}(AB)$ are defined analogously. It should be noted that the formula of specificity gain is very similar to the formula of information gain/mutual information due to the fact that in possibility theory the measure of nonspecificity plays roughly the same role Shannon entropy plays in probability theory.

Alternatively, one may use *possibilistic mutual information* [Borgelt and Kruse 2002]:

$$d_{\mathrm{mi}}(A, B) = \sum_{\substack{a \in \mathrm{dom}(A) \\ b \in \mathrm{dom}(B)}} \pi_{AB}(a,b) \log_2 \frac{\pi_{AB}(a,b)}{\min\{\pi_A(a), \pi_B(b)\}},$$

which is based on a transfer of a different way of writing mutual information to the possibilistic setting. The idea is that by writing mutual information as

$$I_{\mathrm{mut}}(A, B) = \sum_{\substack{a \in \mathrm{dom}(A) \\ b \in \mathrm{dom}(B)}} P(a,b) \log_2 \frac{P(a,b)}{P(a)P(b)}$$

it can be interpreted as measuring the difference between the actual joint distribution $P(a,b)$ and a hypothetical independent distribution $P(a)P(b)$. The transfer to the possibilistic case is achieved by simply inserting the corresponding notion of independence (i.e., that the joint distribution can be constructed as the minimum of the marginal distributions).

Furthermore, a possibilistic version of the χ^2 measure [Borgelt and Kruse 2002] can be employed, which is based on basically the same idea:

$$d_{\chi^2}(A,B) = \sum_{\substack{a \in \mathrm{dom}(A) \\ b \in \mathrm{dom}(B)}} \frac{(\min\{\pi_A(a), \pi_B(b)\} - \pi_{AB}(a,b))^2}{\min\{\pi_A(a), \pi_B(b)\}}.$$

It is worth noting that the optimum weight spanning tree approach has an interesting property in the probabilistic setting: Provided that there is a perfect tree-structured graphical model of the domain of interest and the evaluation measure m that is used satisfies

$$\forall A, B, C: \quad m(C, AB) \geq m(C, B)$$

with equality obtaining only if the attributes A and C are conditionally independent given B and

$$\forall A, C: \quad m(C, A) \geq 0$$

with equality obtaining only if the attributes A and C are (marginally) independent (at least mutual information and the χ^2 measure satisfy these conditions), then the perfect model (directed or undirected) can be found by constructing an optimum weight spanning tree (see [Borgelt and Kruse 2002] for the proof and further details). For mutual information even more can be shown: Constructing an optimum weight spanning tree with this measure yields the best tree-structured approximation of the probability distribution on the domain of interest w.r.t. *Kullback-Leibler information divergence* [Chow and Liu 1968, Pearl 1988].

Unfortunately, these properties do not carry over to the possibilistic setting. Even if there is a perfect graphical model with tree structure, constructing an optimum weight spanning tree with any of the possibilistic measures mentioned above is not guaranteed to find this tree (see [Borgelt and Kruse 2002] for a counterexample). As a consequence there is no analog of the stronger approximation statement either. Nevertheless, the optimum weight spanning tree approach usually leads to good results when constructing possibilistic graphical models, so that the sometimes suboptimal results can be accepted.

3 Extending Spanning Trees

Even if there is no perfect tree-structured graphical model of the domain of interest, constructing an optimum weight spanning tree can be a good starting point for learning a graphical model. The algorithm suggested in [Rebane and Pearl 1987], for example, starts by constructing a(n undirected) spanning tree and then turns it into a (directed) polytree by directing the edges based on the outcomes of conditional independence tests. The advantage of this approach is that it keeps the single-connectedness of the graph and thus allows for a simple derivation of evidence propagation methods. However, by doing so, it does not really restrict the complexity of later inferences, as this complexity depends on the number of parents an attribute has in the polytree. This can be seen by considering the construction of a join tree for the polytree [Castillo et al 1997]. The first step consists in forming a so-called *moral graph* by "marrying" the parents of an attribute (i.e., connecting them with an edge). In this way the set of parents of an attribute together with the attribute itself become a clique in the resulting graph and thus a node in the final join tree. As the size of the nodes in the join tree is a decisive factor of the complexity of inferences,

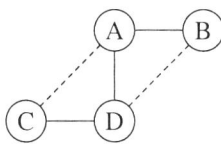

Figure 1: The dotted edges cannot both be the result of "marrying" parents in a directed graph, but may be generated in our algorithm.

the number of parents directly determines this complexity. Unfortunately, there is no way to restrict the number of parents in this algorithm. On the other hand, the restriction to singly connected graphs may be too strong for some learning tasks, as such graphs cannot capture certain rather simple dependence structures.

To amend these drawbacks, we suggest a simple learning algorithm, which also starts from an initial optimum weight spanning tree, but may yield more complex structures than polytrees, while at the same time restricting the size of the nodes in the join tree (which, of course, means that in some respect the structures may also be simpler than a polytree). The basic idea of this algorithm is as follows: First an (undirected) optimum weight spanning tree is constructed. Then this tree is enhanced by edges where a conditional independence statement implied by the tree does not hold. However, we do not check arbitrary conditional independence statements, but only those that refer to edges that connect nodes having a common neighbor in the optimum weight spanning tree. It should be noted that this restriction is similar to directing the edges of the spanning tree, because adding an edge between two nodes having a common neighbor is similar to directing the edges of the spanning tree towards the common neighbor (since the construction of a moral graph would add exactly this edge). However, our approach is more general, since it allows for structures like those shown in figure 1, which cannot result from directing edges alone.

A further restriction of which edges may be added is achieved by the following requirement: If all edges of the optimum weight spanning tree are removed, the remaining graph must be acyclic. This condition is interesting, because it guarantees that the resulting graph has hypertree structure (a precondition for the construction of a join tree, see [Lauritzen and Spiegelhalter 1988, Castillo et al 1997] for details) and that its maximal cliques comprise at most three nodes. Consequently, with this condition we can restrict the size of the join tree nodes and thus the complexity of inferences.

Theorem: If an undirected tree is extended by adding edges only between nodes with a common neighbor in the tree and if the added edges do not form a cycle, then the resulting graph has hypertree structure and its maximal cliques contain at most three nodes.

Proof: Consider first the size of the maximal cliques. Figure 2 shows, with solid edges, the two possible structurally different spanning trees with four nodes. In order to turn these into cliques the dotted edges have to be added. However, in the graph on the left the edge (B, D) connects two nodes not having a common neighbor in the original tree and in the graph on the right the additional edges form a cycle. Therefore it is impossible to get a clique with a size greater than three without breaking the rules for adding edges.

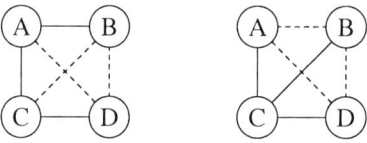

Figure 2: Maximal cliques with four or more nodes cannot be created without breaking the rules for adding edges.

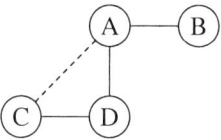

Figure 3: The node A can be bypassed only by an edge connecting the node D to a neighbor of A (which may or may not be B).

In order to show that the resulting graph has hypertree structure, it is sufficient to show that all cycles with a length greater than three have a chord (i.e., an edge connecting two nodes of the cycle that are not adjacent in the considered cycle). This is easily verified with the following argument. Neither the original tree nor the graph without the edges of this tree contain a cycle. Therefore in all cycles there must be a node A at which an edge from the original tree meets an added edge. Let the former edge connect the nodes B and A and the latter connect the nodes C and A. Since edges may only be added between nodes that have a common neighbor in the tree, there must be a node D that is adjacent to A as well as to C in the original tree. This node may or may not be identical to B. If it is identical to B and the cycle has a length greater than three, then the edge (B, C) clearly is a chord. Otherwise the edge (A, D) is a chord, because D must also be in the cycle. To see this, consider Figure 3, which depicts the situation referred to. To close the cycle we are studying there must be a path connecting B and C that does not contain A. However, from the figure it is immediately clear that any such path must contain D, because A can only be bypassed via an edge that has been added between D and a neighbor of A (note that this neighbor may or may not be B). □

In order to test for conditional (in)dependence, we simply use the conditional forms of the marginal dependence measures mentioned above. That is, in the probabilistic case we compute for a measure m

$$m_{\text{ci}}(A, B \mid C) = \sum_{c \in \text{dom}(C)} P(c) \cdot m(A, B \mid c),$$

where $m(A, B \mid c)$ is defined as $m(A, B)$ with all marginal probabilities $P(a)$ and $P(b)$ replaced by their conditional counterparts $P(a \mid c)$ and $P(b \mid c)$. The possibilistic case is analogous. We only have to take into account that the possibility degrees may not add up

net	eds.	pars.	train	test
indep.	0	59	-19921	-20087
orig.	22	219	-11391	-11506
I_{gain}	20	286	-12123	-12340
χ^2	20	283	-12123	-12336
I_{gain}	35	1484	-11454	-12029
χ^2	35	1732	-11441	-12034
I_{gain}	35	1342	-11229	-11818
χ^2	35	1301	-11235	-11805
K2	23	230	-11385	-11511

Table 1: Probabilistic network learning.

be 1, so that a normalization is necessary, i.e.,

$$m_{\text{ci}}(A, B \mid C) = \sum_{c \in \text{dom}(C)} \frac{\pi_C(c)}{s} \cdot m(A, B \mid c),$$

where

$$s = \sum_{c \in \text{dom}(C)} \pi_C(c).$$

Based on these measures we select the additional edges greedily (similar to the selection of edges in the Kruskal algorithm).

As a final remark we would like to point out that this approach is not guaranteed to find the best possible graph with the stated properties, neither in the probabilistic nor in the possibilistic setting. That is, if there is a perfect graphical model of the domain under consideration, which has hypertree structure and the maximal cliques of which have at most size 3, then this approach may not find it. An example of such a case can be found in [Borgelt and Kruse 2002].

4 Experimental Results

We implemented our algorithm in a prototypical fashion as part of the INES program (Induction of NEtwork Structures) [Borgelt and Kruse 2002] and tested it on the well-known Danish Jersey cattle blood group determination problem [Rasmussen 1992].

For our probabilistic tests, we used databases randomly generated from a human expert designed Bayesian network for the Danish Jersey cattle domain. We generated 20 datasets, which where organized into 10 pairs. One dataset of each pair was used for training the network, the other for testing. All networks were evaluated by computing the log-likelihood of the training and the test dataset. Table 1 shows the results, which are averages over the ten trials executed. The first section contains the results for a network without any edges and the original network, followed by results obtained with a pure optimal spanning tree approach. The third section lists the results of the algorithm suggested in this paper and the final section shows the result of greedy parent selection w.r.t. a topological order (the so-called K2-algorithm [Cooper and Herskovits 1992]).

net	eds.	pars.	min.	avg.	max.
indep.	0	80	10.06	10.16	11.39
orig.	22	308	9.89	9.92	11.32
S_{gain}	20	415	8.88	8.99	10.71
d_{χ^2}	20	449	8.66	8.82	10.33
d_{mi}	20	372	8.47	8.60	10.39
S_{gain}	29	2110	8.14	8.30	10.13
d_{χ^2}	35	1672	8.10	8.28	10.18
d_{mi}	31	1353	7.97	8.14	10.25
S_{gain}	31	1630	8.52	8.62	10.29
d_{χ^2}	35	1486	8.15	8.33	10.20
d_{mi}	33	774	8.21	8.34	10.42

Table 2: Possibilistic network learning.

For our possibilistic tests we used a database of 500 real world sample cases, which contains a large number of missing values and is thus well suited for a possibilistic approach. The results are shown in table 2. The meaning of the sections is the same as for table 1, although the evaluation is done differently (details about how we assess the quality of a possibilistic network can be found in [Borgelt and Kruse 2002]).

As was to be expected, in both cases, probabilistic as well as possibilistic, the results are in between those of the pure optimum weight spanning tree algorithm and the greedy parent selection algorithm. However, in comparisons with the latter it should be noted that the greedy parent selection needs a topological order to work on and is thus provided with important additional information, while our algorithm relies on the data alone.

5 Conclusions and Future Work

In this paper we suggested a learning algorithm for graphical models, which extends an optimal spanning tree by adding edges. Due to specific restrictions, which edges may be added, the result is guaranteed to have hypertree structure and maximal cliques of limited size, thus providing for efficient inferences. The experimental results are promising, especially for possibilistic networks.

A drawback of the suggested algorithm is that the size of the maximal cliques is restricted to a fixed value, namely 3. Obviously, it would be more desirable if the size restriction were a parameter. Therefore in our future research we plan to search for conditions that enable us to extend optimal spanning trees in more complex ways, while restricting the model to hypertrees with maximal cliques of at most size 4, 5 etc. Unfortunately, such conditions seem to be much more complex and thus difficult to find.

References

[Borgelt and Kruse 2002] C. Borgelt and R. Kruse. *Graphical Models — Methods for Data Analysis and Mining.* J. Wiley & Sons, Chichester, United Kingdom 2002

[Castillo et al 1997] E. Castillo, J.M. Gutierrez, and A.S. Hadi. *Expert Systems and Probabilistic Network Models.* Springer, New York, NY, USA 1997

[Chow and Liu 1968] C.K. Chow and C.N. Liu. Approximating Discrete Probability Distributions with Dependence Trees. *IEEE Trans. on Information Theory* 14(3):462–467. IEEE Press, Piscataway, NJ, USA 1968

[Cooper and Herskovits 1992] G.F. Cooper and E. Herskovits. A Bayesian Method for the Induction of Probabilistic Networks from Data. *Machine Learning* 9:309–347. Kluwer, Dordrecht, Netherlands 1992

[Gebhardt and Kruse 1995] J. Gebhardt and R. Kruse. Learning Possibilistic Networks from Data. *Proc. 5th Int. Workshop on Artificial Intelligence and Statistics (Fort Lauderdale, FL, USA)*, 233–244. Springer, New York, NY, USA 1995

[Gebhardt 1997] J. Gebhardt. *Learning from Data: Possibilistic Graphical Models.* Habil. thesis, University of Braunschweig, Germany 1997

[Heckerman et al 1995] D. Heckerman, D. Geiger, and D.M. Chickering. Learning Bayesian Networks: The Combination of Knowledge and Statistical Data. *Machine Learning* 20:197–243. Kluwer, Dordrecht, Netherlands 1995

[Jensen 1996] F.V. Jensen. *An Introduction to Bayesian Networks.* UCL Press Ltd. / Springer, London, United Kingdom 1996

[Klir and Mariano 1987] G.J. Klir and M. Mariano. On the Uniqueness of a Possibility Measure of Uncertainty and Information. *Fuzzy Sets and Systems* 24:141–160. North Holland, Amsterdam, Netherlands 1987

[Kruskal 1956] J.B. Kruskal. On the Shortest Spanning Subtree of a Graph and the Traveling Salesman Problem. *Proc. American Mathematical Society* 7(1):48–50. American Mathematical Society, Providence, RI, USA 1956

[Lauritzen and Spiegelhalter 1988] S.L. Lauritzen and D.J. Spiegelhalter. Local Computations with Probabilities on Graphical Structures and Their Application to Expert Systems. *Journal of the Royal Statistical Society, Series B*, 2(50):157–224. Blackwell, Oxford, United Kingdom 1988

[Lauritzen 1996] S.L. Lauritzen. *Graphical Models.* Oxford University Press, Oxford, United Kingdom 1996

[Lopez de Mantaras 1991] R. Lopez de Mantaras. A Distance-based Attribute Selection Measure for Decision Tree Induction. *Machine Learning* 6:81–92. Kluwer, Dordrecht, Netherlands 1991

[Pearl 1988] J. Pearl. *Probabilistic Reasoning in Intelligent Systems: Networks of Plausible Inference.* Morgan Kaufmann, San Mateo, CA, USA 1988 (2nd edition 1992)

[Prim 1957] R.C. Prim. Shortest Connection Networks and Some Generalizations. *The Bell System Technical Journal* 36:1389-1401. Bell Laboratories, Murray Hill, NJ, USA 1957

[Rasmussen 1992] L.K. Rasmussen. *Blood Group Determination of Danish Jersey Cattle in the F-blood Group System (Dina Research Report 8).* Dina Foulum, Tjele, Denmark 1992

[Rebane and Pearl 1987] G. Rebane and J. Pearl. The Recovery of Causal Polytrees from Statistical Data. *Proc. 3rd Workshop on Uncertainty in Artificial Intelligence (Seattle, WA, USA)*, 222–228. USA 1987.

[Shafer and Shenoy 1988] G. Shafer and P.P. Shenoy. *Local Computations in Hypertrees (Working Paper 201).* School of Business, University of Kansas, Lawrence, KS, USA 1988

[Whittaker 1990] J. Whittaker. *Graphical Models in Applied Multivariate Statistics.* J. Wiley & Sons, Chichester, United Kingdom 1990

[Zhou and Dillon 1991] X. Zhou and T.S. Dillon. A statistical-heuristic Feature Selection Criterion for Decision Tree Induction. *IEEE Trans. on Pattern Analysis and Machine Intelligence (PAMI)*, 13:834–841. IEEE Press, Piscataway, NJ, USA 1991

Clustering belief functions based on attracting and conflicting metalevel evidence

Johan Schubert
Department of Data and Information Fusion
Division of Command and Control Systems
Swedish Defence Research Agency
SE-172 90 Stockholm, Sweden
schubert@foi.se
http://www.foi.se/fusion/

Abstract

In this paper we develop a method for clustering belief functions based on attracting and conflicting metalevel evidence. Such clustering is done when the belief functions concern multiple events, and all belief functions are mixed up. The clustering process is used as the means for separating the belief functions into subsets that should be handled independently. While the conflicting metalevel evidence is generated internally from pairwise conflicts of all belief functions, the attracting metalevel evidence is assumed given by some external source.

Kewords: belief functions, Dempster-Shafer theory, clustering.

1 Introduction

In this paper we extend an earlier method within Dempster-Shafer theory [3, 10] for handling belief functions that concern multiple events. This is the case when it is not known a priori to which event each belief function is related. The belief functions are clustered into subsets that should be handled independently.

Previously, we developed methods for clustering belief functions based on their pairwise conflict [2, 7, 8]. These conflicts were interpreted as metalevel evidence about the partition of the set of belief functions [5]. Each piece of conflicting metalevel evidence states that the two belief functions do not belong to the same subset.

The method previously developed is here extended into also being able to handle the case of attracting metalevel evidence. Such evidence is not generated internally in the same way as the conflicting metalevel evidence. Instead, we assume that it is given from some external source as additional information about the partitioning of the set of all belief functions.

For example, in intelligence analysis we may have conflicts (metalevel evidence) between two different intelligence reports about sighted objects, indicating that two objects probably does not belong to the same unit (subset). At the same time we may have information from communication intelligence as an external source (providing attracting metalevel evidence), indicating that the two objects probably do belong to the same unit (subset) as they are in communication.

We begin (Section 2) by giving an introductory problem description. In Section 3 we interpret the meaning of attracting and conflicting metalevel evidence. We assign values to all such pieces of evidence. In Section 4 we combine the metalevel evidence separately for each subset. Here, all attracting metalevel evidence, and all conflicting metalevel evidence are combined as two independent combinations within each subset. At the partition level (Section 5) we combine all metalevel evidence from the subsets, yielding basic beliefs for and against the adequacy of the partition. In Section 6 we compare the information content of attracting metalevel evidence with conflicting metalevel evidence. This is done in order to find a weighting between the basic beliefs for and against the adequacy of the partition in the formulation of a metaconflict function. The order of processing is shown in Figure 1. Finally, the metaconflict function is minimized as the method of finding the best partition of the set of belief functions (Section 7).

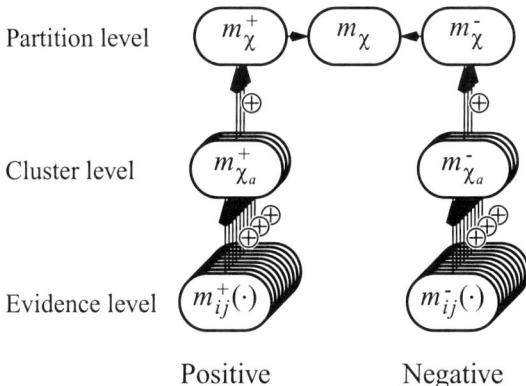

Figure 1: Order of processing.

2 Problem description

When we have several belief functions regarding different events that should be handled independently we want to arrange them according to which event they are referring to. We partition the set of belief function χ into subsets where each subset χ_i refers to a particular event, Figure 2. The conflict of Dempster's rule when all belief functions in χ_i are combined is denoted c_i. In Figure 2, thirteen belief functions e_i are partitioned into four subsets. As these events have nothing to do with each other, they should be analyzed independently.

If it is uncertain whether two different belief functions are referring to the same event we do not know if we should put them into the same subset or not. We can then use the conflict of Dempster's rule when the two belief functions are combined, as an indication of whether belong together. A high conflict between the two functions is an indication of repellency that they do not belong together in the same subset. The higher this conflict is, the less credible that they belong to the same subset. A zero conflict, on the other hand, is no indication at all.

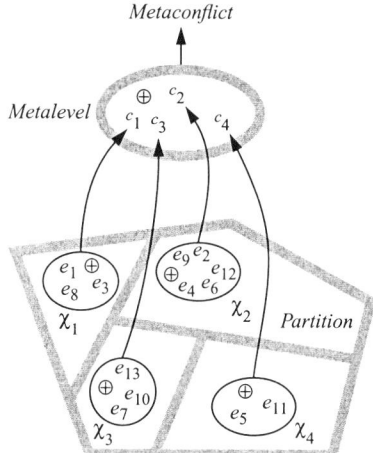

Figure 2: The conflict in each subset is interpreted as evidence at the metalevel.

For each subset we may create a new belief function on the metalevel with a proposition that we do not have an "adequate partition." The new belief functions does not reason about any of the original problems corresponding to the subsets. Rather they reason about the partition of the other belief functions into the different subsets. Just so we do not confuse the two types of belief functions, let us call the new ones "metalevel evidence" and let us say that their combination take place at the metalevel, Figure 2.

On the metalevel we have a simple frame of discernment where $\Theta = \{\text{AdP}, \neg\text{AdP}\}$, where AdP is short for "adequate partition." Let the proposition take a value equal to the conflict of the combination within the subset,

$$m_{\chi_i}(\neg\text{AdP}) \stackrel{\Delta}{=} \text{Conf}(\{e_j | e_j \in \chi_i\})$$

where $\text{Conf}(\{e_j | e_j \in \chi_i\})$ is the conflict of Dempster's rule when combining all basic probability assignments in χ_i.

In [5] we established a criterion function of overall conflict for the entire partition called the metaconflict function (Mcf). The metaconflict is derived as the plausibility of having an adequate partitioning based on $\oplus\{m_{\chi_i}(\neg\text{AdP})\}$ for all subsets χ_i.

DEFINITION. *Let the* metaconflict function,

$$\text{Mcf}(\{e_1, e_2, ..., e_n\}) \stackrel{\Delta}{=} 1 - \prod_{i=1}^{r}(1 - c_i),$$

be the conflict against a partitioning of n belief functions of the set χ into r disjoint subsets χ_i.

Minimizing the metaconflict function was the method of partitioning the belief functions into subsets representing the different events.

However, instead of considering the conflict in each subset we may refine our analysis and consider all pairwise conflicts between the belief functions in χ_i [7], $m^-_{ij}(\cdot) = c_{ij}$, where c_{ij} is the conflict of Dempster's rule when combining e_i and e_j. When $c_{ij} = 1$, e_i and e_j must not be in the same subset, when $c_{ij} = 0$ there simply is no indication of the repellent type. It was demonstrated in [7] that minimizing a sum of logarithmized pairwise conflicts,

$$\sum_i \sum_{\substack{k,l \\ e_k, e_l \in \chi_i}} -\log(1 - c_{kl}),$$

is with a small approximation identical to minimizing the metaconflict function, making it possible the map the optimization problem onto a neural network for neural optimization [2, 9].

In section 3 we will refine the frame of discernment and the proposition of $m^-_{ij}(\cdot)$ in order to make such a refined analysis possible.

In addition to this conflicting metalevel evidence from internal conflicts between belief functions, it is in many applications important to be able to handle attracting metalevel evidence from some external source stating that things do belong together, Figure 3. The analysis of this case is the contribution of the current paper.

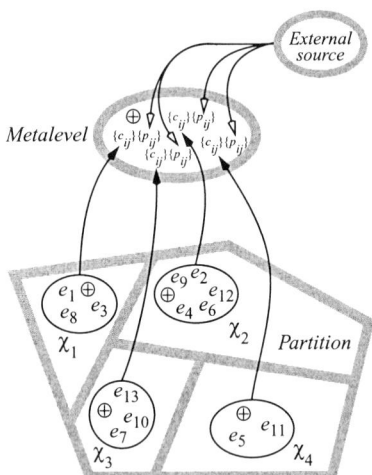

Figure 3: Conflict evidence from subsets and attracting evidence from an external source.

Let $m^+_{ij}(\cdot) = p_{ij}$, where p_{ij} is a degree of attraction, be such an external metalevel evidence. When $p_{ij} = 1$, e_i and e_j must be in the same subset, when $p_{ij} = 0$ we have no indication of the attracting type.

Of course, we can also have external conflicting metalevel evidence. It is then combined with $m_{ij}(\cdot)$, and henceforth we will use $m_{ij}^-(\cdot)$ as the combined result if such external evidence is present.

3 Evidence level

Looking at a series of single subset problems, our frame of discernment for the metalevel of each subset χ_a was initially represented as $\Theta_a = \{\text{AdP}, \neg\text{AdP}\}$, [5]. It is here refined to

$$\Theta_a = \{\forall j . e_j \in \chi_a\} \cup \{e_j \notin \chi_a\}_{j=1}^{|\chi_a|},$$

where "adequate partition" AdP is refined to the proposition $\forall j . e_j \in \chi_a$, that each belief function e_j placed in subset χ_a actually belongs to χ_a. On the other hand, "not adequate partition" $\neg\text{AdP}$ is refined to a set of $|\chi_a|$ propositions $e_j \notin \chi_a$, each stating that a particular belief function is misplaced.
Thus, $|\Theta_a| = 1 + |\chi_a|$, where $|\chi_a|$ is the number of pieces of evidence in χ_a.
Let us assign values to all conflicting and attracting pieces of metalevel evidence. However, we will not combine the attracting and conflicting evidence regarding each pair here on the evidence level as this result is currently not our concern.

3.1 Conflicting evidence: $m_{ij}^-(\cdot)$.

For each pair of belief functions we may receive a conflict. We interpret this as a piece of metalevel evidence indicating that the two belief functions do not belong to the same subset,

$$m_{ij}^-(\forall a . e_i \in \chi_a \Rightarrow e_j \notin \chi_a) = c_{ij},$$
$$m_{ij}^-(\Theta) = 1 - c_{ij}.$$

Here, we simply state that if e_i belongs to a subset χ_a then e_j must not belong to the same subset. Instead, we could have made a disjunction of two different propositions $[(e_i \in \chi_a \Rightarrow e_j \notin \chi_a) \vee (e_j \in \chi_a \Rightarrow e_i \notin \chi_a)]$ where $i \leftrightarrow j$ is permuted in the second term, but this is unnecessary and redundant information because of symmetry.
The metalevel evidence may be simplified to

$$m_{ij}^-(e_i \vee e_j \notin \chi_a) = c_{ij},$$
$$m_{ij}^-(\Theta) = 1 - c_{ij},$$

since

$$\forall a.e_i \in \chi_a \Rightarrow e_j \notin \chi_a = \forall a.\neg e_i \in \chi_a \vee e_j \notin \chi_a = \forall a.e_i \notin \chi_a \vee e_j \notin \chi_a$$
$$= \forall a.e_i \vee e_j \notin \chi_a = e_i \vee e_j \notin \chi_a$$

by implication replacement and dropping universal quantifiers.
We calculate m_{ij}^- for all pairs (ij).

3.2 Attracting evidence: $m_{ij}^+(\cdot)$.

In addition we may also have attracting evidence brought in externally. Such a piece of metalevel evidence is interpreted as the negation of the previous proposition, i.e., that the two pieces of evidence belong to the same cluster,

$$m_{ij}^+(\neg \forall a.e_i \in \chi_a \Rightarrow e_j \notin \chi_a) = p_{ij},$$
$$m_{ij}^+(\Theta) = 1 - p_{ij}.$$

Simplified to

$$m_{ij}^+(e_i \wedge e_j \in \chi_a) = p_{ij},$$
$$m_{ij}^+(\Theta) = 1 - p_{ij},$$

since

$$\neg \forall a.e_i \in \chi_a \Rightarrow e_j \notin \chi_a = \exists a.\neg(e_i \in \chi_a \Rightarrow e_j \notin \chi_a) = \exists a.\neg(\neg e_i \in \chi_a \vee e_j \notin \chi_a)$$
$$= \exists a.\neg(\neg e_i \in \chi_a \vee \neg e_j \in \chi_a) = \exists a.\neg\neg(e_i \in \chi_a \wedge e_j \in \chi_a)$$
$$= \exists a.e_i \in \chi_a \wedge e_j \in \chi_a = \exists a.e_i \wedge e_j \in \chi_a = e_i \wedge e_j \in \chi_a$$

by bringing in negation, implication replacement and dropping of universal quantifiers.
We calculate m_{ij}^+ for all pairs (ij).
Having assigned values to all conflicting and attracting metalevel evidence regarding every pair of belief functions we take the analysis to the cluster level.

4 Cluster level

At the cluster level we use the evidence derived in the previous level. We also use the same frame of discernment. Let us separately combine all conflicting $\{m_{ij}^-\}$ and all attracting evidence $\{m_{ij}^+\}$ for each cluster.

4.1 Combine all conflicting evidence within each cluster

Let us combine $\forall i, j, a. \oplus \{m_{ij}^- | m_{ij}^- \in \chi_a\}$, i.e., all conflicting metalevel evidence within each subset where $m_{ij}^-(e_i \vee e_j \notin \chi_a) = c_{ij}$, Section 3.1.
In [6] we refined the proposition $\neg$AdP separately for each cluster χ_a to $\exists j.e_j \notin \chi_a$, i.e., that there is at least one belief function misplaced in the subset.

Consequently, from the result of the above combination we have,

$$m^-_{\chi_a}(\neg AdP) = m^-_{\chi_a}(\exists j.e_j \notin \chi_a) = 1 - \prod_{(ij) \in \chi_a} (1 - c_{ij}),$$

$$m^-_{\chi_a}(\Theta) = 1 - m^-_{\chi_a}(\neg AdP).$$

We calculate $m^-_{\chi_a}$ for all subsets χ_a. This is the conflicting metalevel evidence derived at the cluster level.

In addition, this piece of evidence $m^-_{\chi_a}$ with proposition $\neg AdP$ may at the next level be refined as $\chi_a \notin \chi$, where χ is the set of all subsets. That is, the same conflict that on the cluster level is interpreted as if there is at least one belief function that does not belong to χ_a, will on the partition level be interpreted as if χ_a (i.e., with all its content) does not belong to χ. This will be useful at the partition level when combining all $m^-_{\chi_a}$ for different subsets χ_a.

4.2 Combine all attracting evidence within each cluster

Similarly to the previous section we begin by combining all attracting metalevel evidence within each individual subset, $\forall i, j, a. \oplus \{m^+_{ij} | m^+_{ij} \in \chi_a\}$, where m^+_{ij} was derived as $m^+_{ij}(e_i \wedge e_j \in \chi_a) = p_{ij}$ in Section 3.2.

For attracting metalevel evidence we refine AdP as the negation of the refinement of $\neg AdP$. We have,

$$AdP = \neg\neg AdP = \neg \exists j.e_j \notin \chi_a = \forall j.\neg e_j \notin \chi_a = \forall j.e_j \in \chi_a.$$

We need to calculate the support for an adequate partition from all attracting evidence $m^+_{\chi_a}(AdP)$ in each subset χ_a. Thus, we will sum up the contribution from all intersections corresponding to a proposition that a conjunction of all pieces of evidence placed in the cluster actually belongs to the subset in question, i.e., $\wedge \{e_j \in \chi_a\}_{j=1}^{|\chi_a|}$.

From the combination of all $\{m^+_{ij}\}$ we have,

$$m^+_{\chi_a}(AdP) = m^+_{\chi_a}(\forall j.e_j \in \chi_a) = \sum_{I \subseteq P_{|\chi_a|} | M_I \equiv N_{|\chi_a|}} \prod_I p_{ij} \prod_{P_{|\chi_a|} - I} (1 - p_{ij}),$$

$$m^+_{\chi_a}(\Theta) = 1 - m^+_{\chi_a}(AdP),$$

where $P_{|\chi_a|} = \{(ij) | 1 \leq i < j \leq |\chi_a|\}$ is a set of all pairs of ordered numbers $\leq |\chi_a|$, $M_I = \{i | \exists p.(ip) \vee (pi) \in I\}$ is the set of all numbers in the pairs of I, and $N_{|\chi_a|} = \{1, ..., |\chi_a|\}$ is the set of all numbers $\leq |\chi_a|$.

We calculate $m^+_{\chi_a}$ for all subsets χ_a. This is the attracting metalevel evidence on the cluster level.

In addition AdP may on the next level be refined as $\chi_a \in \chi$. This will be useful at the partition level when combining all $m^+_{\chi_a}$ from the different subsets.

5 Partition level

The partition level is where all things come together, Figure 1. First, we combine all conflicting metalevel evidence from the subsets, $m^-_{\chi_a}$, Section 5.1. Secondly, we combine all attracting metalevel evidence from the same subsets, $m^+_{\chi_a}$, Section 5.2. Finally, we combine the conflicting and attracting metalevel evidence (in Section 5.3).

However, before we start, let us notice that on the partition level we do not reason about misplaced belief functions. Instead, we reason about the different parts of the partition (i.e., the subsets), and whether each of the subsets can make up part of an adequate partition. For this reason we should represent the frame of discernment differently than on previous levels.

The frame of discernment on the partition level $\Theta = \{\text{AdP}, \neg\text{AdP}\}$ is refined as

$$\Theta = \{\forall a. \chi_a \in \chi\} \cup \{\chi_a \notin \chi\}^{|\chi|}_{a=1},$$

where "adequate partition" AdP is refined to a the proposition $\forall a. \chi_a \in \chi$, stating that every subset χ_a does make up part of an adequate partition. On the other hand, "not adequate partition" $\neg\text{AdP}$ is refined to a set of $|\chi|$ propositions $\chi_a \notin \chi$, each stating that a particular subset does not make up part of an adequate partition.

Thus, the size of the frame is $|\Theta| = 1 + |\chi|$, where $|\chi|$ is the number of subsets in χ.

5.1 Combine all conflicting evidence at the partition level

We begin by combining $\forall a. \oplus \{m^-_{\chi_a}\}$, i.e., all conflicting metalevel evidence from the subsets χ_a that we derived in Section 4.1.

Let us then refine the proposition $\neg\text{AdP}$ of $m^-_{\chi_a}(\neg\text{AdP})$ such that $\neg\text{AdP} = \exists a. \chi_a \notin \chi$, i.e., that there is at least one subset that does not make up part of an adequate partition.

From the combination of all $\{m^-_{\chi_a}\}$ we have,

$$m^-_\chi(\neg\text{AdP}) = m^-_\chi(\exists a. \chi_a \notin \chi) = 1 - \prod_a [1 - m^-_{\chi_a}(\chi_a \notin \chi)]$$

$$= 1 - \prod_a [1 - m^-_{\chi_a}(\neg\text{AdP})] = 1 - \prod_a \prod_{(ij) \in \chi_a} (1 - c_{ij}),$$

$$m^-_\chi(\Theta) = 1 - m^-_\chi(\neg\text{AdP}).$$

This is the conflicting metalevel evidence at the partition level.

5.2 Combine all attracting evidence at the partition level

Let us combine all attracting metalevel evidence $\forall a. \oplus \{m^+_{\chi_a}\}$, derived in Section 4.2.

For attracting metalevel evidence at the partition level we refine the proposition AdP of $m^+_\chi(\text{AdP})$ as the negation of the refinement for $\neg\text{AdP}$ at this level,

$$\text{AdP} = \neg\neg\text{AdP} = \neg\exists a.\chi_a \notin \chi = \forall a.\neg\chi_a \notin \chi = \forall a.\chi_a \in \chi.$$

From the combination of all $\{m^+_{\chi_a}\}$ we find,

$$m^+_\chi(\text{AdP}) = m^+_\chi(\forall a.\chi_a \in \chi) = \prod_a m^+_{\chi_a}(\chi_a \in \chi) = \prod_a m^+_{\chi_a}(\text{AdP})$$

$$= \prod_a \sum_{I \subseteq P_{|\chi_a|}, |M_I| \equiv N_{|\chi_a|}} \prod_I p_{ij} \prod_{P_{|\chi_a|} - I}(1 - p_{ij}),$$

$$m^+_\chi(\Theta) = 1 - m^+_\chi(\text{AdP}).$$

This is the attracting metalevel evidence at the partition level.

5.3 Combine conflicting and attracting evidence

As the final step on the partition level (Figure 1) we combine all already combined conflicting evidence (Section 5.1) with all already combined attracting evidence (Section 5.2), $m_\chi = m^+_\chi \oplus m^-_\chi$. We receive,

$$m_\chi(\text{AdP}) = m_\chi(\forall a.\chi_a \in \chi) = m^+_\chi(\forall a.\chi_a \in \chi)m^-_\chi(\Theta) = m^+_\chi(\text{AdP})m^-_\chi(\Theta),$$
$$m_\chi(\neg\text{AdP}) = m_\chi(\exists a.\chi_a \notin \chi) = m^+_\chi(\Theta)m^-_\chi(\exists a.\chi_a \notin \chi) = m^+_\chi(\Theta)m^-_\chi(\neg\text{AdP}),$$
$$m_\chi(\Theta) = m^+_\chi(\Theta)m^-_\chi(\Theta),$$
$$m_\chi(\varnothing) = m^+_\chi(\forall a.\chi_a \in \chi)m^-_\chi(\exists a.\chi_a \notin \chi) = m^+_\chi(\text{AdP})m^-_\chi(\neg\text{AdP}).$$

With a conflict $m_\chi(\varnothing)$, since

$$(\forall a.\chi_a \in \chi) \wedge (\exists a.\chi_a \notin \chi) = \texttt{False}.$$

This is the amount of support awarded to the proposition that we have an "adequate partition" $m_\chi(\text{AdP})$, and awarded to the proposition that we do not have an "adequate partition" $m_\chi(\neg\text{AdP})$, respectively, when taking everything into account.

6 Weighting by information content

In order to find the best partition we might want to maximize $m_\chi(\text{AdP})$. However, in the special case when there is no positive metalevel evidence then $m_\chi(\text{AdP}) = m^+_\chi(\text{AdP}) = 0$. Alternative, we might like to minimize $m_\chi(\neg\text{AdP})$. This is what was done in [5] when only negative metalevel evidence was available. However, here we also have a special case when there is no negative metalevel evidence. Then, $m_\chi(\neg\text{AdP}) = m^-_\chi(\neg\text{AdP}) = 0$. The obvious solution is to minimize a function of $m_\chi(\neg\text{AdP})$ and $-m_\chi(\text{AdP})$. In doing this, we want to give each term a weighting corresponding to the information content of all conflicting and all attracting metalevel evidence, respectively. This is done in order to let each part have an influence corresponding to its information content.

Thus, let us minimize a metaconflict function

$$Mcf = \alpha[1 - m_\chi(\text{AdP})] + (1-\alpha)m_\chi(\neg\text{AdP}),$$

$0 \leq \alpha \leq 1$, where $\alpha = 0$ when all $p_{ij} = 0$, and $\alpha = 1$ when all $c_{ij} = 0$.
Let

$$\alpha = \frac{H(m^+_{\chi_0})}{H(m^+_{\chi_0}) - H(m^-_{\chi_0})},$$

where $H(m)$ is the expected value of the entropy $-\log_2[m(A)/|A|]$. $H(m)$ is called the average total uncertainty [4], measuring both scattering and nonspecificity, and may be written as the sum of Shannon entropy, $G(\cdot)$, and Hartley information, $I(\cdot)$,

$$H(m) = G(m) + I(m) = -\sum_{A \in \Theta} m(A)\log_2[m(A)] + \sum_{A \in \Theta} m(A)\log_2(|A|).$$

Here, $m^+_{\chi_0}$ and $m^-_{\chi_0}$ are calculated on the cluster level, as if all evidence is put into one large imaginary cluster χ_0.

6.1 Entropy of conflicting metalevel evidence $H(m^-_{\chi_0})$.

First, we combine $\forall i, j. \oplus \{m^-_{ij}\}$, i.e., all conflicting metalevel evidence, taking no account of which subset the different m^-_{ij} actually belongs to.

In this combination all intersections in the combined result are unique. Thus, the number of focal elements are equal to the number of intersections as no two intersections add up. Calculating the average total uncertainty of all conflicting metalevel evidence $H(m^-_{\chi_0}) = G(m^-_{\chi_0}) + I(m^-_{\chi_0})$ is then rather simple,

$$G(m^-_{\chi_0}) = -\sum_{J \subseteq P_n} m^-_{\chi_0}(\wedge\{e_i \vee e_j|(ij) \in J\} \notin \chi_0)\log_2[m^-_{\chi_0}(\wedge\{e_i \vee e_j|(ij) \in J\} \notin \chi_0)],$$

where $P_n = \{(ij)|1 \leq i < j \leq n\}$ is a set of pairs (ij), and n is the number of belief functions, with

$$\forall J \subseteq N ||J| > 1. m^-_{\chi_0}(\wedge\{e_i \vee e_j|(ij) \in J\} \notin \chi_0) = \prod_{I \subseteq P} c_{ij} \prod_{P-I} (1-c_{ij}).$$

The Hartley information is calculated as

$$I(m^-_{\chi_0}) = -\sum_{J \subseteq P_n} m^-_{\chi_0}(\wedge \{e_i \vee e_j | (ij) \in J\} \notin \chi_0)\log_2(|J|).$$

6.2 Entropy of attracting metalevel evidence $H(m^+_{\chi_0})$.

Similarly, we combine $\forall i, j. \oplus \{m^+_{ij}\}$, i.e., regardless of which subset the m^+_{ij}'s actually belongs to.

When calculating $H(m^+_{\chi_0}) = G(m^+_{\chi_0}) + I(m^+_{\chi_0})$ the Shannon entropy may be calculated as

$$G(m^+_{\chi_0}) = -\sum_{J \subseteq N | |J| > 1} m^+_{\chi_0}(\wedge \{e_j | j \in J\} \in \chi_0)\log_2[m^+_{\chi_0}(\wedge \{e_j | j \in J\} \in \chi_0)],$$

where $N = \{1, ..., n\}$, and n is the number of belief functions, and

$$\forall J \subseteq N | |J| > 1. m^+_{\chi_0}(\wedge \{e_j | j \in J\} \in \chi_0) = \sum_{I \subseteq P | M_I \equiv J} \prod_I p_{ij} \prod_{P-I} (1 - p_{ij}),$$

$$m^+_{\chi_0}(\Theta) = \prod_{(ij)} (1 - p_{ij}),$$

where $P = \{(ij) | 1 \leq i < j \leq n\}$.
With Hartley information calculated as

$$I(m^+_{\chi_0}) = -\sum_{J \subseteq N | |J| > 1} m^+_{\chi_0}(\wedge \{e_j | j \in J\} \in \chi_0)\log_2(|N - J| + 1).$$

7 Clustering belief functions

The best partition of all belief functions is found by minimizing

$$Mcf = \alpha[1 - m_\chi(\text{AdP})] + (1 - \alpha)m_\chi(\neg\text{AdP})$$

over all possible partitions. For a small number of belief functions this may be achieved through iterative optimization, but for a larger number of belief functions we need a method with a lower computational complexity, e.g., some neural clustering method similar to what was done in the case with only conflicting metalevel evidence [2].

8 Conclusions

We have extended the methodology for clustering belief function from only being able to manage conflicting information [1, 2, 9] to also being able to handle attracting information. This is important in many practical applications within information fusion.

References

[1] A. Ayoun and P. Smets (2001). Data association in multi-target detection using the transferable belief model. *International Journal of Intelligent Systems*, 16(10), 1167–1182.

[2] M. Bengtsson and J. Schubert (2001). Dempster-Shafer clustering using potts spin mean field theory. *Soft Computing*, 5(3), 215–228.

[3] A.P. Dempster (1968). A generalization of Bayesian inference. *Journal of the Royal Statistical Society B,* 30(2), 205–247.

[4] N.R. Pal, J.C. Bezdek and R. Henasinha (1993). Uncertainty measures for evidential reasoning II: A new measure of total uncertainty. *International Journal of Approximate Reasoning*, 8(1), 1–16.

[5] J. Schubert (1993). On nonspecific evidence. *International Journal of Intelligent Systems*, 8(6), 711–725.

[6] J. Schubert (1996). Specifying nonspecific evidence. *International Journal of Intelligent Systems*, 11(8), 525–563.

[7] J. Schubert (1999). Fast Dempster-Shafer clustering using a neural network structure. In *Information, Uncertainty and Fusion*, B. Bouchon-Meunier, R.R. Yager and L.A. Zadeh, (Eds.), pages 419–430, Kluwer Academic Publishers (SECS 516), Boston, MA.

[8] J. Schubert (1999). Simultaneous Dempster-Shafer clustering and gradual determination of number of clusters using a neural network structure. In *Proceedings of the 1999 Information, Decision and Control Conference*, pages 401–406, Adelaide, Australia, February 1999, IEEE, Piscataway, NJ.

[9] J. Schubert (2000). Managing inconsistent intelligence. In *Proceedings of the Third International Conference on Information Fusion*, volume 1, pages TuB4/10–16, Paris, France, July 2000.

[10] G. Shafer (1976). *A Mathematical Theory of Evidence*. Princeton University Press, Princeton, NJ.

Foundations

B. Bouchon-Meunier, L. Foulloy and R.R. Yager (Editors)
Intelligent Systems for Information Processing:
From Representation to Applications
© 2003 Elsevier Science B.V. All rights reserved.

Models and submodels of fuzzy theories

Vilém Novák

University of Ostrava
Institute for Research and Applications of Fuzzy Modeling
30. dubna 22, 701 03 Ostrava 1, Czech Republic
Vilem.Novak@osu.cz

and

Institute of Information and Automation Theory
Academy of Sciences of the Czech Republic
Pod vodárenskou věží 4, 186 02 Praha 8, Czech Republic

Abstract

This paper is a contribution to the development of the model theory of fuzzy logic in narrow sense with evaluated syntax whose are connectives interpreted in Łukasiewicz algebra. We will define several generalizations of the concept of submodel which have no counterpart in classical logic and demonstrate some properties of them. The main results concern elementary extensions and elementary chains. As a consequence, we will also present a generalization of the well known Craig's interpolation theorem for fuzzy logic.

Keywords: Fuzzy logic in narrow sense, Łukasiewicz logic, fuzzy theory, model theory of fuzzy logic.

1 Introduction

This paper is a contribution to the theory of predicate first-order fuzzy logic in narrow sense with evaluated syntax (FLn), which has been extensively presented in [4]. It is based on the set of truth values, which forms the Łukasiewicz MV-algebra

$$\mathcal{L} = \langle [0,1], \otimes, \oplus, \neg, \mathbf{0}, \mathbf{1} \rangle$$

(see also [2] and elsewhere). Recall that we may define the lattice join $\vee$ and meet $\wedge$ operations in $\mathcal{L}$ as well as the residuation operation $a \rightarrow b = \neg a \oplus b = 1 \wedge (1 - a + b)$. We work with graded concepts of provability and model. Therefore, each formula A is assigned a provability degree being supremum of the values of all its proofs, and a truth degree being infimum of its truth value in all models. The completeness theorem states that for each formula and each fuzzy theory both degrees coincide. Let us remark that classical logic becomes isomorphic to this logic when confining $\mathcal{L}$ to the classical two-valued boolean algebra.

FLn with evaluated syntax possesses a lot of properties which are generalizations of the corresponding properties of classical logic. Their proofs are mostly non-trivial and more complex. Besides other outcomes, they enable us to see classical logic from different point of view.

This paper confirms such claims. Its results are two-fold: we contribute to model theory of FLn and present a theorem on joint consistency of fuzzy theories. The latter is a generalization of the classical Craig-Robinson's one.

2 Preliminaries

We will use notation introduced in [4]. The language of FLn is denoted by J; terms are defined as usual. Formulas are formed using the connectives $\vee$ (disjunction, interpreted by $\vee$), ∇ (Łukasiewicz disjunction, interpreted by $\oplus$), $\wedge$ (conjunction, interpreted by $\wedge$), & (Łukasiewicz conjunction, interpreted by $\otimes$), $\Rightarrow$ (implication, interpreted by $\rightarrow$) and $\neg$ (negation, interpreted by $\neg$). The set of all the well-formed formulas for the language J is denoted by F_J and the set of all the closed terms by M_J. The couple a/A where $a \in L$ and $A \in F_J$ is an *evaluated formula*. Moreover, the language J is supposed to contain *logical constants* $\mathbf{a}$ being names of all the truth values $a \in L$. Similarly as in classical logic, we use the symbol $\top$ instead of $\mathbf{1}$ and $\bot$ instead of $\mathbf{0}$. The symbol A^n denotes n-fold Łukasiewicz conjunction and nA n-fold Łukasiewicz disjunction.

A fuzzy theory T is a fuzzy set $T \subseteq F_J$ of formulas given by a triple

$$T = \langle \text{LAx}, \text{SAx}, R \rangle$$

where $\text{LAx} \subseteq F_J$ is a fuzzy set of logical axioms, $\text{SAx} \subseteq F_J$ is a fuzzy set of special axioms and R is a set of inference rules which includes the rules modus ponens (r_{MP}), generalization (r_G) and logical constant introduction (r_{LC}). If T is a fuzzy theory then its language is denoted by $J(T)$. Note that we may equivalently speak about a *set of evaluated formulas*, or about a *fuzzy set of formulas*. This double view is quite common and makes no harm to definiteness of the explanation. We will usually define a fuzzy theory only by the fuzzy set of its special axioms

Given a fuzzy theory T and a formula A. If w_A is its proof with the value $\text{Val}(w_a)$ then $T \vdash_a A$ means that A is provable in T (a theorem of T) in the degree

$$a = \bigvee \{\text{Val}(w_a) \mid w_A \text{ is a proof of } A \in T\}.$$

If $a = \mathbf{1}$ then we simply write $T \vdash A$.

Let T be a fuzzy theory and $\Gamma \subseteq J(T)$ be a fuzzy set of formulas. Then the extension of T by the special axioms from Γ is a fuzzy theory $T' = T \cup \Gamma$ given by the fuzzy set of special axioms $\text{SAx}' = \text{SAx} \cup \Gamma$.

Theorem 1 (deduction theorem)
Let T be a fuzzy theory, A be a closed formula and $T' = T \cup \{\mathbf{1}/A\}$. Then to every formula $B \in F_{J(T)}$ there is an n such that

$$T \vdash_a A^n \Rightarrow B \quad \text{iff} \quad T' \vdash_a B.$$

Theorem 2 (reduction wrt. consistency)
Let T be a fuzzy theory and $\Gamma \subseteq F_{J(T)}$ a fuzzy set of closed formulas. A fuzzy theory $T' = T \cup \Gamma$ is contradictory iff there are $m_1, \ldots, m_n$ and $A_1, \ldots, A_n \in \text{Supp}(\Gamma)$ such

that
$$T \vdash_c \neg A_1^{m_1} \nabla \cdots \nabla \neg A_n^{m_n}$$
and $a_1^{m_1} \otimes \cdots \otimes a_n^{m_n} > \mathbf{0}$ where $c >^* \neg(a_1^{m_1} \otimes \cdots \otimes a_n^{m_n})$, $a_1 = \Gamma(A_1), \ldots, a_n = \Gamma(A_n)$ (we write $a >^* b$ iff $\mathbf{1} \geq a > b$ or $a = b = \mathbf{1}$).

An extension T' of T is conservative if $T' \vdash_a A$ implies $T \vdash_a A$ for every $A \in F_{J(T)}$. In [4], Theorem 4.11, it has been proved that extension of T by new constants not belonging to $J(T)$ is conservative.

Semantics of FLn is defined by generalization of the classical semantics of predicate logic. The *structure* for J is

$$\mathcal{V} = \langle V, f_\mathcal{V}, \ldots, P_\mathcal{V}, \ldots, u_\mathcal{V}, \ldots \rangle$$

where $f_\mathcal{V} : V^n \to V$ are n-ary[†] functions on V assigned to the functional symbols $f \in J$, $P_\mathcal{V} \subseteq V^n$ are n-ary fuzzy relations on V assigned to the predicate symbols $P \in J$ and $u_\mathcal{V} \in V$ are designated elements assigned to the object constants $\mathbf{u} \in J$.

$\mathcal{V}(t) = v \in V$ is an element being interpretation of the term t in $\mathcal{V}$. To interpret the formula $A(x_1, \ldots, x_n) \in F_J$, we must assign elements of V to its free variables. Therefore, we temporarily extend the language J into the language

$$J(\mathcal{V}) = J \cup \{\mathbf{v} \mid v \in V\}$$

where $\mathbf{v}$ are new names for all the elements of V, i.e. if $\mathbf{v} \in J(\mathcal{V})$ is a name of $v \in V$ then $\mathcal{V}(\mathbf{v}) = v$. Thus

$$\mathcal{V}(A_{x_1,\ldots,x_n}[\mathbf{v}_1, \ldots, \mathbf{v}_n]) \in L \qquad (1)$$

is an interpretation of A (truth value) obtained after assignment of the elements $v_1, \ldots, v_n \in V$ to the respective free variables $x_1, \ldots, x_n$.

Given a fuzzy theory T. The structure $\mathcal{V}$ for $J(T)$ is a model of T, $\mathcal{V} \models T$, if $\mathrm{SAx}(A) \leq \mathcal{V}(A)$ holds for all A. Then we say that A is true in T, $T \models_a A$ if

$$a = \mathcal{V}(A) = \bigwedge \{\mathcal{V}(A) \mid \mathcal{V} \models T\}.$$

If $a = \mathbf{1}$ then we simply write $T \models A$.

A fuzzy theory T is *consistent* if $T \vdash_a A$ and $T \vdash_b \neg A$ implies $a \otimes b = \mathbf{0}$; otherwise it is *inconsistent*. It can be proved that $T = F_J$ for an inconsistent T (F_J is seen as a fuzzy set with all the membership degrees equal to $\mathbf{1}$).

As mentioned, the completeness theorem states that

$$T \vdash_a A \qquad \text{iff} \qquad T \models_a A$$

holds for each fuzzy theory T and a formula $A \in F_{J(T)}$ (for the details see [4]).

[†] The arity n, of course, depends on the symbol in concern. However, we will not explicitly stress this.

3 Model theory in FLn

In [4], the concepts of submodel (substructure), elementary submodel, elementary equivalence, homomorphism and isomorphism of models have been defined. We will slightly modify the definition of substructure since a more subtle classification is necessary.

Definition 1
Let $\mathcal{V}$, $\mathcal{W}$ be two structures for the language J and let $\Gamma \subseteq F_J$.

(i) The $\mathcal{V}$ is a *weak substructure* of $\mathcal{W}$, in symbols
$$\mathcal{V} \subseteq \mathcal{W},$$
if $V \subseteq W$, $f_\mathcal{V} = f_\mathcal{W}|V^n$ holds for every n-ary function assigned to a functional symbol $f \in J$, $u_\mathcal{V} = u_\mathcal{W}$ for every object constant $\mathbf{u} \in J$ and
$$P_\mathcal{V} \leq P_\mathcal{W}|V^n \tag{2}$$
holds for the n-ary fuzzy relations $P_\mathcal{V}$ and $P_\mathcal{W}$ assigned to all the predicate symbols $P \in J$ in $\mathcal{V}$ and $\mathcal{W}$, respectively.

(ii) The $\mathcal{V}$ is a *substructure* of $\mathcal{W}$, in symbols
$$\mathcal{V} \subset \mathcal{W},$$
if the equality holds in (2) for all the predicate symbols $P \in J$.

(iii) The $\mathcal{V}$ is a *strong Γ-substructure* of $\mathcal{W}$ ($\mathcal{W}$ is a *strong Γ-extension* of $\mathcal{V}$), in symbols
$$\mathcal{V} \leq_\Gamma \mathcal{W},$$
if $V \subseteq W$ and
$$\mathcal{V}(A_{x_1,\ldots,x_n}[\mathbf{v}_1,\ldots,\mathbf{v}_n]) \leq \mathcal{W}(A_{x_1,\ldots,x_n}[\mathbf{v}_1,\ldots,\mathbf{v}_n]) \tag{3}$$
holds for every formula $A \in \Gamma$ and $v_1,\ldots,v_n \in V$. If $\mathcal{V} \leq_\Gamma \mathcal{W}$ for $\Gamma = F_J$ then $\mathcal{V}$ is a *strong substructure* of $\mathcal{W}$ ($\mathcal{W}$ is a strong extension of $\mathcal{V}$) and we write $\mathcal{V} \leq \mathcal{W}$.

(iv) Let $V \subseteq W$. The *expanded structure* is
$$\mathcal{W}_\mathcal{V} = \langle \mathcal{W}, \{\mathbf{v} \mid v \in V\} \rangle$$
where $\mathbf{v}$ are names for all the elements $v \in V$ taken in $\mathcal{W}_\mathcal{V}$ as new object constants, which, however, are interpreted by the same elements, i.e. $\mathcal{W}_\mathcal{V}(\mathbf{v}) = v$ for all $v \in V$.

(v) The $\mathcal{V}$ is an *elementary Γ-substructure* of $\mathcal{W}$ ($\mathcal{W}$ is an *elementary Γ-extension* of $\mathcal{V}$), in symbols $\mathcal{V} \prec_\Gamma \mathcal{W}$, if $V \subseteq W$ and
$$\mathcal{V}(A_{x_1,\ldots,x_n}[\mathbf{v}_1,\ldots,\mathbf{v}_n]) = \mathcal{W}(A_{x_1,\ldots,x_n}[\mathbf{v}_1,\ldots,\mathbf{v}_n]) \tag{4}$$
where $A(x_1,\ldots,x_n) \in \Gamma$ and $v_1,\ldots,v_n \in V$. If $\Gamma = F_J$ then $\mathcal{V}$ is an *elementary substructure* of $\mathcal{W}$ ($\mathcal{W}$ is an *elementary extension* of $\mathcal{V}$) and write $\mathcal{V} \prec \mathcal{W}$.

Lemma 1
(a) If $\mathcal{V}_1 \subseteq \mathcal{V}_2$ ($\mathcal{V}_1 \subset \mathcal{V}_2$) and $\mathcal{V}_2 \subseteq \mathcal{V}_1$ ($\mathcal{V}_2 \subset \mathcal{V}_1$) then $\mathcal{V}_1 = \mathcal{V}_2$.
(b) If $\mathcal{V}_1 \leq \mathcal{V}_2$ then $\mathcal{V}_1 \subseteq \mathcal{V}_2$.

Note, however, that if $\mathcal{V}_1 \subseteq \mathcal{V}_2$ then there is no explicit relation between $\mathcal{V}_1(A)$ and $\mathcal{V}_2(A)$ for an arbitrary formula A. Note also that the distinction between strong and weak substructures has no sense in classical logic.

Definition 2
Two structures $\mathcal{V}$ and $\mathcal{W}$ are *isomorphic*, $\mathcal{V} \cong \mathcal{W}$, if there is a bijection $g : V \to W$ such that the following holds for all $v_1, \ldots, v_n \in V$:

(i) For each couple of functions $f_\mathcal{V}$ in $\mathcal{V}$ and $f_\mathcal{W}$ in $\mathcal{W}$ assigned to a functional symbol $f \in J$,
$$g(f_\mathcal{V}(v_1, \ldots, v_n)) = f_\mathcal{W}(g(v_1), \ldots, g(v_n)).$$

(ii) For each predicate symbol $P \in J$,
$$P_\mathcal{V}(v_1, \ldots, v_n) = P_\mathcal{W}(g(v_1), \ldots, g(v_n)). \tag{5}$$

(iii) For each couple of constants u in $\mathcal{V}$ and w in $\mathcal{W}$ assigned to a constant symbol $\mathbf{u} \in J$,
$$g(u) = u'.$$

Note that Definitions 1 and 2 are simplified with respect to the corresponding ones introduced in [4].

Lemma 2
Let $\mathcal{V} \cong \mathcal{W}$ be two isomorphic structures. Then
$$g(\mathcal{V}(t)) = \mathcal{W}(t)$$
holds for every closed term t and
$$\mathcal{V}(A) = \mathcal{W}(A)$$
holds for every formula $A \in F_J$.

PROOF: By induction on the complexity of terms and formulas. □

The Γ-*diagram* $D_\Gamma(\mathcal{V})$ of $\mathcal{V}$ is a fuzzy theory with the special axioms
$$\mathrm{SAx}_{D_\Gamma(\mathcal{V})} = \{a/A_{x_1,\ldots,x_n}[\mathbf{v}_1, \ldots, \mathbf{v}_n] \mid A \in \Gamma,$$
$$a = \mathcal{V}(A_{x_1,\ldots,x_n}[\mathbf{v}_1, \ldots, \mathbf{v}_n]), v_1, \ldots, v_n \in V\}. \tag{6}$$

If $\Gamma = F_J$ then we speak about *diagram* of $\mathcal{V}$ and write simply $D(\mathcal{V})$. Obviously, $\mathcal{V} \models D_\Gamma(\mathcal{V})$.

Lemma 3
Let $\mathcal{V}, \mathcal{W}$ be two structures for the language J and $\mathcal{V} \subseteq \mathcal{W}$. Then $\mathcal{V}$ is a Γ-substructure of $\mathcal{W}$ iff $\mathcal{W}_\mathcal{V} \models D_\Gamma(\mathcal{V})$.

PROOF: Let $A \in \Gamma$ and $v_1, \ldots, v_n \in V$. Then

$$\mathcal{V}(A_{x_1,\ldots,x_n}[\mathbf{v}_1,\ldots,\mathbf{v}_n]) \leq \mathcal{W}(A_{x_1,\ldots,x_n}[\mathbf{v}_1,\ldots,\mathbf{v}_n])$$

which means that $\mathcal{W}_\mathcal{V} \models D_\Gamma(\mathcal{V})$ since $\mathbf{v}_1, \ldots, \mathbf{v}_n$ are, at the same time, object constants of $\mathcal{W}_\mathcal{V}$. The converse is obvious. $\square$

A set Γ of formulas is called *regular* if $x = y \in \Gamma$ or $x \neq y \in \Gamma$ and if $A' \in \Gamma$ is an instance of A then also $A \in \Gamma$.

Theorem 3
Let T be a (consistent) fuzzy theory, $\mathcal{V}$ a structure for $J(T)$ and $\Gamma \subseteq F_{J(T)}$ be a regular set of formulas. Then the following is equivalent.

(a) There is a Γ-extension of $\mathcal{V}$, $\mathcal{V} \leq_\Gamma \mathcal{W}$, being a model of T, i.e. $\mathcal{W} \models T$.

(b) Let $B := \neg A_1^{m_1} \nabla \cdots \nabla \neg A_n^{m_n}$ be a formula where $A_1, \ldots, A_n \in \Gamma$ and $m_1, \ldots, m_n \in \mathbb{N}^+$ such that
$$T \vdash_b B.$$
Then
$$b \leq \mathcal{V}(\neg A_1^{m_1} \nabla \cdots \nabla \neg A_n^{m_n}).$$

PROOF:
(a)$\Rightarrow$(b): Let $\mathcal{V} \leq_\Gamma \mathcal{W}$ where $\mathcal{W} \models T$. Let $A_1, \ldots, A_n \in \Gamma$ and $T \vdash_b \neg(A_1^{m_1} \& \cdots \& A_n^{m_n})$ for some $m_1, \ldots, m_n \in \mathbb{N}^+$. Then there are $v_1, \ldots, v_p$ such that

$$T \vdash_{b'} \neg((A'_1)^{m_1} \& \cdots \& (A'_n)^{m_n}), \qquad b' \geq b, \tag{7}$$

where $(A'_i)^{m_i} := A_i^{m_i}[\mathbf{v}_1, \ldots, \mathbf{v}_p]$ is a $\mathcal{V}$-instance of $A_i^{m_i}$, $i = 1, \ldots, n$.

It follows from the assumption, the substitution axiom and monotonicity of $\otimes$ that

$$\mathcal{V}(A_1^{m_1} \& \cdots \& A_n^{m_n}) = \mathcal{V}(A_1)^{m_1} \otimes \cdots \otimes \mathcal{V}(A_n)^{m_n} \leq$$
$$\leq \mathcal{V}(A'_1)^{m_1} \otimes \cdots \otimes \mathcal{V}(A'_n)^{m_n} \leq \mathcal{W}(A'_1)^{m_1} \otimes \cdots \otimes \mathcal{W}(A'_n)^{m_n} =$$
$$= \mathcal{W}((A'_1)^{m_1} \& \cdots \& (A'_n)^{m_n}).$$

Since $\mathcal{W} \models T$, we obtain

$$b \leq b' \leq \mathcal{W}(\neg((A'_1)^{m_1} \& \cdots \& (A'_n)^{m_n})) \leq \mathcal{V}(\neg(A_1^{m_1} \& \cdots \& A_n^{m_n}))$$

by the properties of negation operation.

(b)$\Rightarrow$(a): Let us extend conservatively T by constants for elements of V into the fuzzy theory T' and, furthermore, put

$$T'' = T' \cup D_\Gamma(\mathcal{V})$$

where $D_\Gamma(\mathcal{V})$ is the Γ-diagram (6) of $\mathcal{V}$. By contradiction, we may show that T'' is consistent. Hence, it has a model $\mathcal{W}'' \models T''$. Moreover, since $\mathcal{W}'' \models D_\Gamma(\mathcal{V})$, we have $\mathcal{V}(A) \leq \mathcal{W}''(A)$ for every $A \in \Gamma$. This means that $\mathcal{V} \leq_\Gamma \mathcal{W}''$.

We must now show that there is an isomorphic structure W such that $\mathcal{W}_\mathcal{V} \models D_\Gamma(\mathcal{V})$. The problem is that there may be constants $\mathbf{v}_1, \mathbf{v}_2$ in $\mathcal{W}''$ assigned to only a single element $v \in V$. Thus, let $v_1, v_2 \in V$ be different elements. Since Γ is regular, $\mathcal{V}(\mathbf{v}_1 \neq \mathbf{v}_2) = \mathbf{1}$ and so, $D_\Gamma(\mathcal{V}) \vdash \mathbf{v}_1 \neq \mathbf{v}_2$. Therefore also $\mathcal{W}''(\mathbf{v}_1 \neq \mathbf{v}_2) = \mathbf{1}$ and consequently, we can consider an isomorphic structure $\mathcal{W}'$ instead of the original $\mathcal{W}''$. When considering its restriction $\mathcal{W}$ by names of elements from V, we realize that $\mathcal{W}' = \mathcal{W}_\mathcal{V}$. Since $\mathcal{W}_\mathcal{V} \models T''$, we have $\mathcal{W}_\mathcal{V} \models D_\Gamma(\mathcal{V})$ and thus, $\mathcal{W}$ is a Γ-extension of $\mathcal{V}$ such that $\mathcal{W} \models T$. $\square$

Corollary 1
Let $\mathcal{V} \models T$. Then there is a strong extension $\mathcal{W}$, $\mathcal{V} \leq \mathcal{W}$ such that $\mathcal{W} \models T$.

PROOF: The existence of such structure follows from Theorem 3 by setting $\Gamma = F_{J(T)}$. Then for every formula $T \vdash_a A$ implies $a \leq \mathcal{V}(A)$ since $\mathcal{V} \models T$. $\square$

Theorem 4
Let T be a fuzzy theory of the language J, $\mathcal{V}$ be a structure and $\Gamma \subseteq F_J$. Then there is a model $\mathcal{W} \models T$ which is an elementary Γ-extension of $\mathcal{V}$.

PROOF: Similarly as in the proof of Theorem 3, we will consider a fuzzy theory $T'' = T' \cup D(\mathcal{V})$ where T' is a conservative extension of T by constants for the elements of V and $D(\mathcal{V})$ is the diagram of $\mathcal{V}$.

The fuzzy theory T'' is consistent since otherwise there should exist formulas $A_1, \ldots, A_n$ and $m_1, \ldots, m_n \in \mathbb{N}^+$ such that $T' \vdash_d \neg((A'_1)^{m_1} \& \cdots \& (A'_n)^{m_n}))$ (A'_i are $\mathcal{V}$-instances of A_i) so that
$$d > \neg(a_1^{m_1} \otimes \cdots \otimes a_n^{m_n})$$
where $a_i = \mathcal{V}_\mathcal{V}(A'_i)$. Since $\mathcal{V}_\mathcal{V} \models T'$, we obtain
$$d \leq \mathcal{V}(\neg((A'_1)^{m_1} \& \cdots \& (A'_n)^{m_n}))) < d$$
which is a contradiction. Hence, there is a model $\mathcal{W}'' \models T''$. We will now replace $\mathcal{W}''$ by an isomorphic structure $\mathcal{W}$ and by induction on the complexity of formula, we show that $\mathcal{W}(A) = \mathcal{V}(A)$ holds for every $\mathcal{V}$-instance of a formula A. Thus, we obtain an elementary extension of $\mathcal{V}$. Finally, since $\Gamma \subseteq F_J$, $\mathcal{W}$ is an elementary Γ-extension. $\square$

Definition 3
Let $\mathcal{V}_1 \subset \mathcal{V}_2 \subset \ldots \subset \mathcal{V}_\alpha \subset \ldots$ (or $\mathcal{V}_1 \subseteq \mathcal{V}_2 \subseteq \ldots \subseteq \mathcal{V}_\alpha \subseteq \ldots$) be a chain of structures, $\alpha < \xi$ for some ordinal number ξ. Put $V = \bigcup_{\alpha < \xi} V_\alpha$ and each $f_\mathcal{V} = \bigcup_{\alpha < \xi} f_{\mathcal{V}_\alpha}$. Furthermore, we extend $P_{\mathcal{V}_\alpha}$ into V by putting

$$P_\mathcal{V}(v_1, \ldots, v_n) = \bigvee_{\beta \leq \alpha < \xi} P_{\mathcal{V}_\alpha}(v_1, \ldots, v_n) \qquad (8)$$

for each predicate symbol P and each sequence $v_1, \ldots, v_n \in V$, where β is the first ordinal such that $v_1, \ldots, v_n \in V_\beta$. Then the union of the above chain of structures is the structure
$$\mathcal{V} = \bigcup_{\alpha < \xi} \mathcal{V}_\alpha = \langle V, P_\mathcal{V}, \ldots, f_\mathcal{V}, \ldots, u_\mathcal{V}, \ldots \rangle.$$

Theorem 5
Let $\mathcal{V}_1 \subset \mathcal{V}_2 \subset \ldots \subset \mathcal{V}_\alpha \subset \ldots$, $\mathcal{V}_\alpha \models T$, be a chain of models of some fuzzy theory T, $\alpha < \xi$. Then $\mathcal{V} \models T$ where $\mathcal{V}$ is the union of this chain.

PROOF: We may confine only to formulas in prenex form. Let A be a formula and $T \vdash_a A$. We will show that $a \leq \mathcal{V}(A)$.

First, let $A := (\forall x_1) \cdots (\forall x_n) B(x_1, \ldots, x_n)$ for some formula B. Then

$$\mathcal{V}(A) = \bigwedge_{v_1, \ldots, v_n \in V} \mathcal{V}(B_{x_1, \ldots, x_n}[\mathbf{v}_1, \ldots, \mathbf{v}_n]) =$$

$$= \bigwedge_{\alpha < \xi} \bigwedge_{v_1, \ldots, v_n \in V_\alpha} \mathcal{V}_\alpha(B_{x_1, \ldots, x_n}[\mathbf{v}_1, \ldots, \mathbf{v}_n]) =$$

$$= \bigwedge_{\alpha < \xi} \mathcal{V}_\alpha((\forall x_1) \cdots (\forall x_n) B) \geq a$$

by the assumption on $\mathcal{V}_\alpha$.

It remains to show that if A does not contain quantifiers then $a \leq \mathcal{V}(A)$. This will be done by induction on the complexity of A.

(a) Let $A := P_{x_1, \ldots, x_n}[t_1, \ldots, t_n]$ be a closed atomic formula and $\mathcal{V}(t_i) = v_i \in V$, $i = 1, \ldots, n$. Then there is $\beta < \xi$ such that $v_1, \ldots, v_n \in V_\alpha$ for all $\beta \leq \alpha$. By (8), the assumed equality in (2) and the assumption on $\mathcal{V}_\alpha$ we have

$$a \leq \mathcal{V}_\alpha(A) = P_{\mathcal{V}_\alpha}(v_1, \ldots, v_n) = \bigvee_{\beta \leq \alpha < \xi} P_{\mathcal{V}_\alpha}(v_1, \ldots, v_n) = P_\mathcal{V}(v_1, \ldots, v_n) = \mathcal{V}(A).$$

(b) The case when $A := B \Rightarrow C$ is a closed formula without quantifiers follows from the induction assumption. □

The following theorem has been proved in [4].

Theorem 6
Let $\mathcal{V}_1 \prec \mathcal{V}_2 \prec \ldots \prec \mathcal{V}_\alpha \prec \ldots$ be an elementary chain of models, $\alpha < \xi$ for some ordinal number ξ. Then
$$\mathcal{V} = \bigcup_{\alpha < \xi} \mathcal{V}_\alpha$$
is an elementary extension of each $\mathcal{V}_\alpha$, i.e., $\mathcal{V}_\alpha \prec \mathcal{V}$ for every α.

The following is a fuzzy logic version of the classical Craig-Robinson's theorem on simultaneous consistency of theories (cf. [7]) whose proof is based on the application of the previous theorems and can be found in [5].

Theorem 7 (simultaneous consistency)
Let T and T' be consistent fuzzy theories. Then $T \cup T'$ is contradictory iff there is a closed formula $A \in F_{J(T)} \cap F_{J(T')}$ and $a, b \in L$ such that

$$T \vdash_a A \quad \text{and} \quad T' \vdash_b \neg A \quad \text{and} \quad a \otimes b > \mathbf{0}.$$

Using Theorem 7 we can, analogously as in classical logic, prove also generalization of the classical Craig's interpolation theorem.

Theorem 8 (interpolation theorem)
Let T and T' be fuzzy theories, $A \in F_{J(T)}$ and $B \in F_{J(T')}$. Let

$$T \cup T' \vdash_a A \Rightarrow B$$

where $a > \mathbf{0}$. Then there is a closed formula $C \in F_{J(T)} \cap F_{J(T')}$ and $m, n \in \mathbb{N}^+$ such that

$$T \vdash_c A^m \Rightarrow C \quad \text{and} \quad T' \vdash_d C \Rightarrow nB$$

for some c and d such that $c \otimes d > \mathbf{0}$.

PROOF: It is sufficient to consider only closed formulas because for open formulas we can use the theorem on constants (which is valid in fuzzy logic, as well).

Let A, B be closed formulas and let us consider the fuzzy theory

$$\bar{T} = (T \cup \{\mathbf{1}/A\}) \cup (T' \cup \{\mathbf{1}/\neg B\}).$$

This fuzzy theory is contradictory since $\bar{T} \vdash_{a'} A \Rightarrow B$ where $a' \geq a$, $\bar{T} \vdash A$, $\bar{T} \vdash \neg B$ and thus, $\bar{T} \vdash_{a''} B \& \neg B$ where $a'' \geq a > \mathbf{0}$. By Theorem 7, there is a closed formula $C \in F_{J(T)} \cap F_{J(T')}$ such that

$$T \cup \{\mathbf{1}/A\} \vdash_c C \quad \text{and} \quad T' \cup \{\mathbf{1}/\neg B\} \vdash_d \neg C \quad \text{and} \quad c \otimes d > \mathbf{0}.$$

By the deduction theorem there are $m, n \geq 1$ such that

$$T \vdash_c A^m \Rightarrow C \quad \text{and} \quad T' \vdash_d (\neg B)^n \Rightarrow \neg C$$

where the right part implies $T' \vdash_d C \Rightarrow nB$. □

Corollary 2
Let T be a fuzzy theory and $A, B \in F_{J(T)}$ formulas. Let

$$T \vdash_a A \Rightarrow B$$

where $a > \mathbf{0}$. Then there is a closed formula $C \in F_{J(T)}$ and $p \in \mathbb{N}^+$ such that

$$T \vdash_c A^p \Rightarrow C \quad \text{and} \quad T \vdash_d C \Rightarrow pB \qquad (9)$$

for some c and d such that $c \otimes d > \mathbf{0}$.

PROOF: Take $T = T'$ in the above theorem and put $p = \max(m, n)$. □

The above interpolation theorem raises a question what should be the numbers m, n. It is not possible to derive them syntactically because the proof uses the deduction theorem in which the exponent n depends on the number of usages of the modus ponens rule. Therefore, we offer a semantic criterion.

Corollary 3
Let T be a fuzzy theory and $A, B \in F_{J(T)}$ formulas. Let
$$T \vdash_a A \Rightarrow B$$
where $a > 0$. Let $C \in F_{J(T)}$ be a closed formula such that
$$T \cup \{\mathbf{1}/A\} \vdash_c C \quad \text{and} \quad T \cup \{\mathbf{1}/\neg B\} \vdash_d \neg C \quad \text{and} \quad c \otimes d > \mathbf{0}. \tag{10}$$
Then there are the least $m, n \in \mathbb{N}^+$ such that
$$T \vdash_c A^m \Rightarrow C \quad \text{and} \quad T \vdash_d C \Rightarrow nB. \tag{11}$$

PROOF: The existence of m, n follows from the deduction theorem using the reasoning in the proof of Theorem 8. Let $T \vdash_f A$, $T \vdash_b B$ and $T \vdash_e C$ for some $f, b, e \in L$ (this assumption is sound since in fuzzy logic, every formula is provable in some degree). Then using the completeness theorem we obtain $c \otimes f^m \leq e$ and $d \otimes e \leq nb$ for some m, n (in the proof we must use the identities $\bigwedge_i a_i^n = (\bigwedge_i a_i)^n$ and $n \bigwedge_i a_i = \bigwedge_i na_i$ which hold in the Łukasiewicz MV-algebra). These inequalities enable us to find the least m, n fulfiling them. □

It follows from this corollary that the (least) exponents m, n depend on the provability degrees of the formula C and its negation in the extensions (10) of the fuzzy theory T. Note that since c, d are some general degrees, the formulas in (11) are nontrivial (i.e. they do not collapse into $\top$).

4 Conclusion

In this paper, we have discussed some properties of model theory of fuzzy logic in narrow sense with evaluated syntax. It can be seen that when dealing with degrees, we can introduce a lot of special properties which have no counterpart in classical logic. This concerns various kinds of relations among models of fuzzy theories. Note also, that they can be generalized to hold in some general degree only. Such generalization can be useful, e.g. in areas such as modeling of natural language semantics, or theories dealing with prototypes (cf. e.g. [6]).

References

[1] Chang, C. C. and Keisler, H. J. (1973). *Model Theory*, North–Holland, Amsterdam.

[2] Hájek, P. (1998). *Metamathematics of fuzzy logic*. Kluwer, Dordrecht.

[3] Novák, V. and I. Perfilieva (eds.) (2000). *Discovering World With Fuzzy Logic*. Springer-Verlag, Heidelberg, 555 pp. (Studies in Fuzziness and Soft Computing, Vol. 57)

[4] Novák, V., Perfilieva I. and J. Močkoř (1999). *Mathematical Principles of Fuzzy Logic*. Kluwer, Boston.

[5] Novák, V. (2002). Joint consistency of fuzzy theories. *Mathematical Logic Quaterly* **48**, 4, 563–573.

[6] Novák, V. (2002). Intensional theory of granular computing. *Soft Computing* (to appear).

[7] Shoenfield, J. R. (1967). *Mathematical Logic*. New York: Addison–Wesley.

Numerical Representations of Fuzzy Relational Systems

Sergei Ovchinnikov
Mathematics Department
San Francisco State University
San Francisco, CA 94132, USA
sergei@sfsu.edu

Abstract

We show that there exists a continuous fuzzy quasi–order $\widetilde{R}$ on the set of real numbers $\mathbb{R}$ such that for any fuzzy quasi–order R on a finite set X there is a mapping $f : X \to \mathbb{R}$ such that $R(x,y) = \widetilde{R}(f(x), f(y))$ for all $x, y \in X$.

Keywords: Fuzzy relational system, Fuzzy quasi–order

1 Introduction

The paper is concerned with numerical representations of fuzzy relational systems. A *fuzzy relational system* is a pair (X, R) where X is a set and R is a fuzzy binary relation on X. In addition, we assume that X is a topological space and the membership function $R(x, y)$ of R is a continuous function on $X \times X$. (Finite sets are assumed to be endowed with the discrete topology.) In the case when $X = \mathbb{R}$, the set of real numbers, $(\mathbb{R}, R)$ is said to be a *numerical fuzzy relational system*.

Given two fuzzy relational systems (X, P) and (Y, Q), a *homomorphism* from (X, P) to (Y, Q) is a continuous mapping $f : X \to Y$ such that
$$P(x, y) = Q(f(x), f(y)), \quad \text{for all } x, y \in X.$$

A homomorphism from (X, R) to a numerical fuzzy relational system $(\mathbb{R}, \widetilde{R})$ is said to be a *numerical representation* of (X, R) in $(\mathbb{R}, \widetilde{R})$.

In this paper, all fuzzy relations R on a given space X are assumed to be *fuzzy quasi–orders*, i.e., fuzzy binary relations satisfying conditions:

(i) $R(x, x) = 1$, (*reflexivity*),

(ii) $R(x, y)R(y, z) \leq R(x, z)$, (*max–product transitivity*),

for all $x, y \in X$.

The following theorem is the main result of the paper.

Main Theorem. *There exists a numerical fuzzy relational system $(\mathbb{R}, \widetilde{R})$ such that any fuzzy relational system (X, R) with finite X and positive R ($R(x, y) > 0$ for all $x, y \in X$) admits a numerical representation in $(\mathbb{R}, \widetilde{R})$.*

Similar results for arbitrary fuzzy transitive relations are established in [3].

2 Preliminaries

First, we show that a continuous fuzzy quasi–order R on $\mathbb{R}$ must have a positive membership function $R(x, y)$. This fact explains why we need the positivity condition in the Main Theorem. We define

$$x \sim_0 y \quad \Leftrightarrow \quad R(x, y) > 0 \text{ and } R(y, x) > 0,$$

for all $x, y \in \mathbb{R}$. Thus defined $\sim_0$ is obviously a reflexive and symmetric relation. By the transitivity of R, the relation $\sim_0$ is an equivalence relation on $\mathbb{R}$. Clearly, $\{(x, y) : x \sim_0 y\}$ is an open subset of $\mathbb{R}^2$. Since equivalence classes of $\sim_0$ are open sets and $\mathbb{R}$ is connected, $x \sim_0 y$ for all $x, y \in \mathbb{R}$. Hence, $R(x, y)$ is positive.

A fuzzy quasi–order R on X is said to be *strict* if

$$\min\{R(x, y), R(y, x)\} < 1, \quad \forall x \neq y \text{ in } X.$$

We show now that in order to prove the Main Theorem it suffices to consider only the case of strict quasi–orders. We define

$$x \sim_1 y \quad \Leftrightarrow \quad R(x, y) = 1 \text{ and } R(y, x) = 1,$$

for all $x, y \in X$. Thus defined $\sim_1$ is an equivalence relation on X. Let us define

$$(R/\sim_1)(u, v) = R(x, y),$$

for $x \in u \in X/\sim_1$, $y \in v \in X/\sim_1$. The relation $R/\sim_1$ is well–defined on $X \sim_1$. Indeed, suppose that $x' \in u$ and $y' \in v$. Then

$$R(x, y) \geq R(x, x')R(x', y) = R(x', y)$$
$$\geq R(x', y')R(y', y) = R(x', y').$$

By symmetry, $R(x, y) = R(x', y')$. Clearly, $R/\sim_1$ is a strict fuzzy quasi–order on $X/\sim_1$.

The canonical mapping $\varphi : X \to X/\sim_1$ defines a homomorphism from (X, R) to

$$(X/\sim_1, R/\sim_1).$$

Let f be a numerical representation of $(X/\sim_1, R/\sim_1)$. Then the mapping $f \circ \varphi$ defines a numerical representation of (X, R).

3 Proof of the Main Theorem

According to the previous section, we may restrict our attention to numerical representations of fuzzy relational systems (X, R) with positive strict fuzzy quasi–orders R. We also assume that X is a finite set.

Let

$$M_n = \{\overline{x} : \frac{1}{n} \leq x_i \leq 1,\ 1 \leq i \leq n\} = \left[\frac{1}{n}, 1\right]^n$$

where $\overline{x} = \{x_1, \ldots, x_n\}$. We define

$$R_n(\overline{x}, \overline{y}) = \min_i \min\left\{\frac{x_i}{y_i}, 1\right\},$$

for all $\overline{x}, \overline{y} \in M_n$. Clearly,

$$R_n(\overline{x}, \overline{y}) R_n(\overline{y}, \overline{z}) \leq R_n(\overline{x}, \overline{z}).$$

Thus, R_n is a strict fuzzy quasi–order on M_n.

Let $I_n = [n-1, n]$ for $n > 0$. There exists a continuous surjection f_n from I_n onto M_n (a "Peano curve" [1, IV(4)]). We define a fuzzy relation $\widetilde{R}_n$ on I_n by

$$\widetilde{R}_n(x, y) == \min\{\max\{\min\{e^{y-x}, 1\}, 1 - \frac{1}{n}\}, R_n(f_n(x), f_n(y))\}. \tag{1}$$

Clearly, $\widetilde{R}_n$ is a continuous function.

Lemma 1. $\widetilde{R}_n$ *is a strict fuzzy quasi–order on* I_n.

Proof. First, we show that $\widetilde{R}_n$ is transitive. Note, that $P(x, y) = \min\{e^{y-x}, 1\}$ defines a transitive relation on I_n. Let Q be defined by

$$Q(x, y) = \max\{P(x, y), 1 - \frac{1}{n}\}$$

on I_n. Thus defined Q is transitive. Indeed,

$$Q(x, y)Q(y, z) =$$
$$\max\{P(x, y), 1 - \frac{1}{n}\} \cdot \max\{P(y, z), 1 - \frac{1}{n}\}$$
$$\leq \max\{P(x, y)P(y, z), 1 - \frac{1}{n}\}$$
$$\leq \max\{P(x, z), 1 - \frac{1}{n}\} = Q(x, z).$$

Finally, a simple calculation shows that

$$\widetilde{R}_n(x, y)\widetilde{R}_n(y, z) = \widetilde{R}_n(x, z).$$

It follows from (1) that $\widetilde{R}_n(x, y) = 1$ implies $x \geq y$. Hence, $\widetilde{R}_n$ is a strict fuzzy quasi–order. □

Lemma 2. *Let Y be a finite subset of M_n such that $R_n(\overline{x}, \overline{y}) \leq 1 - \frac{1}{n}$ for all $\overline{x} \neq \overline{y}$ in Y. There is a homomorphism from (Y, R_n) to $(I_n, \widetilde{R}_n)$.*

Proof. Since f_n is a surjection, for any $\overline{z} \in Y$, there is $x_{\overline{z}} \in I_n$ such that $f_n(x_{\overline{z}}) = \overline{z}$. Let $Z = \{x_{\overline{z}} : \overline{z} \in Y\} \subseteq I_n$. Then $g_n : \overline{z} \mapsto x_{\overline{z}}$ is a bijection from Y to Z. Since $R_n(\overline{z}, \overline{z}') \leq 1 - \frac{1}{n}$, we have, by (1), $R_n(\overline{z}, \overline{z}') = \widetilde{R}_n(g_n(\overline{z}), g_n(\overline{z}'))$. □

Let R be a positive strict fuzzy quasi–order on a finite set $X = \{x_1, \ldots, x_m\}$ and let $\alpha = \min\{R(x,y)\}$ and $\beta = \min\{1 - R(x,y) : x \neq y\}$. We define

$$n(R) = \max\{m, \lfloor \alpha^{-1} \rfloor, \lfloor \beta^{-1} \rfloor\}.$$

Clearly,
$$\frac{1}{n(R)} \leq R(x,y) \leq 1 - \frac{1}{n(R)}, \tag{2}$$

for $x \neq y$.

The following lemma is a special case of Theorem 1 in [2].

Lemma 3. *Let $n = n(R)$. There exists a homomorphism from (X, R) to (M_n, R_n).*

Proof. We define $f : X \to M_n$ by

$$f(x_i) = (R(x_i, x_1), \ldots, R(x_i, x_m), \underbrace{1, \ldots, 1}_{(n-m)\ 1\text{'s}}),$$

for $1 \leq i \leq m$. We have, by the transitivity property,

$$R_n(f(x_i), f(x_j)) = \min_k \min\left\{\frac{R(x_i, x_k)}{R(x_j, x_k)}, 1\right\}$$
$$\geq R(x_i, x_j),$$

and, for $k = j$, $\dfrac{R(x_i, x_k)}{R(x_j, x_k)} = R(x_i, x_j)$.

Hence, $R(x_i, x_j) = R_n(f(x_i), f(x_j))$ for all $1 \leq i, j \leq m$. $\square$

Let R be a positive strict fuzzy quasi–order on X and $n = n(R)$. It follows from Lemma 3, inequalities (2) and Lemma 2 that there exists a homomorphism from (X, R) to $(I_n, \widetilde{R}_n)$.

It remains to show that there exists a continuous fuzzy quasi–order $\widetilde{R}$ on $\mathbb{R}$ such that its restriction on each interval I_n coincides with $\widetilde{R}_n$.

Lemma 4. *Let R_1 and R_2 be two continuous fuzzy quasi–orders defined on intervals $[a, b]$ and $[b, c]$, respectively. Then R defined by*

$$R(x,y) = R_1(x,y),\ on\ T_1 = [a,b] \times [a,b],$$
$$R(x,y) = R_2(x,y),\ on\ T_2 = [b,c] \times [b,c],$$
$$R(x,y) = R_1(x,b)R_2(b,y),$$
$$on\ T_3 = [a,b] \times [b,c],$$
$$R(x,y) = R_2(x,b)R_1(b,y),$$
$$on\ T_4 = [b,c] \times [a,b]$$

is a continuous fuzzy quasi–order on $[a, c]$.

Proof. Clearly, $R(x,y)$ is well–defined on the rectangle $[a,c] \times [a,c]$. Since $R(x,y)$ is continuous on each $T_1, \ldots, T_4$, it is continuous on $[a,c] \times [a,c]$.

To show that
$$R(x,y)R(y,z) \leq R(x,z),$$
for all $x,y,z \in [a,c]$, we first consider the following three cases:

1) $x \in [a,b]$, $y, z \in [b,c]$. Then
$$R(x,y)R(y,z) = R_1(x,b)R_2(b,y)R_2(y,z) \leq$$
$$R_1(x,b)R_2(b,z) = R(x,z).$$

2) $y \in [a,b]$, $x, z \in [b,c]$. Then
$$R(x,y)R(y,z) =$$
$$R_2(x,b)R_1(b,y)R_1(y,b)R_2(b,z) \leq$$
$$R_2(x,z) = R(x,z).$$

3) $z \in [a,b]$, $x, y \in [b,c]$. Then
$$R(x,y)R(y,z) =$$
$$R_2(x,y)R_2(y,b)R_1(b,z) \leq$$
$$R_2(x,b)R_1(b,z) = R(x,z).$$

The remaining alternatives are similar to these three and omitted. □

Note that the previous proof is valid in the case when one (or both) of the intervals $[a,b]$ and $[b,c]$ is unbounded.

We define $\widetilde{R}_0(x,y) = \min\{e^{y-x}, 1\}$ on $I_0 = \{t : t \leq 0\}$. It follows from the previous lemma that there exists a fuzzy quasi–order $\widetilde{R}$ on $\mathbb{R}$ which coincides with each $\widetilde{R}_n$ on I_n for $n = 0, 1, \ldots$.

This completes the proof of the Main Theorem.

References

[1] J.Dugunji, Topology (Wm. C. Brown Publishers, Dubuque, Iowa, 1989).

[2] S.Ovchinnikov, Representations of transitive fuzzy relations, in: Aspects of Vagueness (D.Reidel Publ. Co., Dordrecht, Holland, 1984).

[3] S. Ovchinnikov, Numerical representation of transitive fuzzy relations, *Fuzzy Sets and Systems* **126** (2002) 225–232.

B. Bouchon-Meunier, L. Foulloy and R.R. Yager (Editors)
Intelligent Systems for Information Processing:
From Representation to Applications
© 2003 Elsevier Science B.V. All rights reserved.

Normal forms for fuzzy relations and their contribution to universal approximation*

Irina Perfilieva

University of Ostrava

Institute for Research and Applications of Fuzzy Modeling

30. dubna 22, 701 03 Ostrava 1

Czech Republic

Irina.Perfilieva@osu.cz

Abstract

This paper continues the investigation of approximating properties of generalized normal forms in fuzzy logic. The problem is formalized and solved algebraically. Normal forms are considered in two variants: infinite and finite. It is proved that infinite normal forms are universal representation formulas whereas finite normal forms are universal approximation formulas for extensional functions. The estimation of the quality of approximation is suggested. Moreover, functions which can be precisely represented by the discrete normal forms are considered.

Keywords: Fuzzy relations, disjunctive and conjunctive normal forms, BL-algebra, universal approximation, extensional functions

1 Introduction

This paper is a contribution to the theory of representation/approximation of fuzzy relations by special formulas. By a fuzzy relation we mean a function which is defined on an arbitrary set and takes values from some set of truth values L. Normally, $L = [0, 1]$ and in this case we do not distinguish between fuzzy relations and bounded real valued functions. Based on this remark, we will further use term "L-valued function" instead of "fuzzy relation".

The theory of representation has many practical applications because it provides a formal description (precise or approximate) of functions which can be used for different purposes. What makes this theory general is the fact that one formal description given by a certain formula with a number of parameters can represent (precisely or approximately) a class of functions. The respective formulas are therefore, called *universal formulas*.

The following examples serve us as prototypes of universal formulas. In boolean algebra of logical functions, each function can be represented by disjunctive and conjunctive normal forms. In this connection we say that normal forms in this boolean algebra play the role of universal representation formulas. In the algebra of many-valued logical functions,

---

*This paper has been partially supported by grant IAA1187301 of the GA AV ČR

the Rosser-Turquette formula generalizes the disjunctive normal (boolean) form and also serves as universal representation formula. From the theory of approximation, we know that e.g., Lagrange polynomials or Fourier series are examples of universal approximation formulas for continuous functions defined on bounded intervals.

Fuzzy logic also contributed to this particular problem. It offers formulas of a first order calculus with fuzzy predicates which formalize a linguistic description given by a set of IF-THEN rules and thus, serve as universal approximation formulas for continuous functions defined on a compact set (see e.g.,[1, 3, 6, 7]). Later, it has been shown that these formulas can be regarded as generalizations of boolean normal forms ([5, 7, 8, 9]). Moreover, in [9] this fact has been proved using formal logical means.

This paper continues the investigation of approximation properties of generalized normal forms in fuzzy logic. The problem is formalized and solved algebraically. To be more general, an arbitrary BL-algebra has been chosen as an algebra of fuzzy logic operations (Section 2). This implies that the construction of the normal forms remains unchanged independently on which concrete operations we use: Gödel, Łukasiewicz, product or others based on a choice of a t-norm. Normal forms are considered in two variants: infinite and finite (Sections 4, 5). It is proved that infinite normal forms are universal representation formulas whereas finite normal forms are universal approximation formulas for extensional functions (Section 3). The estimation of the quality of approximation is suggested. The proofs of the statements are not included due to the limitation of space, they can be found in [10].

2 BL-algebra

BL-algebra has been introduced in [2] as the algebra of logic operations which correspond to connectives of basic logic (BL). In the same sense as BL generalizes boolean logic we can say that BL-algebra generalizes boolean algebra. This appears in the extension of the set of boolean operations by two semigroup operations which constitute so called adjoined couple. The following definition summarizes definitions which have been introduced in [2].

Definition 1
A *BL-algebra* is an algebra

$$\mathcal{L} = \langle L, \vee, \wedge, *, \rightarrow, \mathbf{0}, \mathbf{1} \rangle$$

with four binary operations and two constants such that

(i) $(L, \vee, \wedge, \mathbf{0}, \mathbf{1})$ is a lattice with $\mathbf{0}$ and $\mathbf{1}$ as the least and largest elements w.r.t. the lattice ordering,

(ii) $(L, *, \mathbf{1})$ is a commutative semigroup with the unit $\mathbf{1}$, such that multiplication $*$ is associative, commutative and $\mathbf{1} * x = x$ for all $x \in L$,

(iii) $*$ and $\rightarrow$ form an adjoint pair, i.e.
$z \leq (x \rightarrow y)$ iff $x * z \leq y$, $\quad x, y, z \in L$,

(iv) and moreover, for all $x, y \in L$
$x * (x \rightarrow y) = x \wedge y$,
$(x \rightarrow y) \vee (y \rightarrow x) = \mathbf{1}$.

Another two operations of $\mathcal{L}$: unary $\neg$ and binary $\leftrightarrow$ can be defined by

$$\neg x = x \to \mathbf{0},$$
$$x \leftrightarrow y = (x \to y) \wedge (y \to x).$$

The following property will be widely used in the sequel:

$$x \leq y \quad \text{iff} \quad (x \to y) = \mathbf{1}.$$

Note that if a lattice $(L, \vee, \wedge, \mathbf{0}, \mathbf{1})$ is given, then BL-algebra is completely defined by the choice of multiplication operation $*$. In particular, $L = [0,1]$ and $*$ is known as a t-norm (see examples below).

The following are examples of BL-algebras.

Example 1 (Gödel algebra)

$$\mathcal{L}_G = \langle [0,1], \vee, \wedge, \to_G, 0, 1 \rangle$$

where the multiplication $* = \wedge$ and

$$x \to_G y = \begin{cases} 1 & \text{if } x \leq y, \\ y & \text{if } y < x. \end{cases} \tag{1}$$

Example 2 (Goguen algebra)

$$\mathcal{L}_P = \langle [0,1], \vee, \wedge, \odot, \to_P, 0, 1 \rangle$$

where the multiplication $\odot = \cdot$ is the ordinary product of reals and

$$x \to_P y = \begin{cases} 1 & \text{if } x \leq y, \\ \frac{y}{x} & \text{if } y < x. \end{cases} \tag{2}$$

Example 3 (Łukasiewicz algebra)

$$\mathcal{L}_Ł = \langle [0,1], \vee, \wedge, \otimes, \to_Ł, 0, 1 \rangle \tag{3}$$

where

$$x \otimes y = 0 \vee (x + y - 1), \tag{4}$$
$$x \to_Ł y = 1 \wedge (1 - x + y). \tag{5}$$

3 BL-algebras of L-valued Functions and Extensional Functions

Let X be a nonempty set, $\mathcal{L}$ a BL-algebra on L, and P_L a set of all L-valued functions $f(x_1, \ldots, x_n)$, $n \geq 0$, which are defined on X and take values from L. To shorten the denotation we will write $\tilde{\mathbf{x}}^{(n)}$ instead of $(x_1, \ldots, x_n)$. Let us extend the operations from $\mathcal{L}$ on P_L so that for each $f(\tilde{\mathbf{x}}^{(n)}), g(\tilde{\mathbf{x}}^{(n)}) \in P_L$

$$(f \vee g)(\tilde{\mathbf{x}}^{(n)}) = f(\tilde{\mathbf{x}}^{(n)}) \vee g(\tilde{\mathbf{x}}^{(n)}),$$

$$(f * g)(\tilde{\mathbf{x}}^{(n)}) = f(\tilde{\mathbf{x}}^{(n)}) * g(\tilde{\mathbf{x}}^{(n)}),$$
$$(f \wedge g)(\tilde{\mathbf{x}}^{(n)}) = f(\tilde{\mathbf{x}}^{(n)}) \wedge g(\tilde{\mathbf{x}}^{(n)}),$$
$$(f \to g)(\tilde{\mathbf{x}}^{(n)}) = f(\tilde{\mathbf{x}}^{(n)}) \to g(\tilde{\mathbf{x}}^{(n)}).$$

Furthermore, let $\mathbb{U}$ and $\mathbb{O}$ be constant functions from P_L taking the values $\mathbf{1}$ and $\mathbf{0}$, respectively. Then
$$P_{\mathcal{L}} = \langle P_L, \vee, \wedge, *, \to, \mathbb{U}, \mathbb{O} \rangle$$
is a BL-algebra.

Note that

(i) in case $n = 0$, L-valued functions degenerate to constants (elements of L),

(ii) in case $n = 1$, L-valued functions are usually considered as membership functions of L-valued fuzzy sets,

(iii) in other cases, L-valued functions are identified with fuzzy relations.

We will be concerned with the subclass of L-valued functions formed by so called extensional functions. The reason comes from the fact that extensional functions have properties similar to continuity and therefore, they can be represented (precisely or approximately) by special formulas over $P_{\mathcal{L}}$ (see [8, 9]). But before we give the definition of extensional functions we will introduce the similarity relation on X which helps to describe a neighborhood of a point.

Definition 2
A binary fuzzy relation E on X given by L-valued function $E(x, y)$ is called a *similarity* if for each $x, y, z \in X$ the following properties hold true

$$E(x, y) = \mathbf{1}, \quad \text{(reflexivity)}$$
$$E(x, y) = E(y, x), \quad \text{(symmetry)}$$
$$E(x, y) * E(y, z) \leq E(x, z). \quad \text{(transitivity)}$$

In our text we will formally distinguish between the similarity relation (denoted by E) and the (membership) function representing it (denoted by $E(x, y)$), although they are closely connected. The value $E(x, y)$ can be interpreted as the *degree of similarity* of x and y or the degree which characterizes that x belongs to a neighborhood of y.

Let us consider some examples of similarity. Note that each crisp equivalence on X is also a similarity on X in any BL-algebra $P_{\mathcal{L}}$. If $L = [0, 1]$ and the multiplication $*$ is a continuous Archimedian t-norm with continuous generator $g : [0, 1] \longrightarrow [0, \infty]$, then for any pseudo-metric $d : X^2 \longrightarrow [0, \infty]$ on X the fuzzy relation E_d on X given by $E_d(x, y) = g^{(-1)}(d(x, y))$ is a similarity (see [4]). For example, if $*$ is Łukasiewicz conjunction $\wedge$ with generator $1 - x$ then $E_d(x, y) = \max(0, 1 - d(x, y))$ defines the similarity on X in BL-algebra $P_{\mathcal{L}_{\text{L}}}$.

The following definition of extensional function is taken from [2].

Definition 3
An L-valued function $f(x_1,\ldots,x_n)$ is *extensional* w.r.t. a similarity relation E on X if for all $x_1,\ldots,x_n, y_1,\ldots,y_n \in X$

$$E(x_1,y_1) * \cdots * E(x_n,y_n) * f(x_1,\ldots,x_n) \leq f(y_1,\ldots,y_n).$$

It is easy to see that if $f(x,y) = E(x,y)$ or $f(x,y) = E(x,y) * a$ where a is an arbitrary element of L then $f(x,y)$ is extensional w.r.t. similarity E. A general characterization of extensional functions can be given in the BL-algebras where the operation $*$ is a continuous Archimedian t-norm. This characterization is based on the property which is analogous to Lipschitz continuity.

Theorem 1
Let $L = [0,1]$ and multiplication $*$ be a continuous Archimedian t-norm with the continuous generator $g : [0,1] \longrightarrow [0,\infty]$. Let $d : X \longrightarrow [0,\infty]$ be a pseudo-metric on X and the fuzzy relation E_d on X given by $E_d(x,y) = g^{(-1)}(d(x,y))$ be a similarity. Then any extensional function $f(x_1,\ldots,x_n)$, w.r.t. the similarity E_d fulfils the inequality

$$|g(f(x_1,\ldots,x_n)) - g(f(y_1,\ldots,y_n))| \leq$$
$$\leq \min(g(0), d(x_1,y_1) + \cdots + d(x_n,y_n)) \quad (6)$$

where $n \geq 2$ or (in case $n = 1$)

$$|g(f(x)) - g(f(y))| \leq d(x,y).$$

Particularly, if the underlying BL-algebra is the Łukasiewicz algebra $\mathcal{L}_L$ then extensional functions are Lipschitz continuous in classical sense.

Corollary 1
Suppose that the above given conditions are fulfilled, and $*$ is the Łukasiewicz t-norm with generator $g(x) = 1 - x$. Then any extensional function $f(x_1,\ldots,x_n)$ fulfils the inequality

$$|f(x_1,\ldots,x_n) - f(y_1,\ldots,y_n)| \leq \min(1, d(x_1,y_1) + \cdots d(x_n,y_n)).$$

It is worth noticing that a class of extensional functions w.r.t. one specific similarity do not form a subalgebra of $P_\mathcal{L}$. This can be illustrated by the following example: the function $f(x,y) = E^2(x,y)$ (this is a short for $E(x,y) * E(x,y)$) is not extensional w.r.t. similarity E. A class of extensional functions w.r.t. one specific similarity is a subalgebra of a weaker algebra than $P_\mathcal{L}$.

Lemma 1
Let $E(x,y)$ define a similarity on X, $n \geq 1$ a natural number. A class of extensional w.r.t. E functions depending on n variables, forms a sublattice of $\langle P_L, \vee, \wedge, \mathbb{U}, \mathbb{O} \rangle$.

4 Normal Forms for L-valued Functions

In this section, we will introduce three normal forms for L-valued functions analogously to that, how it is done for boolean two-valued functions. We will construct one generalized disjunctive normal form and *two* generalized conjunctive normal forms. The latter fact is due to the absence of the law of double negation in BL-algebra.

Let us fix some complete BL-algebra $\mathcal{L}$ and a similarity E on a set X. The following lemma introduces so called constituents of normal forms and shows their relation to the original function.

Lemma 2

Let $c_1, \ldots, c_n$ where $n \geq 1$, be arbitrary elements of X and $d \in L$. Then functions represented by

$$E(x_1, c_1) * \cdots * E(x_n, c_n) * d \tag{7}$$

and

$$E(x_1, c_1) * \cdots * E(x_n, c_n) \to d, \tag{8}$$

are extensional w.r.t. similarity E.

Let moreover, $f(x_1, \ldots, x_n)$ be an extensional w.r.t. E function such that

$$f(c_1, \ldots, c_n) = d.$$

Then for all $x_1, \ldots, x_n \in X$

$$E(x_1, c_1) * \cdots * E(x_n, c_n) * d \leq f(x_1, \ldots, x_n), \tag{9}$$
$$f(x_1, \ldots, x_n) \leq E(x_1, c_1) * \cdots * E(x_n, c_n) \to d. \tag{10}$$

The functions represented by the formula

$$(E(x_1, c_1) * \cdots * E(x_n, c_n) * d)$$

or

$$(E(x_1, c_1) * \cdots * E(x_n, c_n) \to d)$$

will be called *lower* and *upper constituents* of the function $f(x_1, \ldots, x_n)$, respectively provided that $f(c_1, \ldots, c_n) = d$. Using them, we can introduce the disjunctive normal form (DNF) of f as a supremum of all its lower constituents and the second conjunctive normal form of f as an infimum of all its upper constituents. The first conjunctive normal form of f will be introduced as the formula dual to DNF.

Definition 4

Let $f(x_1, \ldots, x_n) \in P_\mathcal{L}^n$ be an L-valued function. The following formulas over $P_\mathcal{L}$ are called the *disjunctive normal form* of f

$$f_{\mathrm{DNF}}(x_1, \ldots, x_n) = \bigvee_{c_1, \ldots, c_n \in X} (E(x_1, c_1) * \cdots * E(x_n, c_n) * f(c_1, \ldots, c_n)) \tag{11}$$

and the *conjunctive normal forms* of f

$$f_{\mathrm{CNF}_I}(x_1, \ldots, x_n) = \neg \bigvee_{c_1, \ldots, c_n \in X} (E(x_1, c_1) * \cdots * E(x_n, c_n) * (\neg f(c_1, \ldots, c_n)))$$
$$\tag{12}$$

and
$$f_{\mathrm{CNF}_{II}}(x_1,\ldots,x_n) = \bigwedge_{c_1,\ldots,c_n \in X} (E(x_1,c_1) * \cdots * E(x_n,c_n) \to f(c_1,\ldots,c_n)). \quad (13)$$

Remark 1

Two conjunctive normal forms are generally not equivalent except for the case of Łukasiewicz algebra.

At first, we will investigate whether the functions represented by normal forms are extensional w.r.t. similarity E?

Theorem 2

The functions

$$\bigvee_{c_1,\ldots,c_n \in X} (E(x_1,c_1) * \cdots * E(x_n,c_n) * d_{c_1\ldots c_n}) \quad (14)$$

$$\neg \bigvee_{c_1,\ldots,c_n \in X} (E(x_1,c_1) * \cdots * E(x_n,c_n) * (\neg d_{c_1\ldots c_n})) \quad (15)$$

$$\bigwedge_{c_1,\ldots,c_n \in X} (E(x_1,c_1) * \cdots * E(x_n,c_n) \to d_{c_1\ldots c_n}) \quad (16)$$

where $d_{c_1\ldots c_n} \in L$, are extensional w.r.t. similarity E.

Now, we will investigate the relation between the original function f and the functions represented by normal forms of f.

Theorem 3

Let $f(x_1,\ldots,x_n)$ be an extensional function w.r.t. similarity E. Then

$$f(x_1,\ldots,x_n) = f_{\mathrm{DNF}}(x_1,\ldots,x_n), \quad (17)$$
$$f(x_1,\ldots,x_n) = f_{\mathrm{CNF}_{II}}(x_1,\ldots,x_n), \quad (18)$$
$$f(x_1,\ldots,x_n) \leq f_{\mathrm{CNF}_I}(x_1,\ldots,x_n). \quad (19)$$

Let us stress that Theorem 3 demonstrates very convincing results, asserting that even in the fuzzy case, normal forms can be equal to the original function. Moreover, the construction of normal forms gives us the idea of how they can be simplified without significant loss of their ability to represent (at least approximately) the original function.

5 Discrete Normal Forms

The normal forms introduced above can hardly be used in practice because they are based on the full knowledge of the represented function in all its points. Having on mind practical applications, we have to simplify normal forms. We will do this by removing some of its elementary terms. Of course, after such removing we cannot expect that thus obtained simplified formulas will represent the original function precisely. But we expect an approximate representation which in many cases is sufficient. Aiming at this, we will introduce discrete normal forms which are based on partial knowledge of the represented function in some nodes.

Let BL-algebra $\mathcal{L}$ and a similarity E on the set X be given.

Definition 5
Let $f(x_1, \ldots, x_n)$ be an L-valued function defined on X, $c_1, \ldots, c_k$ some chosen elements (*nodes*) from X. Let elements $d_{i_1 \ldots i_n} \in L$ be chosen so that

$$d_{i_1 \ldots i_n} = f(c_{i_1}, \ldots, c_{i_n})$$

for each collection of indices $(i_1 \ldots i_n)$ where $1 \leq i_1, \ldots, i_n \leq k$. Then the following formulas over $P_\mathcal{L}$ are called the *discrete disjunctive normal form* of f

$$f_{\text{DNF}}(x_1, \ldots, x_n) = \bigvee_{i_1, \ldots, i_n = 1}^{k} (E(x_1, c_{i_1}) * \cdots * E(x_n, c_{i_n}) * d_{i_1 \ldots i_n}) \quad (20)$$

and the *discrete conjunctive normal forms* of f

$$f_{\text{CNF}_I}(x_1, \ldots, x_n) = \neg \bigvee_{i_1, \ldots, i_n = 1}^{k} (E(x_1, c_{i_1}) * \cdots * E(x_n, c_{i_n}) * (\neg d_{i_1 \ldots i_n})) \quad (21)$$

and

$$f_{\text{CNF}_{II}}(x_1, \ldots, x_n) = \bigwedge_{i_1, \ldots, i_n = 1}^{k} (E(x_1, c_{i_1}) * \cdots * E(x_n, c_{i_n}) \to d_{i_1 \ldots i_n}). \quad (22)$$

5.1 Universal Approximation of Functions by Discrete Normal Forms

We will show that the discrete normal forms do approximate the original function if the nodes are chosen properly. It is interesting that, independently on the choice of nodes, DNF and both CNF's can be considered as lower and upper approximate representations of the given function, respectively. This fact is proved in the theorem given below.

Theorem 4
Let $f(x_1, \ldots, x_n)$ be an extensional function w.r.t. similarity E defined on X, $c_1, \ldots, c_k$ be nodes from X. Let elements $d_{i_1 \ldots i_n} \in L$ be chosen so that

$$d_{i_1 \ldots i_n} = f(c_{i_1}, \ldots, c_{i_n})$$

for each collection of indices $(i_1 \ldots i_n)$ where $1 \leq i_1, \ldots, i_n \leq k$. Then discrete normal forms given by (20)–(22) fulfil the following inequalities:

$$f_{\text{DNF}}(x_1, \ldots, x_n) \leq f(x_1, \ldots, x_n), \quad (23)$$
$$f(x_1, \ldots, x_n) \leq f_{\text{CNF}_I}(x_1, \ldots, x_n), \quad (24)$$
$$f(x_1, \ldots, x_n) \leq f_{\text{CNF}_{II}}(x_1, \ldots, x_n). \quad (25)$$

The following theorem exposes the approximation property of discrete normal forms. By this we mean, that the equivalence between an extensional function and the functions represented by each of its discrete normal forms can be estimated from below. Thus, in the language of BL-algebra $P_\mathcal{L}$, the approximation means the *conditional equivalence*.

Theorem 5
Let $f(x_1,\ldots,x_n)$ be an extensional function w.r.t. similarity E defined on X, $c_1,\ldots,c_k$ be nodes from X. Let elements $d_{i_1\ldots i_n} \in L$ be chosen so that

$$d_{i_1\ldots i_n} = f(c_{i_1},\ldots,c_{i_n})$$

for each collection of indices $(i_1\ldots i_n)$ where $1 \leq i_1,\ldots,i_n \leq k$. Then functions represented by the discrete normal forms given by (20) and (22) are conditionally equivalent to the original function f which means

$$f_{\text{DNF}}(x_1,\ldots,x_n) \leftrightarrow f(x_1,\ldots,x_n) \geq \bigvee_{i_1\ldots i_n} (E^2(x_1,c_{i_1}) * \cdots * E^2(x_n,c_{i_n})) \quad (26)$$

$$f_{\text{CNF}_{II}}(x_1,\ldots,x_n) \leftrightarrow f(x_1,\ldots,x_n) \geq \bigvee_{i_1\ldots i_n} (E^2(x_1,c_{i_1}) * \cdots * E^2(x_n,c_{i_n})) \quad (27)$$

Note, that the condition $\bigvee_{i_1\ldots i_n} (E^2(x_1,c_{i_1}) * \cdots * E^2(x_n,c_{i_n}))$ of equivalences (26) and (27) describes the *quality of approximation* which is the degree of similarity between two points $(x_1,\ldots,x_n)$ and $(c_{i_1},\ldots,c_{i_n})$. The direct estimation of thus expressed quality of approximation is given below for special BL-algebras.

Corollary 2
Let the conditions of the above theorem be fulfilled. Moreover, let $L = [0,1]$ and $*$ be a continuous Archimedian t-norm with a continuous additive generator $g : [0,1] \longrightarrow [0,\infty]$. Let $d : X \longrightarrow [0,\infty]$ be a pseudo-metric on X and the similarity E on X be given by $E(x,y) = t^{(-1)}(d(x,y))$. Then we can estimate the quality of approximation of the original function f by functions represented by the discrete normal forms given by (20) or (22)

$$|g(f(x_1,\ldots,x_n)) - g(f_{\text{DNF}}(x_1,\ldots,x_n))| \leq \min(t(0), \min_{i_1,\ldots,i_n} \sum_{j=1}^n 2d(x_j,c_{i_j})), \quad (28)$$

$$|g(f(x_1,\ldots,x_n)) - g(f_{\text{CNF}_{II}}(x_1,\ldots,x_n))| \leq \min(t(0), \min_{i_1,\ldots,i_n} \sum_{j=1}^n 2d(x_j,c_{i_j})). \quad (29)$$

Figures 1–4 illustrate how the extensional function can be approximated by the corresponding normal forms. In fact, the function represented by DNF coincides with the original one while the function represented by CNF really approximates the original one. The approximation error is shown on Figure 4.

Remark 2
It is worth noticing that the discrete normal forms model what is known in fuzzy literature as fuzzy systems. Usually, fuzzy systems are developed with the help of IF–THEN rules which describe a behaviour of a dynamic system in a language close to the natural one. The proof that fuzzy systems are really able to do this job is given in a number of papers unified by the key words "universal approximation" (see e.g.,[1, 3, 7]). The result proved in Theorem 5 also belongs to this family. However, the method of a proof suggested here, differs from the other ones by its algebraic origin.

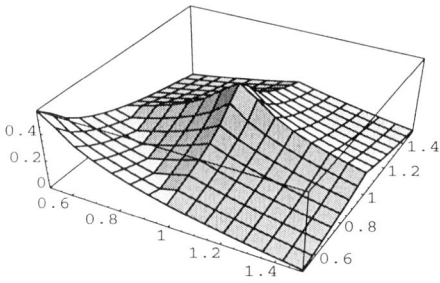

Figure 1: Original function

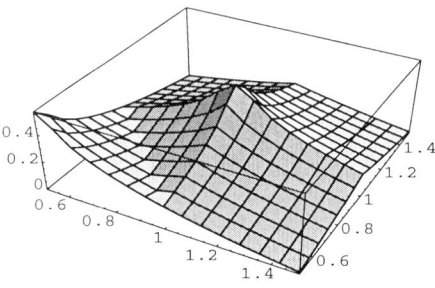

Figure 2: The corresponding d.n.f..

5.2 Representation of Bounded Functions with One Variable by Discrete Normal Forms

In this subsection we will consider the problem whether there exist functions admitting a precise representation by the discrete normal forms. It would be nice to find a full characterization of all such functions, but the problem seems to be very difficult. We suggest a partial answer for the case of functions of one variable defined on a closed interval of the real line.

Theorem 6
Let L be a support of some BL-algebra $\mathcal{L}$ and an L-valued function $f(x)$ be defined on the interval $[a, b]$. Moreover, let $f(x)$ have its largest value at point $c \in [a, b]$. Then $f(x)$ can be precisely represented by the respective discrete disjunctive normal form.

An analogous representation of f by the discrete conjunctive normal form requires a stronger algebra than BL. Without going into the details, we give the following definition.

Definition 6
A BL-algebra $\mathcal{L} = \langle L, \vee, \wedge, *, \rightarrow, \mathbf{0}, \mathbf{1} \rangle$ where the double negation law $x = \neg\neg x$ is valid for all $x \in L$, is called an MV-algebra.

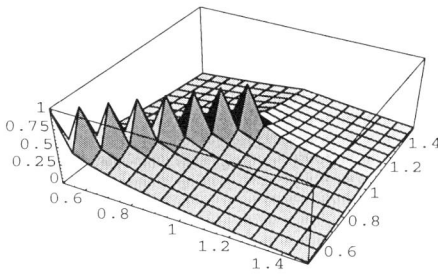

Figure 3: The corresponding c.n.f..

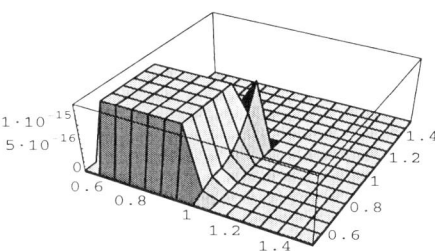

Figure 4: The difference between $g(x,y)$ and the corresponding c.n.f..

Now, using the means of MV-algebra we can prove the second representation theorem.

Theorem 7
Let L be a support of some MV-algebra $\mathcal{L}$ and an L-valued function $f(x)$ be defined on the interval $[a,b]$. Moreover, let $f(x)$ have its least value at point $c \in [a,b]$. Then $f(x)$ can be precisely represented by the respective discrete conjunctive normal form.

6 Conclusions

In this paper we have introduced the disjunctive and conjunctive normal forms as special formulas of BL-algebra of functions. On the one side, the expressions for the normal forms generalize boolean ones which are used for the representation of logical functions. On the other side, they generalize formulas known in fuzzy literature as universal approximation formulas. The latter has been used for approximate description of continuous functions defined on compact domains.

References

[1] Castro J.L., M. Delgado (1996) Fuzzy systems with defuzzification are universal approximators. IEEE Trans. Syst., Man, Cybern., 26(1), 149–152

[2] Hájek P. (1998) Metamathematics of fuzzy logic, Kluwer, Dordrecht

[3] Kosko B. (1992) Fuzzy systems as universal approximators. Proc. IEEE Int. Conf. on Fuzzy Syst., San Diego, CA, 1153–1162

[4] Klement P., Mesiar R., L. Papp (2001) Triangular norms, Kluwer, Dordrecht

[5] Kreinovich V., Nguyen H.T., D.A. Sprecher (1996) Normal forms for fuzzy logic – an application of Kolmogorov's theorem. Int. Journ. of Uncert., Fuzz. and Knowl.-Based Syst., 4, 331–349

[6] Novák V., Perfilieva I. and J. Močkoř (1999) Mathematical Principles of Fuzzy Logic, Kluwer, Boston- Dordrecht

[7] Perfilieva I. (1999) Fuzzy logic normal forms for control law representation. In: Verbruggen H., Zimmermann H.-J., Babuska R. (eds.) Fuzzy Algorithms for Control, Kluwer, Boston, 111–125

[8] Perfilieva I. (2001) Normal Forms for Fuzzy Logic Functions and Their Approximation Ability. Fuzzy Sets and Syst., Vol. 124, No 3, pp. 371–384

[9] Perfilieva I. (2002) Logical Approximation. Soft Computing, 2002, vol.7, N=2, 73–78

[10] Perfilieva I. (2003) Normal Forms in BL-algebra and their Contribution to Universal Approximation of Functions, Fuzzy Sets and Syst., submitted.

B. Bouchon-Meunier, L. Foulloy and R.R. Yager (Editors)
Intelligent Systems for Information Processing:
From Representation to Applications
© 2003 Elsevier Science B.V. All rights reserved.

Associative operators based on t-norms and t-conorms

M. Mas[1], R. Mesiar[2], M. Monserrat[1] and J. Torrens[1]*

[1]Dpt. de Ciències Matemàtiques i Informàtica
Universitat de les Illes Balears
07122 Palma de Mallorca. Spain
dmi{mmg, mma, jts}0@clust.uib.es

[2] Department of Mathematics and Descriptive Geometry
Faculty of Civil Engineering
Slovak Technical University
Bratislava, Eslovakia
mesiar@cvt.stuba.sk

Abstract

This work presents several alternative definitions of uninorm and it is proved that all of them are equivalent to the initial one given by Yager-Rybalov. Special cases are studied for uninorms in $\mathcal{U}_{\min}$ and $\mathcal{U}_{\max}$, from this study some new kinds of operators appear which are a generalization of both: t-operators and uninorms. Moreover, the same kinds of operators are obtained, from an alternative approach, as a kind of median operators following the notation in [6].

Keywords: Uninorm, t-operator, associativity, bisymmetry, boundary conditions.

1 Introduction

Associative operators acting on intervals were studied already by Abel [1], compare also Aczél [2]. A prominent role among these operators play t-norms and t-conorms introduced by Schweizer and Sklar [14] and extensively studied in [145] and [10]. A specific class of associative operators on [0,1] containing t-norms and t-conorms as border cases is the class of uninorms. From their introduction by Yager-Rybalov in [16], uninorms have been extensively studied by many authors from both: the applicational and the theoretical point of view. The first study has proved that uninorms are useful in many fields like aggregation, expert systems, neural networks or fuzzy system modelling. On the other hand, the theoretical study of uninorms has allowed to obtain a complete description of their structure in some distinguished cases in [8], their classification in some classes in [4], the characterization of idempotent uninorms in [3] and also some generalizations of uninorm have appeared, e.g. in [13]. Recall also that uninorms have been studied already by Golan [9] as multiplications U in distributive semirings $([0,1], \sup, U)$ and $([0,1], \inf, U)$.

---

*Correspondence author

Another kind of operators closely related to t-norms and t-conorms are t-operators studied in [11]. The structure of both t-operators and uninorms is similar enough with the particularity that t-operators have a null element instead of a neutral one. However, the existence of this null element for t-operators is derived from their definition whereas for uninorms the existence of the neutral element is one of the conditions in the definition. Observe that nullnorms introduced in [5] turn out to be exactly the earlier introduced t-operators.

Can we give an alternative definition of uninorm in such a way that the existence of the neutral element is derived from? Or also, can any other condition on the definition be changed by some other obtaining an equivalent definition? In this work we present positive answers to these questions. In particular we prove that associativity and commutativity can be substituted by bisymmetry. Moreover, in cases of uninorms in $\mathcal{U}_{\min}$ and $\mathcal{U}_{\max}$ it is proved that the existence of the neutral element can be derived from some border conditions plus a condition on continuity. From this study two new classes of operators appear which we have called S-uninorms in $\mathcal{U}_{\min}$ and T-uninorms in $\mathcal{U}_{\max}$. The first class presents a generalization of t-operators and uninorms in $\mathcal{U}_{\min}$ and the second one a generalization of t-operators and uninorms in $\mathcal{U}_{\max}$. Moreover, both classes are dual one of the other. Finally, we introduce a common generalization of all classes of associative operators discussed above.

2 Preliminaries

We assume the reader to be familiar with basic notions concerning t-norms and t-conorms (see e.g. [10]). On the other hand, we recall here some definitions and results about uninorms and t-operators.

Definition 1 *A uninorm is a two-place function $U : [0,1]^2 \longrightarrow [0,1]$ which is associative, commutative, non-decreasing in each place and such that there exists an element $e \in [0,1]$ called the neutral element such that $U(e, x) = x$ for all $x \in [0, 1]$.*

It is clear that, when $e = 1$, the function U becomes a t-norm and, when $e = 0$, a t-conorm. For any uninorm we have $U(0, 1) \in \{0, 1\}$. Uninorms satisfying that both functions $U(x, 0)$ and $U(x, 1)$ are continuous except perhaps at the point e are characterized in [8] as follows:

Theorem 1 *(see Theorem 1 in [8]) Let U be a uninorm with neutral element $e \neq 0, 1$ and such that both functions $U(x, 1)$ and $U(x, 0)$ are continuous except at the point $x = e$.*

i) *If $U(0, 1) = 0$, then*

$$U(x,y) = \begin{cases} eT\left(\frac{x}{e}, \frac{y}{e}\right) & \text{if } 0 \leq x, y \leq e \\ e + (1-e)S\left(\frac{x-e}{1-e}, \frac{y-e}{1-e}\right) & \text{if } e \leq x, y \leq 1 \\ \min(x, y) & \text{otherwise} \end{cases}$$

ii) *If $U(0, 1) = 1$, then*

$$U(x,y) = \begin{cases} eT\left(\frac{x}{e}, \frac{y}{e}\right) & \text{if } 0 \leq x, y \leq e \\ e + (1-e)S\left(\frac{x-e}{1-e}, \frac{y-e}{1-e}\right) & \text{if } e \leq x, y \leq 1 \\ \max(x, y) & \text{otherwise} \end{cases}$$

In both formulas T is a t-norm and S is a t-conorm. It is denoted by $\mathcal{U}_{\min}$ the class of uninorms as in i) and by $\mathcal{U}_{\max}$ the class of uninorms as in ii).

Note that in both cases the uninorm U is an ordinal sum of semigroups in the sense of Clifford [7].

Remark 1 *Without the assumption of continuity of the functions $U(x, 1)$ and $U(x, 0)$, the same structure holds but the values of $U(x, y)$ when $\min(x, y) < e < \max(x, y)$ are not determined and it is only known that they are somewhere between $\min$ and $\max$.*

From now on, given any binary operator $U : [0, 1]^2 \to [0, 1]$, and any element $a \in [0, 1]$, we will denote by U_a the section $U_a : [0, 1] \to [0, 1]$ given by $U_a(x) = U(a, x)$.

Definition 2 *A t-operator is a two-place function $F : [0, 1]^2 \longrightarrow [0, 1]$ which is associative, commutative, non-decreasing in each place and such that:*

- $F(0, 0) = 0; F(1, 1) = 1$.

- *The sections F_0 and F_1 are continuous on $[0, 1]$.*

Remark 2 *When F is a t-operator, the element $k = F(1, 0)$ is a null (or absorbent) element in the sense that $F(x, k) = k$ for all $x \in [0, 1]$ and therefore, the function F becomes a t-norm when $k = 0$ and a t-conorm when $k = 1$.*

The proof of the following theorem can be found in [11].

Theorem 2 *Let $F : [0, 1]^2 \longrightarrow [0, 1]$ be a t-operator with $F(1, 0) = k \neq 0, 1$. Then:*

$$F(x, y) = \begin{cases} kS\left(\frac{x}{k}, \frac{y}{k}\right) & \text{if } 0 \leq x, y \leq k \\ k + (1 - k)T\left(\frac{x-k}{1-k}, \frac{y-k}{1-k}\right) & \text{if } k \leq x, y \leq 1 \\ k & \text{if } \min(x, y) \leq k \leq \max(x, y) \end{cases}$$

where S is a t-conorm and T a t-norm.

Note that following [6] each t-operator F with null-element k can be represented in the form $F = med(k, T, S)$ for some t-norm T and some t-conorm S. Viceversa, given any t-norm T and any t-conorm S, each operator F defined by $F = med(k, T, S)$ is a t-operator. Moreover, for the genuine n-ary extension of F (due to its associativity) we have:

$$F(x_1, ..., x_n) = med(k, T(x_1, ..., x_n), S(x_1, ..., x_n)).$$

3 New look on uninorms

We begin with equivalent definitions of uninorm in the general case and we deal with the case of uninorms in $\mathcal{U}_{\min}$ and $\mathcal{U}_{\max}$ later. The first modification involves the bisymmetric property which we recall here.

Definition 3 *An operator $U : [0, 1]^2 \longrightarrow [0, 1]$ is said to be bisymmetric if*

$$U(U(x, y), U(u, v)) = U(U(x, u), U(y, v))$$

for all $x, y, u, v \in [0, 1]$.

Proposition 1 *Let $U : [0,1]^2 \to [0,1]$ be a binary operator. Then U is a uninorm if and only if U is non-decreasing in each place, bisymmetric and has a neutral element.*

Proof: It is well known that associativity and commutativity imply bisymmetry. Conversely, if U is bisymmetric and has a neutral element $e \in [0,1]$, then for all $x, y, z \in [0,1]$ we have

$$U(x, U(y, z)) = U(U(x, e), U(y, z)) = U(U(x, y), U(e, z)) = U(U(x, y), z)$$

and so U is associative. Moreover, for all $x, y \in [0,1]$,

$$U(x, y) = U(U(e, x), U(y, e)) = U(U(e, y), U(x, e)) = U(y, x)$$

and commutativity follows. ∎

The next proposition shows that the existence of a neutral element can be replaced by the existence of an idempotent element with onto section.

Proposition 2 *Let $U : [0,1]^2 \to [0,1]$ be a binary operator. Then U is a uninorm if and only if U is non-decreasing in each place, associative, commutative and has an idempotent element e whose section U_e is an onto map.*

Proof: It is obvious that if U is a uninorm then its neutral element e is idempotent and its section is the identity map.

Conversely, let us suppose that e is an idempotent element of U for which U_e is onto. Thus, given $x \in [0,1]$ there exists $y \in [0,1]$ such that

$$x = U_e(y) = U(e, y) = U(y, e).$$

and therefore

$$U(e, x) = U(e, U(e, y)) = U(U(e, e), y) = U(e, y) = x$$

consequently e is a neutral element of U and U is a uninorm. ∎

From now on, let us deal with uninorms in $\mathcal{U}_{\min}$ and in $\mathcal{U}_{\max}$. For these classes of uninorms, the sections U_0 and U_1 are such that one is continuous and the other is also continuous except perhaps at the neutral element. Recall on the other hand that, for a t-operator F, the existence of the null element follows precisely from the continuity of its sections F_0 and F_1. So the question is: Are there some additional conditions on sections U_0 and U_1 in such a way that the existence of the neutral element of U can be derived from these additional conditions?

Unfortunately the answer is negative as the following example shows.

Example 1 *Let T be a t-norm, $e \in (0,1)$ and let us consider the operator $U : [0,1]^2 \to [0,1]$ defined by*

$$U(x,y) = \begin{cases} eT\left(\frac{x}{e}, \frac{y}{e}\right) & \text{if } x, y \in [0, e]^2 \\ 1 & \text{if } x, y \in [e, 1]^2 - \{(e, e)\} \\ \min(x, y) & \text{otherwise} \end{cases}$$

It is easy to see that U is associative, commutative and non-decreasing in each place. Moreover, the sections U_0 and U_1 are exactly those of a uninorm in $\mathcal{U}_{\min}$ although U is not a uninorm since it has no neutral element. However, U is an ordinal sum of semigroups in the sense of Clifford [7] with one t-norm summand T and one t-superconorm summand constant 1. This operator can be viewed on figure 1.

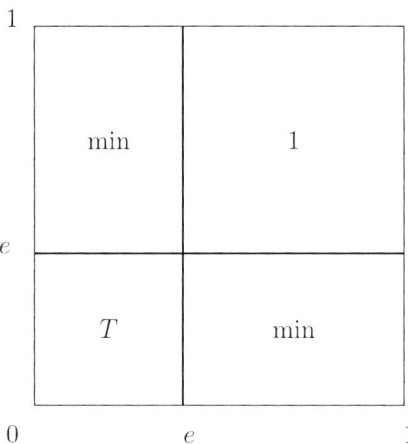

Figure 1: Operator given in Example 1

From the example above, if we want to derive the existence of the neutral element from conditions on the sections U_0 and U_1 we will need at least one additional condition. We will reduce now our reasonings to uninorms in $\mathcal{U}_{\min}$ and similar results for uninorms in $\mathcal{U}_{\max}$ will follow later.

The following theorem is essential in the paper for two reasons. On one hand, it introduces a new kind of operators which are a generalization of both; t-operators and uninorms in $\mathcal{U}_{\min}$ and, on the other hand, an immediate corollary will give us another equivalent definition of uninorm in $\mathcal{U}_{\min}$.

Theorem 3 *Let $U : [0,1]^2 \to [0,1]$ be a binary operator. Then: U satisfies the properties*

- *associative, commutative, non-decreasing in each place*
- *U_0 is continuous with $U_0(0) = 0$*
- *$U_1(x) \geq x$ for all $x \in [0,1]$ and U_1 is continuous except perhaps at one point e but then $U_1(e) = 1$ and the section U_e is continuous*

if and only if there exist t-conorms S, S' and a t-norm T' such that U is given by

$$U(x,y) = \begin{cases} kS\left(\frac{x}{k}, \frac{y}{k}\right) & \text{if } 0 \leq x, y \leq k \\ k & \text{if } \min(x,y) \leq k \leq \max(x,y) \\ k + (e-k)T'\left(\frac{x-k}{e-k}, \frac{y-k}{e-k}\right) & \text{if } k \leq x, y \leq e \\ e + (1-e)S'\left(\frac{x-e}{1-e}, \frac{y-e}{1-e}\right) & \text{if } e \leq x, y \leq 1 \\ \min(x,y) & \text{otherwise} \end{cases} \quad (1)$$

In expression (1) it is understood that: The part involving S must be omitted when $k = 0$ and analogously the part involving S' when $e = 1$. Moreover, the case when U_1 is continuous corresponds in expression (1) with $e = 1$.

Before prove this theorem let us introduce a name for this kind of operators and let us also give the promised corollary.

Definition 4 *Any binary operator satisfying the properties stated in the theorem above will be called, according to its structure, an S–uninorm in $\mathcal{U}_{\min}$.*

From their structure, which can be viewed in Figure 2, it is clear that S–uninorms in $\mathcal{U}_{\min}$ are a generalization of t-operators (cases when $e = 1$, that is, when U_1 is also continuous) and they are also a generalization of uninorms in $\mathcal{U}_{\min}$ (cases when $k = 0$, that is, when $U_1(0) = 0$). As an immediate consequence of this observation we have the following corollary.

Corollary 1 *Let $U : [0,1]^2 \to [0,1]$ be a binary operator. Then U is a uninorm in $\mathcal{U}_{\min}$ if and only if U is*

- *associative, commutative, non-decreasing in each place,*
- *$U_1(x) \geq x$ for all $x \in [0,1]$, $U_1(0) = 0$ and U_1 is continuous except perhaps at one point e but then $U_1(e) = 1$ and the section U_e is continuous*

Let us now deal with the proof of the theorem.

Proof of Theorem 3: It is a straightforward computation to prove that all binary operators given by expressions (1) satisfy the properties stated in the theorem.

Conversely, let U be a binary operator satisfying these properties and let us say $k = U(0,1)$. Then we have by associativity

$$U(k,1) = U(U(0,1),1) = U(0,1) = k$$

and similarly $U(0,k) = k$. Moreover,

$$U(k,k) = U(U(0,1),k) = U(0,U(1,k)) = U(0,k) = k$$

and consequently $U(x,y) = k$ for all (x,y) such that $\min(x,y) \leq k \leq \max(x,y)$. Now we continue our proof by distinguishing some cases:

- If U_1 is continuous then, since $U_1(x) \geq x$, we have $U_1(1) = 1$ and consequently U is a t-operator.

- If there exists $e \in [0,1]$ such that U_1 is continuous except at point $x = e$ then we necessarily have $k < e$, because $U_1(k) = k$ whereas $U_1(e) = 1$. Thus, for any $k \leq x < e$ there exists y such that $U(1,y) = x$, again by continuity and the fact that $U_1(x) \geq x$, but then

$$U(1,x) = U(1,U(1,y)) = U(1,y) = x.$$

Moreover

$$U(e,x) = U(e,U(1,x)) = U(U(e,1),x) = U(1,x) = x$$

for all x such that $k \leq x < e$ and since U_e is continuous we obtain that e is an idempotent element. This proves that $U(x,y) = \min(x,y)$ for all (x,y) such that $k < \min(x,y) < e \leq \max(x,y)$. Finally, for all $x > e$ there exists an element y such that $U(e,y) = x$ and then

$$U(e,x) = U(e,U(e,y)) = x.$$

To end the proof let us define S, T' and S' by

$$S(x,y) = \frac{U(kx,ky)}{k}, \qquad T'(x,y) = \frac{U((e-k)x+k,(e-k)y+k)-k}{e-k},$$

$$S'(x,y) = \frac{U((1-e)x+e,(1-e)y+e)-e}{1-e}.$$

It is a simple computation to verify that T' is a t-norm, S and S' are t-conorms and they are such that U is given by equation (1). ∎

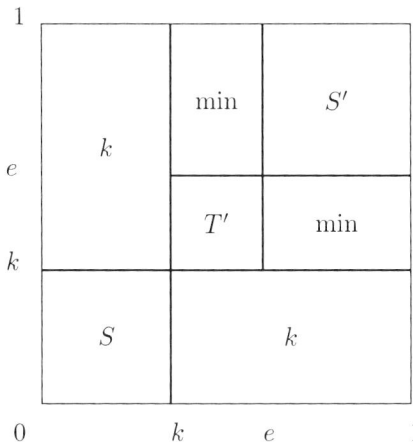

Figure 2: General structure of S-uninorms in $\mathcal{U}_{\min}$

Similarly, we can give an analogous study for uninorms in $\mathcal{U}_{\max}$. Following the same line of reasonings let us give an example analogous to Example 1 proving that the existence of a neutral element can not be derived from boundary conditions.

Example 2 *Let S be a t-conorm, $e \in (0,1)$ and let us consider the operator $U : [0,1]^2 \to [0,1]$ defined by*

$$U(x,y) = \begin{cases} 0 & \text{if } x,y \in [0,e]^2 - \{(e,e)\} \\ e + (1-e)S\left(\frac{x-e}{1-e}, \frac{y-e}{1-e}\right) & \text{if } x,y \in [e,1]^2 \\ \max(x,y) & \text{otherwise} \end{cases}$$

It is easy to see that U is associative, commutative and non-decreasing in each place. Moreover, the sections U_0 and U_1 are exactly those of a uninorm in $\mathcal{U}_{\max}$. However U is not a uninorm since it has no neutral element.

On the other hand, we have the following results that we present without proofs since they are quite similar to those given for uninorms in $\mathcal{U}_{\min}$.

Theorem 4 *Let $U : [0,1]^2 \to [0,1]$ be a binary operator. Then: U satisfies the properties*

- *associative, commutative, non-decreasing in each place*
- *U_1 is continuous with $U_1(1) = 1$*
- *$U_0(x) \leq x$ for all $x \in [0,1]$ and U_0 is continuous except perhaps at one point e but then $U_0(e) = 0$ and the section U_e is continuous*

if and only if there exist t-norms T, T' and a t-conorm S' such that U is given by

$$U(x,y) = \begin{cases} eT'\left(\frac{x}{e}, \frac{y}{e}\right) & \text{if } 0 \leq x, y \leq e \\ \max(x,y) & \text{if } \min(x,y) \leq e < \max(x,y) \\ e + (k-e)S'\left(\frac{x-e}{k-e}, \frac{y-e}{k-e}\right) & \text{if } e \leq x, y \leq k \\ k + (1-k)T\left(\frac{x-k}{1-k}, \frac{y-k}{1-k}\right) & \text{if } k \leq x, y \leq 1 \\ k & \text{otherwise} \end{cases} \quad (2)$$

In expression (2) it is understood that: The part involving T' must be omitted when $e = 0$ and analogously the part involving T when $k = 1$. Moreover, the case when U_0 is continuous corresponds in expression (2) with $e = 0$.

Definition 5 *Any binary operator satisfying the properties of the theorem above will be called, attending to its structure, a T–uninorm in $\mathcal{U}_{\max}$.*

From their structure, which can be viewed in Figure 3, it is clear that T–uninorms in $\mathcal{U}_{\max}$ are a generalization of t-operators (cases when $e = 0$, that is, when U_0 is also continuous) and they are also a generalization of uninorms in $\mathcal{U}_{\max}$ (cases when $k = 1$, that is, when $U_0(1) = 1$). As an immediate consequence of this observation we have the following corollary.

Corollary 2 *Let $U : [0,1]^2 \to [0,1]$ be a binary operator. Then U is a uninorm in $\mathcal{U}_{\max}$ if and only if U is*

- *associative, commutative, non-decreasing in each place,*
- *$U_0(x) \leq x$ for all $x \in [0,1]$, $U_0(1) = 1$ and U_0 is continuous except perhaps at one point e but then $U_0(e) = 0$ and the section U_e is continuous*

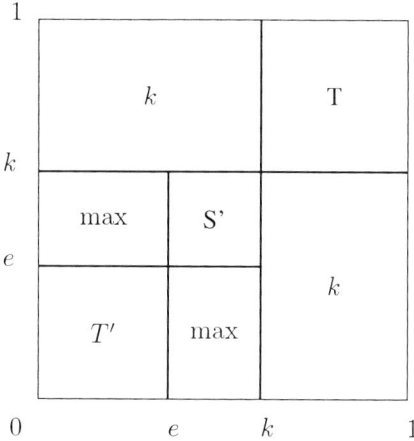

Figure 3: General structure of T–uninorms in $\mathcal{U}_{\max}$

Finally, let us point out that, given any strong negation N, the two kinds of operators introduced in this paper, S–uninorms in $\mathcal{U}_{\min}$ and T–uninorms in $\mathcal{U}_{\max}$, are N–dual one of each other. Namely, given any strong negation N and any binary operator $U : [0,1]^2 \to [0,1]$, let us denote by U_N the N–dual operator of U, that is,

$$U_N(x,y) = N(U(N(x), N(y)))$$

for all $x, y \in [0, 1]$.

Then, we have the following theorem.

Theorem 5 *Let N be a strong negation and $U : [0,1]^2 \to [0,1]$ a binary operator. Then U is an S–uninorm in $\mathcal{U}_{\min}$ if and only if U_N is a T–uninorm in $\mathcal{U}_{\max}$.*

Proof: It is well known that if U is associative, commutative and non-decreasing in each place then so is U_N. Moreover it is trivial that the boundary sections of U_N are given by

$$(U_N)_0 = N \circ U_1 \circ N \quad \text{and} \quad (U_N)_1 = N \circ U_0 \circ N$$

and then the boundary conditions also follow. ∎

4 Alternative approach

As already observed, S-uninorms in $\mathcal{U}_{\min}$ as well as T-uninorms in $\mathcal{U}_{\max}$ generalize both uninorms and t-operators. Taking into account the representation $F = med(k,T,S)$ of t-operators from [6], we may try to define new type of binary operators on the unit interval [0,1]

$$F = med(k, G, H) \qquad (3)$$

where $k \in [0,1]$ is a constant, and G, H are from the class $\mathcal{A} = \mathcal{T} \cup \mathcal{S} \cup \mathcal{U}$, where $\mathcal{T}$ is the class of all t-norms, $\mathcal{S}$ is the class of all t-conorms and $\mathcal{U}$ is the class of all uninorms in $\mathcal{U}_{\min} \cup \mathcal{U}_{\max}$.

Obviously, any operator $F : [0,1]^2 \to [0,1]$ given by (3) is nondecreasing, commutative and $F(0,0) = 0, F(1,1) = 1$. However, F is not associative and k need not be a

null-element of F, in general. To ensure that k is the null-element of F, it is enough to suppose that
$$G(k,x) \leq k \leq H(k,x) \quad \text{for all } x \in [0,1] \qquad (4)$$
but still associativity is not guaranteed.

Example 3 *Just taking $F = med(1/2, T, U)$ where T is the Hamacher t-norm with parameter $\lambda = 0$ and U is the uninorm given by*
$$U(x,y) = \begin{cases} 1 & \text{if } x = 0, y = 1 \text{ or } x = 1, y = 0 \\ \frac{xy}{xy + (1-x)(1-y)} & \text{otherwise} \end{cases}$$
we obtain an operator satisfying (4) that is not associative. Indeed, we have
$$F(F(1/5, 2/5), 3/10) = 6/53 \quad \text{while} \quad F(1/5, F(2/5, 3/10)) = 2/13$$
violating the associativity of F.

A property that can be extended to this kind of operators is the one given in Theorem 5 concerning duality. Namely, the following result holds trivially.

Theorem 6 *Let $k \in [0,1]$ be a fixed constant, let G, H be two binary operators from $\mathcal{A}$ and N a strong negation. The N-dual operator of $F = med(k, G, H)$, denoted by F_N is given by $F_N = med(N(k), H_N, G_N)$. Moreover, F satisfies condition (4) if and only if F_N satisfies it.*

Let us now deal with associativity of operators given by (3). First we need the following definition.

Definition 6 *Let $F : [0,1]^2 \to [0,1]$ be a binary operator. F will be called a bi-uninorm if and only if there exist $e_1, e_2 \in [0,1]$ such that $e_2 < k < e_1$ and two uninorms, U_1 and U_2 with neutral elements e_1 and e_2 respectively, such that F is given by:*
$$F(x,y) = \begin{cases} kU_2\left(\frac{x}{k}, \frac{y}{k}\right) & \text{if } 0 \leq x, y \leq k \\ k & \text{if } \min(x,y) \leq k \leq \max(x,y) \\ k + (1-k)U_1\left(\frac{x-k}{1-k}, \frac{y-k}{1-k}\right) & \text{if } k \leq x, y \end{cases}$$

Theorem 7 *Let $k \in (0,1)$ be a fixed constant, let G, H be two binary operators from $\mathcal{A}$ with $G(0,1) = 0, H(0,1) = 1$ and let $F = med(k, G, H)$. Then the following conditions are equivalent:*

i) *F is associative*

ii) *F is either a t-operator, a T-uninorm in $\mathcal{U}_{\max}$, a S-uninorm in $\mathcal{U}_{\min}$ or a bi-uninorm with underlying uninorms U_1 in $\mathcal{U}_{\min}$ and U_2 in $\mathcal{U}_{\max}$.*

iii) *There exist two binary operators from $\mathcal{A}$, G', H' with $G' \leq H'$, $G'(k,x) \leq k \leq H'(k,x)$ for all $x \in [0,1]$ such that $F = med(k, G', H')$.*

Moreover, in this case, k is always a null-element for F.

Proof: Full proof of this result can be found in [12] ∎

Example 4 Take $k = 1/2$ and U_1, U_2 the following uninorms:

$$U_1(x,y) = \begin{cases} \max(x,y) & \text{if } \min(x,y) \geq 4/5 \\ \min(x,y) & \text{else} \end{cases}$$

$$U_2(x,y) = \begin{cases} \min(x,y) & \text{if } \max(x,y) \leq 1/5 \\ \max(x,y) & \text{else} \end{cases}$$

Note that U_1 and U_2 are dual one to another with respect to the usual negation $N = 1 - Id$. Then

$$F = med(1/2, U_1, U_2)$$

is a commutative, nondecreasing, associative operator, with 0, 1 idempotent elements and 1/2 null-element. Moreover, F is self-dual and it is given by:

$$F(x,y) = \begin{cases} \min(x,y) & \text{if } \max(x,y) \leq 1/5 \text{ or } 1/2 \leq \min(x,y) < 4/5 \\ \max(x,y) & \text{if } 1/5 < \max(x,y) \leq 1/2 \text{ or } \min(x,y) \geq 4/5 \\ 1/2 & \text{otherwise} \end{cases}.$$

This operator can be viewed in Figure 4.

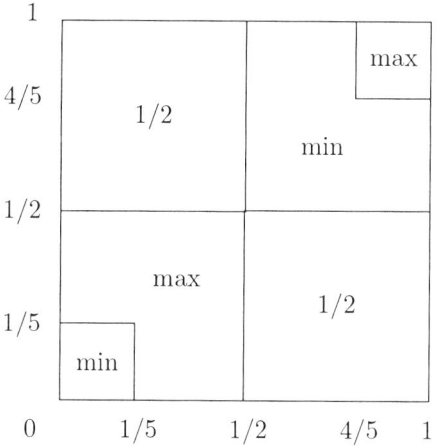

Figure 4: The bi-uninorm given in example 4

Acknowledgements The authors M. Mas, M. Monserrat and J. Torrens have been partially supported by the Spanish DGI project BFM2000-1114, while R. Mesiar was partially supported by grant VEGA 1/0273/03.

References

[1] N.H. Abel (1826). Untersuchungen der funktionen zweier unabhängigen veränderlichen Größen x und y wie $f(x, y)$, welche die eigenschaft haben, daß $f(z, f(x, y))$ eine symmetrische funktion von x, y und z ist, *J. Reine Angew. Math.*, Vol. 1, pp. 631-642.

[2] J. Aczél (1966). *Lectures on functional equations and their applications*, Academic Press, New York.

[3] B. De Baets (1999). Idempotent uninorms, *European J. Oper. Res.*, Vol. 118(3), pp. 631-642.

[4] B. De Baets (1988). Uninorms: the known classes, in *Proc. Third International FLINS Workshop on Fuzzy Logic and Intelligent Technologies for Nuclear Science and Industry* (Antwerp, Belgium), World Scientific, 1988.

[5] T. Calvo, B. De Baets and J. Fodor (2001). The functional equations of Frank and Alsina for uninorms and nullnorms, *Fuzzy Sets and Systems*, Vol 120, pp. 385-394.

[6] T. Calvo, A. Kolesárová, M. Komorníková and R. Mesiar (2002). Aggregation operators: basic concepts, issues and properties, in T. Calvo, G. Mayor, R. Mesiar Editors. *Studies in Fuzziness and Soft Computing* 97: "Aggregation operators. New trends and applications", Physica-Verlag 2002.

[7] A.H. Clifford (1954) Naturally totally ordered commutative semigroups, *Amer. J. Math.*, Vol. 76, pp. 631-646.

[8] J.C. Fodor, R.R. Yager and A. Rybalov (1997). Structure of Uninorms, *Int. J. of Uncertainty, Fuzziness and Knowledge-based Systems*, Vol. 5, N.4, pp. 411-427.

[9] J.S. Golan (1992). *The theory of semirings with applications in mathematics and theoretical computer science*. Longman, Essex.

[10] E.P. Klement, R. Mesiar and E. Pap (2000). *Triangular norms*, Kluwer Academic Publishers, Dordrecht.

[11] M. Mas, G. Mayor and J. Torrens (1999). t-operators, *Int. J. of Uncertainty, Fuzziness and Knowledge-Based Systems*, Vol. 7, No 1, pp. 31-50.

[12] M. Mas, R. Mesiar, M. Monserrat and J. Torrens. Aggregation operators with annihilator, in preparation.

[13] M. Mas, M. Monserrat and J. Torrens (2001). On left and right uninorms, *Int. J. of Uncertainty, Fuzziness and Knowledge-Based Systems*, Vol. 9, No 4, pp. 491-507.

[14] B. Schweizer and A. Sklar (1960). Statistical metric spaces, *Pacific J. of Math.*, Vol. 10, pp. 313-334.

[15] B. Schweizer and A. Sklar (1983). *Probabilistic metric spaces*, North-Holland, New York.

[16] R.R. Yager and A. Rybalov (1996). Uninorm aggregation operators, *Fuzzy Sets and Systems*, Vol. 80, pp. 111-120.

Applications

B. Bouchon-Meunier, L. Foulloy and R.R. Yager (Editors)
Intelligent Systems for Information Processing:
From Representation to Applications
© 2003 Elsevier Science B.V. All rights reserved.

Non-Analytical Approaches to Model-Based Fault Detection and Isolation

Paul M. Frank
Universität Duisburg – Essen
Bismarckstr. 81, 47057 Duisburg, Germany
e-mail: p.m.frank@uni-duisburg.de

Abstract
The paper deals with the treatment of modeling uncertainties in model-based fault detection and isolation (FDI) systems using different kinds of non-analytical models which allow *accurate* FDI under even imprecise observations and at reduced complexity.

Keywords: Fault detection and isolation, modeling uncertainty, robustness, qualitative models.

1 Introduction
The classical approach to model-based fault detection and isolation FDI makes use of functional models in terms of *analytical* (*"parametric"*) mathematical models. A fundamental difficulty with analytical models is that there are always modeling uncertainties due to unmodeled disturbances, simplifications, idealizations and parameter mismatches which are basically unavoidable in the mathematical modeling of a real system. They may be subsumed under the term *unknown inputs*. They are not mission-critical, but they can obscure small faults, and if they are misinterpreted as faults they cause false alarms which can make an FDI system totally useless. Hence, the most essential requirement for an analytical model-based FDI algorithm is *robustness* w. r. t. the different kinds of modeling uncertainties. Analytical approaches to robust FDI schemes that enable the detection and isolation of faults with low false alarm rates in the presence of modeling uncertainties have attracted increasing research attention in the past two decades and there is a great number of solutions to this problem [7, 11, 12, 14, 16, 30, 31, 32 ,33 , 38].

Surprisingly, much less attention has been paid to the use of *qualitative* models in FDI systems, in which case the parameter uncertainty problem does inherently not appear at all. The appeal of the qualitative approaches lies in the fact that qualitative models permit accurate FDI decision making even under imperfect system modeling and imprecise measurements. Moreover, qualitative models may be less complex than comparably powerful analytical models. At present, increased research is going on in the field of FDI using qualitative modeling and computational intelligence, and there is a good deal of publications with most encouraging results, see, for example, [1, 10, 11, 13, 15, 21, 23, 24, 25, 26, 33, 37, 39, 46].

In this paper, we focus our attention on how to cope with modeling uncertainties and imprecise measurements by using non-analytical, i.e., qualitative, structural, data-based and computationally intelligent models. Our intention is to stress the fact that modeling abstraction enables us to make accurate decisions for FDI with less complexity even in the face of large modeling uncertainty, measurement imprecision and lack of system knowledge.

2 The model-based approach to FDI

2.1. Diagnostic strategy

The basic idea of the model-based approach to FDI is to compare the behavior of the actual system with that of its functional model. The diagnostic strategy can follow either of the two policies:

1) If the measurements of outputs are inconsistent with those of a fault-free model with the same input, this indicates that a fault has occurred.

2) If the measurements are consistent with the model behavior corresponding to a certain fault scenario, f_i, then the fault scenario, f_i is declared.

In general, the FDI task is accomplished by the following two-step procedure (Figure 1):

1) *Residual/symptom generation*. This means to generate residuals/symptoms that reflect the faults of interest from the measurements or observations of the actual system. If the individual faults in a set of faults are to be *isolated*, one has to generate properly *structured* residuals or *directed* residual vectors.

2) *Residual/symptom evaluation*. This is a logical decision making process to determine the time of occurrence of faults (fault *detection*) and to localize them (fault *isolation*). If, in addition, faults are to be identified, this requires the determination of the type, size and cause of a fault (fault *analysis*).

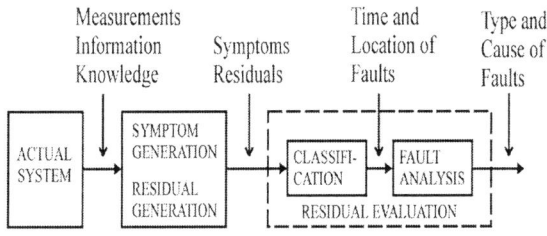

Figure 1: The two-step process of residual generation and evaluation.

2.2. Types of models for residual generation

It has been mentioned earlier that any kind of model that reflects the faults can be used for residual generation. The most appropriate model is the one which allows a acurrate fault decision at a minimum false alarm rate and low complexity. There is a variety of different kinds of non-analytical models that can be used for this task. The types of models can roughly be classified into four categories, namely analytical (quantitative), qualitative, knowledge-based (statistical, fuzzy, computationally intelligent), data-based (fuzzy, neural), structural. The classification of the corresponding *residual generation* methods is shown in Figure 2.

Analytical models are problematic unless one can renounce those parts of the model which carry substantial uncertainty, knows as robust FDI strategies. This means that one concentrates on the certain part of the model, which reflects the faults of interest.

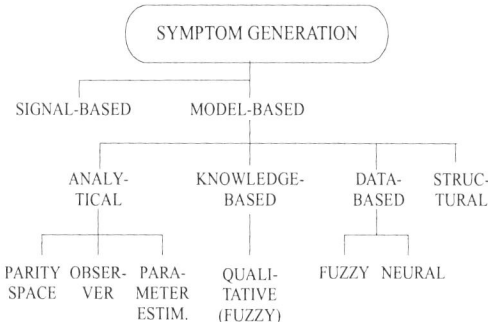

Figure 2: Classification of different model-based approaches to residual generation

3 FDI with non-analytical models
3.1. The power of abstraction

The best way to overcome model uncertainties is to avoid them from the very beginning. That is to say, to use such kinds of models that are not precisely (analytically) defined in terms of parameters. The use of non-analytical, such as qualitative or structural models associated with the dealing with symptoms rather than signals means an increase of the degree of *abstraction*, which plays a fundamental role in reaching accurate results. Logically, achieving accurateness in FDI implies that the check of the reference model must be accurate, i.e., it must be in agreement with the observations of the fault-free system even if the observations are imprecise. This is possible with an according degree of *abstraction* of the model. In addition, abstraction may reduce the complexity of the model and consequently of the FDI system.

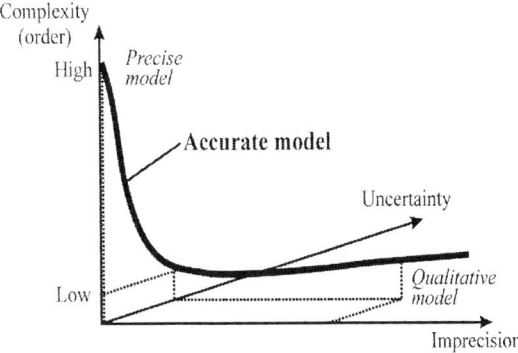

Figure 3: Complexity of an accurate model for FDI versus uncertainty and imprecision.

Figure 3 shows the typical relationship between model complexity, measurement imprecision and modelling uncertainty of an *accurate* model for FDI depending on different kinds of modeling. It can be seen that, to reach accuracy, the required complexity is maximum for precise, i.e. quantitative analytical models, and it decreases considerably with the degree of abstraction obtained by the use of non-analytical models. This means that accurate decisions are possible even in case of imprecise observations if abstract (non-analytic) modeling is applied, or, in other words:

A reduction of complexity of robust FDI algorithms can be obtained by increasing the degree of abstraction of the model.

3.2. FDI based on qualitative models
3.2.1. Qualitative approaches to FDI

Qualitative models reduce the resolution of the representations by introducing tolerances in order to emphasize relevant distinctions and ignore unimportant or unknown details. Under imprecise observations this description represents the systems accurately if a *set* of values rather than single values become the primitives of representation.

In the last decade, the study of applying qualitative models to system monitoring and FDI received much attention, see, e.g., [10, 21, 22, 23, 34], and the concept of *qualitative (knowledgebased) observer* was introduced [13]. Typical qualitative descriptions of variables are signs [9], intervals [20], [23] or fuzzy sets [35]. As a fuzzy set can be divided into a series of intervals, the use of the α-cut identity principle proposed by Nguyen [29] allows to reduce fuzzy mappings into interval computations. Therefore, intervals are the fundamental representations in qualitative modeling. The rough representation of variables leads to the imprecision of the qualitative model which relates the variables to each other.

According to the available information about a system, there are different possibilities for a qualitatively representation of the information of the dynamic process. Basically, a qualitative simulation method should be responsible for retaining the accuracy of the represented system behavior (so called soundness property following the definition of Kuipers [20]), so that the FDI approaches based on them can avoid false alarms. The most important types of representation known are:

- Qualitative differential equations (QDE) [20, 35]
- Envelope behavior (e.g., [5, 18]
- Stochastic qualitative behavior [23, 46].

Other relevant methods to qualitative models for fault diagnosis are, e.g., signed directed graphs [22], logical based diagnosis [24] and structural analysis [36]. Dynamic behaviors are not emphasized in these methods, their main concern is the causality or correlativity among various parts of the systems, which are useful for performing fault isolation and fault analysis.

3.2.2. FDI using qualitative observers based on QDE

Conceptually, a qualitative differential equation can be considered as the extension of an ordinary differential equation

$$\dot{x} = g(x, u, \theta), \qquad (1)$$

where x, u and θ denote the vectors of state variables, known inputs and parameters with the dimension of n, r and s, respectively. However, in a QDE, the variables take intervals as their values and the variant of the non-linear function g(.) is allowed to include various imprecise representations: e.g., interval parameters, non-analytical functions empirically represented by IF-THEN rules and even, in the algorithm QSIM of Kuipers [18], unknown monotonic functions. If the non-linear function g(.) is rational, its corresponding QDE can be readily derived from it by using the natural interval extension of the real function [28]. Qualitative simulation procedures that are composed of the two main steps "generation" and "test/exclusion" are basically different from the numerical ones. The behavior of continuous variables is discretely represented by a branching tree of qualitative states.

The resulting *qualitative* observer (QOB) based on QDE is an extension of a qualitative simulator, and it functions in further reducing the number of irrelevant behaviors (including spurious solutions) to the system under consideration [39] as illustrated in Fig. 4. The principle of observation filtering is that the simulated qualitative behavior of a variable must cover its counterpart of the measurements obtained from the system itself; otherwise the simulated behavioral path is inconsistent and can be eliminated. Since these procedures do not lead to the violation of the accuracy of the qualitative behavior under fault free condition, the output of QOB is the refined prediction behavior in this case.

However, when a fault occurs which causes a significant deviation of the system output such that no consistent predicted counterpart of the output could be generated, the output of the QOB becomes an empty set, which indicates the fault occurrence. Following this principle, fault detection and sensor fault isolation can be implemented [39]. It is important to note that, in exchange with the advantage of requiring weaker process knowledge in this method, one has to put up with an increase in computational complexity and less sensitivity to small faults.

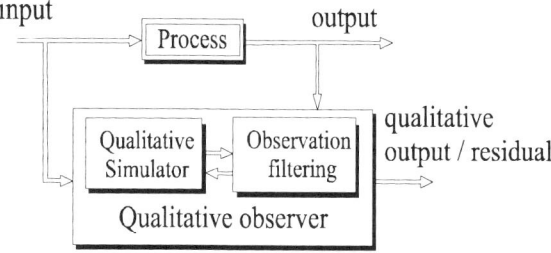

Figure 4: Qualitative observer

3.2.3. Fault detection based on envelope behaviors

A key issue of improving the small fault detectability when applying qualitative methods is that the qualitative system behavior should be predicted as precisely as possible. Different from the qualitative model and the simulation method presented above, the model considered in this and the next sections is of less ambiguity. In other words, imprecision in Equ. (1) is caused only by interval parameters and interval initial states, the structure of g(.) is considered to be fully known. While qualitative behaviors here are interval values of system variables against time, qualitative simulation aiming at producing all possible dynamic behaviors means the generation of their envelope. Once the envelope is generated, the fault detection task is a direct comparison between the envelope and the measurements. In fault-free case, the measurements are contained in the envelope; otherwise, it indicates a fault.

Recently, many efforts have been made to increase the efficiency of classical qualitative simulation, i.e., to avoid unnecessary conservativeness. More quantitative information is brought into the model representation [3], and simulation methods tend to be more constructive. Kay and Kuipers [18] and Verscovi et al. [38] propose approaches based on standard numerical methods to obtain the bounding behavior. In [5, 19] Bonarini et al. and Keller et al. treat the interval parameters and the state variables as a super-cube, whose evolution at any time is specified by its external surface. Armengo et al. [1] present the computation of envelopes making use of modal interval analysis.

3.2.4. Residual generation via stochastic qualitative behaviors

Another qualitative representation of system behaviors is the stochastic distribution under partitioned state and output spaces. Beginning with the similar model assumptions as in section 3.3.1, the parameter vector is in θ and the initial state is uniformly distributed within a prescribed area, say cell 0. $X_i(t)$ and $Y_i(t)$ denote the probabilities that the trajectories of the respective state and output variables, which start from all initial states in cell 0, fall into the i-th cell at any time t. The behavior can be approximately represented by a Markov chain [46]. It turns out that the new state and output variables X and Y can be described by the following discrete hidden Markov model (HMM):

$$X(k+1) = A(u,\theta) X(k) + V(k) \tag{2}$$

$$Y(k+1) = C(\theta) X(k+1), \tag{3}$$

where V represents the influence of spurious solutions.

A fault detection scheme based on the HMM is shown in Fig. 5 [46]. A qualitative observer (QOB) aiming at attenuating the effect of V and watching over the possible abnormal behavior of measurements is applied. The residual r and its credibility v can be calculated, the latter reflects the degree of spurious solutions.

3.3. Residual generation employing computational intelligence

In the case of fault diagnosis in complex systems, one is faced with the problem that no or insufficiently accurate mathematical models are available. The use of data-model-based (neural) diagnosis expert systems or in combination with a human expert, is then a much more appropriate way to proceed. The approaches presented in the following section employ computational intelligence techniques such as neural networks, fuzzy logic, genetic algorithms and combinations of them in order to cope with the problem of uncertainty, lacking analytical knowledge and non-linearity [15].

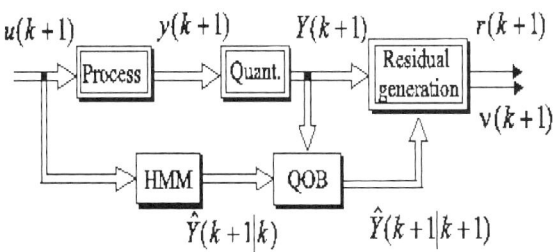

Figure 5: Observer-based residual generation using HMM

3.3.1. Neural observer-based residual generation

Neural networks can be used as non-linear multiple-input single-output (MISO) models of ARMA type to set up different kinds of observer schemes [15, 27]. The neural networks replace the analytical models of observer-based FDI. If instead of a single multiple-input multiple-output structure a separate neural network is taken for each output, a set of smaller neural networks can be used for each class of system behavior.

The type of neural network employed for this task is of a mixed structure called *dynamic multi-layer perceptron (DMLP-MIX)* integrating three generalized structures of a DMLP [25]. These three are: the DMLP with synaptic generalized filters, which have each synapse represented by an ARMA filter with different orders for denominator and numerator, the DMLP with internal generalized filters [2] integrating an ARMA filter within the neurons before the activation function, and the DMLP with a connectionist hidden layer, which has a partially recurrent structure interconnecting only the hidden units. The mixed structure is implemented selecting either a basic architecture or a combination of them. The training of the DMLP-MIX neural network is performed by applying dynamic back propagation, the problem of structural optimization is solved with the help of a genetic algorithm [26]. Two types of observer schemes for actuator, component and instrument fault detection have been proposed by Marcu et al. [27]: the neural single observer scheme (NSOS) and the neural dedicated observer scheme (NDOS).

3.3.2. Fuzzy observer-based residual generation

There are many ways of using fuzzy logic to cope with uncertainty in observer-based residual generation [15]. The resulting type of fuzzy observer depends upon the type of the fuzzy model used. Fuzzy modelling can roughly be classified into four categories: fuzzy rule-based, fuzzy qualitative, fuzzy relational and fuzzy functional (Tagaki-Sugeno type).

3.3.3. Residual generation with hierarchical fuzzy neural networks

Here the fault diagnosis system is designed by a knowledge-based approach and organized as a hierarchical structure of fuzzy neural networks (FNN) [6]. FNNs combine the advantage of fuzzy reasoning, i.e. being capable of handling uncertain and imprecise information, with the advantage of neural networks, i.e. being capable of learning from examples. The neural nets consist of a fuzzification layer, a hidden layer and an output layer. Fault detection is performed through the knowledge-based system, where the detection rules are generated from knowledge obtained from the structural decomposition of the overall system into subsystems and operational experience. After detecting a fault the diagnostic module is triggered which consists of a hierarchical structure (usually three layers) of FNNs. The number of FNNs is determined by the number of faults considered. The lower level only contains one FNN, which processes all measured variables. The FNNs on the medium level are fed by all measurements but also by the outputs of the previous level. The upper level consists of an OR operation on the outputs of the medium level. This hierarchical structure can cope with multiple simultaneous faults under highly uncertain conditions.

3.3.4. Fuzzy residual evaluation

Fuzzy logic is especially useful for decision making under considerable uncertainty. The three main categories of current residual evaluation methods are: classification (clustering) or pattern recognition, inference or reasoning, and threshold adaptation. Although all approaches employ fuzzy logic, the first one is actually data-based while the other two are knowledge-based.

Fuzzy clustering

The approach of fuzzy clustering actually consists of a combination of statistical tests to evaluate the time of occurrence of the fault and the fuzzy clustering to provide isolation of the fault [8]. The statistical tests are based on the analysis of the mean and the variance of the residuals, e.g., the CUSUM test [17]. The subsequent fault isolation by means of fuzzy clustering consists of the two following steps: In an online phase the characteristics of the different classes are determined. A learning set which contains residuals for all known faults is necessary for this online phase. In the online phase the membership degree of the current residuals to each of the known classes is calculated. A commonly used algorithm is the fuzzy C-means algorithm [4].

Fuzzy reasoning

The basic idea behind the application of fuzzy reasoning for residual evaluation is that each residual is declared as *normal, high* or *low* with respect to the nominal residual value [8, 37]. These linguistic attributes are defined in terms of fuzzy sets, and the rules among the fuzzy sets are derived from the dynamics of the system. For fault detection, the only relevant information is whether or not the residual has deviated from the fault free value, and hence it is only necessary to differentiate between normal and abnormal behavior. However, if isolation of faults is desired, it may be necessary to consider both the direction and magnitude of the deviation.

Fuzzy threshold adaptation

Fuzzy reasoning has been applied with great success to threshold adaptation [13, 33]. In the case of poorly defined systems it is difficult or even impossible to determine adaptive thresholds. In such situations the fuzzy logic approach is much more efficient. The relation for the adaptive threshold can be defined as a function of input u and output y by

$$T(u,y) = T_0 + \Delta T(u,y). \tag{4}$$

Here $T_o = T_0(u_0, y_0)$ denotes a constant threshold for nominal operation at the operational point (u_o, y_0) where only the effects of the stationary disturbances including measurement noise are taken into account. The increment $\Delta T(u, y)$ represents the effects of $u(t)$ and $y(t)$ caused by the modeling errors. These effects are described in terms of IF-THEN rules and the variables by fuzzy sets (e.g. SMALL, MIDDLE, LARGE, etc.) that are characterized by proper membership functions.

As a typical example of an industrial application we consider the residual evaluation via fuzzy adaptive threshold of a six-axis industrial robot (Manutec R3) [13, 33]. Let the goal be to detect a collision of the robot by checking the moments of the drives. A model of the robot is available, but without knowledge of the friction of the bearings, which is highly uncertain. It is known, however, that the residual of the moment is heavily distorted by the friction which strongly depends on the arm acceleration. This knowledge can be formulated by rules. For example for the third axis the following rules apply:

- IF*{speed small}*, THEN*{threshold middle}*
- IF*{acceleration high}*, THEN*{threshold large}*
- IF*{acceleration very high}*, THEN *{threshold very large}*
- IF *{acceleration of any other axis very high}*, THEN *{threshold middle}*.

The linguistic variables *small, middle, high, very high, large, very large* are defined by proper membership functions [33], they are assigned intuitively based on the experience of the operators or the manufacturers of the robot.

Figure 6 shows the time shape of the threshold together with the shape of the residual of axis 3 for a particular manoeuvre of the robot. Note that at $t = 4,5$ sec the heavy robot which can handle 15 kg objects in its gripper, hits an obstacle which causes a momentum of about 5 Nm. As can be seen, this small fault can be detected at high robustness to the uncertainty caused by the neglected unknown friction.

3.4. FDI based on structural models

The use of structural system models together with structural analysis is another way of abstraction of the modelling of the system behavior in order to increase the robustness of the FDI algorithm to model uncertainties. Here we only consider the *structure* of the constraints, i.e., the existence of links between variables and parameters rather than the constraints themselves [36]. The links are usually represented by a *bi-partite graph*, which is independent of the nature of the constraints and variables (quantitative, qualitative, equations, rules, etc.) and of the values of the parameters. Structural properties are true almost everywhere in the system parameter space.

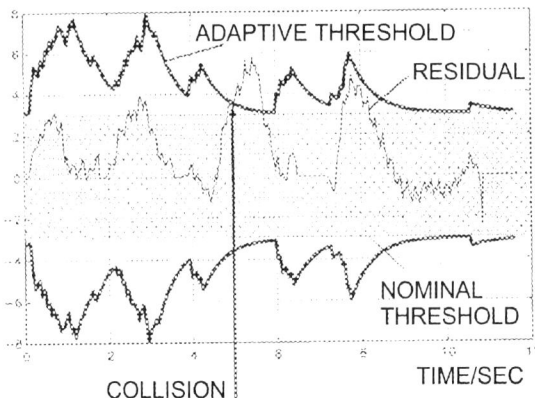

Figure 6: Obstacle detection of a robot with fuzzy adaptive threshold.

This represents indeed a very low-level easy-to-obtain model of the system behavior, which is logically extremely insensitive to changes in the system parameters but, of course, also to parametric faults. The important tasks of structural analysis are solved with the aid of the analysis of the system structural graph and its canonical decomposition. An important factor in the canonical decomposition is the property of causality which complements the bi-partite graph with an orientation. FDI is performed with the aid of analytical redundancy relations based on a structural analysis and the generation of structured residuals.

Note that the use of structural models together with the strong decoupling approach solves automatically the robustness problem in structurally observable systems.

4   Conclusion

The paper reviews the methods of handling modelling uncertainties, incomplete system knowledge and measurement imprecision in model-based fault detection and isolation by using non-analytical models. It is pointed out that abstract non-analytical models may be superior over analytical models with respect to uncertainty, impression and complexity. The paper outlines the state of the art and relevant on-going research in the field approaching the modeling uncertainty and measurement imprecision problem in FDI by various types of non-analytical models.

References

[1] Armengo, J., L. Travé-Massuyès, J. Vehi and M. Sáinz (1999). Semi-qualitative simulation using modal interval analysis. In: *Proc. of the 14th IFAC World Congress 99*. Beijing, China.

[2] Ayoubi, M. (1994). Fault diagnosis with dynamical neural structure and application to a turbo-charger. In: *IFAC Symp. SAFE-PROCESS'94*, Espoo, Finland. Vol. 2.

[3] Berleant, D. and B. Kuipers (1992). Qualitative-numeric simulation with q3. In: *Recent Advances in Qualitative Physics* (B. Faltings and P. Struss, Eds.). MIT Press Cambridge., Mass..

[4] Bezdek, J. C. (1991). Pattern Recognition with Fuzzy Objective Functions Algorithms. *Plenum Press*. New York.

[5] Bonarini, A. and G. Bontempi (1994). A qualitative simulation approach for fuzzy dynamical models. *ACM transactions on Modeling and Computer Simulation* 4(4), pp. 285-313.

[6] Calado, J. M. F. and J. M. Sa da Costa (1999). Online fault detection and diagnosis based on a coupled system. In: *Proc. ECC'99*, Karlsruhe.

[7] Chen, J and R. Patton (1999). Robust model-based fault diagnosis for dynamic systems. *Kluwer Academic Publishers*.

[8] Dalton, T., N. Kremer and P. M. Frank (1999). Application of fuzzy logic based methods of residual evaluation to the three tank benchmark. In: *ECC'99*, Karlsruhe.

[9] de Kleer, J. and J.S. Brown (1984). Qualitative physics based on confluences. *Artificial Intelligence* 24, pp. 7-83.

[10] Dvorak, D. and B. Kuipers (1989). Model-based monitoring of dynamic systems. In: *Proc. 11th Intern. Joint Conf. on Art. Intell.*, Detroit, MI. pp. 1238-1243.

[11] Frank, P. M. (1990). Fault diagnosis in dynamic systems using analytical and knowledge-based redundancy - A survey and some new results. *Automatica* 26, pp. 459-474.

[12] Frank, P.M. (1994). Enhancement of robustness in observer-based fault detection. *Int. J. Control*. 59, pp. 955-981.

[13] Frank, P.M. (1996). Analytical and qualitative model-based fault diagnosis – A survey and some new results. *Europ. J. Control* 2, pp. 6-28.

[14] Frank, P.M., G. Schreier and E. Alcorta Garcia (1999). Nonlinear observers for fault detection and isolation. In: *New Directions in Nonlinear Observer Design* (Ed.: H. Nejmeijer et al.). Springer, pp. 399-422.

[15] Frank, P.M. and T. Marcu (2000). Diagnosis strategies and systems: Principles, Fuzzy and neural approaches. In: *Teodorescu, H.-N. et al. (Eds.). Intelligent Systems and Interfaces*. Kluwer Akademic Publishers

[16] Gertler, J.J. (1998). Fault Detection and Diagnosis in Engineering Systems. *Marcel Dekker*.

[17] Gomez, R. E. (1995). Modellbasierte Fehlererkennung und -diagnose in Mehrgroessensystemen mit Hilfe statistischer Methoden. PhD Thesis Univ. Bochum Germany.

[18] Kay, H. and B. Kuipers (1993). Numerical behavior envelopes for qualitative models. In: *Proc. of the 7th National Conf. on Artificial Intelligence*. pp. 606-613.

[19] Keller, U., T. K. Wyatt and R. R. Leitch (1999). Frensi - a fuzzy qualitative simulator. In: *Proc. of Workshop on Applications of Interval Analysis to Systems and Control*. Girona, Spain. pp. 305-313.

[20] Kuipers, B. (1986). Qualitative simulation. *Artificial Intelligence* 66(29), pp. 289-338.

[21] Leitch, R., Q. Shen, G. Conghil, M. Chantler and A. Slater (1994). Qualitative model-based diagnosis of dynamic systems. In: *Colloquium of the Institution of Measurement and Control*. London.

[22] Leyval, L., J. Montmain and S. Gentil (1994). Qualitative analysis for decision making in supervision of industrial continuous processes. *Mathematics and computers in simulation* 36, pp. 149-163.
[23] Lunze, J. (1994). Qualitative modelling of linear dynamical systems with quantised state measurements. *Automatica* 30(3), pp. 417-431.
[24] Lunze, J. (1995). Künstliche Intelligenz für Ingenieure, Band 2: Technische Anwendungen. *Oldenbourg Verlag*. München.
[25] Marcu, T., L. Mirea and P. M.Frank (1998). Neural observer schemes for robust detection and isolation of process faults. In: *UKACC Int. Conf. CONTROL'98*, Swansea, UK. Vol. 2.
[26] Marcu, T., L. Ferariu and P. M. Frank (1999a). Genetic evolving of dynamical neural networks with application to process fault diagnosis. In: *ECC'99*, Karlsruhe.
[27] Marcu, T., M. H. Matcovschi and P. M. Frank (1999b). Neural observer-based approach to fault detection and isolation of a three-tank system. In: *ECC'99*, Karlsruhe.
[28] Moore, R. (1979). Methods and applications of Interval analysis. *SIAM*. Philadelphia.
[29] Nguyen, H.T. (1978). A note on the extension principle for fuzzy sets. *Journal of Mathematical Analysis and Applications* 64, pp.369-380.
[30] Patton, R., P. M. Frank and R. N. Clark (1989). Fault diagnosis in dynamic systems. *Prentice Hall*.
[31] Patton, R., P. M. Frank and R. N. Clark (2000). Issues of fault diagnosis for dynamic systems. *Springer Verlag*.
[32] Rambeaux, F. F. Hammelin and D. Sauter (1999). Robust residual generation via LMI. In: *Proc. of the 14th IFAC World Congress 99*, Beijing, China.
[33] Schneider, H. Implementation of fuzzy concepts for supervision and fault detection of robots. In: *Proc .of EUFIT'93*, Aachen. pp. 775-780.
[34] Shen, L. and P. Hsu (1998). Robust design of fault isolation observers. *Automatica, 34, pp. 1421-1429*.
[35] Shen, Q and R. Leitch (1993). Fuzzy qualitative simulation. *IEEE Trans. SMC* 23(4).
[36] Staroswiecki, M., J. P. Cassar and P. Declerck (2000). A structural framework for the design of FDI system in large scale industrial plants. In: *Issues of Fault Diagnosis for Dynamic Systems* (Patton, Frank and Clark, Eds.). Springer. pp. 453-456.
[37] Ulieru, M. (1994). Fuzzy reasoning for fault diagnosis. *2nd Int. Conf. on Intelligent Systems Engineering..*
[38] Verscovi, M., A. Farquhar and Y. Iwasaki (1995). Numerical interval simulation. In: *International Joint Conference on Artificial Intelligence IJCAI* 95. pp. 1806-1813.
[39] Zhuang, Z. and P.M. Frank (1997). Qualitative observer and its application to fault detection and isolation systems. *J. of Systems and Contr. Eng., Proc. of Inst. of Mech. Eng.*, Pt. I 211(4), pp. 253-262.
[40] Zhuang, Z. and P.M Frank (1999). A fault detection scheme based on stochastic qualitative modeling. In: *Proc. of the 14th IFAC World Congress 99*, Beijing, China.

A Hybrid Fuzzy-Fractal Approach for Time Series Analysis and Prediction and Its Applications to Plant Monitoring

Oscar Castillo and Patricia Melin
Dept. of Computer Science
Tijuana Institute of Technology
P.O. Box 4207, Chula Vista CA 91909, USA
Email: ocastillo@tectijuana.mx

Abstract

We describe in this paper a new hybrid fuzzy-fractal approach for plant monitoring. We use the concept of the fractal dimension to measure the complexity of a time series of observed data from the plant. We also use fuzzy logic to represent expert knowledge on monitoring the process in the plant. In the hybrid fuzzy-fractal approach a set of fuzzy if-then rules is used to classify different conditions of the plant. The fractal dimension is used as input linguistic variable in the fuzzy system to improve the accuracy in the classification. An implementation of the proposed approach is shown to describe in more detail the method.

Keywords: Fuzzy Logic, Fractal Theory, Time Series Analysis, Plant Monitoring

1 Introduction

Diagnostic systems are used to monitor the behavior of a process and identify certain pre-defined patterns that are associated with well-known problems [10]. These problems, once identified, imply suggestions for specific solutions. Most diagnostic systems are in the form of a rule-based expert system: a set of rules is used to describe certain patterns. Observed data are collected and used to evaluate these rules. If the rules are logically satisfied, the pattern is identified, and the problem associated with that pattern is suggested. In general, the diagnostic systems are used for consultation rather than replacement of human expert [15].

Most current plant monitoring systems only check a few variables against individual upper and lower limits, and start an audible alarm should each variable move out of its predefined range [9]. Other more complicated systems normally involve more sensors that provide more data but still follow the same pattern of independently checking individual sets of data against some upper and lower limits. The warning alarm from these systems only carries a meaning that there is something wrong with the process in the plant [12].

In this paper a new fuzzy-fractal approach for plant monitoring is proposed. The concept of the fractal dimension is used to measure the complexity of the time series of relevant variables for the process [3]. A set of fuzzy rules is used to represent the knowledge for monitoring the process [11]. In the fuzzy rules, the fractal dimension is used as a linguistic variable to help in recognizing specific patterns in the measured data. The fuzzy-fractal approach has been applied before in problems of financial time series prediction [4, 5, 7] and for other types of problems [6, 14], but now it is proposed to the monitoring of plants.

This paper is structured as follows. First, the problem of plant monitoring and diagnosis is described in more detail. Second, some basic concepts of fractal theory are given. Third, the problem of fuzzy modelling for monitoring and diagnosis is described in more detail. Then, the fuzzy-fractal approach for plant monitoring is proposed and explained. Finally, some experimental results are given and future work is proposed.

2 Monitoring and Diagnosis

Monitoring means checking or regulating the performance of a machine, a process, or a system [10]. Diagnosis, on the other hand, means deciding the nature and the cause of a diseased condition of a machine, a process, or a system by examining the symptoms. In other words, monitoring is detecting suspect symptoms, whereas diagnosis is determining the cause of the symptoms.

The importance of monitoring and diagnosis of plant processes now is widely recognized because it results in increased productivity, improved product quality and decreased production cost. As a result, in the past decade, a large number of research and development projects have been carried and many monitoring and diagnosis methods have been developed [10, 11]. The commonly used monitoring and diagnosis methods include modeling-based methods, pattern recognition methods, fuzzy systems methods, knowledge-based systems methods, and artificial neural networks [8]. It is interesting to note that even though these methods are rather different, they share a very similar structure as shown in Figure 1.

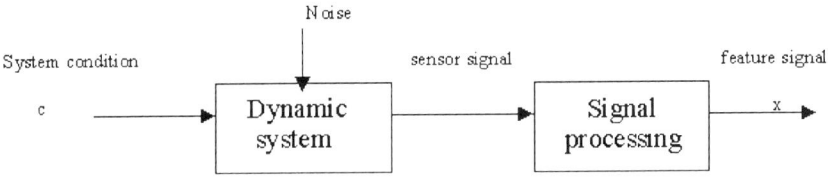

Figure 1: Plant monitoring and diagnosis.

The "health" of a machine, a process, or an engineering system (which will be referred to as system condition and denoted by $c \in \{c_1, c_2,..., c_m\}$) can be considered as the "input", the system working conditions and noises (including system noise and sampling noise) can be considered as the "noise", and the sensor signals are the "outputs" from the system. Typically, the sensor signals are processed by a computer, after which the signals are transformed into a set of features called feature signals, denoted as $\mathbf{x} = \{x_1, x_2,..., x_n\}$. In general, the systems conditions are predefined, such as normal, critical, etc. On the other hand, the features may be the mean of a temperature signal, the variance of a displacement signal, etc. Sensing and signal processing are very important to the success of plant monitoring and diagnosis [9].

More formally, the goal of monitoring is to use the feature signals, x, to determine whether the plant is in an acceptable condition(s) (a subset of $\{c_1, c_2,..., c_m\}$). On the other hand, the objective of diagnosis is to use the feature signals, x, to determine the system condition, $c \in \{c_1, c_2,..., c_m\}$. No matter how monitoring and diagnosis methods may differ, monitoring and diagnosis always consist of two phases:

training and decision making. Training is to establish a relationship between the feature signals and the systems conditions. Without losing generality, this relationship can be represented as

$$\mathbf{x} = F(\mathbf{c}). \qquad (1)$$

It should be pointed out that $F(\mathbf{c})$ represents a fuzzy system, a neural network or another method that could be used to obtain this relationship. In fact, it is the form of the relationship that determines the methods of monitoring and diagnosis, as well as the performance of the methods. The relationship $F(\mathbf{c})$ is established based on training samples, denoted by $\mathbf{x}_1, \mathbf{x}_2,..., \mathbf{x}_k,..., \mathbf{x}_N$, where the system condition for each training sample is known [and denoted as $\mathbf{c}(\mathbf{x}_k)$].

After the relationship is established, when a new sample is given (from an unknown system condition), its corresponding condition is estimated based on the inverse relationship

$$\mathbf{c} = F^{-1}(\mathbf{x}). \qquad (2)$$

This is called decision-making, or classifying. Whereas it is not likely that the training samples will cover all possible cases, decision making often involves reasoning or inferencing.

In particular, when a fuzzy system is used, the relationship is given by a set of fuzzy rules as shown in Figure 2. The input to the fuzzy system is the feature signal and the output of the fuzzy system is the estimated plant condition(s) [i.e., $\mathbf{z} = (z_1, z_2,..., z_m)$ is an estimate of $\mathbf{c} = (c_1, c_2,..., c_m)$]. In other words, the fuzzy system models the inverse relationship between the system conditions and the feature signals.

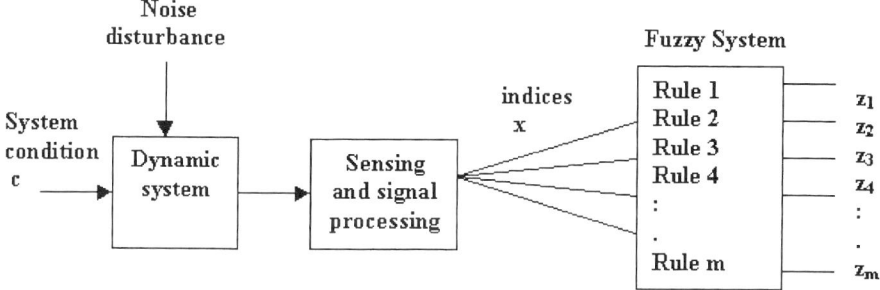

Figure 2: Fuzzy system for plant monitoring and diagnosis.

3 Fractal Dimension of a Geometrical Object

Recently, considerable progress has been made in understanding the complexity of an object through the application of fractal concepts [13], and dynamic scaling theory. For example, financial time series show scaled properties suggesting a fractal structure [2, 5, 7]. The fractal dimension of a geometrical object can be defined as follows:

$$d = \lim_{r \to 0} [\ln N(r)] / [\ln(1/r)] \qquad (3)$$

where $N(r)$ is the number of boxes covering the object and r is the size of the box. An approximation to the fractal dimension can be obtained by counting the number of boxes covering the boundary of the object for different r sizes and then performing a logarithmic regression to obtain d (box counting algorithm). In Figure 3, we illustrate the box counting algorithm for a hypothetical curve C. Counting the number of boxes for different sizes of r and performing a logarithmic linear regression, we can estimate the box dimension of a geometrical object with the following equation:

$$\ln N(r) = \ln \beta - d \ln r \tag{4}$$

this algorithm is illustrated in Figure 4.

The fractal dimension can be used to characterize an arbitrary object. The reason for this is that the fractal dimension measures the geometrical complexity of objects. In this case, a time series can be classified by using the numeric value of the fractal dimension (d is between 1 and 2 because we are on the plane xy). The reasoning behind this classification scheme is that when the boundary is smooth the fractal dimension of the object will be close to one. On the other hand, when the boundary is rougher the fractal dimension will be close to a value of two.

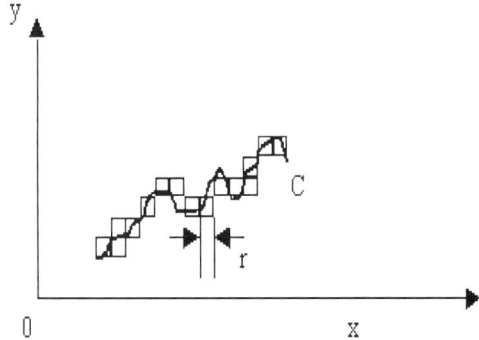

Figure 3: Box counting algorithm for a curve C.

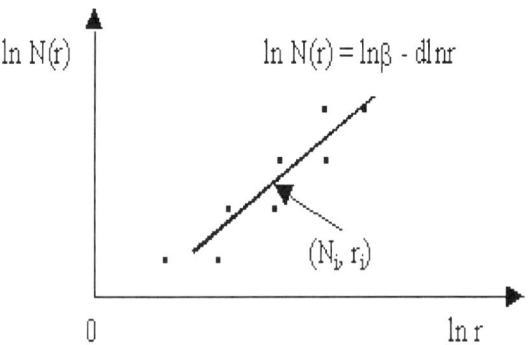

Figure 4: Logarithmic regression to find dimension.

4 Fuzzy Estimation of the Fractal Dimension

The fractal dimension of a geometrical object is a crisp numerical value measuring the geometrical complexity of the object. However, in practice it is difficult to assign a unique numerical value to an object. It is more appropriate to assign a range of numerical values in which there exists a membership degree for this object. For this reason, we will assign to an object O a fuzzy set μ_o, which measures the membership degree for that object.

Lets consider that the object O is in the plane xy, then a suitable membership function is a generalized bell function:

$$\mu_o = 1 / [1 + | (x-c) / a |^{2b}] \tag{5}$$

where a, b and c are the parameters of the membership function. Of course other types of membership functions could be used depending on the characteristics of the application.

By using the concept of a fuzzy set [17, 18, 19, 20] we are in fact generalizing the mathematical concept of the fractal dimension. In fact, our definition of the fuzzy fractal dimension is as follows.

Definition 1: Let O be an arbitrary geometrical object in the plane xy. Then the fuzzy fractal dimension is the pair: (d_o, μ_o)
where d_o is the numerical value of the fractal dimension calculated by the box counting algorithm, and μ_o is the membership function for the object.

With this new definition we can account for the uncertainty in the estimation of the fractal dimension of an object. Also, this new definition enables easier pattern recognition for objects, because it is not necessary to match an exact numerical value to recognize a particular object [1, 16].

5 Plant Monitoring using the Fuzzy-Fractal Approach

In this section, we show how to implement a fuzzy rule-based expert monitoring system with two basic sensors: temperature, and pressure. We also use as input the fuzzy fractal dimension of the time series of the measured variables. Individual sensors can identify three linguistic values (normal, high, and low). The three inputs can be combined to give 9 different scenarios. With the perfectly normal case (where all three input variables have normal values), there are additionally 9 more cases where combinations of abnormal readings can be observed.

Let x_1 be the temperature, x_2 the pressure, x_3 the fuzzy fractal dimension, and y the diagnostic statement. Let L_i, N_i, and H_i, represent the three sets of low range, normal range, and high range for input data x_i, where i = 1, 2, or 3. Furthermore, let C_1, C_2,..., C_9 be the individual scenarios that could happen for each combination of the different data sets. The fuzzy rules have the general form:

$$R^{(0)}: \text{IF } x_1 \text{ is } N_1 \text{ AND } x_2 \text{ is } N_2 \text{ AND } x_3 \text{ is } N_3 \text{ THEN } y \text{ is } C_1$$

...

$$R^{(i)}: \text{IF } x_1 \text{ is } V_1 \text{ AND } x_2 \text{ is } V_2 \text{ AND } x_3 \text{ is } V_3 \text{ THEN } y \text{ is } C_i \tag{6}$$

...

$$R^{(26)}: \text{IF } x_1 \text{ is } H_1 \text{ AND } x_2 \text{ is } H_2 \text{ AND } x_3 \text{ is } H_3 \text{ THEN } y \text{ is } C_9$$

In this case, V_i represents L_i, H_i, or N_i, depending on the condition for the plant. Experts have to provide their knowledge in plant monitoring to label the individual cases Ci for i = 1, 2,..., 9. Also, the membership functions for the linguistic values of variables have to be defined according to historical data of the problem and expert knowledge.

We can use the Fuzzy Logic Toolbox of the MATLAB programming language to implement the fuzzy monitoring system described above. In this case, we need to specify the particular fuzzy rules and the corresponding membership functions for the problem. We show bellow a sample implementation of a health monitoring system using MATLAB. First, we show in Figure 5 the general architecture of the fuzzy monitoring system. In this figure, we can see the input linguistic variables (temperature, pressure, and fractal dimension) and the output linguistic variable (condition of the plant) of the fuzzy monitoring system. Of course, in this case the fractal dimension is estimated using the box counting algorithm, which was implemented also as a computer program in MATLAB.

In Figure 6 the implementation of the fuzzy rule base in MATLAB is shown. The actual 27 rules were defined according to expert knowledge on the process. In Figure 7, the Gaussian membership functions for the output variable (condition of the plant) are shown. In Figure 8 the non-linear surface for the problem of plant monitoring is shown. In this figure, we show two ways of viewing the surface because there are four linguistic variables in the fuzzy system.

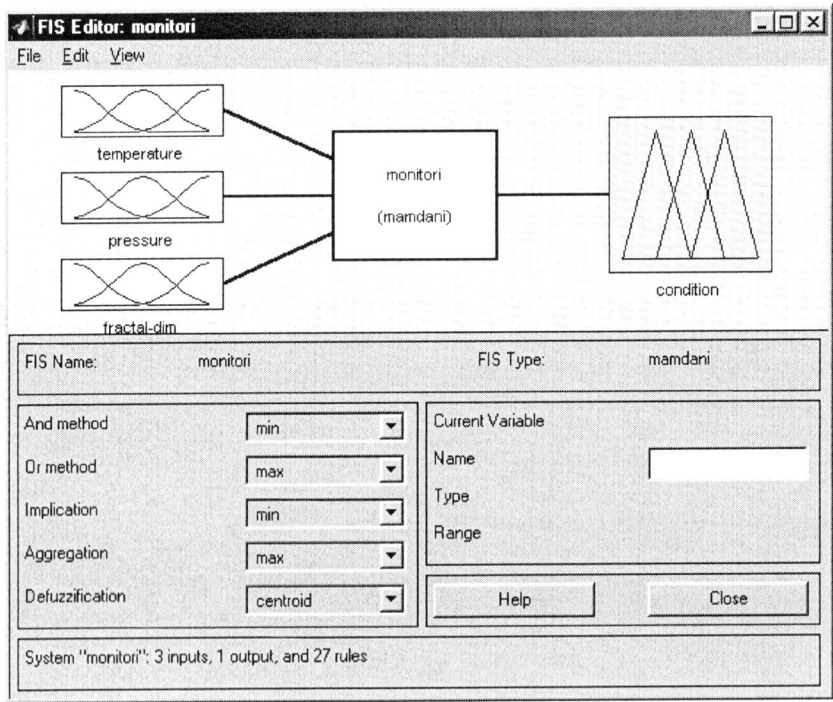

Figure 5: Architecture of the fuzzy system for plant monitoring.

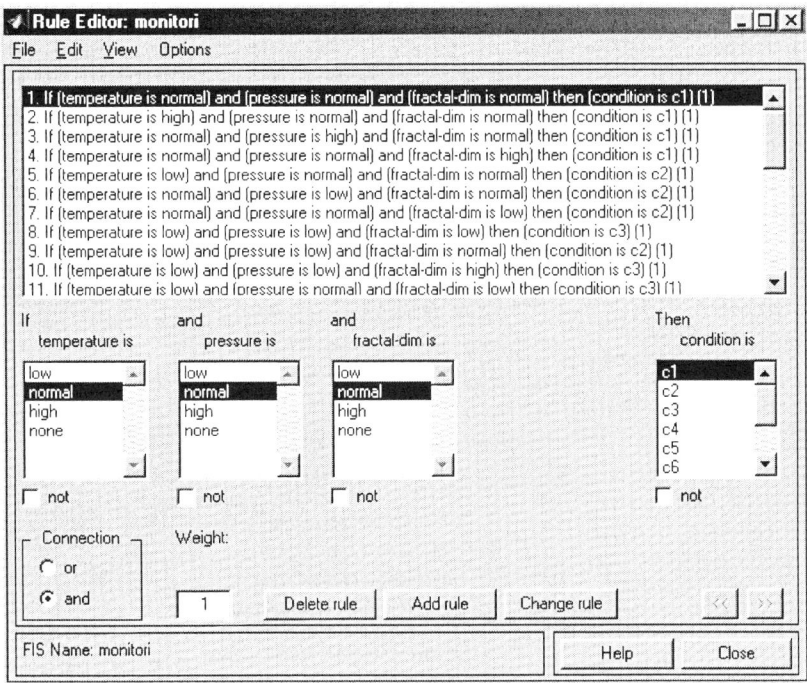

Figure 6: Fuzzy rule base for plant monitoring.

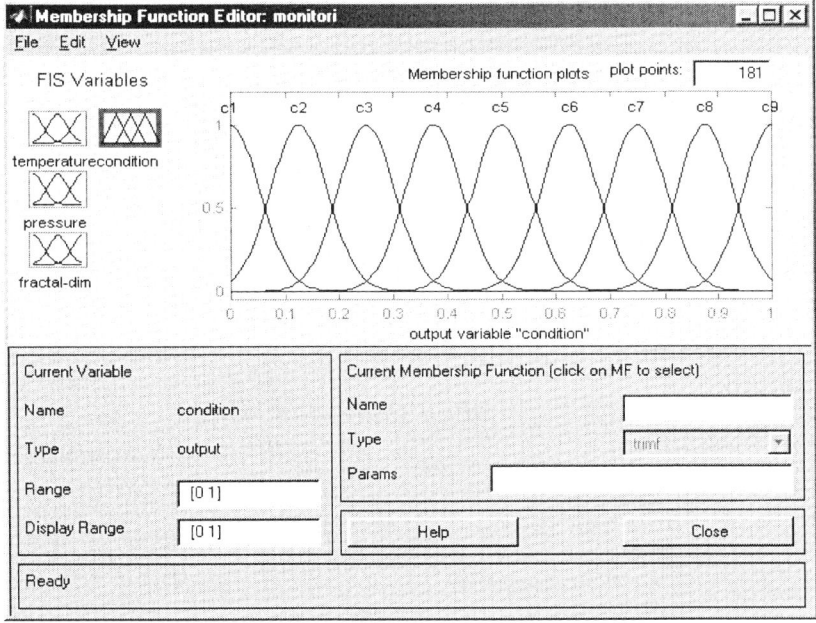

Figure 7: Membership functions for the output variable.

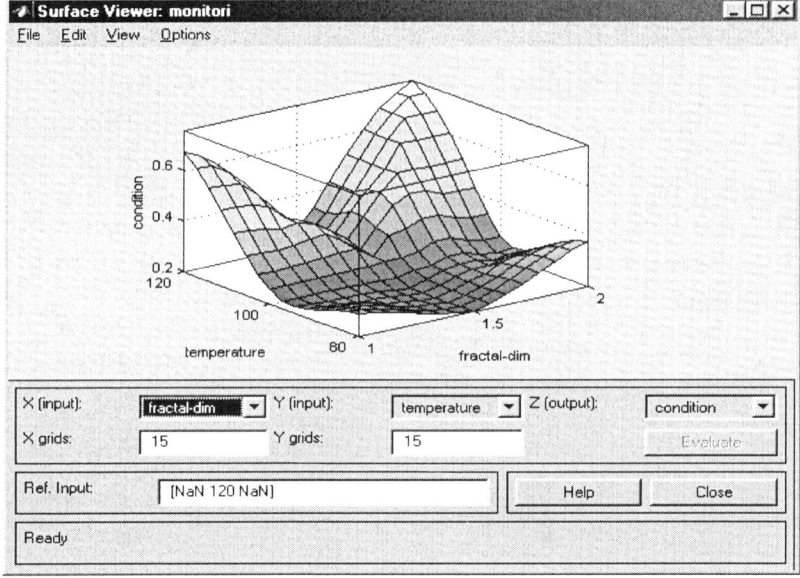

(a)

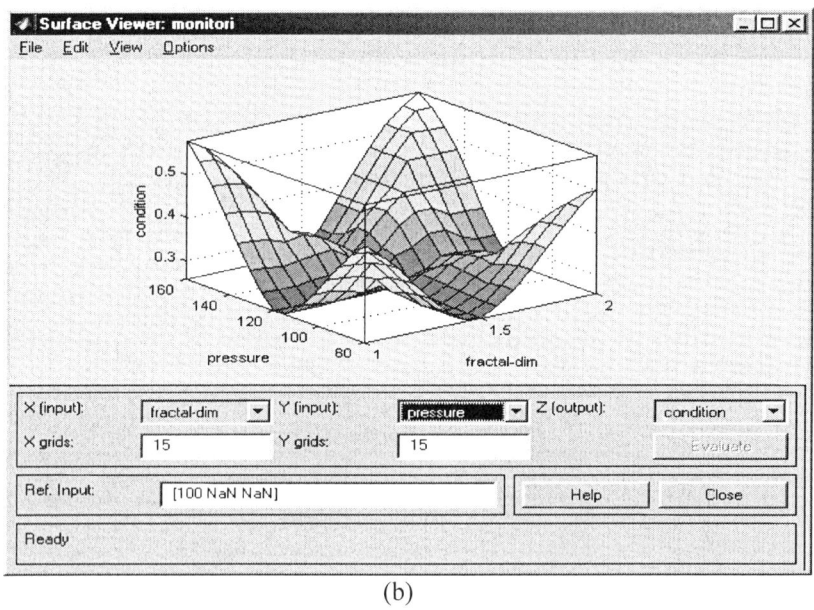

(b)

Figure 8: Non-linear surface for plant monitoring: (a) three-dimensional view of temperature and fractal dimension, (b) three-dimensional view of pressure and fractal dimension.

6 Experimental Results

In this section, numerical simulations are given to demonstrate the workability of the 27 rules given in the previous section representing a monitoring expert system.

Case 1. In this case, the first rule for normal condition is tested. The temperature profile $x_1(t)$, and pressure profile $x_2(t)$ are generated according to the following equations:

$$x_1(t) = 100 + 4\sin(\pi t/20)e(-t/20) + n(t) \qquad (7)$$

$$x_2(t) = 120 + 40\sin(\pi t/20)e(-(t-20)/20) + n(t) \qquad (8)$$

Where the function $n(t)$ is random white noise and $e(t)$ the exponential function of the form:

$$e(t) = e^t \text{ if } 0 < t < 100, e(t) = 0 \text{ otherwise.} \qquad (9)$$

The $x_3(t)$ variable is calculated from the time series of the $x_1(t)$ and $x_2(t)$ variables using the fractal dimension concept. The above formulas create three separate time intervals, $I_1 = [7, 14]$, $I_2 = [27, 34]$, and $I_3 = [47, 54]$. These data profiles will result in the first rule being valid in I_1, the second rule being valid in I_2, and the third rule being valid in I_3. Thus, the corresponding membership function should yield the value 1 for $t \in I_1$ and the value 0 for everywhere else. Similarly, for the other two intervals.

Case 2. In this case, the rule $R^{(20)}$ is tested. The temperature profile $x_1(t)$, and pressure profile $x_2(t)$ are generated according to the following equations:

$$x_1(t) = 100 + 4\sin(\pi t/40)e(-t/20) + n(t) \qquad (10)$$

$$x_2(t) = 120 - 40\sin(\pi t/40)e(-t/20) + n(t) \qquad (11)$$

Where the function $n(t)$ is random white noise and $e(t)$ the exponential function described in the previous case. These formulas create an interval $I_4 = [10, 30]$ where the rule $R^{(20)}$ will be valid, i.e., the corresponding membership function should yield the value 1 for $t \in I_4$ and the value 0 for everywhere else.

Case 3. In this case, the rule $R^{(26)}$ is tested. The temperature profile $x_1(t)$, and pressure profile $x_2(t)$ are generated according to the following equations:

$$x_1(t) = 100 + 4\sin(\pi t/40)e(-t/20) + n(t) \qquad (12)$$

$$x_2(t) = 120 + 40\sin(\pi t/40)e(-t/20) + n(t) \qquad (13)$$

Where the function $n(t)$ is random white noise and $e(t)$ the exponential function described in case 1. These formulas create an interval $I_4 = [10, 30]$ where the rule $R^{(26)}$ will be valid, i.e., the corresponding membership function should yield the value 1 for $t \in I_5$ and the value 0 for everywhere else.

Figures 9 and 10 show the plot of the data $x_1(t)$ for temperature, and $x_2(t)$ for pressure for Case 1. In the first 20 seconds, $x_2(t)$ is in the normal range, and $x_1(t)$ is gradually moving toward the high range, causing the membership function of condition C_2 to rise toward 1. Similarly, during the time interval [20,40], $x_1(t)$ is in the normal range, and $x_2(t)$ is gradually moving toward the high range, causing the membership function of condition C_3 to rise toward 1; and during the time interval [40,60], $x_1(t)$ and $x_2(t)$ are in the normal range, causing the membership function of condition C1 to rise toward 1.

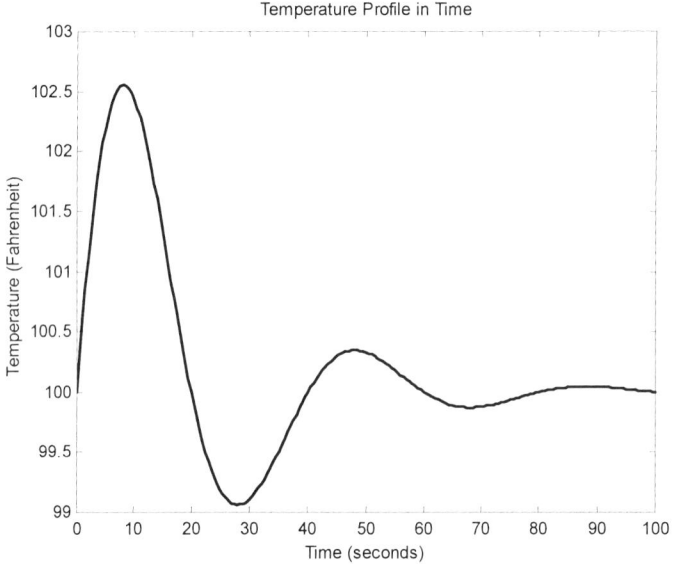

Figure 9: Temperature profile in time for case 1.

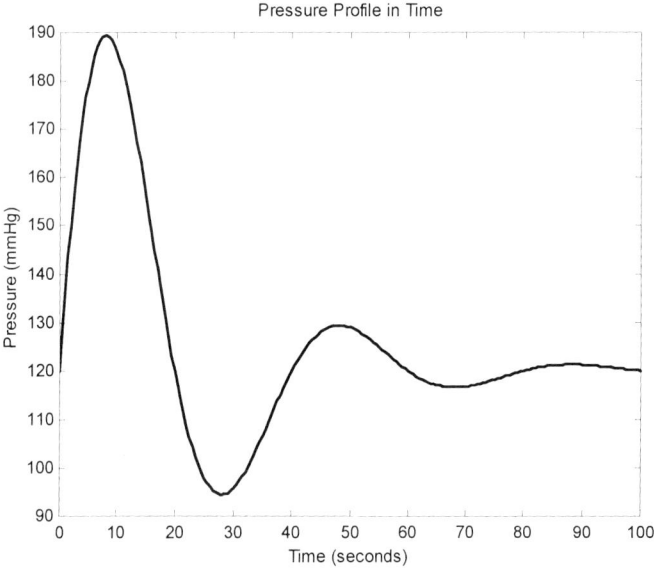

Figure 10: Pressure profile in time for case 1.

Based on the examples presented in this section, we see that using fuzzy logic in monitoring and diagnostics always results in improved performance. Also, the use of the fractal dimension improves the accuracy of the method. We have compared the success rate of the fuzzy-fractal approach against the use of only fuzzy logic, using the simulated data described before. The results are shown in Table 1. We are using in all of the cases a specific biochemical reactor. The reactor is considered to be in a different condition in each of the three cases. The comparison is between the results of the intelligent system using the fuzzy-fractal approach, and a computer program using only fuzzy logic with the Mamdani approach.

Table 1: Success Rate of the fuzzy-fractal approach for monitoring.

Applications	Fuzzy-Fractal	Fuzzy Logic
condition monitoring in a biochemical reactor (case 1)	98%	82%
condition monitoring in a biochemical reactor (case 2)	86%	73%
condition monitoring in a biochemical reactor (case 3)	90%	79%

It should be pointed out that no matter what techniques are used, there is no guarantee of success because monitoring and diagnosis is a process of abduction. First, the training samples may not represent all the patterns of different system conditions. An effective solution to this problem is to add more training samples. Second the patterns of different system conditions overlap and/or are inseparable owing to the definition of system conditions and the use of monitoring indices.

Finally, it is interesting to compare the performance of the fuzzy-fractal approach with that of using only fuzzy logic (see Table 1). We see that the fuzzy-fractal approach outperforms the fuzzy logic approach by at least 10% in all the cases. This demonstrates that the fuzzy-fractal approach is indeed a more effective method and, in general outperforms the use of fuzzy logic alone.

7 Conclusions

In this paper a hybrid fuzzy-fractal approach for plant monitoring has been proposed. An implementation in MATLAB has been shown, to describe in more detail the advantages of the new approach. The hybrid fuzzy-fractal approach combines the advantages of fuzzy logic (expert knowledge representation) with the advantages of the fractal dimension concept (ability to measure object complexity), to achieve efficient monitoring and diagnostics. A problem yet to be considered, is how to automatically learn (or adapt) the membership functions and rules of the fuzzy system using real data for the problem. A genetic approach could be used to evolve the fuzzy system (including the fractal dimension).

Acknowledgements

We would like to thank the Research Grant Committee of COSNET for the financial support given to this project (under grant 487.02-P). We would also like to thank the Department of Computer Science of Tijuana Institute of Technology for the time and resources given to this research project.

References

[1] Bezdek, J.C. (1981). Pattern Recognition with Fuzzy Objective Function Algorithms, Plenum Press.
[2] Brock, W.A., Hsieh, D.A., and LeBaron, B. (1991). "Nonlinear Dynamics, Chaos and Instability", MIT Press, Cambridge, MA, USA.
[3] Castillo, O. and Melin, P. (1994). Developing a New Method for the Identification of Microorganisms for the Food Industry using the Fractal Dimension, *Journal of Fractals*, **2**, No. 3, pp. 457-460.
[4] Castillo, O. and Melin, P. (1996). "Automated Mathematical Modelling for Financial Time Series Prediction using Fuzzy Logic, Dynamical System Theory and Fractal Theory", Proceedings CIFEr'96, IEEE Press, USA, pp. 120-126.
[5] Castillo, O. and Melin P. (1998). "A New Fuzzy-Genetic Approach for the Simulation and Forecasting of International Trade Non-Linear Dynamics", Proceedings of CIFEr'98, IEEE Press, New York, NY, USA, pp. 189-196.
[6] Castillo, O. and Melin, P. (1998). "A new Fuzzy-Fractal-Genetic Method for Automated Mathematical Modelling and Simulation of Robotic Dynamic Systems", Proceedings of FUZZ'98, IEEE Press, USA, Vol. 2, pp. 1182-1187.
[7] Castillo, O. and Melin, P. (2001). Soft Computing for Control of Non-Linear Dynamical Systems, Springer-Verlag, Heidelberg, Germany.
[8] Castillo, O. and Melin, P. (2003). Soft Computing and Fractal Theory for Intelligent Manufacturing, Springer-Verlag, Heidelberg, Germany.
[9] Chen, G. And Pham, T. T. (2001). "Introduction to Fuzzy Sets, Fuzzy Logic, and Fuzzy Control Systems", CRC Press, Boca Raton, Florida, USA.
[10] Du, R. (1998). "Engineering Monitoring and Diagnosis", Edited Book " Industrial and Manufacturing Systems", Academic Press, San Diego, California, USA, pp. 119-144.
[11] Du, R., Elbestawi, M. A., and Wu, S.M. (1993). "Automated Monitoring of Manufacturing Processes", Transactions ASME J. Eng. Ind., pp. 245-252.
[12] Jang, J.R., Sun, C.T. and Mizutani E. (1997). Neuro-Fuzzy and Soft Computing, Prentice Hall.
[13] Mandelbrot, B. (1987). "The Fractal Geometry of Nature", W.H. Freeman and Company.
[14] Melin, P. and Castillo, O. (1998). "An Adaptive Model-Based Neuro-Fuzzy-Fractal Controller for Biochemical Reactors in the Food Industry", Proceedings of IJCNN'98, IEEE Computer Society Press, Alaska, USA, Vol. 1, pp. 106-111.
[15] Melin, P. and Castillo, O. (2002). Modelling, Simulation and Control of Non-Linear Dynamical Systems, Taylor and Francis, London, Great Britain.
[16] Yager, R. and Filev, D. (1994). Generation of Fuzzy Rules by Mountain Clustering, Intelligent and Fuzzy Systems, Vol. 2, No. 3, pp. 209-219.
[17] Zadeh, L. A. (1965). "Fuzzy Sets", *Information and Control*, 8, pp. 338-353.
[18] Zadeh, L. A. (1971). "Quantitative Fuzzy Semantics", *Information Sciences*, 3, pp. 159-176.
[19] Zadeh, L. A. (1973). "Outline of a New Approach to the Analysis of Complex Systems and Decision Processes", *IEEE Transactions on Systems, Man, and Cybernetics*, 3(1), pp. 28-44.
[20] Zadeh, L. A. (1975). "The Concept of a Linguistic Variable and its Application to Approximate Reasoning", *Information Sciences*, 8, pp. 43-80.

Linguistic Modeling of Physical Task Characteristics

Sofia Visa[1], Anca Ralescu
ECECS Dept. ML 0030
Univ. of Cincinnati
Cincinnati, OH 45221, USA
{svisa,aralescu}@ececs.uc.edu

Simon Yeung
Dept. of Rehab. Sci.
HK Polytech. Univ.
Hong Kong
rssyeung@polyu.edu.hk

Ash Genaidy
MINE Dept. ML 0072
Univ. of Cincinnati
Cincinnati, OH 45221, USA
Ash.Genaidy@uc.edu

Abstract

This paper describes an approach to linguistic modeling based on fuzzy sets applied to modeling the meaning of linguistic descriptions for lifting task variables. The model aims at capturing the correspondence between numerical values and their qualitative/linguistic descriptions as perceived by human experts (workers performing lifting). The performance of the model obtained is compared to that using a neural network approach.

Keywords: fuzzy models, neural networks, lifting task variables.

1 Introduction

The current study is motivated by the desire to model the perception of the lifting task variables by manual workers whose activity consists in frequent lifting. The objective is to obtain a correspondence between the actual physical characteristics of the lifting task and its perception, as expressed in words. Modeling this correspondence is part of a larger study in the field of manual labor safety, in which the relationships between workers' abilities, tasks characteristics are studied with the goal to achieve better task design, so as to minimize the risks (and therefore injuries) associated with performing of such tasks. So far, research in this field has concentrated on developing an equation with seven variables associated with lifting (the NIOSH lifting task equation), [6], using knowledge from labor safety experts. No perception of the task difficulty from the actual workers has been included in that approach. Yet, it seems natural to assume that an actual worker's experience with lifting tasks should be useful in describing the perception of the effort required for such a task. Moreover, it is also reasonable to assume that the individual worker characteristics, not all of which can be explicitly captured, should affect this perception.

The values of the lifting task variables are described by words rather than exact numbers. For example, weight of load (W) is a variable which determines the difficulty of lifting. Verbal descriptions of measurements, such as *light, medium light, heavy, medium heavy, very heavy*, etc. are natural to people. Yet, these descriptions are

---

[1] Partial support of this work was received by a Graduate Fellowship from Ohio Board of Regents

subjective, the quantities which one description applies to vary, and different such descriptions may describe, eventually to a *different degree*, the same value.

If a computer based approach is to capture the correspondence between the actual values of a variable and their linguistic description, a paradigm that accommodates both the numeric and symbolic/linguistic descriptions of values, that allows for categories to be both distinct and somewhat overlapping, that distinguishes between members of the same category is needed. Fuzzy sets theory [9] is one such paradigm which has proved successful in many similar problems [8] and is adopted in this study. Although the use of fuzzy sets for the purpose described above is not new, their use in this particular application is.

The remainder of this paper is organized as follows: Section 2 presents a brief introduction to fuzzy sets, their definition, the relation of fuzzy sets to probability distributions and the use of this to determining the membership function of a fuzzy set. The actual application, the lifting task variables data and the use of Section 2 to this application are described in Section 3. For comparison purposes, a feed forward neural network, trained on the same data is described in Section 4. A data adjustment step and its effects both on the fuzzy model and the neural network are presented in Section 5.

2 Elements of fuzzy sets theory

Fuzzy sets were introduced in 1965 [9]. They aim at capturing and describing concepts that do not have sharp boundaries. Such natural concepts abound in the real world and are to be distinguished from man-made, technical (mathematical) concepts defined in terms of necessary and sufficient conditions. Given a universe of discourse, S, a (classical) set A is identified by an indicator function, $I_A : S \rightarrow \{0,1\}$ such that $I_A(x) = 1$ if $x \in A$ and $I_A(x) = 0$ otherwise. Similarly, a fuzzy set, $\overline{A}$ is identified by a membership function, $\mu_{\overline{A}} : S \rightarrow [0,1]$ such that $\mu_{\overline{A}}(x)$ is the degree to which x is in $\overline{A}$. Fuzziness arises in the process of assigning linguistic descriptions to numerical values, and is due what can be called lack of definition. Fuzzy sets and classical sets do share certain similarities but they are also different in what they can capture. Table 1 summarizes the features of these two representations.

Table 1: Fuzzy sets versus classical sets to model linguistic labels.

Features	Fuzzy Sets	Classical Sets
Overlapping	✓	x
Exhaustive	✓	✓
Between-label distinction	✓	✓
Within-label distinction	✓	x

2.1 Determining the membership function

An important issue in connection with fuzzy sets is that of determining their membership function. It can be seen that, given a label, e.g. *tall* for the variable height, the membership function that describes it is not necessarily unique. More precisely, while it is accepted that for heights under a certain value, α, the membership function is 0, for heights exceeding a value $\beta > \alpha$ it is 1, and we expect it to be nondecreasing between α and β, the exact values as well as exact definition in the interval $[\alpha, \beta]$ are not fixed, their choices reflecting various conditions under which the concept *tall* may be considered. Without doubt there is a *statistical* aspect underlying the notion of a fuzzy set, in the sense that several different in- stances must have been given as examples of the concept to be represented as a fuzzy set (alternatively, the notion of degree of membership seems to make little sense, when all that is known about such a concept is only one instance). It is therefore natural to relate the notion of fuzziness to the notion of frequency and, moreover, to do so in a formal manner. This idea is at the basis of work done by [1], in which a formal correspondence between fuzzy sets on a universe of discourse and a probability distribution on it is established. The same was further studied in [2] and was the basis for an approach to derive subjective membership functions [3], [5]. For the current problem the basic mechanism for converting a relative frequency distribution (discrete probability distribution) into fuzzy sets is used. This mechanism can be presented independently of the more advanced aspects of the general theory.

From this point on A denotes a discrete fuzzy set with membership function $\mu_A(x_i)$, $i = 1,...,n$. For $0 < \alpha \leq 1$ the α-level, A_α is the crisp set defined as $A_\alpha = \{x \mid \mu_A(x) \geq \alpha\}$. The representation theorem [4] shows how a fuzzy set and its level sets are connected. In particular, $\{A_\alpha\}_{\alpha \in (0,1]}$ is a collection of nested sets such that $A = \bigcup_\alpha A_\alpha$. Let $0 < \alpha_i \leq 10$ denote a sequence of non-increasing levels, $\alpha_1 \geq \alpha_2 \geq ,..., \geq \alpha_n$, and A_i their corresponding level sets. It can be verified that $\sum_{i=1}^{n}(\alpha_i - \alpha_{i+1}) = \alpha_1$, and therefore, it is equal to 1, when $\alpha_1 = 1$.

Let $\mu_{(i)}$ denote the values of $\mu_A(x_i)$ arranged in nondecreasing order and A_i the level set corresponding to $\mu_A(x_i)$. For simplicity, it is assumed that $\mu_{(i)} = \mu_A(x_i)$. Then $A_i = \{x_1,...,x_i\}$. Let p_{ik} denote a selection rule from the level A_i as defined in [5]. That is, $p_{ik} = p_i(x_k)$, $k = 1,...,i$, denotes the probability of selecting the element x_k from the level set A_i such that $0 \leq p_{ik} \leq 1$ and $\sum_{i=1}^{n} p_{ik} = 1$. Let $m : \wp(S) \to [0,1]$ be defined as

$$m(A) = \begin{cases} \alpha_i - \alpha_{i+1} & \text{if } A = A_i \\ 0 & \text{otherwise} \end{cases} \tag{1}$$

and as in [3], let $f_k : S \to [0,1]$ be defined as

$$f_k(x) = \sum_{i=1}^{n} m(A_i) \times p_{ik}(x) \tag{2}$$

It can be easily verified that $0 \le f_k(x) \le 1$ and that $\sum_{k=1}^{n} f_k(x) = 1$. That is, $\{f_k\}_k$ is a probability distribution on $\{x_1,\ldots,x_n\}$. It can also be seen easily that $\{x \mid f_k(x) \ge f_k(x_i)\} = A_i$. As shown in [5] (2) can be used to derive the membership function μ_A when the frequency f_k and the selection rules p_{ik} are known. The special case, $p_{ik} = 1/|A_i| = 1/i$, usually referred to as the *least prejudiced distribution (lpd)*, was known for quite some time in the literature of fuzzy sets (see for example [1]). A discussion of the general case can be found in [3], [2]. For the (lpd) case it follows from (2) that the membership function μ_A is determined (setting $f_{n+1} = 0$) by (3).

$$\mu_{(k)} = k f_{(k)} + \sum_{i=k+1}^{n} f_{(i)} \quad \text{for } k = 1,\ldots,n \tag{3}$$

In (3) $f_{(k)}$ and $\mu_{(k)}$ denote the kth largest value of the frequency distribution and membership function respectively.

3 Fuzzy sets for the lifting task variables

Using the mechanism outlined above the data for seven lifting task variables were assigned to three linguistic labels on their corresponding universe of discourse, represented as fuzzy sets.

3.1 Data collection

The data set used in this study, obtained from a questionnaire of 217 workers carried out at the Hong Kong Polytechnic, consists of 4,557 lines of input: each individual assessed the meaning of three linguistic labels for each of the seven lifting task variables. A typical data item consists of a numeric range for the value of the variable and the linguistic description assigned by a worker. Table 2 shows the labels assigned to values of the variable *horizontal distance* (HD). It can be seen from this table the variability of values within labels as well as the variability of labels for the same value.

Table 2: Between-label and within-label variability for the variable horizontal distance (HD).

Label	Value
S	30-40
S	40-45
S	28-53
M	40-65
M	50-60
M	35-45
B	33-45
B	50-59
B	48-63

A mild preprocessing step was applied to data which eliminated what was deemed as incorrect: entries in which one of the endpoints of the interval for values of the variable was not specified, or the endpoints were specified in reverse order. The final data set, shown in Table 3, for each variable and, for the fuzzy set approach, by labels within each variable, has 4,232 entries. Of these, half, selected randomly, are used to model the fuzzy sets and train the neural network. The remaining are used with each model for testing purposes. The corresponding domains of the fuzzy sets are contiguous intervals of the real line obtained from the initial data (intervals which may overlap) through a step of refinement. For two intervals I and J the refinement operation is defined by (4).

$$r(I,J) = \begin{cases} \{I,J\} & \text{if } I \cap J = \emptyset \\ \{I \setminus J, I \cap J, J \setminus I\} & \text{otherwise} \end{cases} \quad (4)$$

The frequency distribution is then calculated on the newly obtained set of intervals. Example 1 illustrates this step on a small data set.

Example 1 *Let* {[2,6],[1,3],[4,7],[5,6]} *be the data provided by different subjects for a linguistic label l. Successive overlapping of these intervals, results in the following collection of non overlapping intervals* {[1,2],[2,3],[3,4],[4,5],[5,6],[6,7]} *with the corresponding frequencies* {1,2,1,2,3,1}.

Table 3:
Breakdown of the data by variables and variable labels; floor level (FW), waist level (W), horizontal distance(HD), twisting angle (TA), frequency(F), work duration(WD), vertical distance(VD).

.Variable	Total data points	Data for fuzzy sets per label	Data sets for the NN
FW	512	S:176 M:168 B:168	256 train 256 test
W	480	S:160 M:160 B:160	240 train 240 test
HD TA F WD VD	648 for each variable	S:216 M:216 B:216	324 train 324 test

Table 4 lists the seven lifting tasks variables along with dictionaries of linguistic labels, and statistics for these labels for the samples used. Figures 1 and 2 show the frequency distributions and the fuzzy sets for the two variables highlighted in Table 4.

3.2 Testing the fuzzy model

The fuzzy sets obtained are used for classification of a given data point as follows. Given the value x, a point or interval value, for a lifting variable the following steps are applied:

1. <u>Match x to each label</u>: Calculate the degrees $\mu_S(X), \mu_M(X), \mu_B(X)$.
2. <u>Assign x to a label</u>: Based on the degrees computed at the preceding point assign x to that category for which its degree is highest. More precisely, if pred_l denotes the label predicted for the data x, then

$$\text{pred\_l} = \arg\{\max_{L \in \{S,M,B\}} \mu_L(X)\} \quad (5)$$

Table 4: Characteristics of data.

Variable name/symbol	Range within labels	Mean within labels	Standard Deviation	Dictionary of labels
floor weight/FW	10-45 4-25 2.27-14	18.2-24 9.8-15.9 2.7-7.4	3-3.6 2.2-2.9 0.9-2.1	
waist weight/W	10-40 5-25 2.27-17	19.2-24.2 11.1-17.2 3-8.7	2.87-2.82 2.5-2.9 1.5-2.6	heavy/big/**B**
horizontal distance/HD	33-75 25-70 25-55	52.7-65.3 38.3-49.6 25.5-35.2	8.1-9.4 5.5-7.9 1.7-5.2	moderate/**M**
twist angle/TA	21-180 11-80 10-51	56-78.3 32.3-51.4 10.8-27.1	13.6-16.9 8.6-12.5 2.6-7.6	
frequency/F	4-45 1-40 0-16	11.5-16.5 5.9-10.5 0.6-4.8	3.7-4.8 2.6-3.9 1.3-2.4	light/small/**S**
duration/D	1-8 0.5-7 0.3-3	4.3-6.4 2.2.-3.8 0.3-1.6	1.4-1.6 0.9-1.3 0.3-0.8	
vertical distance/VD	55-200 36-175 25-205	106.8-146.7 54.3-98.7 27.4-55.7	23.9-30.1 17.2-22.8 7.4-14.8	

3.3 Classification error

Due to the variability of descriptions among various subjects, and to the overlap between various labels, a given value x may be misclassified. To track the errors of classification a simple error model (6) is used.

$$\text{error}(x) = \begin{cases} 0 & \text{pred\_l}(x) = \text{actual\_l}(x) \\ 1 & \text{otherwise} \end{cases} \quad (6)$$

For a collection of data, V, the overall error is simply the average number of errors (7).

$$\text{error}(V) = \frac{\sum_{x \in V} \text{error}(x)}{|V|} \quad (7)$$

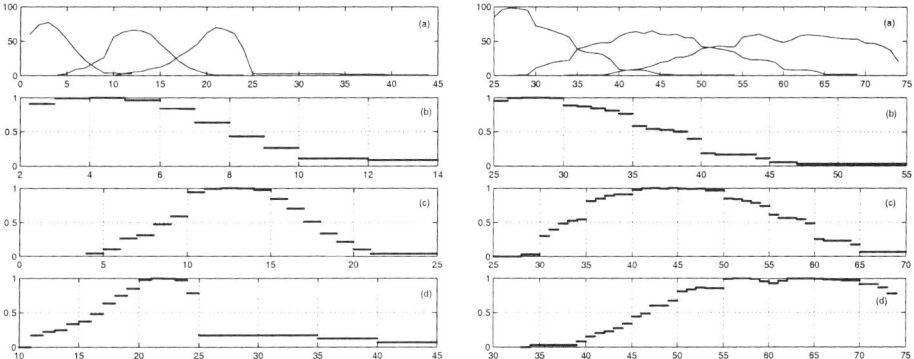

Figure 1: Frequency distribution (a) and fuzzy sets (b, c, d) for the FW variable.

Figure 2: Frequency distribution (a) and fuzzy sets (b, c, d) for the HD variable.

3.4 Modeling error versus predictive error

For fuzzy models (as well as in other approaches, e.g. neural networks) the error of classification cannot be excluded even for the training data. This is known as the *modeling error* and the corresponding correct classification of the training data is known as *modeling power*. For the test data these are called *generalization error* and *generalization power* respectively. The idea behind modeling approaches such the one used in this study is to *give up some of the modeling power, in order to achieve acceptable levels of generalization power of the model.*

Table 5 shows the generalization error of the fuzzy sets based model for the seven variables considered. The modeling error for the two variables highlighted in Table 5, FW (lowest error) and HD (largest error), are 5.5% and 15.75% respectively. Note that these are cumulative errors for each variable, and that within each label, the errors are much smaller. It should also be noted that the fuzzy model loses very little from its power when it is applied to test data.

The overlap between data evident from Table 4 indicates how fuzzy the labels and therefore, the underlying fuzzy sets are. It can be seen that the extent to which data corresponding to different labels overlap affects the accuracy of the prediction.

Table 5: Generalization error of the fuzzy model. For the variables FW and HD corresponding to best and worst case respectively, the error is shown for each label.

Variable	% error for test data
FW	5.5 $\begin{cases} 2.9 - B \\ 1.9 - M \\ 0.7 - S \end{cases}$
W	6.7
TA	15.75
F	15.13
D	15.2
VD	13.28
HD	15.75 $\begin{cases} 8.93 - B \\ 5.42 - M \\ 1.4 - S \end{cases}$

4 Neural network modeling of the lifting task variables

For comparison purposes, neural networks were trained to capture the meaning of the linguistic labels for each variable. The training and test data are shown in Table 3. Feed forward neural networks, with one hidden layer of ten neurons, sigmoidal activation function, back-propagation learning algorithm, training error 0:1, two input values and one output were used. The input values correspond to the endpoints of the interval given as data to be assigned to a linguistic label. The output, corresponding to the class label, is 1 for *Small*, 0 for *Medium* and +1 for *Big* respectively.

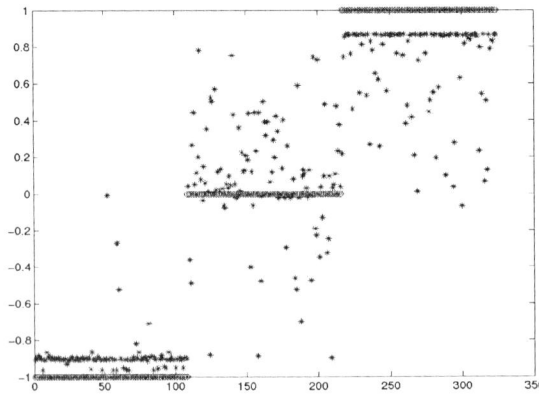

Figure 3: Neural networks results.

4.1 Testing the feed forward neural network model

After training, for each input data the neural network outputs a value in the interval $[-1,1]$. A particular test data, the endpoints of an interval, is classified to the label **S**, **M**, **B** according to whether the corresponding output is *close enough* to $-1, 0$, or 1 respectively. This rule of thumb is implemented simply by selecting a threshold such that if $NN_V(x)$ denotes the neural network output for the variable V, at x, x is assigned the label L_0 if $|NN_V(x) - L_0| \leq \gamma$.

Figure 3 shows the neural network prediction results for the twisting angle (TA) variable. These results are typical for all variables. It can be seen from these that the neural network model distinguishes better between the labels **S** and **M** than between **M** and **B**. With the error model of (6) and (7) and $\gamma = 0.5$ for FW, W, HD, TA, F, and $\gamma = 0.4$, for D and VD the best prediction power of the neural network approach compared to that of the fuzzy approach, is shown in Figure 4 and Table 6 which show that fuzzy set and neural network approaches have comparable performance.

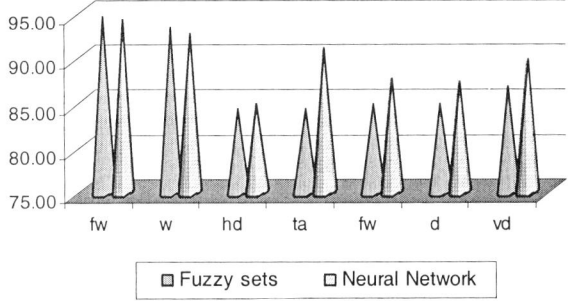

Figure 4: Comparison of the prediction power for fuzzy sets and neural network.

The selection of the threshold values requires *an added step after the training of the network*. No such step is required by the fuzzy model. From a different point of view, once an input data is to a label is done, the neural network cannot distinguish between different instances of the same label, implementing in effect, each label as a crisp category. Alternatively, one could come up with a way of differentiating between the elements of the same label. However, such a step would be yet another additional step and, more importantly, it would be *ad-hoc* without a clear meaning.

Table 6: Complete results for fuzzy sets and neural networks.

Variable name	Fuzzy sets	Neural networks	γ
FW	**94.5**	**94.15**	**0.5**
W	93.30	92.5	0.5
TA	84.25	91.04	0.5
F	84.87	87.65	0.5
D	84.80	87.3	0.4
VD	86.72	89.81	0.4
HD	**84.25**	**84.87**	**0.5**

5 Data adjustment for improving the fuzzy model

Although the fuzzy model performance is satisfactory, both from the point of view of the modeling and prediction power, and by comparison to the neural network selected, it is interesting to investigate if indeed this performance cannot be improved. The performance of the fuzzy model depends on the variability within labels (as illustrated in Table 2). This variability is further augmented by the procedure of extracting the unit intervals from each data item.

Table 7: Adjustment of variable intervals for the variable HD.

Label	Initial value → Adjusted value
S	30-40 → **25**-40
	40-45 → **25**-**55**
	28-**53** → **25**-**32**
M	40-65 → 40-65
	50-60 → 50-60
	35-45 → 35-45
B	33-**45** → 33-**75**
	50-**59** → 50-**75**
	48-**63** → 48-**75**

To reduce this variability a *rationality assumption* is introduced as follows: When a subject estimates the interval $[a,b]$ as his/her perception of the concept *small/light* assume that this subject will also accept any interval $[x,b]$ for any $x \le a$. It is possible that the reason for which the subject produced the interval $[a,b]$ is that for values $x < a$ this subject would have associated another label, e.g. very small. Yet in the experiments which produced the data for this study there was no other option except that of using labels from the set $\{S, M, B\}$. A similar adjustment, of the upper limit of the interval, is done for intervals which are used to estimate the label *big/large*. Intervals corresponding to the label *medium* are not adjusted. Equation (8) describes the above adjustment procedure.

$$\mathrm{adj}([a,b],L) = \begin{cases} [m_l, b] & \text{if } L = S \\ [a, b] & \text{if } L = M \\ [a, M_l] & \text{if } L = B \end{cases} \quad (8)$$

where $m_l = \min_L\{a;[a,b]\}$ and $M_l = \max_L\{b;[a,b]\}$.

Table 7 illustrates the result of the adjustment for the data in Table 2. The recognition procedure remains the same as for the case of the original data.

The neural network described in Section 4 was also trained (with training error 0.1) and tested (with $\gamma = 0.5$) on the adjusted data. The complete results for the fuzzy model and the neural network are in Table 8.

Table 8: Prediction results for fuzzy sets and neural networks on the adjusted data.

Variable name	Fuzzy sets	Neural networks
FW	**95.4**	**98.28**
W	95.25	98.33
TA	91.66	96.29
F	87	87.34
D	87.4	87.6
VD	91.95	90.74
HD	**91.5**	**98.14**

6 Conclusions

This study addressed the problem of modeling human perception of the linguistic descriptions of numerical values. Two approaches, a fuzzy set and neural network approach, were considered and their results evaluated. From these results it can be concluded that the two approaches give similar results and that the choice of one over the other may depend on several factors, including the goal of the modeling, further use of the results, etc. Unlike the neural network approach, the fuzzy model requires no additional steps beyond training in order to classify given data. However, more importantly it was also found that it converges much faster. Not included in this paper are results showing the learning curves for the fuzzy model: the generalization power of the model exceeds 84% for training sets as small as 30% of the data set. An application in the domain of manual labor workers' safety was considered. From this point of view the use of both approaches is novel. Although the fuzzy approach is more transparent and easier to link with the actual data, further research in the application domain is needed to decide which of these two approaches is more suitable.

Reference

[1] J. Baldwin, T. Martin and B. Pilsworth. Fuzzy and Evidential Reasoning in Artificial Intelligence, John Wiley and Sons Inc., 1995. 1.
[2] A. Inoue and A. L. Ralescu. The Associative Nature of Perceptual Information Processing. In *Proceedings of Artificial Neural Networks In Engineering (ANNIE)*, November 1999.
[3] A. Inoue and A. L. Ralescu. Generation of Mass Assignment with Nested Focal Elements. In *Proceedings of 18th International Conference of the North American Fuzzy Information Processing Society (NAFIPS-99)*, 208-212, June 1999.
[4] C. V. Negoita and D. A. Ralescu. Representation Theorems for Fuzzy Concepts. *Kybernetes*, 4, pages 169-174, 1975.
[5] A. L. Ralescu. Quantitative Summarization of Numerical Data Using Mass Assignment Theory. Invited lecture at the SOFT Kansai Meeting, May 1997.

[6] T. Waters, V. Putz-Anderson, A. Garg and L. Fine. Revised NIOSH equation for the design and evaluation of manual lifting tasks. *Ergonomics*, 36, 749-776, 1993.

[7] S. Yeung, A. Genaidy, W. Karwowski, R. Huston and J. Beltran. Assessment of manual lifting activities using worker expertise: A comparison of two workers population. *Asian Journal of Ergonomics*, 2, 11-24, 2001.

[8] L. A. Zadeh. Fuzzy Logic = Computing with Words. *IEEE Transactions of Fuzzy Systems*, 4.2, 103-111, 1996.

[9] L. A. Zadeh. Fuzzy Sets. *Information Control*, 8, 338-353, 1965.

B. Bouchon-Meunier, L. Foulloy and R.R. Yager (Editors)
Intelligent Systems for Information Processing:
From Representation to Applications
© 2003 Elsevier Science B.V. All rights reserved.

Validation of Diagnostic Models Using Graphical Belief Networks

Oscar Kipersztok[1]   Haiqin Wang

Mathematics and Computing Technology
Phantom Works, The Boeing Company
P.O.Box 3707, M/C: 7L-44
Seattle, WA 98124
oscar.kipersztok@boeing.com   haiqin.wang@boeing.com

Abstract

A system that facilitates airplane maintenance provides decision support for finding the root-cause of a failure from observed symptoms and findings. Such system provides diagnostic advice listing the most probable causes and recommending possible remedial actions. Furthermore, its goal is to reduce the number of delays and cancellations and unnecessary parts removal, which add significant costs to airplane maintenance operations. Bayesian belief networks, a model-based approach, are presently being used for building such diagnostic models. The paper describes the pertinent issues in using such models.

Keywords: Bayesian networks, diagnostics, model validation, sensitivity analysis.

1   Decision Support Methods for Diagnosis

Delays and cancellations add significant cost to airline maintenance operations. Unnecessary part removals, in addition, compound the problem. Much of this operational costs are attributed to a decrease in diagnostics ability of airline mechanics as a result of lack of experience with an increasing variety of airplane types in the fleet and the increasing practice of outsourcing maintenance operations by airlines.
The critical factors in commercial airline maintenance operations are airplane safety, dispatch reliability and turn-around time. To ensure safety and reliability, airline operators must adhere to government regulatory agencies' standards, which require a Minimum Equipment List (MEL) with the minimal set of Line Replaceable Units (LRUs) that must be in working order before dispatch is approved. In response to a reported fault and under time pressure to meet scheduled departures, the tendency of operators is to replace suspect parts unnecessarily. Seasoned mechanics can quickly narrow the list of possible causes to a small number of replaceable units. The challenge is to disambiguate between the most probable parts by further performing troubleshooting tests, before departure time. To avoid costly delays and potential cancellations, the mechanics have to decide what action to take before departure, which suspect LRUs should be replaced and what remedial actions can be deferred to the next destination.

---

[1] Correspondence author

A diagnostic decision support system for airplane maintenance should be designed to facilitate the decision process in such a way as to improve the accuracy of airplane diagnosis without compromising safety and reliability. This paper describes the basis for such system.

2 Decision Support Methods for Diagnosis

There are several methods for building diagnostic models, and tools to help build them. Model-based reasoning systems rely on physical models that describe the input/output relations between system sub-components and the fault-propagation dependencies between them [3,5,4]. Case-based reasoning systems, rely on historical references to associations between feature problem descriptions and actions taken to correct them [15,6]. Although several approaches incorporate measures of uncertainty to help resolve ambiguities, there are methods that are inherently probabilistic. One such method uses Bayesian belief networks to encode probabilistic dependencies between the variables of a diagnostic problem into the structure of a directed acyclic graph [16,12]. Such graph is capable of updating probabilities of component failures when evidence of a fault is observed. Other approaches to diagnosis include the use of rule-based expert systems, fuzzy logic, and neural networks, etc.[1,13,11].

In this paper, we suggest to define a diagnostic model as a transfer function between the causes of a problem and their observed effects. In airplane maintenance, the causes are LRUs, and the observed effects are either Flight Deck Effects (FDEs), which are failure triggered events visible to pilots in the cockpit, or other perceived anomalies such as unordinary sounds, smells or visible cues (e.g, smoke in the cabin). Once such function is defined, the diagnostic problem is reduced to that of computing the problem root-causes given the observed effects. In this manner, a diagnostic model is directly built to simulate the way a system fails, rather than to simulate the way the system deviates from its normal behavior. Airplane diagnosticians do not rely only on their systemic knowledge of the system, just as medical diagnosticians do not only rely on their understanding of the physiology and bio-chemistry of the body when seeing a patient. Beyond systemic knowledge, diagnosis-sis is also reliant on experiential knowledge accumulated over repeated exposure to similar problems and associations between causes and effects.

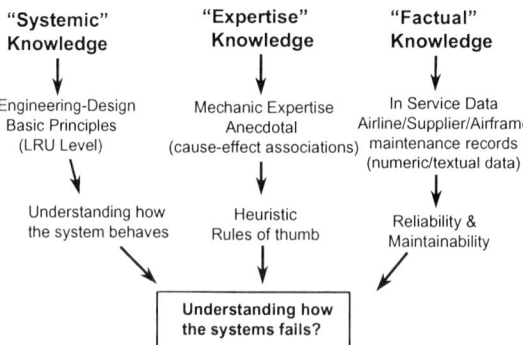

Figure 1. Three types of knowledge needed to diagnose complex airplane systems.

Figure 1 shows the three sources of knowledge that are critical for diagnosis of a complex system such as an airplane. First, the "systemic" knowledge, which entails the understanding of how the sub-components of the system relate to each other and operate under normal conditions, so it is possible to understand the different operational pathways conducive to failures. This is the type of knowledge possessed mostly by engineers responsible for designing and building the various systems. Secondly, the "experiential" knowledge, which entails the cause-and-effect associations learned over long periods of maintenance exposure and familiarity with the system. Mechanics and engineers who are the maintenance operators of the systems mostly possess this type of knowledge. And thirdly, the "factual" knowledge, which is a combination of text and numeric records that capture the actual field experience, i.e., the history of the actions taken in the field, and the component reliability data for each replaceable component. The latter is usually in the form of Mean Times Between Failures (MTBFs) or Unscheduled Removals (MTBURs). These three essential sources of knowledge provide the required information content for any comprehensive airplane diagnosis decision support system.

3 The Preflight Troubleshooting Process at the Airport Gate

Diagnosis at an airport gate is done as part of a decision support process to determine:
a) which LRUs, if possible, should be fixed on the ground before scheduled departure,
b) which LRUs should be replaced before scheduled departure, c) whether scheduled departure should be delayed to support either a or b and if so, for how long, and d) whether the flight should be cancelled all together.

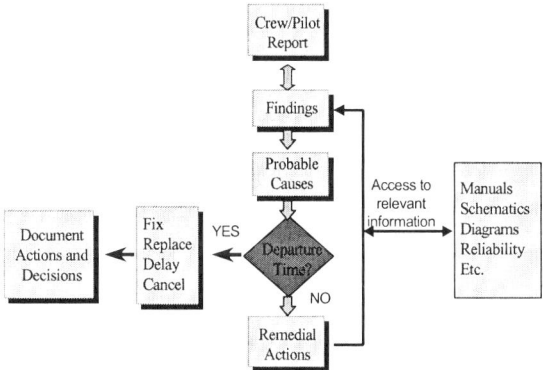

Figure 2. Maintenance process at the airport gate.

Described in Figure 2 is the maintenance cycle process that takes place at the airport gate. Preflight troubleshooting begins when the aircraft arrives and is scheduled to depart on an outgoing flight. If a failure is detected by the pilot or flight crew in the preceding flight, or by the maintenance crew while on the ground, troubleshooting begins to ensure safe and timely airplane dispatch. The deadline for decisions is the departure time of the next scheduled flight. Troubleshooting is the responsibility of several decision makers including Airline Maintenance Operation Control (MOC), the ground maintenance staff, and the airplane flight crew.

4   Building Bayesian Belief Networks for Airplane Diagnosis

Bayesian belief networks are directed acyclic graphs that capture probabilistic dependencies between the variables of a problem. Bayesian networks approximate the joint probability distribution over the variables of the diagnosis problem using the chain rule of probability, which, subject to simplifying conditional independence assumptions, results in the product of probabilities of the variables conditioned on their parents:

$$p(x_1, x_2, ..., x_n) = \prod_{k=1}^{n} p(x_k \mid pa(x_k)) \qquad (1)$$

where a set $X = \{ X_1, ..., X_n \}$ denotes the domain variables of interest, and $Pa(X_i)$ denotes the parents of a node X_i (the nodes pointing to X_i in the graph).

The general approach to building Bayesian networks is to map the fault causes (LRUs) to the observed effects (FDEs), keeping in mind that what is being modeled is not the normal behavior of the system but rather the behavior of the system when one or more of its parts fail.

The construction of a Bayesian network requires: first, the identification of nodes with associated discrete or continuous states, secondly, connecting the nodes with arcs according to their dependence relationships, and thirdly, establishing probability distributions for each node conditioned on the states of the parents. Figure 3 shows a segment of a Bayesian network in a model built for air-conditioning system diagnosis. From the figure, we can see that the connection channels between LRUs and FDEs often go through intermediate nodes.

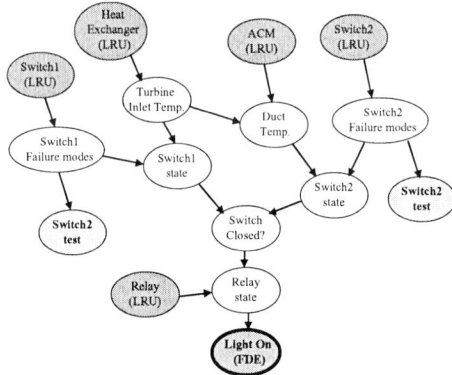

Figure 3. Section of a Bayesian network connecting causes (LRUs) to effects (FDEs).

The process of building such networks requires the elicitation of knowledge from domain experts. In the case of an airplane system diagnosis model the experts should represent the three types of knowledge shown in Figure 1.To improve knowledge elicitation in the creation of an airplane diagnostic model, we have found that the modeler must become familiar with the functionality and terminology of the system, and understand its behavior well enough to be conversant about it with the experts. One can achieve such level of understanding from system manuals (also used for training mechanics), maintenance manuals, and system schematics. Building the network requires to start from a list of the most problematic system faults.

The faults are not necessarily those that occur the most frequently, but rather, they are the most difficult to troubleshoot. Parentless nodes in the network are populated with prior probabilities derived from component reliability data. The data is available from various sources in the form of Mean Time Between Unscheduled Removals (MTBUR). These estimates can be converted into probability estimates using the exponential distribution assuming a Poisson process,

$$F(x) = 1 - e^{-\lambda x} \qquad (2)$$

where $1/\lambda$ is the long-term average life-time of the LRU and x can be interpreted as a single cycle of operation, equivalent, for example, to the average duration of the last flight leg. Typical values of MTBURs are of order greater than 10^5 hours. Since $\lambda \ll 1$, Equation 2 can be approximated by $F \sim \lambda$. These probability estimates constitute the priors for the LRU components. For nodes with parents, the conditional probability tables (CPTs) are elicited from expert opinion during the knowledge acquisition and building of the network.

The probability model of Equation 2 for the parentless priors of the network can certainly be improved to better reflect the true lifetime of the replaceable parts. Furthermore, better estimates can also be obtained by replacing MTBURs with the Mean Time Between Failures (MTBF). This will require a much higher cost since MTBF is a more difficult quantity to obtain. Whether it is necessary to improve the estimates of the priors can be addressed by conducting sensitivity analysis. With sensitivity analysis, one can assess how much noise can be tolerated for the prior estimates before it can significantly impact the diagnosis outcome of the Bayesian models.

Although the process of building Bayesian networks by hand is as much an art as it is a science, there are several reported methods and techniques that can help improve the efficiency and accuracy of the process by addressing knowledge acquisition issues [18,19,8–10]. Methods are also available to acquire the parameters of the network or even to derive its structure directly from data [2,7]. The latter approach currently being an active research area with results that offer significant promise [7].

5 Validation of Diagnostic Models

Three important validations that need to be performed when building Bayesian networks are, a) validation for knowledge accuracy, b) sensitivity analysis for robustness of the network's performance, and c) network evaluation for the quality of diagnostic recommendations obtained using the models.

5.1. Knowledge Accuracy

Bayesian networks are built using an iterative process that includes repeated knowledge testing as part of the process. The networks are tested for their ability to pro-vide accurate diagnosis, which for an airplane diagnosis system, it constitutes a partial-ordered list of probable causes ranked by probability. During knowledge validation of the models the experts must agree on the recommended diagnosis for each test-case scenario. The more test cases used, the better.

Knowledge validation is key to the performance of the network and should be an ongoing process throughout the life and use of the network. Techniques have been developed to evaluate elicitation schemes when building probabilistic models based on knowledge acquisition from domain experts [18]. Learning network parameters from numerical or textual data could be complementary to direct validation from the experts.

5.2. Sensitivity Analysis

In probabilistic models, robustness of the output is measured by sensitivity, i.e., how inaccurate probability values affect diagnostic performance of the models. For networks with high sensitivity to noise, the nominal diagnosis could advise the airplane maintainer to follow a series of irrelevant actions that could result in un-necessary and costly part removals, delays or cancellations.

Empirical study of sensitivity analysis on a Bayesian network examines the effects of varying the network's probability parameters on the posterior probabilities of the true hypothesis. One approach to modeling the variation of the probability parameters is to add normal noise to the log-odds of the nominal probabilities.

Equation 3 illustrates the probability density of the log-odds normal distribution, in terms of the nominal probability p and noise ε:

$$\log \frac{1-p'}{p'} = \log \frac{1-p}{p} + \varepsilon \qquad (3)$$

where p' is the noisy value of p after imparted noise ε, and $\varepsilon \sim N(0,\sigma)$. The relation between the noisy odds and the nominal odds is then:

$$\frac{1-p'}{p'} = \frac{1-p}{p} 10^{\varepsilon} \qquad (4)$$

or Odds' = Odds $*10^{\varepsilon}$, where Odds' = (1-p')/p' and Odds = (1-p)/p.

This indicates that error introduced by the log-odds normal noise ε reflects the scale of the change of odds by a factor of 10^{ε}. Therefore, for example, an approximate increase in odds of 25% due to noise in the priors corresponds to the noise level with a standard deviation of 0.1. Note that this model for probability noise is only valid for values of σ < 1, which allows for a unimodal of log-odds normal distribution. For values of σ > 1, the distribution becomes bimodal with values of zero and one. Using the distribution in that range is not valid to describe the noise on the probability values. As an expert is unlikely to err in estimate and misjudges a probability value as one, but in fact it is near zero. In [14], we give a more detailed explanation on how to model the variation of probability estimates in empirical sensitivity analysis.

Typically, for airplane diagnosis, the reliability estimates of most airplane parts is of order greater than 10^5 hours for the mean time between part failures. The corresponding prior probabilities are therefore approximately of order smaller than 10^{-5}. At such low probabilities the log-odds normal distribution is very asymmetric and the average rank does not adequately represent the effect that the noise imparts on the network. As an example, Figure 4 illustrates the lower-bound confidence estimates for confidence levels from 50% to 99% of the rank changes of the most probable failed parts in a production Bayesian network built for a particular airplane system diagnosis. In the figure, five levels of prior noise are used to affect the nominal network. As expected, performance degrades as the noise increases. The rank of the most probable failed part drops about one position when noise is distributed with standard deviation (std) of $\sigma = 0.1$. There is only a 10% chance that the most probable suspect part may drop in rank by more than two positions, due to inaccuracies in prior estimates. That is a reasonably small risk for that level of noise. Note that 50th percentile lower bounds are smaller than rank averages, further indicating the asymmetry of the noise distribution.

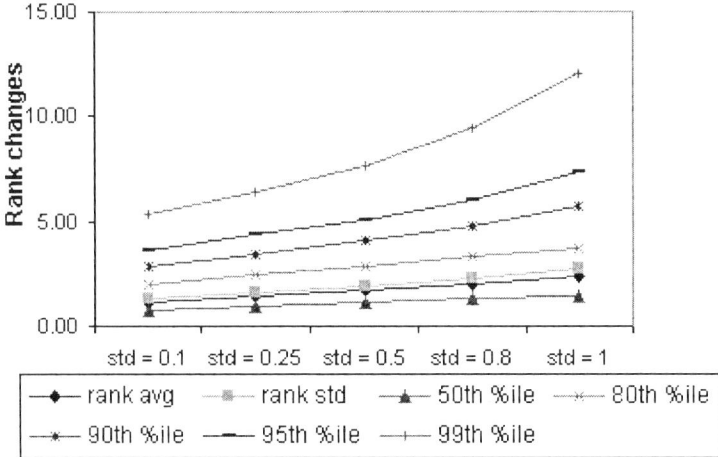

Figure 4. Posterior rank changes of the most probable failed parts in an airplane diagnosis network using 100 run cases across different test scenarios and prior noise.

The suggested diagnosis, then, should be considered according to the partial ordering of probable causes resulting from the update of the posterior probabilities given a set of findings. While the most probable cause is often given the highest consideration, typically, in multiple-fault diagnostic systems, it is the set of top causes (e.g. the top five) and their partial ordering that is most informative. Since very seldom the diagnosis singles out a particular cause, the partial ordering provides guidance for subsequent actions. The effect that noise has on the posterior partial ordering of the causes is, therefore, a significant measure of the network sensitivity. Hence, a suitable lower confidence bound on the posterior rank changes of the partial ordering is an adequate measure to assess the effect of noise on network probabilities. Figure 5 plots the posterior rank changes for the top five most probable failed parts at the noise level $\varepsilon \sim N(0, 1.0)$ in the same experimental test as Figure 4.

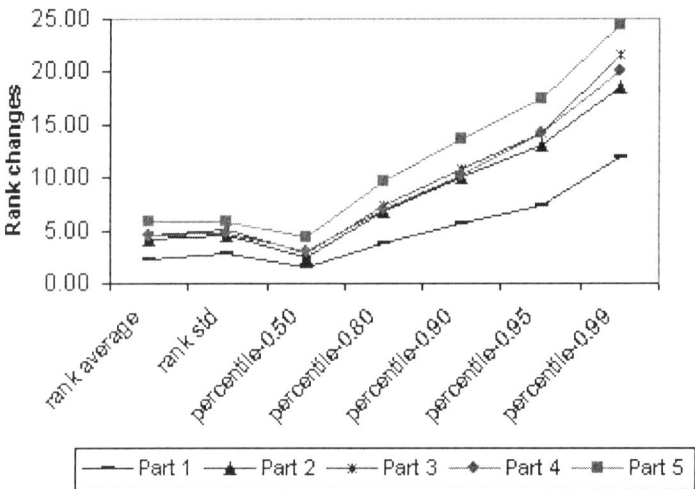

Figure 5. Posterior rank changes of the top five most probable failed parts in an airplane diagnosis network using 100 run cases across different test scenarios and prior noise
$\varepsilon \sim N(0, 1.0)$.

5.3. Network Evaluation

Prior to their use, Bayesian network models need to be tested for their diagnostic performance. In addition to their knowledge accuracy and sensitivity evaluations, it is also important to validate Bayesian networks for the quality of their recommended diagnosis, because the system being modeled may have diagnostic limitations such as missing or improperly located sensors. It happens often that failures of the components cannot be diagnosed convincingly, or implicate each other with high ambiguity, or are consistently misinterpreted as failures of other components. Techniques were recently developed to help identify the critical elements of Bayesian network models that are responsible for incorrect diagnosis [17].

6 Conclusion

A diagnosis decision support approach was described using Bayesian belief networks for facilitating airplane maintenance at an airport gate. The approach com-bines engineering and mechanic knowledge with statistical component reliability data. It is argued that Bayesian network framework contains a rich representation language that permits to encode the different types of knowledge needed for airplane diagnosis. The high degree of system integration in an airplane typically results in ambiguous diagnoses. The inference engine of a Bayesian network provides a consistent probability update mechanism to help disambiguate between the possible causes of a failure. Sensitivity analysis of the networks to noisy priors justifies the use of simple probability models from MTBUR data, and also shows reason-able robustness of the network diagnosis resulting from limited sensitivity of the network to noise in the priors.

Acknowledgements

The authors would like to thank several individuals who have provided various types and levels of support and who have contributed in various capacities to the work presented in this paper. Special thanks go to Cathy Kitto, Dick Shanafelt, Susan Chew, Glenn Dildy, Nick Walker, Chris Esposito, Karl Rein-Weston, Dave Naon, Dan Goldenberg, Brian Wood, Jeff Spiro, Hai Tran, Andrew Booker, Paul Jackson and John Bremer.

References

[1] Bruce G. Buchanan and Edward H. Shortliffe, editors. *Rule-Based Expert Systems:The MYCIN Experiments of the Stanford Heuristic Programming Project*. Addison-Wesley, Boston, MA, 1984.

[2] G. Cooper and E. Herskovits. A Bayesian method for the induction of probabilistic networks from data. *Machine Learning*, 9:309–347, 1992.

[3] Randall Davis and Walter Hamscher. Model-based reasoning: Troubleshooting. In Howard E. Shrobe, editor, *Exploring Artificial Intelligence: Survey Talks from the National Conferences on Artificial Intelligence*, chapter 8, pages 297–346. Morgan Kaufmann Publishers, San Mateo, CA, 1988.

[4] Johan de Kleer. Using crude probability estimates to guide diagnosis. *Artificial Intelligence*, 45(3):381–391, October 1990.

[5] Johan de Kleer and Brian C. Williams. Diagnosing multiple faults. *Artificial Intelligence*, 32(1):97–130, April 1987.

[6] Kalyan Moy Gupta. Case-based troubleshooting knowledge management. In *AI in Equipment Maintenance Service and Support*, 1999 AAAI Spring Symposium, Technical Report SS-99-04, Menlo Park, CA, 1999. AAAI Press.

[7] D. Heckerman, D. Geiger, and D. Chickering. Learning Bayesian networks: The combination of knowledge and statistical data. *Machine Learning*, 20:197–243,1995.

[8] David Heckerman. Similarity networks for the construction of multiple-fault belief networks. In P.P. Bonissone, M. Henrion, L.N. Kanal, and J.F. Lemmer, editors, *Uncertainty in Artificial Intelligence* 6, pages 47–64. Elsevier Science Publishing Company, New York, NY, 1991.

[9] Max Henrion. Some practical issues in constructing belief networks. In L.N. Kanal, T.S. Levitt, and J.F. Lemmer, editors, *Uncertainty in Artificial Intelligence* 3, pages 161–173. Elsevier Science Publishing Company, New York, NY, 1989.

[10] Max Henrion, John S. Breese, and Eric J. Horvitz. Decision Analysis and Expert Systems. *AI Magazine*, 12(4):64–91, Winter 1991.

[11] Donna L. Hudson and Maurice E. Cohen. *Neural Networks and Artificial Intelligence for Biomedical Engineering*. IEEE Press Series on Biomedical Engineering. IEEE Press, Piscataway, NJ, 2000.

[12] Finn Verner Jensen. *Bayesian Networks and Decision Graphs*. Springer Verlag, New York, 2001.

[13] Oscar Kipersztok. Fault propagation using fuzzy cognitive maps. In *Proceedings of the Ninth International Workshop on Principles of Diagnosis*, pages 204–208. Cape Cod, 1998.

[14] Oscar Kipersztok and Haiqin Wang. Another look at sensitivity of Bayesian networks to imprecise probabilities. In *Proceedings of the Eighth International*

Workshop on Artificial Intelligence and Statistics (AISTAT-2001), pages 226–232, San Francisco, CA, 2001. Morgan Kaufmann Publishers.

[15] Janet L. Kolodner. *Case-Based Reasoning*. Morgan Kaufmann Publishers, San Francisco, CA, 1993.

[16] Judea Pearl. *Probabilistic Reasoning in Intelligent Systems: Networks of Plausible Inference*. Morgan Kaufmann Publishers, San Mateo, CA, 1988.

[17] Wojtek K. Przytula, Denver Dash, and Don Thompson. Evaluation of Bayesian networks used for diagnostics. In *Proceedings of the Twenty-fourth Annual IEEE Aerospace Conference*, Big Sky, MN, 2003. IEEE Press.

[18] Haiqin Wang, Denver Dash, and Marek Druzdzel. A method for evaluating elicitation schemes for probabilistic models. *IEEE Transactions on Systems, Man, and Cybernetics–Part B:Cybernetics*, 32(1):38–43, February 2002.

[19] Haiqin Wang and Marek J. Druzdzel. User interface tools for navigation in conditional probability tables and graphical elicitation of probabilities in Bayesian networks. In *Uncertainty in Artificial Intelligence: Proceedings of the Sixteenth Conference (UAI-2000)*, pages 617–625, San Francisco, CA, 2000. Morgan Kaufmann Publishers.

B. Bouchon-Meunier, L. Foulloy and R.R. Yager (Editors)
Intelligent Systems for Information Processing:
From Representation to Applications
© 2003 Elsevier Science B.V. All rights reserved.

Adjustment of parallel queuing processes by multi-agent control [1]

Rainer Palm   Thomas A. Runkler

Siemens AG, Corporate Technology, Information and Communications
81730 Munich, Germany
e-Mails: Rainer.Palm@siemens.com
Thomas-Runkler@siemens.com

Abstract

Multi-agent control is an efficient optimization and control method for local systems in large decentralized environments. Market-oriented algorithms work in multi-agent scenarios in which producer and consumer agents compete and cooperate on a virtual market of commodities. The method is applied to a set of parallel birth and death processes where the local death rates can be influenced by some control strategy. The goal of the control strategy is to change the individual death rates so that, after some finite time, the distributions of all queuing processes are adjusted. The death rates of the local queuing processes are changed on the basis of the probabilities of the occurence of a selected event in each of the local queuing processes.

Keywords: Multi-agent control, market-based control, decentralized systems, queuing process, birth and death process, Markov process

1. Introduction

The control of distributed systems is difficult especially in the case when the system consists of a large number of local systems. Normally, this control is done in a centralized way which can lead to problems with respect to the exchange of information and to find an appropriate optimum. A decentralized option is multi agent control which may be able to cope better with the problems arising with centralized control and optimization methods. Multi agent control methods are applied in congestion control of traffic networks [1] and manufacturing systems [3,13,15]. In [7] a multi-agent framework is presented that deals with logistic operations in a distributed network environment which is another growing field of application. This approach is used both for analysis and for the design of a distributed intelligent agent architecture. In [5] a multi-level warehouse hierarchy with its market mechanisms is applied to real-timetransportation, dynamic freight allotment, depot agents, scheduling problems, and production planning. An interesting and promising approach is *market-based control*. Market-based control imitates the behavior of economic systems. In this framework so-called producer and consumer agents both compete and cooperate on a market of certain

---

[1] This work is supported by the German BMBF under Contract No. 13N7906

commodities. Several market-based control and optimization strategies are presented in [2, 4]. In [6,11,12] market-based control algorithms are presented in more detail. On the basis of cost functions used by producer and consumer agents the optimization of distributed coupled linear systems is shown. The present paper mainly adopts ideas from [6, 11] (see also [8,9]). The method is applied to a set of local Markov processes, e.g. queuing processes, respectively. In the case of several queues of customers it might be of interest to keep all queue lengths approximately the same in order to occupy the whole system equally. In contrast to the local birth rates it is assumed that the local death rates can be influenced by some external control. The main goal is to change the individual death rates so that, after some finite time, the distributions of all queuing processes are adjusted. The death rates of the local queuing processes are changed on the basis of the probabilities of the occurence of a selected event in each of the local queuing processes. The paper is organized as follows. Section 2 gives an introduction into the problem of decentral queuing (birth and death) processes acting in parallel. In Section 3 the market-based algorithm is applied to a set of birth and death processes. In Section 4 simulation results are presented. Section 5 concludes with a short summary of the methods presented.

2. Decentral queuing processes

Decentral queuing processes acting in parallel can be explained by the following simple application example. Given a company producing n types of commodities (e.g. computers). Furthermore, let an unknown number of customers order different numbers of each type of commodity. Some of the customers may, for example, order a number of computers of type 1, and others may order another number of computers of type 2 etc. Each computer needs for its production different materials (elements) some of them may be used for both types of computers.

Let both the stream of demands by the customers and the times for handling the commissions by the company be exponentially distributed (see Fig 1). Furthermore, for simplicity we assume that a customer orders only one piece of a special type of commodity. We also assume that at a time instant t the change in the number of demands is 0 or 1.

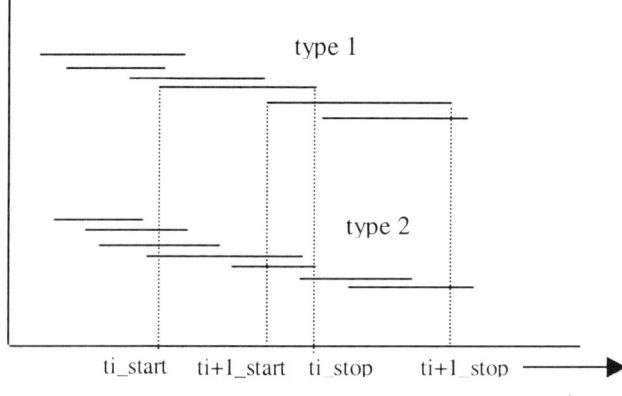

Figure 1: Occurence of demands and handling times of the commissions

Let the time difference between two demands be $\Delta t_{i\,start} = t_{i+1\,start} - t_{i\,start}$. Let further the difference between the stops of orders be $\Delta t_{i\,stop} = t_{i+1\,stop} - t_{i\,stop}$. The corresponding probability functions for $t, \tau > 0$ are

$$P(\Delta t_{i\,start} < t) = 1 - e^{-\lambda t} \qquad t > 0 \qquad (1)$$

and

$$P(\Delta t_{i\,stop} < \tau) = 1 - e^{-\mu \tau} \qquad \tau > 0 \qquad (2)$$

respectively. $\lambda, \mu > 0$ are the parameters of the exponential processes (1) and (2). This type of stochastic process can be modeled by a homogeneous Markov process [10, 14]. Since the demands are stochastically "born" and the corresponding handlings of commissions "die out" the process can be modeled by a so-called birth and death process. In the differential equations for the probability of the occurence of the states $j = 1, 2, \ldots$

$$\begin{aligned}\dot{P}_0(t) &= -\lambda_0 P_0(t) + \mu_1 P_1(t) \\ \dot{P}_j(t) &= \lambda_{j-1} P_{j-1}(t) - (\lambda_j + \mu_j) P_j(t) + \mu_{j+1} P_{j+1}(t)\end{aligned} \qquad (3)$$

are λ_j's the birth rates and the μ_j's are the death rates. $P_j(t)$ is the probability to reach the state j at time t when having started in state i at time t_0. In (3) the birth and death rates are different for every state j. When assuming the rates to be constant for all states one obtains instead

$$\begin{aligned}\dot{P}_0(t) &= -\lambda P_0(t) + \mu P_1(t) \\ \dot{P}_j(t) &= \lambda P_{j-1}(t) - (\lambda + \mu) P_j(t) + \mu P_{j+1}(t)\end{aligned} \qquad (4)$$

We call a process *ergodic* or *stable*, respectively, if $0 < \lambda/\mu < 1$. In this case the process converges. That is, for a specific state $\tilde{j}$ the probability to reach a higher state becomes increasingly unlikely $P_j < P_{j-1}$ for $j > \tilde{j}$. For our example the states j are the number of handled commissions at time t. $P_j(t)$ is the probability to have j commissions in the queue at time t. $\dot{P}_j(t)$ is the change in the probability $P_j(t)$ at time t. Equation (4) can therefore be rewritten

$$\begin{aligned}\dot{P}_0^{\,k}(t) &= -\lambda^k P_0^{\,k}(t) + \mu^k P_1^{\,k}(t) \\ \dot{P}_j^{\,k}(t) &= \lambda^k P_{j-1}^{\,k}(t) - (\lambda^k + \mu^k) P_j^{\,k}(t) + \mu^k P_{j+1}^{\,k}(t) \\ j &= 1, 2, \ldots, \qquad k = 1, 2, \ldots, M\end{aligned} \qquad (5)$$

Observe here that for each type of commodity the λ's and μ's are supposed to be different. Equations (4) can be considered as a system of equations concerning all types of commodities. Equations (4) correspond to the graph depicted in Fig.2 which deals with the total Markov process of all events with n different states. That is, in (4) and Fig. 2 we do not distinguish between different types of commodities.

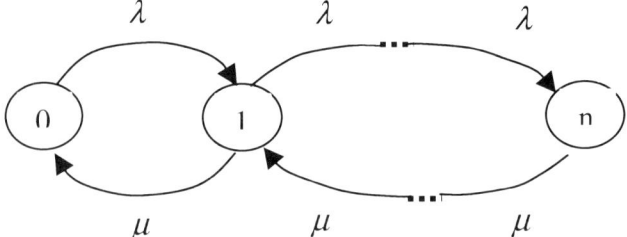

Figure 2: Graph of the centralized Markov process

State j denotes "j active commissions" of an arbitrary commodity. When we split the process into subprocesses each of which regarding a special type of commodity we obtain M different Markov processes (5) and the corresponding graphs (see Fig 3) each of which having n^k states

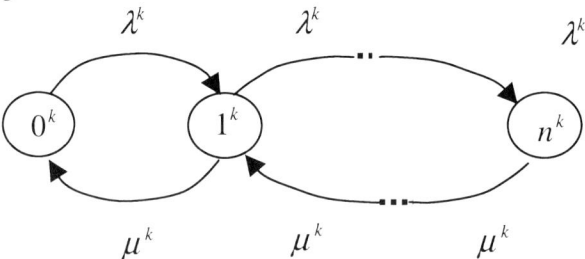

Figure 3: Graph of the decentralized Markov processes

State j^k denotes "j^k active commissions" of commodity type k. The relation between the the λ's and μ's of the centralized process and the decentralized processes can be found by

$$\lambda = \sum_{k=1}^{M} \lambda^k; \quad \mu = \sum_{k=1}^{M} \mu^k \qquad (6)$$

because of the OR-combinations of probabilities. That is, the individual graphs are connected via (6). In addition the total number m of active commissions and the numbers m^k of individual active commissions are connected by

$$m = \sum_{k=1}^{M} m^k; \qquad (7)$$

However, since m can be obtained in many ways the number of possible combinations to get the same total number m can be very high.

3. Market-based algorithms and process control

In general, a queuing system consists of a buffer or a queue and a service station (see Fig. 4). Each type of order k arrives with an interarrival rate λ^k and is handled at the service station with a service rate μ^k. The orders leave the system with the interdeparture rate μ^k_{dep}. The process is said to be stable if for large times t the service rate μ^k is greater than the interarrival rate λ^k : $\mu^k > \lambda^k$. In this case the interdeparture rate μ^k_{dep} of the process is equal to the interarrival rate λ^k independent of the servicerate μ^k : $\mu^k_{dep} = \lambda^k$. It can further be shown that not only the rates but also the distributions of the interdeparture and the interarrival process are equal if the service distribution is exponential [10].

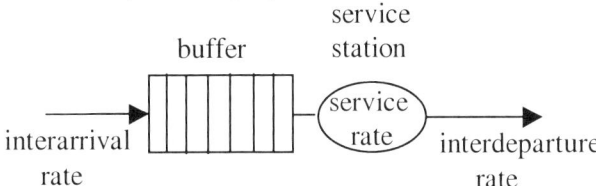

Figure 4: Block scheme of a queuing process

In the following a decentral control method for both the subprocesses and the complete process is investigated. The birth (interarrival) and death (service) processes (5) are completely determined by their initial conditions and their parameters λ^k and μ^k. Since the birth rates λ^k completely depend on the customer the λ^k's cannot be changed or influenced in some way. Therefore, only the death (service) rates μ^k can be changed or influenced by some control or optimization strategy. Increasing μ^k means that a commodity can be delivered faster (earlier). One way of increasing μ^k is, e.g., the exchange or loaning of materials (elements), respectively, from other types of commodities. This, however, leads to a decrease of the μ^k for these other types. To reach a compromise for all subsystems a market-based algorithm is applied to find a so-called Pareto-optimum for the complete set of subprocesses. One may argue that it may be sufficient to make the queues as fast as possible. But under the condition of restricted recources of the suppliers this control policy may lead to longer queues, larger buffer capacities, and lower quality of service. Instead, the control policy applied here is to change the death (service) rates of the individual queues in a way that their dynamical behaviors become equal. This is identical with the requirement for using the capacity of all subprocesses in a well-balanced way which in turn leads to a synchronization of the queues. This can be obtained by minimizing a cost /energy function of the total process via the local cost/energy functions. An appropriate calculation of the death rates is done by a market-based approach that is presented in the following. In order to avoid misinterpretations one has to note that the below-mentioned terms "producer" and "consumer" should not be confused with real customers/consumers and suppliers/producers acting in the commisioning process described in the previous section. Subsequently, producer and consumer agents are virtual actors trading with virtual commodities like, e.g., death rates. Let now be defined a producer agent Pag_j^k and a consumer agent Cag_j^k, respectively, for each state j^k belonging to a commodity of type k. Pag_j^k "produces" a certain death rate μ_p^k, and tries to maximize a local profit function $\rho_j^k > 0$. Cag_j^k "demands" for a certain death rate μ_c^k, and tries to maximize a local utility function $U_j^k > 0$. The trade between producer and consumers agents takes place on the basis the local profit and utility functions ρ_j^k and U_j^k, and common prices p_j. The prices p_j can be calculated from the equlibrium of the whole "economy" at every time step. Then, an equilibrium is reached as the sum over all supplied death rates μ_p^k is equal to the sum over all utilized death rates μ_c^k.

$$\sum_{l=1}^{M} \mu_p^l = \sum_{l=1}^{M} \mu_c^l \qquad (8)$$

Then, from (8) the common prices p_j can be calculated. The quadratic utility function for the consumer agent Cag_j^k is now defined by

Utility = benefit - expenditure
$$U_j^k = \tilde{b}_j^k \mu_c^k - \tilde{c}_j^k p_j (\mu_c^k)^2 \qquad (9)$$

where $\tilde{b}_j^k, \tilde{c}_j^k > 0$ will be determined later in connection with the queue dynamics (5). The quadratic profit function for the producer Pag_j^k is defined by

profit = income - costs
$$\rho_j^k = g_j^k p_j \mu_p^k - e_j^k (\mu_p^k)^2 \qquad (10)$$

where $g_i^j, e_i^j > 0$ are free parameters that determine the average price level. p_j is the actual price that has to be payed for μ_p^k by each consumer Cag_j^k. According to [6] the individual terms in U_j^k and ρ_j^k are chosen so that benefit and income increase with μ_c^k and μ_p^k, respectively. On the other hand, the expenditure per μ_c^k and costs per

μ_p^k, respectively, are assumed to increase with μ_c^k and μ_p^k, respectively. The parameters $\tilde{b}_j^k, \tilde{c}_j^k$ in (9) can be determined by using local energy functions based on the dynamics (5) where, for simplicity, the argument t is eliminated

$$\dot{P}_j^k = \lambda^k P_{j-1}^k - (\lambda^k + \mu_c^k) P_j^k + \mu_c^k P_{j+1}^k \qquad (11)$$

From (11) a local energy function for each subprocess is defined as

$$J_{j_c}^k = (\dot{P}_j^k)^2 = \alpha_j^k + \beta_j^k \mu_c^k + \gamma_j^k (\mu_c^k)^2 \qquad (12)$$

where

$$\alpha_j^k = (\lambda^k)^2 \cdot (P_{j-1}^k - P_j^k)^2 \geq 0$$

$$\beta_j^k = -2\lambda^k \cdot (P_{j-1}^k - P_{j+1}^k)(P_j^k - P_{j+1}^k) \leq 0 \quad \text{for small } |\dot{P}_j^k| \qquad (13)$$

$$\gamma_j^k = (P_j^k - P_{j+1}^k)^2 \geq 0$$

It can be shown that for the stationary case $\dot{P}_j^k = 0$, and for $p_j = 1$, the energy function (12) reaches its minimum at the maximum of the utility function (9), independently of the parameter α_j^k. Therefore, a comparison of (9) and (12) leads to the intuitive choice

$$\tilde{b}_j^k = |\beta_j^k|, \quad \tilde{c}_j^k = \gamma_j^k \qquad (14)$$

in order to guarantee $\mu_c^k \geq 0$. The optimization of the whole process takes place by individual maximization of the local utility and profit functions, respectively. Maximization of the utility function (9) yields

$$\frac{\partial U_j^k}{\partial \mu_c^k} = \tilde{b}_j^k - 2\tilde{c}_j^k p_j \mu_c^k = 0 \qquad (15)$$

from which optimum "demanded" μ_c^k's are obtained

$$\mu_c^k = \frac{\tilde{b}_j^k}{2\tilde{c}_j^k} \cdot \frac{1}{p_j} \qquad (16)$$

Maximization of the profit function (10) yields

$$\frac{\partial \rho_j^k}{\partial \mu_p^k} = -g_j^k p_j + 2e_j^k \mu_p^k = 0 \qquad (17)$$

from which optimum "produced" μ_p^k's are obtained

$$\mu_p^k = \frac{p_j}{2\eta_j^k}; \quad \eta_j^k = \frac{e_j^k}{g_j^k} \qquad (18)$$

The requirement for an equilibrium between the sums of the "produced" μ_p^k's and the "demanded" μ_c^k's led to the balance equation (8). Substituting (16) and (18) into (8) gives the price p_j for μ_p^k and μ_c^k, respectively.

$$p_j = \sqrt{\sum_{l=1}^{M} \tilde{b}_j^l / \tilde{c}_j^l \bigg/ \sum_{l=1}^{M} 1/\eta_j^l} \qquad (19)$$

Substituting (19) into (16) yields the final equation for the death rate to be implemented in each subsystem. η_j^k can be chosen as a constant which gives reasonable results. A better choice, however, is $\eta_j^k = \frac{\mu^k}{\lambda^k}$ to let η_j^k be dependent on μ^k and λ^k. For the *stationary case* we obtain

$$P_j^k = \frac{\lambda^k}{\mu^k} P_{j-1}^k \qquad (20)$$

For β_j^k, and γ_j^k we obtain with (13) and (20)

$$\beta_j^k = -2\mu^k \cdot (1 + \frac{\lambda^k}{\mu^k})(1 - \frac{\lambda^k}{\mu^k})^2 (P_j^k)^2 \leq 0 \quad \gamma_j^k = (1 - \frac{\lambda^k}{\mu^k})^2 (P_j^k)^2 \geq 0 \quad (21)$$

The determination of β_j^k and γ_j^k for the *nonstationary case*, however, requires the computation/measurement of the probabilities $P_j^k(t)$ that can be done by constructing histograms for every point in time. The probability of a state $P_j^k(t)$ is approximated by the number of events j^k divided by the total number of events

$$P_j^k(t) = n(j^k) \bigg/ \sum_{i=1}^{N} n(i^k) \qquad (22)$$

4. Simulation examples

Example 1
The first example deal with 3 different Markov processes with 11 states for each process. The continuous processes and the optimization strategy are implemented as discrete models. The simulations have been done on a small time scale which can be changed for a real process accordingly. The initial values are

Process P1:
P0_1=1; P1_1=0; P2_1=0; P3_1=0; P4_1=0; P5_1=0; P6_1=0;
P7_1=0; P8_1=0; P9_1=0; P10_1=0;
Process P2:
P0_2=1; P1_2=0; P2_2=0; P3_2=0; P4_2=0; P5_2=0; P6_2=0;
P7_2=0; P8_2=0; P9_2=0; P10_2=0;
Process P3:
P0_3=1; P1_3=0; P2_3=0; P3_3=0; P4_3=0; P5_3=0; P6_3=0;
P7_3=0; P8_3=0; P9_3=0; P10_3=0;
The corresponding parameters are
lambda_1=1; lambda_2=2.5; lambda_3=3.8; mu_1=1.4; mu_2=1.7; mu_3=1.9;
Process P1 is ergodic since `lambda_1< mu_1`. The processes P2 and P3 are non-ergodic since `lambda_2 > mu_2` and `lambda_3 > mu_3`.

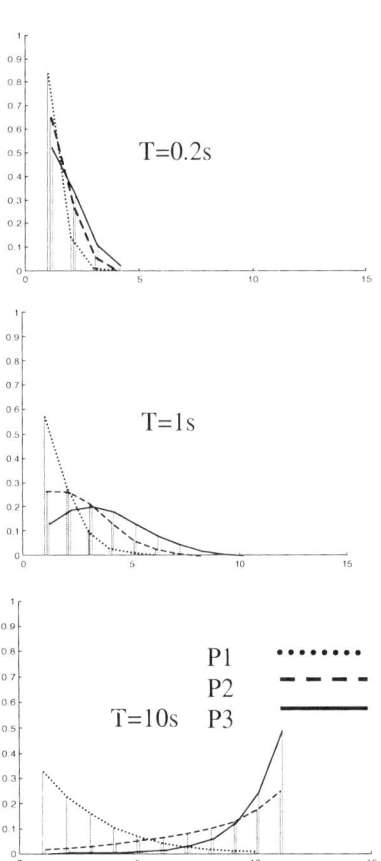

Figure 5. Process evolution P1- P3 (no optimization)

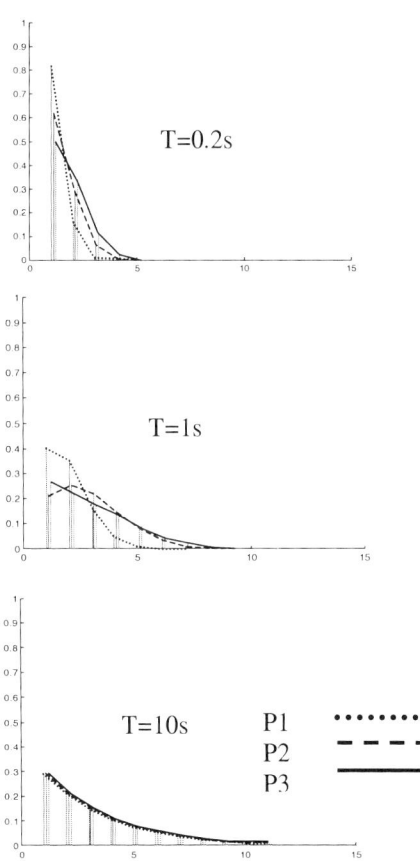

Figure 6. Process evolution P1-P3 (with optimization)

Figure 5 shows the evolution of P1-P3 without any cooperation or interconnection between the processes. After 10s the processes are almost stationary. For this case P1 shows the characteristic ergodic feature with $P_k > P_{k+1}$ where, on the other hand, for the processes P2 and P3 we have the non-ergodic feature $P_k < P_{k+1}$. Despite of this, P2 and P3 are not unstable but come to rest at a stationary distribution depicted in Fig. 5. The reason is that the number of states is restricted. Figure 6 presents the case in which the processes cooperate (compete) with each other. The corresponding parameters for the market-based algorithm are

```
e_1=2; g_1=1; eta_1=e_1/g_1;
e_2=2; g_2=1; eta_2=e_2/g_2;
e_3=2; g_3=1; eta_3=e_3/g_3;
```

The initial distribution is the same as in the previous case. The μ 's are changed at each time step according to (16). Already after 1s it can be observed that the process distributions approach to each other. After 10s all processes exhibit ergodic features which means that, finally, the corresponding μ 's take larger values than the corresponding λ 's

```
lambda_1 = 1;   lambda_2= 2.5;   lambda_3 = 3.8;
mu_1 = 1.3799;     mu_2 = 3.5065 ;    mu_3 = 5.3328.
```

Example 2

In the previous example the distributions shown in Figs. 5 and 6 were generated directly by the differential equations (5) and (10), respectively. Instead, in the following example the distributions are the result of stochastic processes generated by noise generators that produce both the birth processes of demands and the death processes of handling the commissions.

We assume to have only one time series available, and we make also the assumption that all birth and death processes are stochastically independent, stationary, and ergodic. The whole time series is divided into l time intervals N1. Then the time intervals N1 are divided into smaller time intervals T. For these intervals we calculate histograms and, by normalization, corresponding distributions. Figure 7a shows an example for t=10s and no optimization from the market-based algorithm. and Fig. 8b shows the same with optimization. We observe that the results are comparable with the results from the differential equations.

The birth and death rates λ and μ in Fig.7a,b are the final rates after the experiment. Figure 8 shows the evolution of the price. It can be noticed that after 30 cycles (3s) the price and the final μ values have already been established.

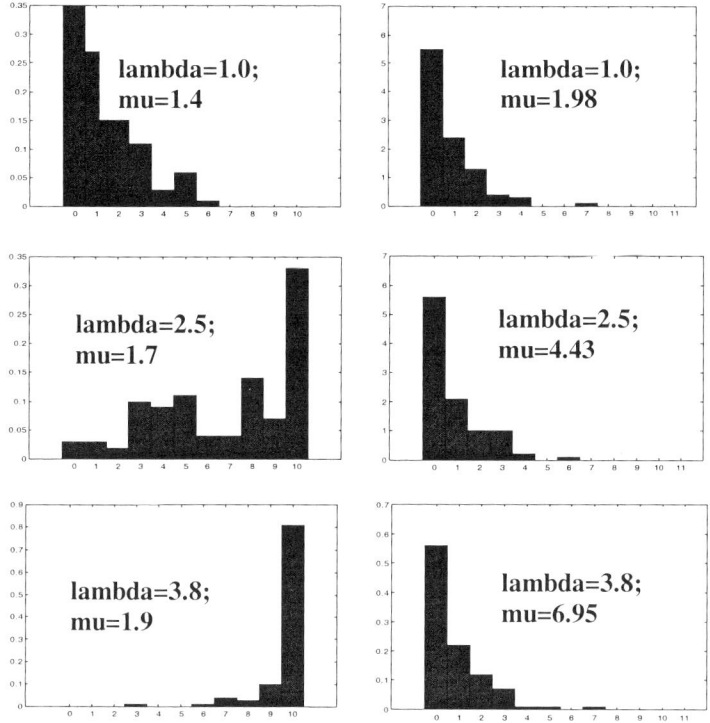

Figure 7: Processing of the death rates; a/b no/with optimization

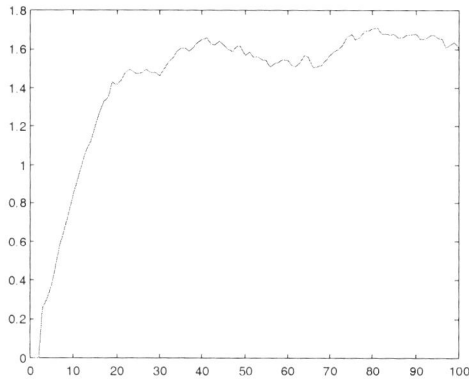

Fig. 8 Evolution of the price

5. Conclusions

In the paper a set of local birth and death processes was presented whose local birth rates are fixed but the local death rates are influenced by market-based control algorithms. The control goal is to change the local death rates so that the distributions of all queuing processes are adjusted. In the paper an introduction into distributed birth and death processes is given, and the problem of their centralized or decentral representation is discussed. For the market-based algorithm applied here local utility and profit functions, respectively, are defined on the basis of which the particular Markov processes are optimized. The result of the optimization is a so-called Pareto-optimum which represents a compromise between the competing processes. The simulation experiments show that the algorithm leads to excellent results for the adjustment of local queuing processes.

References

[1] E. Altman; T.Basar, R. Srikant (1999). A team-theoretic approach to congestion control. 14th IFAC World Congress 1999, Beijing, P.R.China., Paper J-3d-02-2
[2] A.A.Berlin ; T.Hogg, W. Jackson (1998). Market-based control of active surfaces. Proceedings of the SPIE, Boston, MA, USA, 4-5 Nov. 1998, pp. 118-129
[3] S. Bussmann (1998). Agent-oriented programming of manufacturing control tasks. Proceedings International Conference on Multi Agent Systems , Paris, France, 3-7 July 1998, pp.57-63
[4] S.H.Clearwater (ed.) (1996). Market-based Control: A Paradigm for Distributed Resource Allocation. World Scientific, Singapore, 1996.
[5] J. Falk; S.Spieck; M.Weigelt; P. Mertens (1993). Cooperative and competitive approaches with agents in logistics and production. Thirteenth International Conference. Artificial Intelligence, Expert Systems, Natural Language Proceedings, Avignon, France, 24-28 May 1993, pp.519-528
[6] O.Guenther; T.Hogg; B. Huberman (1997). Controls for unstable structures. Proceedings of the SPIE, San Diego, CA, USA, 3-6 March 1997, pp. 754-763
[7] M.L .Moore; R.B. Reyns; S.R.T. Kumara; J.R.Hummel (1997). Distributed intelligent agents for logistics (DIAL). 1997 IEEE International Conference on Systems, Man, and Cybernetics. Computational Cybernetics and Simulation , vol.3, Orlando, FL, USA, 12-15 Oct. 1997. pp.2782-2787
[8] R.Palm (2002). Synchronization of decentralized multiple-model and fuzzy systems by market-based optimization. Proceedings IFAC2002, Barcelona, Session T-Tu-E04
[9] R.Palm, Th. Runkler (2002). Adjustment of parallel queuing processes by multi-agent control. IPMU'2002, Annecy, July 2002, Vol. III, pp.1495-1502
[10] S. M. Ross (1996): Stochastic Processes. 2nd edition, John Wiley & Sons, Inc. New York, 1996
[11] H.Voos; L.Litz (1999). A new approach for optimal control using market-based algorithms. Proceedings of the European Control Conference ECC'99, Karlsruhe, 1999.
[12] H.Voos (1999). Market-based control of complex dynamic systems. Proceedings of the 1999 IEEE International Symposium on Intelligent Control, Intelligent Systems and Semiotics, Cambridge, MA, USA,15-17 Sept. 1999. pp.284-289.
[13] A.Wallace(1998). Flow control of mobile robots using agents. 29th International Symposium on Robotics Birmingham, UK, 27 April -1 May 1998, pp. 273-27
[14] R.W .Wolff (1989). Stochastic Modeling and the Theory of Queues. Prentice-Hall London 1989
[15] M.B. Zaremba; Z.A. Banaszak; P. Majdzik; K.J.Jedrzejek (1999). Distributed flow control design for repetitive manufacturing processes.
14th IFAC World Congress 1999, Beijing, P.R.China. Paper A-1a-01-1

Is she gonna like it? Automated Inspection System Using Fuzzy Aggregation

A. Soria-Frisch and M. Köppen and Th. Sy

Fraunhofer IPK Berlin, Dept. Security and Inspection Tech.
Pascalstr. 8-9, 10587 Berlin, Germany
E-Mail:{aureli|mario.koeppen|thorsten.sy}@ipk.fhg.de

Abstract

A framework for the automated visual inspection of texture surfaces was implemented and used in an industrial application for the evaluation of fault perceptual relevance. This new paradigm for automated visual inspection systems, whose appearance is due to demands from the industry, allows attaining the objective reproduction of the end user's subjective decisions on texture surface quality of a particular product. The here presented paper analyzes in detail the fuzzy fusion methodologies used in this framework in order to predict the end user's opinion. Finally the paper presents the results obtained in a real industrial application, namely the inspection of organic material plates.

Keywords: Automated Visual Inspection, Fuzzy Aggregation, Order Weighted Averaging, Choquet Fuzzy Integral, Pattern Recognition.

1 Introduction

The utilization of automated visual inspection systems for quality control on textured surfaces has been an outstanding field of research in computer vision. In this context industrial visual inspection is leaving behind the traditional application goal of finding functional faults in textured materials trying to conquer new and more complex goals. Nowadays the classical approach of detecting a fault, assigning it to a fault class, and delivering the crisp results to the superposed production system has lost importance due to economic demands. The possibility to reuse some parts of the produced goods, to redirect its selling to alternative channels of distribution or to take advantage of the long experience of human operators in inspection tasks encourage the establishment of new paradigms for automated visual inspection systems. The paradigm proposed in this paper is denoted as perceptual relevance evaluation.

The systems developed within perceptual relevance evaluation basically attain the detection of perceptually relevant anomalies of any kind, which might disturb the total visual appearance of the surface texture. Furthermore such a system attempts to predict the response of an end user in front of a particular object. In this case the quality control is preventing a product end user of eventually rejecting some items due to their aesthetic value. The resolution of such problems are characterized by the following main features:

- The items subject to inspection are in its majority end consumer goods.

- Only part of such faults are of functional nature, or at least malfunction plays a secondary role.

- It is desired that the inspection system delivers a bit more complex description of the faultiness degree than the two-valued decision fault or not.

Summarizing, where traditional visual inspection systems detect, modern ones have to interpret (see Fig. 1). Wooden and textile surfaces have been a long-termed research target in this direction. However, a general attempt to treat this class of inspection problems has not been considered so far.

Figure 1: Challenge for automated visual inspection systems of textured surfaces: from the detection of faults eventual producing malfunctioning (a) to the the interpretation of "ugliness" (b). (a): Mechanical piece. (b): Example of organic material plate analyzed with the here presented framework.

Perceptual relevance evaluation is mainly characterized by its subjective nature. Also, perceptual faults are stuck on both, local and global image properties. A floor tile might successfully pass all fault tests by itself, whereas it may fail in a pavement due to its perceptual relevance (e.g. a large-scale unintended regularity). So far the consideration of fuzzy methodologies [18] in the prediction process seems unavoidable.

Taking all these facts into consideration a framework for fuzzy evaluation of perceptually relevant faults was developed [9]. The components of the framework treat specific texture detection problems with different binarization approaches. The evaluation of the binary images delivered by the binarization modules is realized through a hierarchical network of fuzzy fusion operators. The purpose of the here presented paper is to analyze in detail the fuzzy strategies employed in that framework [9].

The paper is organized as follows. In section 2 a brief overview on fuzzy fusion strategies is given. The framework is presented in section 3. The usage of the hierarchical network of fuzzy aggregation operators is described in section 4. The framework was applied in the evaluation of organic material plates. In section 5 the perceptual relevance evaluation of the faults in a test set and the results obtained by the final industrial system are presented. Finally, the reader can find the conclusions in section 6.

2 Fuzzy Fusion Approaches

Sugeno's work on fuzzy integrals [12] incorporate the subjectivity, which was already considered by Zadeh [18] in the case of the classical set theory, into the mathematical concept of integration. The final goal was the approximation of such an operation to different aggregation processes undertaken by human beings. The successful usage of the fuzzy measures theoretical framework [14] in problems of fuzzy classification in pattern recognition [4], multicriteria decision making [5], and multiattribute subjective evaluation [13] constitute good antecedents for the problem of perceptual relevance evaluation.

In the here presented framework, an OWA and a Choquet Fuzzy Integral are used as fusion operators. OWAs [16] used a weighting of the different factors being aggregated. The OWAs weighting scheme is based on the numerical ranking of the channel values presented to the aggregation function. Formally, being X the set of values to be aggregated, $X = \{x_1, x_2, \ldots, x_n\}$, and W the set of weights, $W = \{w_1, w_2, \ldots, w_n\}$, the result O of the fusion operation can be expressed as:

$$O = \sum_{i=1}^{n} w_i x_{(i)}, \tag{1}$$

where the weights satisfy I. $w_i \in [0, 1]$ and II. $\sum_{i=1}^{n} w_i = 1$. Moreover the enclosed subindex express the result of the sorting operation in ascending order: $x_{(1)} \leq x_{(2)} \leq \cdots \leq x_{(n)}$.

In the Choquet Fuzzy Integral the weighting procedure is made through so-called fuzzy measures μ, which are functions on fuzzy sets defined as $\mu : \mathcal{P}(X) \longrightarrow [0,1]$ for $\mathcal{P}(X)$ being the set of all subsets of X. Fuzzy measures satisfy the following conditions when being considered on sets of a finite number of elements: I. $\mu\{\emptyset\} = 0; \mu\{X\} = 1$ and II. $A \subset B \rightarrow \mu(A) \leq \mu(B) \forall A, B \in \mathcal{P}(X)$.

The quantification of the *a priori* importance is made assigning values to the fuzzy measure coefficients. Different types of fuzzy measures are used in different theoretical frameworks. In the λ-fuzzy measures [12] only the fuzzy measure coefficients of the individual components and a parameter λ, which characterizes the computation of component coalitions coefficients as $\mu(A \cup B) = \mu(A) + \mu(B) + \lambda \mu(A) \mu(B)$, have to be determined.

The formal expression of the Choquet Fuzzy Integral, $(C) \int f d\mu$, is:

$$(C) \int f d\mu = \sum_{i=1}^{n} x_{(i)} \cdot [\mu(A_{(i)}) - \mu(A_{(i-1)})], \tag{2}$$

with $A_{(i)} = \{x_{(i)}, \ldots, x_{(n)}\}$ and $A_{(0)} = \emptyset$.

The utilization of fuzzy fusion operators in the here presented framework allows the flexible representation of subjective reasoning present in the human task of perceptual relevance evaluation. Moreover, the admission of several operators and the possibility of considering different types of weighting make the fuzzy fusion framework the adequate one for the resolution of the application on hand. Finally the deliverance of the result to the production system in form of fuzzy membership degrees allows its flexible utilization.

3 Framework for Perceptual Relevance Evaluation

A system for the automated visual inspection of texture surfaces was implemented [9]. The purpose of the visual inspection system is the evaluation of the perceptual relevance of different fault types (e.g. holes, cracks) on plates of an organic material (see Fig. 1b). The framework is basically composed of a processing chain of alternating Binary Pattern Processing Modules ($BPPM_i$), each of them having the same internal structure. These modules reduce the grayvalue domain of the input images to a common binary one by detecting a specific type of faults. Thus all fault types can be further analyzed in the same way. Some additional testing modules can be found between such modules (see Fig. 2). The testing modules, where a fast testing routine based on reduced information is undertaken, are optional design components. Their main purpose is to bypass a BPPM computations, if there is no evidence for the faults, which are processed by that BPPM.

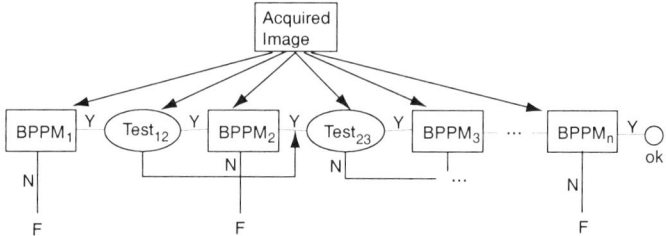

Figure 2: Framework for perceptual relevance evaluation. Different fault types, whose detection presents increasing complexity from the left to the right, are detected through different Binary Pattern Processing Module ($BPPM_i$). Different Test modules ($Test_{ij}$) are interpolated among them in order to break the analysis and thus to optimize the performance of the system in terms of detection time. The BPPMs deliver a decision on the relevance of the analyzed fault (F).

Each BPPM has access to the acquired image and to the evaluation of the foregoing testing module. Hence, the processing of each BPPM is independent of the processing of the others, but may refer to the results of the foregoing modules. Modules for the detection of the more frequent faults, or of the more simple to detect faults should come first in the chain. Since the appearance of a sufficient relevant fault in foregoing modules can interrupt the more time consuming computation in later ones, the system is computationally optimized.

Each single BPPM is designed for a particular fault type, which is defined with respect to the fault's morphological and contrast structure. Each module is composed by a preprocessing, a binarization, a fuzzy feature extraction and a fuzzy evaluation stages (see Fig. 3). In the binarization stage, a set of k binarizing procedures is performed in parallel, giving k resulting binary images. These images act as complementary pieces of evidence in the analysis of the object under inspection. So far, four basic designs of a BPPM for four different fault types have been implemented (for a detailed explanation the reader is referred to [9]). A taxonomy of these types is given following:

Strong-contrast localized faults appear as an area presenting a grayvalue very different from the image background. Only one binarization operation is applied, which may be an interval thresholding.

Long-range faults are not related to a strong local contrast, but to a distortion within the global distribution of grayvalues within the image. Such faults may be detected by employing the auto-lookup procedure [9] based on the co-occurrence matrix [7] of a subset

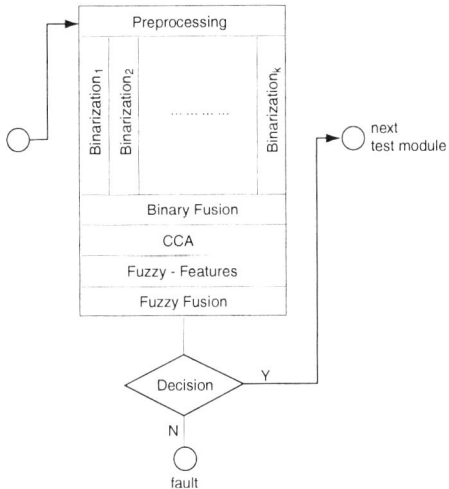

Figure 3: Structure of a Binary Pattern Processing Module. Although the faults to be treated are different, the BPPM maintain the same structure up to the number of parallel binarizations. CCA: Connected Component Analysis.

of pixels from the image. The procedure acts as a novelty filter (see Fig. 4).

Low-contrast faults are characterized by its low contrast in the image grayvalues. For their treatment, the framework called Lucifer2 [8] may be used. The purpose of the framework is the automated generation of texture filters given the original and the expected goal images.

Frequency related faults appear in form of a disruption of the uniformity resulting from the repetition of a basic element. For its detection a Gabor image decomposition [2] for different spectral bands are calculated following the schemata presented in [10]. Finally, the total image energy in the analyzed spectral bands is computed and thresholded.

A fusion procedure derives a binary image by applying logical operators on the k images delivered by the binarization stage. The foreground (or Black pixels) of that binary image are candidates for perceptual faults which will be fuzzy processed in the following stages. The CCA module connects components from the binary fused image, which are perceptual fault regions, and removes noise that remained after the binary fusion.

The fuzzy evaluation of the binary image is undertaken by a network of fusion operators (see Sect. 4). Thus, the network achieves the path between binary pixel information and the relevance of the defect under inspection. As a result each BPPM delivers a two-tuple of fuzzy membership degrees upon two classes ("no relevant faultiness", "relevant faultiness"). For instance, the vector $(0.3, 0.7)$ would indicate an inspected objet where the analyzed texture fault are perceptually relevant.

The final decision for the rejection of a piece upon each defect type is undertaken by the production system based on this information. In case of a very complex interaction between the different defect types another fuzzy integral could be used in the highest level of abstraction for the integration of their ("no relevant faultiness","relevant faultiness") membership degrees.

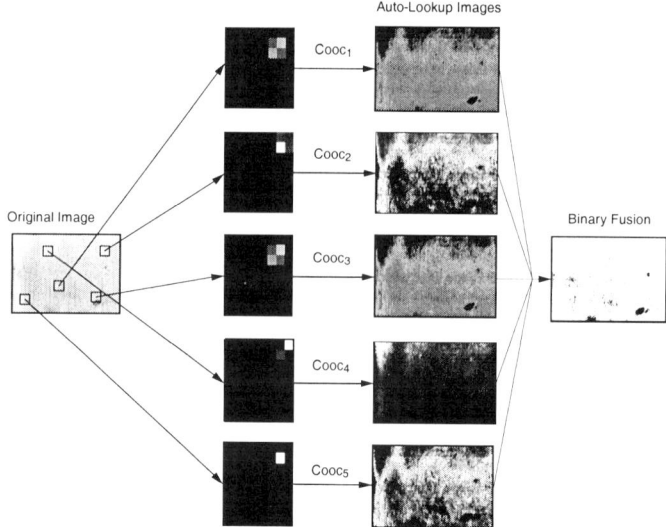

Figure 4: Auto-Lookup based procedure. From the acquired image, five co-occurrence matrices [7] are derived at five randomly located windows. Then, for each co-occurrence matrix, the auto-lookup procedure [9] is performed on the whole image. As a result dark regions stand for "atypical" regions.

4 Fuzzy Aggregation for Perceptual Relevance Evaluation

The binary images extracted by the parallel binarization submodules (see Fig. 3) have to be fuzzy processed in order to better approximate the desired subjective evaluation (see Sect. 2).

The analysis of the binary images obtained is undertaken at this point by a network of fuzzy aggregation operators (see Fig. 5), where the complexity of the problem attached in each following stage needs the usage of operators of also increasing complexity. This concept has been analyzed in detail in [11]. With each fusion operator in the network a new abstraction level in the way from pixel information to the quantification of the perceptual relevance of the defects is achieved. The different stages will be analyzed in the following.

At first the remaining connected components are measured (see Fig. 6) through classical operators as the statistical first moment of the black pixel positions or the sum of pixel values (1 or 0). As a result, local geometric features for each detected fault are extracted. E.g. height, width, area, perimeter, roundness.

4.1 Fuzzy Meta-Feature Extraction through OWAs

A holistic description of the local features is needed, in order to get global description of the elements under analysis. For that end, traditional fusion operators on the one hand and Ordered Weighted Averaging (OWA) operators on the other are applied on them. The employment of traditional fusion operators is trivial, e.g. computation of the number of faults in an item. On the other hand it is worth detailing the employment of OWAs. The

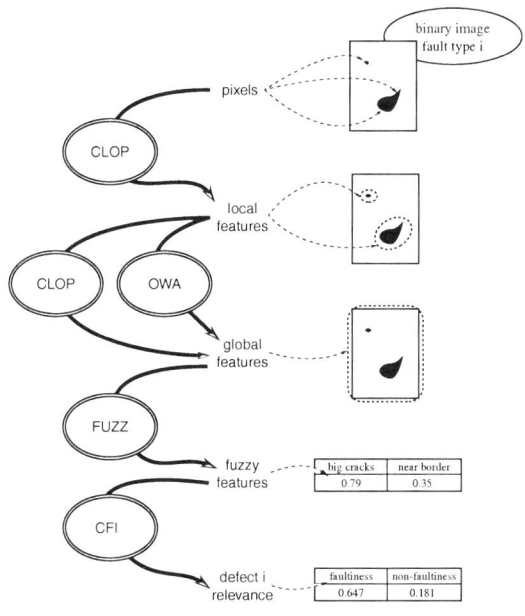

Figure 5: Hierarchically organized network of fusion operators used in the here presented paper for the fuzzy evaluation of binary images. The operators of increasing softness [11] are in charge of tasks of increasing complexity. These correspond to increasing levels of abstraction in the evaluation from the consideration of pixel information, $value(x, y)$, to the result in form of a double fuzzy membership degree expressing "faultiness". CLOP: Classical Operators; OWA: Ordered Weighted Averaging operators; FUZZ: fuzzification stage; CFI: Choquet Fuzzy Integral.

global descriptor obtained up to them can be considered as a meta-feature, resulting from the fuzzy aggregation of the local fault features.

The weighting configuration of the OWAs, which are softer aggregation operators than traditionally used ones, increase the flexibility. Since the weighting is done by taking into consideration the numerical ranking of the features, the result can be biased for giving preference to a determined range of values or for reinforcing the presence of coincident ones [17]. Furthermore, the usage of the OWAs allows the comparison of vectors with different number of components without suffering the low-pass filtering effect of traditionally used ones, e.g. average. An example of the usage of the OWA weights and its effect on the result is shown in Fig. 7b.

The global descriptors till this stage are fuzzified when necessary by defining linguistic terms with trapezoidal fuzzy membership functions (see Fig. 7c). The fuzzification of the *meta-features*, which apply a linguistic descriptor on them, eases the conceptual development and parameterization of the following stages. Moreover the usage of two opposed fuzzy features increase the robustness of the evaluation system.

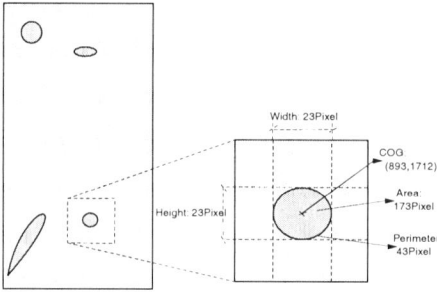

Figure 6: Measuring of binary patterns after CCA. Geometric features for each detected fault are extracted. COG: Center of gravity.

4.2 Decision Making over Perceptual Relevance through the Choquet Fuzzy Integral

The last stage of the hierarchical network of fuzzy aggregation operators (see Fig. 5) is implemented through a Choquet Integral. This operator attains binding the different *fuzzy meta-features* in order to make a decision over the goodness of the item. Thus for each perceptual fault class the *fuzzy meta-features* are fused into a value from $[0, 1]$ by using the Choquet fuzzy integral in order to characterize the faultiness perceptual relevance in form of a membership degree. The application of an aggregation operator can produce such a membership function.

The most important reason for the application of the fuzzy integral is the capability of this fuzzy operator for fusing information taking into consideration the a priori importance of both individual and groups of attributes. The fusion of the different fuzzy features is needed in order to find the joint perceptual relevance of the faults in the object under inspection. In such a process the interaction between the different fuzzy features has to be considered. The fuzzy integral is the only fuzzy fusion operator to allow such a characterization [5].

The kind of analysis undertaken reflects in the result the different possibilities of interaction. E.g., if the presence of defects on the plate border is very important, the result of the relevance quantification should increase; if such a presence is not so important but coincides with a very big defect, the relevance should also increase; if the defects are small and there are not so much of them, the relevance decreases. Such a characterization could have been undertaken also with a system of fuzzy rules. However, the fuzzy integral approach is more synthetic and understandable from the developer point of view. When the kind of interactions to be characterized are very complex or numerous, the number of rules increase so much that the problem is no longer tractable. Furthermore, in many cases the descriptions delivered by the inspection experts are difficult to collect, when not full of contradictions. The automated finding of the fuzzy measures helps avoiding this problem and allows an easier redesign of the feature extraction stage.

Furthermore the Choquet Fuzzy Integral has been already used as classificator in pattern recognition problems, where its performance was superior to that of other classificators, e.g. multilayer perceptron, bayesian independent classification [4]. Moreover the parameterization of the fuzzy integral was taken into consideration. Diverse algorithms

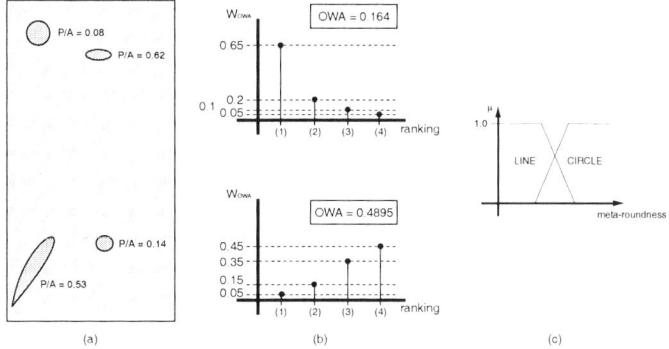

Figure 7: Example of the employment of OWAs in the computation of *fuzzy meta-features*. The image shows the exemplary computation of the meta-roundness of a plate based on the fuzzy aggregation of the local features Perimeter/Area (P/A) (a). (b) A weight configuration very sensitive to the presence of just one fault with circular form is displayed at the top. On the contrary if a detection of more than one elongated fault is desired, a configuration like the one at the bottom would be used. (c) Example of fuzzification functions for the computation of *fuzzy meta-features* (in this case, meta-roundness). The result of the OWAs is fuzzified with trapezoidal fuzzifying functions.

have been presented for the parameterization of the Choquet integral [4][5][15], while the Sugeno integral lacks of such a diversity.

4.3 Automated Parameterization of the Fuzzy Integral

It is worth making a detailed analysis of the automated parameterization of the Choquet Fuzzy Integral. In such an industrial application as the one presented here the expert knowledge is difficult to collect from the system end-users. The subjective characterization of the "ugliness" of an object with respect to its *fuzzy meta-features* is not exact, if not even presenting some contradictions. Moreover the importance of the different object *meta-features* and of its interaction is neither easy to characterize. Since the automated parameterization is based on examples of "ugly" plates, it facilitates the characterization of the *fuzzy meta-features*' importance and that of their interactions. Finally the automated parameterization increases the flexibility of the system in front of changes in such specification.

So far, some experiments were done on the problem at hand for the automated construction of the fuzzy measures for the Choquet integral by using an optimization technique based on quadratic programming [4]. Those experiments were not successful. This may lie on the necessity for the optimization technique to have a clear distinction among the memberships of the classes. In case of unclear definition of them the quadratic programming delivers trivial solutions, i.e. a fuzzy integral equivalent to a maximum operator.

Some previous works [5] take into consideration histograms and two-scatter plots for the importance assessment. This approach was not implemented in the final system because it does not result in a full automation of the parameterization.

Genetic Algorithms [3] constitute a good, though not trivial, alternative for the purpose under consideration [15]. This approach makes possible the consideration of a search in the parameter space in form of a trend instead of the comparison with an expected final value of traditional supervised approaches. This type of setting was considered in the final system. The applied GA uses a real codification of the fuzzy measure coefficients, linear scaling for fitness, roulette wheel as selection operator and and a two-point crossover operator [3]. All of them are GA's standard operators.

One important discussion point in the parameterization through GAs is which fitness function to use. The implemented fitness function takes two factors into consideration. On the one hand the minimization of the error between the fault perceptual relevance established by the experts on the plates belonging to the training set and the result delivered by the fuzzy integral is attained. On the other hand the result of the fuzzy integral is biased in order to present a maximal difference between the classes ("no relevant faultiness", "relevant faultiness").

A λ-fuzzy measure was used for the computation of the fuzzy integral. Such an election avoids the consideration of the monotonicity condition of general fuzzy measures, since only the coefficients for the individual components have to be determined. Moreover, it reduces the search space of the GA, what leads to shorter computation times.

5 Results

The framework for the evaluation of fault perceptual relevance is implemented in an automated visual inspection system for the industrial evaluation of organic material plates (see Fig. 1b). The results for a particular fault type are presented following. The training of the fuzzy integral was undertaken for this BPPM based on a set of 39 plates. The industry experts had determined the expected output membership degrees in advance. The obtained results on the training set after determining the fuzzy measure coefficients is depicted in Fig. 8a. Taking into consideration the maximal difference between classes within the fitness function improved the results in terms of false accepted/rejected rates and of approximation to the relevance established by the experts. Taking into consideration the larger membership degree as a crisp result the system achieved a recognition rate of 84.61% on the training set.

The results obtained on a test set of 84 plates are depicted in Fig. 8b. The recognition rate with the test set was of 70.23%. On hand of these results the generalization capability of the evaluation framework can be analyzed.

The implemented industrial system was tested with 11 different evaluation data sets, which are composed by 100 plates taken directly from the production line. The results obtained on these evaluation sets can be observed in Tab. 1. Three parameters characterize the goodness of the system. First the percentage of plates with relevant faultiness that present a false fault type as the principal factor influencing its final classification (EFT). Finally the false rejectance (FR) and acceptance (FA) rates are given.

6 Conclusions

The here presented framework has been successfully implemented. The slight differences between the recognition rates of the experimental results and those of the industrial system lays on the fact that the training and test sets just take faulty plates into consideration.

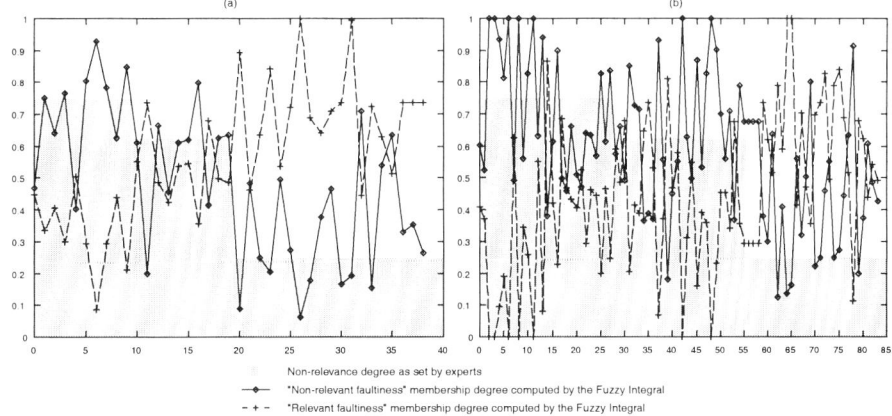

Figure 8: (a) Results obtained on a 39 plates training set. (b) Results obtained on a 84 plates test set with the automatically determined fuzzy measure coefficients.

Table 1: Results (in %) of automated industrial system in 11 evaluation data sets from production line. ET: Error in fault type. FR: False rejectance. FA: False acceptance.

	1	2	3	4	5	6	7	8	9	10	11	ε_m	$\bar{\varepsilon}$	σ_ε	ε_M
ET	1	1	7	13	9	2	5	11	13	4	4	1	6.4	4.5	13
FR	1	3	1	0	3	9	2	0	4	4	4	0	2.8	2.6	9
FA	24	9	21	17	11	3	5	5	6	5.3	11	3	10.7	7.1	24

After analyzing the results on hand of the binary images entering the fuzzy evaluation subsystem it could be stated that a great part of the errors were due to external factors. Among them is worth mentioning the presence of contradictions in the expected output membership degrees established by the experts. The interactive assessment of these output membership degrees taking into consideration the results of the automated system should be undertaken in order to minimize such errors.

The utilization of general fuzzy measures in the fuzzy integral, the employment of a more balanced (in terms of expected membership values) training set, and the characterization of GAs for the determination of fuzzy measure coefficients are some of the lines of future research that is expected to improve the performance of the industrial system.

References

[1] D. Dubois, H. Fargier, H. Prade (1997). Beyond min aggregation in multicriteria decision: (ordered) weighted min, discri-min, lexmin. In: Yager, R. R., Kacprzyk, J. (eds.), *The ordered weighted averaging operators — Theory and applications*, Kluwer Academic Publishers, Dordrecht.

[2] D. Gabor (1946). Theory of Communication, *J. Inst. Elec. Eng.*, vol. 93, pp. 429-457.

[3] D.E. Goldberg (1989). *Genetic algorithms in search, optimization & machine learning*, Addison-Wesley, Reading, MA.

[4] M. Grabisch, J.-H. Nicolas (1994). Classification by fuzzy integral: Performance and tests, *Fuzzy Sets and Systems* (65), pp.255-271, 1994.

[5] M. Grabisch (1997). Fuzzy Measures and Integrals for Decision Making and Pattern Recognition. In *Tatra Mountains: Fuzzy Structures Current Trends*, vol.13, pp. pp. 7-34.

[6] M. Grabisch, T. Murofushi, M. Sugeno (1999). *Fuzzy Measures and Integrals*, Physica-Verlag.

[7] R. Haralick, K. Shanmugam, I. Dinstein (1973). Textural features for image calssification, *IEEE Trans. SMC*, 3, (6), pp. 610-621.

[8] M. Köppen, M. Teunis, B. Nickolay (1998). A framework for the evolutionary generation of 2D-Lookup based texture filters, *Proc. IIZUKA'98*, Iizuka, Japan, pp.965-970.

[9] M. Köppen, A. Soria-Frisch, Th. Sy (2001). Binary Pattern Processing Framework for Perceptual Fault Detection, *Proc. 12th Scandinavian Conf. Image Analysis, SCIA 2001*, Bergen, Norway, pp.363-370.

[10] O. Nestares, *et al.* (1998). Efficient spatial-domain implementation of a multiscale image representation based on Gabor functions, *J. Electronic Imaging*, vol. 7, pp. 166-173.

[11] A. Soria-Frisch (2002). Soft Data Fusion in Image Processin. In *Soft-Computing and Industry: Recent Advances*, R. Roy et al. eds., pp. 423-444, Springer-Verlag.

[12] M. Sugeno (1974). *Theory of fuzzy integrals and its applications*, Ph.D. thesis, Tokyo University.

[13] K. Tanaka, M. Sugeno (1991). A study on subjective evaluation on printed color images, *Int. J. Approximate Reasoning*, vol. 5, No. 3, pp 213-222.

[14] Z. Wang, G.J. Klir (1992). *Fuzzy Measure Theory*, Plenum Press: New York.

[15] D. Wang *et al.* (1997). Determining Fuzzy Integral densities using a Genetical Algorithm for Pattern Recognition, *Proc. 1997 NAFIPS Annual Meeting*, pp. 263-267.

[16] R.R. Yager (1988). On ordered weighted averaging aggregation operators in multicriteria decision making, *IEEE Trans. SMC*, 18, pp.183-190.

[17] R.R. Yager, A. Kelman (1996). Fusion of Fuzzy Information with considerations for compatibility, partial aggregation and reinforcement,*International Journal of Approximate Reasoning*, 1996.

[18] L. Zadeh (1965) Fuzzy Sets, *Information and Control*, vol. 8, No. 3, pp. 338-353.

Author Index

Abásolo C. 129
Andreasen T. 107
Baldwin J.F. 303
Beaubouef T. 37
Beringer J. 327
Bloch I. 47
Bodenhofer U. 59
Borgelt C. 339
Bosc P. 141
Castillo O. 419
Cayrol C. 179
Cholvy L. 223
Coletti G. 191
Cordón O. 315
De Caluwe R. 117
De Cooman 277
De Tré G. 117
Del Jesus M.J. 315
Dempster A.P. 213
Denoeux T. 291
Fernández F. 71
Frank P.M. 407
Genaidy A. 431
Gómez M. 129
González Rodríguez I. 303
Guadarrama S. 71
Gutiérrez J. 71
Haenni R. 203
Hallez A. 117
Herrera F. 81, 315
Herrera-Viedma E. 81
Hüllermeier E. 327
Inuiguchi M. 93
Kacprzyk J. 153
Kipersztok O. 443
Köppen M. 465
Kruse R. 339
Lagasquie M.C. 179
Lawcewicz K. 153
Lawry J. 303
Liétard L. 141
Magdalena L. 315
Martínez L. 81
Mas M. 393
Masson M. 291
Melin P. 419
Mendel J.M. 233
Mesiar R. 393
Monserrat M. 393
Novák W. 363
Ovchinnikov S. 375
Palm R. 453
Pawlak Z. 243
Perfilieva I. 381
Petry F.E. 37
Pivert O. 141
Ralescu A. 431
Ramer A. 253
Runkler T.A. 453
Saffiotti A. 47
Sánchez A.M. 315
Sánchez P.J. 81
Schubert J. 349
Scozzafava R. 191
Smets Ph. 265
Soria-Frisch A. 465
Sy Th. 465
Tanino T. 93
Torrens J. 393
Trillas E. 71
Troffaes M.C.M. 277
Vantaggi B. 191
Verstraete J. 117
Villar P. 315
Visa S. 431
Wang H. 443

Yager R.R. 167
Yeung S. 431

Zadeh L.A. 3
Zadrozny S. 153